Editorial Board Members' Collection Series: "Global Food Security and Sustainability"

Editorial Board Members' Collection Series: "Global Food Security and Sustainability"

Guest Editors

Theodoros Varzakas
Costas Stathopoulos

MDPI

Basel • Beijing • Wuhan • Barcelona • Belgrade • Novi Sad • Cluj • Manchester

Guest Editors

Theodoros Varzakas
Department of Food Science and Technology
University of the Peloponnese
Antikalamos
Greece

Costas Stathopoulos
School of Medical, Molecular, and Forensic Sciences
Murdoch University
Perth
Western Australia

Editorial Office
MDPI AG
Grosspeteranlage 5
4052 Basel, Switzerland

This is a reprint of the Special Issue, published open access by the journal *Foods* (ISSN 2304-8158), freely accessible at: https://www.mdpi.com/journal/foods/special_issues/9GYMYELBG8.

For citation purposes, cite each article independently as indicated on the article page online and as indicated below:

Lastname, A.A.; Lastname, B.B. Article Title. *Journal Name* **Year**, *Volume Number*, Page Range.

ISBN 978-3-7258-5239-0 (Hbk)
ISBN 978-3-7258-5240-6 (PDF)
https://doi.org/10.3390/books978-3-7258-5240-6

Contents

About the Editors

Theodoros Varzakas

Theodoros Varzakas serves as Full Professor at the Department of Food Science and Technology, University of the Peloponnese, Greece since 2019, specializing in issues of food technology, food processing/engineering, and food quality and safety. He serves as Section Editor in Chief Journal Foods in Food Security and Sustainability (2020-), ex Editor in Chief Current Research in Nutrition and Food Science (2015–2019), and reviewer and member of the editorial board in many international journals such as (*International Journal of Food Science & Technology, Journal of Food Engineering, Waste Management, Critical Reviews in Food Science and Nutrition, Italian Journal of Food Science, Journal of Food Processing and Preservation, Journal of Culinary Science and Technology, Journal of Agricultural and Food Chemistry, Journal of Food Quality, Foods, Microorganisms*, and *Current Research in Food Science*). He has written more than 300 research papers and chapters in books and has presented more than 160 papers and posters at national and international conferences. He has written and edited 14 books in Greek, and 16 in English on sweeteners, biosensors, food engineering and food processing, published by CRC. He has participated in many European and national research programs as a coordinator or scientific member. His h-index is 45 and his no. of citations is approx. 7600.

According to September 2021, September 2022 data, October 2023, and August 2024 and September 2025-update for "Updated science-wide author databases of standardized citation indicators", Varzakas Theodoros is in the top 2% of scientists worldwide.

Costas Stathopoulos

Costas Stathopoulos is a food scientist and technologist with more than 20 years' experience of working in research and academia in 8 countries. He is an expert in food waste utilisation, and his research has been extensively published in top quality peer reviewed journals (120+ publications, 70+ conference presentations). Prof Stathopoulos has secured competitive research funding in excess of EUR 20 million, and has supervised a number of PhD students to completion. Prof Stathopoulos is a Fellow of IFST (UK), a Professional Member of AIFST (Australia), and a member of the scientific committee of PETET (Greek IFST).

Preface

The reprint belongs to the Special Issue Editorial Board Members' Collection Series: "Global Food Security and Sustainability. It addresses issues of food security or insecurity and sustainability. Seventeen contributions have been discussed thoroughly, addressing issues of food sustainability, food policy, nutrition, sustainable supply management, food safety, environmental footprint, sustainable food systems (SFSs), as connected to the Sustainable Development Goals (SDGs), and others.

As already described, it is important to take into account a holistic dimension of these systems in order to safeguard food security and sustainability, health, and more specifically, a One Health approach.

In this context, all stakeholders will benefit and our planet will eventually be saved.

Theodoros Varzakas and Costas Stathopoulos
Guest Editors

foods

Editorial

Global Food Security and Sustainability

Theodoros Varzakas [1,*] and Costas Stathopoulos [2]

1 Department Food Science and Technology, University of the Peloponnese, Antikalamos, 24100 Kalamata, Greece
2 Food Futures Institute, School of Medical Molecular and Forensic Sciences, Murdoch University, Food Innovation Precinct WA, Room G.05, 37 Dollyup Street, Perth, WA 6181, Australia; costas.stathopoulos@murdoch.edu.au
* Correspondence: t.varzakas@uop.gr; Tel.: +30-2721045279

1. Main Text

This Editorial introduces the Special Issue "Editorial Board Members' Collection Series: "Global Food Security and Sustainability"", and highlights key topics on this subject. Sixteen papers have been published, amongst which ten were original research and six were reviews.

They are all within the wide area of food security/insecurity and sustainability.

In the first contribution by Xiao et al. [Contribution 1], the spatial and temporal patterns of the rate of grain self-sufficiency (RSSG) at the county level in Guangdong Province from 2014 to 2023 were reported, proposing policy recommendations such as the prioritization of farmland preservation; the implementation of the "storing grain in the land" strategy to ensure food supply; the adoption of advanced agricultural technologies for the improvement of grain yield; the reduction in grain loss by the improvement of storage management and the enhancement of storage efficiency; and the facilitation of interdepartmental coordination.

The second contribution refers to the transition toward sustainable food, which requires integrated policies and collaborative action. Key commodity/supply chain groups need to be established. "Net-Map Analysis of the Italian Wheat Supply Chain" is well evidenced, analyzing the different stakeholder relationships and their complexities.

Sustainable food supply chains need to be taken into account as discussed by Dania et al. [1], depicting both vertical and horizontal collaboration approaches.

The third contribution by Ayuso et al. [Contribution 3] deals with dietary fiber (DF) and, more specifically, soluble dietary fiber, a key characteristic of nutrition that affects food security and sustainability. DF is considered an essential component of a healthy diet, and has been associated with several health benefits, reducing mortality from diseases. The authors reported that higher-value-added ingredients could arise from the revalorization of carob, artichoke, apple, and broccoli byproducts after enzymatic treatment, thus improving their nutritional and physicochemical properties.

The fourth contribution by Mastos et al. [Contribution 4] investigated the performance of sustainable supply chain management (SSCM) in the Greek food industry by using structural equation modeling. They concluded that society and the environment are affected by supply chain strategic orientation, and that they both affect economics. Moreover, firm-level critical sustainability factors play a determining role.

The fifth contribution by Loi et al. [Contribution 5] addressed the issue of mycotoxins as a key factor in global food security through the Horizon 2020 project. The authors referred to risk assessment and management strategies through the adoption of the harmonization of methods and marketable solutions.

 https://doi.org/10.3390/foods14111983

The sixth contribution by Lima et al. [Contribution 6] deals with nanoemulsions as a means of food security in order to enhance essential oils' dispersion and protection. Natural preservatives instead of synthetic ones are a key priority in food systems.

Consumers show preference for low-greenhouse gas (GHG)-emission produce [2], as well as the declaration of a carbon footprint (e.g., the European Commission's Product Environmental Footprint). Hence, in the seventh contribution by Simmons et al. [Contribution 7], strategies for reducing the carbon footprint of cherry and orange production were examined, including the avoidance of emissions by sequestering C from the atmosphere.

The eighth contribution by Lee et al. [Contribution 8] considered global food safety systems and standards, addressing many case studies and issues related to food safety culture in developed and developing countries.

In the ninth contribution by Ajayi et al. [Contribution 9], the issue of bioactive peptides from plant-based proteins was explored, which has been gaining attention due to their promotion of human health and alleviation of diseases. In this direction, quinoa protein hydrolysates (QPHs) have been considered as an alternative therapy for the treatment of hypercholesterolemia.

The tenth contribution by Kotsou et al. [Contribution 10] dealt with edible insects as well as *Tenebrio molitor*, and their enhancement through the addition of albedo orange peel waste.

The eleventh contribution by Varzakas and Antoniadou [Contribution 11] dealt with food ethics and ethically informed food choices. The spiritual and physical significance of food were determined through the perspective of oral health aiding holistic aspects of well-being. Ethical practices and transparency will work their way towards more sustainable and equitable food systems.

The twelfth contribution by Varzakas and Smaoui [Contribution 12] brought to the forefront the whole dimension of sustainable food systems (SFSs), as connected to the Sustainable Development Goals (SDGs). A plant-based protein case study was well evidenced.

The thirteenth contribution examined food products developed using *T. molitor* larvae, addressing issues of food security and environmental sustainability linked to nutrition.

The fourteenth contribution by Vinas-Ospino et al. [Contribution 14] addressed the issue of green solvent technology offering significant advantages in terms of extracting carotenoids from fruit and vegetable byproducts. This is a major aspect of food security and sustainability.

The fifteenth contribution by Carvalho et al. [Contribution 15] employed a holistic way of considering sustainability, uniting healthy eating, personal health, and well-being.

Finally, Lara et al. [Contribution 16] described issues of food biodiversity and underutilized and neglected food crops, leading to food and nutrition insecurity. Indigenous crops, landrace varieties, and diversification are some of the proposed strategies underlined.

2. Conclusions

Global food security and sustainability issues need to be addressed in a holistic way, taking into account all stakeholders' needs and expectations. All of the issues raised above will aid in safeguarding our health and the health of our planet; this being the case, the mission is to achieve this, and this will involve a coordinated effort from all of us. It is in our hands.

Author Contributions: Writing—review and editing, T.V. and C.S. All authors have read and agreed to the published version of the manuscript.

Funding: This research received no external funding.

Data Availability Statement: Not applicable.

Acknowledgments: The Guest Editors would like to acknowledge the support given by *Foods* and all of the authors and editors who contributed to the success of this Special Issue.

Conflicts of Interest: The authors declare no conflicts of interest.

List of Contributions:

1. Xiao, Y.; Zhao, J.; Zhang, Y.; Wang, S. Grain Self-Sufficiency in Guangdong, China: Current Trends and Future Perspectives. *Foods* **2025**, *14*, 1126. https://doi.org/10.3390/foods14071126.
2. Ciasca, B.; Cito, N.M.; Lattanzio, V.M.T. Getting Insights in Stakeholder Collaboration in the Transition Toward Safe and Sustainable Food Production: Net-Map Analysis of the Italian Wheat Supply Chain. *Foods* **2025**, *14*, 786. https://doi.org/10.3390/foods14050786.
3. Ayuso, P.; Peñalver, R.; Quizhpe, J.; Rosell, M.d.l.Á.; Nieto, G. Broccoli, Artichoke, Carob and Apple By-Products as a Source of Soluble Fiber: How It Can Be Affected by Enzymatic Treatment with Pectinex® Ultra SP-L, Viscozyme® L and Celluclast® 1.5 L. *Foods* **2025**, *14*, 10. https://doi.org/10.3390/foods14010010.
4. Mastos, T.D.; Gotzamani, K.; Ieromonachou, P.; Tsiotras, G. A Structural Equation Model for Sustainable Supply Chain Management in the Food Industry. *Foods* **2024**, *13*, 3713. https://doi.org/10.3390/foods13233713.
5. Loi, M.; Moretti, A.; Lippolis, V.; Özer, H.; PembeciKodolbas, C.; Yener, E.; Demirtaş, İ.; Vila-Donat, P.; Manyes, L.; Lattanzio, V.M.T. MycoTWIN Working Group Discussion: A Multi-Actor Perspective on Future Research Directions for Mycotoxins and Toxigenic Fungi Along the Food and Feed Chain. *Foods* **2024**, *13*, 3582. https://doi.org/10.3390/foods13223582.
6. Lima, M.A.; Carusi, J.; Rocha, L.d.O.; Tonon, R.V.; Cunha, R.L.; Rosenthal, A. Physicochemical Characterization, Rheological Properties, and Antimicrobial Activity of Sodium Alginate-Pink Pepper Essential Oil (PPEO) Nanoemulsions. *Foods* **2024**, *13*, 3090. https://doi.org/10.3390/foods13193090.
7. Simmons, A.T.; Simpson, M.; Bontinck, P.-A.; Golding, J.; Grant, T.; Fearnley, J.; Falivene, S. Emissions Reduction Strategies for the Orange and Cherry Industries in New South Wales. *Foods* **2023**, *12*, 3328. https://doi.org/10.3390/foods12183328.
8. Lee, J.C.; Neonaki, M.; Alexopoulos, A.; Varzakas, T. Case Studies of Small-Medium Food Enterprises around the World: Major Constraints and Benefits from the Implementation of Food Safety Management Systems. *Foods* **2023**, *12*, 3218. https://doi.org/10.3390/foods12173218.
9. Ajayi, F.F.; Mudgil, P.; Jobe, A.; Antony, P.; Vijayan, R.; Gan, C.-Y.; Maqsood, S. Novel Plant-Protein (Quinoa) Derived Bioactive Peptides with Potential Anti-Hypercholesterolemic Activities: Identification, Characterization and Molecular Docking of Bioactive Peptides. *Foods* **2023**, *12*, 1327. https://doi.org/10.3390/foods12061327.
10. Kotsou, K.; Chatzimitakos, T.; Athanasiadis, V.; Bozinou, E.; Adamaki-Sotiraki, C.; Rumbos, C.I.; Athanassiou, C.G.; Lalas, S.I. Waste Orange Peels as a Feed Additive for the Enhancement of the Nutritional Value of *Tenebrio molitor*. *Foods* **2023**, *12*, 783. https://doi.org/10.3390/foods12040783.
11. Varzakas, T.; Antoniadou, M. A Holistic Approach for Ethics and Sustainability in the Food Chain: The Gateway to Oral and Systemic Health. *Foods* **2024**, *13*, 1224. https://doi.org/10.3390/foods13081224.
12. Varzakas, T.; Smaoui, S. Global Food Security and Sustainability Issues: The Road to 2030 from Nutrition and Sustainable Healthy Diets to Food Systems Change. *Foods* **2024**, *13*, 306. https://doi.org/10.3390/foods13020306.
13. Kotsou, K.; Chatzimitakos, T.; Athanasiadis, V.; Bozinou, E.; Athanassiou, C.G.; Lalas, S.I. Innovative Applications of *Tenebrio molitor* Larvae in Food Product Development: A Comprehensive Review. *Foods* **2023**, *12*, 4223. https://doi.org/10.3390/foods12234223.
14. Viñas-Ospino, A.; López-Malo, D.; Esteve, M.J.; Frígola, A.; Blesa, J. Green Solvents: Emerging Alternatives for Carotenoid Extraction from Fruit and Vegetable By-Products. *Foods* **2023**, *12*, 863. https://doi.org/10.3390/foods12040863.
15. Carvalho, G.P.; Costa-Camilo, E.; Duarte, I. Advancing Health and Sustainability: A Holistic Approach to Food Production and Dietary Habits. *Foods* **2024**, *13*, 3829. https://doi.org/10.3390/foods13233829.

16. Lara, S.W.; Tsiami, A.; Cross, P. Discovering and Mapping Colloquial Terminologies Describing Underutilized and Neglected Food Crops—A Comprehensive Review. *Foods* **2023**, *12*, 2428. https://doi.org/10.3390/foods12122428.

References

1. Dania, W.A.P.; Xing, K.; Amer, Y. Collaboration behavioural factors for sustainable agri-food supply chains: A systematic review. *J. Clean. Prod.* **2018**, *202*, 193–208. [CrossRef]
2. Thøgersen, J.; Nielsen, K.S. A better carbon footprint label. *J. Clean. Prod.* **2016**, *125*, 86–94. [CrossRef]

foods

MDPI

Article

Grain Self-Sufficiency in Guangdong, China: Current Trends and Future Perspectives

Yi Xiao [1], Jianya Zhao [2], Yanglan Zhang [3] and Shu Wang [2,3,*]

1 Department of Science and Technology Research, Jinan University, 855 Xingye E. Ave., Guangzhou 510632, China
2 Jinan University-University of Birmingham Joint Institute, Jinan University, 855 Xingye E. Ave., Guangzhou 511436, China
3 School of Economics, Jinan University, 601 W. Huangpu Ave., Guangzhou 510632, China
* Correspondence: shuwang@jnu.edu.cn

Abstract: The rising number of disturbances caused by natural hazards, epidemics, and international conflicts seriously threatens global agriculture and food systems. In order to combat the increasing uncertainty, the Chinese government proposed a "dual circulation" strategy, which puts forward a new requirement of promoting self-sufficiency. Among various food categories, grain serves as a cornerstone for sustaining human life during public crises. Guangdong Province is one of the most economically developed and highly densely populated regions in China; therefore, it faces significant pressure to enhance food security. This study investigates the spatial and temporal patterns of the rate of grain self-sufficiency (RSSG) at the county levels in Guangdong Province from 2014 to 2023. The findings reveal the following: (1) rice remains the predominant grain crop, with its sown area and production consistently accounting for over 75%; (2) the RSSG is 28.14% on average, which is lower than the governmental goal of 30%; (3) the RSSG has significant regional disparities and distinct agglomeration patterns; and (4) scenario simulations indicate that grain yield improvement and grain loss reduction have the potential to promote the RSSG to 34.14%. Based on the research findings, this paper proposes the following policy recommendations: (1) prioritize farmland preservation and implement the "storing grain in the land" strategy to ensure food supply; (2) adopt advanced agricultural technologies under the "storing grain in the technology" strategy to improve grain yield; (3) reduce grain loss by strengthening disaster prevention, improving storage management, and enhancing storage efficiency; and (4) strengthen interdepartmental coordination. The integrated application of these strategies will help Guangdong Province enhance its grain self-sufficiency, ensure food security, and contribute to the achievement of national food security goals.

Keywords: grain self-sufficiency; food security; China

1. Introduction

Since 2020, the rising number of internal and external shocks from natural hazards, epidemics, and international conflicts demonstrate the frailty and vulnerability of the global agriculture and food system. Extreme weather events such as drought and heavy rain, favored by climate change, have a negative impact on crop yields [1,2]. Since 2020, the world has grappled with the coronavirus 2019 (COVID-19) pandemic and the Russia–Ukraine war, both of which have had significantly adverse influences on food accessibility [3]. From early 2020, with the widespread transmission of COVID-19, extensive disruptions to food systems

 https://doi.org/10.3390/foods14071126

were noted [4]. Agricultural production had been disrupted by the shortage of input, such as labor and agricultural materials [5]. Ultimately, the vulnerable food supply chain during the shutdown period resulted in household food insecurity [6–8]. Just as pressures from the COVID-19 crisis began to wane, the Russia–Ukraine conflict disrupted grain exports from Russia's Black Sea region and caused a considerable rise in food prices [9,10]. Apart from the direct military conflict as the most significant actor in the global agriculture system [11], the China–U.S. trade disputes have also caused severe damage to agriculture and the food supply [12]. Such conflicts have been suggested to become more tense because Donald Trump's government in the U.S. has adopted an approach of weaponizing food supplies [13]. Therefore, these factors exacerbate the vulnerability of agriculture and the food system and put a heavy threat on global food security.

In order to combat rising uncertainties, the Chinese government proposed the spirit of "holding the rice bowl of the Chinese people in their own hands firmly", which emphasizes the importance of maintaining food self-sufficiency as the central theme of China's food security [14]. As one of the largest agricultural countries globally, China grapples with multifaceted challenges in agricultural production. These challenges include cropland scarcity, water scarcity, non-point source pollution from chemical fertilizers, and the looming impacts of climate change. The challenges outlined pose significant barriers to the sustainable development of agriculture, potentially endangering national food security and the welfare of farmers in the long term. Therefore, China's central government issued the No.1 document in 2014, declaring a break away from self-sufficiency to a new food security strategy—one that relies on domestic production with moderate imports [15]. China was a net exporter of maize until the 1990s, but since 2009, it has become a constant importer, with the import volume reaching 28 million tons in 2021 [15]. Consequently, China has become the worlds' largest importer of maize [15]. In 2022, China's total crop imports amounted to 164 million tons, representing an effective import of 0.8 billion hectares of arable land [16].

However, the "import substitution" policy should establish the stable international environment and free trade. Because of the increasing complexity of the international environment and the globalization of environmental governance, China's agricultural trade is facing an increasing number of difficulties and challenges [17]. Crop failure exacerbated by extreme weather [18], food processing, and logistics suspended by public health emergencies [4], trade conflicts [12], and even worse, the war [9,10], is threatening stable and smooth food import. China has foreseen that "The world is undergoing profound changes unseen in a century". In May 2020, China proposed the domestic and international "dual circulation" strategy, which refers to the domestic cycle as the mainstay, while domestic and international dual circulation can mutually promote and benefit each other [19,20]. With such a movement, the urgency and importance of self-sufficiency is reinforced again. For the purposes of sustaining human life, food variety has to be limited [21], and grain supply should have high priority [22–24].

As the one of the largest developing countries in the world, China also suffers from strict regulations for ensuring national food security [24–28]. China's vast territory is characterized by significant regional disparities [29–32], with Guangdong Province being particularly representative. Guangdong is one of the most economically developed regions in China, with rapid urbanization and industrialization [24,25]. Since the reformation and the opening up of the province, urbanization and economic growth in Guangdong have been rapidly promoted (see Figure 1). However, rapid urbanization and industrialization have severe negative impacts on food security and environmental sustainability. Industrialization is attracting large numbers of migrants to Guangdong, and the demand for food surges dramatically, while the agricultural endowments (farmland, water, etc.) are limited. The increase in agriculture cannot keep pace with the population growth [33–35]. Urbanization leads to significant land use changes, including the destruction of farmland, which affects its carrying capacity and food security [36]. Meanwhile, Guangdong faces serious internal regional economic disparities [37]. Guangdong is divided into four parts: east (Shantou, Shanwei, Chao Yang, Jieyang), west (Yangjiang, Zhanjiang, Maoming), north (Shaoguan, Heyuan, Meizhou, Qingyuan, Yunfu) and Pearl River Delta (Guangzhou, Zhuhai, Foshan, Huizhou, Jiangmen, Zhaoqing, Dongguan, Zhongshan). Located in the heart position of Guangdong, the Guangdong–Hong Kong–Macau Big Bay Area (Pearl River Delta plus Hong Kong SAR and Macau SAR) is a major economic hub in China. It is famous around the world, with two major international cities: Hong Kong and Macau [38]. On the other hand, the varied topographies, ranging from mountainous areas to the Pearl River Delta [39], involve certain endowments that result in different economic development and agricultural practices. Although the remaining parts of Guangdong have great gaps with the Big Bay Area, the regions' reliance on the other parts for resources, such as water and food, makes great contributions to the achievement of the Big Bay Area for the remarkable and impressive economic development. In addition, as one of the pioneers of reform and opening up, Guangdong has built up well-established statistical systems, and the available data are comparatively comprehensive.

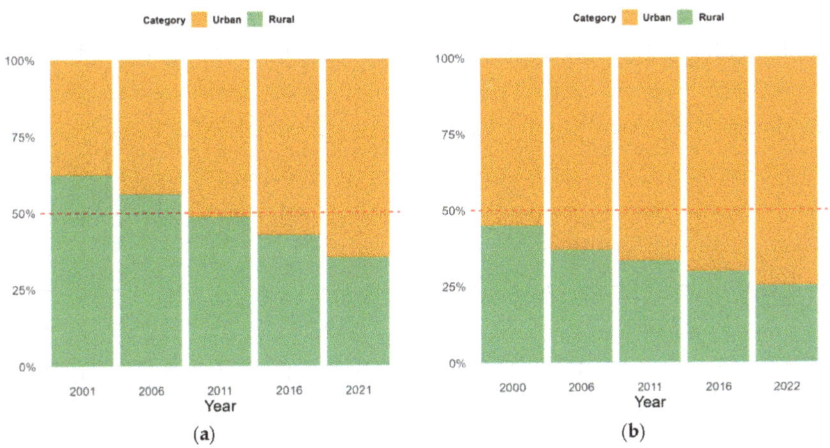

Figure 1. Ratios of urbanization: (**a**) China; (**b**) Guangdong province. Data Source: China Statistical Yearbook [40].

Previous studies have independently considered several problems and solutions of food security from different perspectives. Yang et al. (2024) [39] presents an "element coupling–function synergy" analytical framework to assess the multifunctionality of cultivated land resources in Guangdong, China, revealing trade-offs and synergies among various functions such as food production, ecological regulation, and social security, and proposing multi-objective management plans for sustainable land use. Mutelo et al. (2023) [41] employs high-resolution remote sensing to monitor the destruction of high-quality farmland in Guangdong, revealing significant conversion of farmland to industrial and residential uses, which poses a threat to food security due to the loss of fertile land. Fang et al. (2023) [36] evaluates the carrying capacity of cultivated land in Guangdong, using the entropy weight method; they found that ecological and environmental pressures have a significant impact on capacity, while economic and social factors remain stable. This observation highlights the need for green recovery strategies to improve land productivity and support sustainable agriculture. Chen et al. (2022) [42] quantifies the net carbon sequestration value of the rice production system in Guangdong, finding that compensating for this value could alleviate farmland abandonment and enhance food security and carbon neutrality. The study also points out that the self-sufficiency rate of rice in Guangdong was only about 22% in 2019, highlighting a significant food security crisis.

However, the potential risks of food shortages arising from public crises have yet to be fully explored and should not be overlooked. In 2019, Guangdong Province released a plan for the rural revitalization strategy, and a the goal of ensuring that grain self-sufficiency exceeds 30% [43]. Therefore, in this study, we collected data on grain production and consumption in Guangdong, assessing the self-sufficiency of grain; accordingly, we propose a practical pathway for further improvement. In this study, we aim to investigate the spatial and temporal patterns of the rate of grain self-sufficiency (RSSG) at the county level in Guangdong Province from 2014 to 2022; additionally, we analyze how grain yield improvement and grain loss reduction can promote the self-sufficiency of grain in Guangdong. Addressing Guangdong's food security issues and ensuring the goal of keeping "the Chinese rice bowl firmly in their own hands" are important contributions to achieving the stable and sustainable development of countries and regions around the world.

2. Data and Methods

2.1. Method

Grain self-sufficiency can be assessed using various metrics, such as caloric value, production volume, or monetary value [44,45]. To evaluate the balance between grain production and consumption, the most intuitive and straightforward approach is to compare actual grain production with the region's standard per capita grain consumption and the population it supports. By calculating either the gap (difference) or the ratio (division) between production and consumption, the local grain supply can be effectively assessed, and potential risks to grain self-sufficiency can be identified. This method [45,46] provides a practical framework for analyzing regional food security dynamics and identifying areas where localized interventions may be needed to address imbalances.

This study analyzes the grain self-sufficiency at the county/district level in Guangdong Province. For the assessment of regional food security, a micro-scale approach (at the county level) is more accurate than a macro-scale approach (at the provincial or municipal levels) [47]. County-level units play a key role in agricultural production and food security management. County governments are responsible for implementing policies and directly influencing grain planting and yield management [48]. Given China's vast territory and significant regional disparities, studying county-level units helps to identify unique challenges and potentials in different areas, contributing to the improvement of grain production capacity across regions and thus promoting overall food security. The role of county-level units is particularly critical in responding to disasters and climate change. Therefore, studying grain self-sufficiency at the county level provides important support for the national food security strategy [49]. The analysis is based on the fundamental concepts of division and subtraction between grain production and consumption:

$$RSSG_{i,y} = \frac{Y_{i,y}}{S \times P_{i,y}} \tag{1}$$

$$GSSG_{i,y} = Y_{i,y} - (S \times P_{i,y}) \tag{2}$$

where $RSSG_{i,y}$ (unit: %) indicates the ratio of self-sufficiency of grain in region, i, of year, y; $Y_{i,y}$ (unit: kg/ha) is grain production in county, i, of year, y; S (unit: kg grain/person*year) is the standard of individual grain consumption per year; $P_{i,y}$ (unit: person) is the population in region, i, in year, y. $GSSG_{i,y}$ (unit: %) indicates the gap of self-sufficiency of grain in region, i, of year, y. If $GSSG_{i,y}$ is positive, it indicates that the grain production in the region exceeds the local consumption needs, ensuring self-sufficiency. Conversely, a negative $GSSG_{i,y}$ implies a grain deficit, requiring imports from other regions (via domestic allocation or international trade). Several factors influence the calculation of actual grain production and consumption, including grain loss during seeding, harvesting, and storage. Apart from direct consumption for food, grain is also consumed indirectly for forage and food processing. These factors act as additional constraints and are often estimated using conversion coefficients. For example, grain loss accounts for approximately 12% of total production [50,51]. The security line for grain has been set to 400 kg per capita annually [52]. This methodological framework provides a structured approach to quantifying grain self-sufficiency and identifying regional disparities in grain supply, which are critical for ensuring food security.

2.2. Statistical Analytical Framework

The analyses in this study were conducted in four distinct stages (Figure 2). First, a descriptive statistical analysis was performed to identify the structure and trends in grain production in Guangdong.

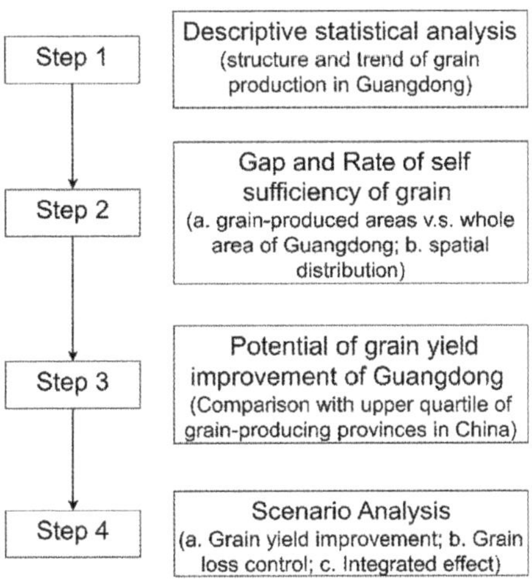

Figure 2. Data analysis procedure in this study.

Second, the study examined the gap and rate of grain self-sufficiency (*GSSG*, *RSSG*) in both grain-producing areas and the entire province of Guangdong Province, based on current yields. This analysis provides valuable insights into the current state of grain self-sufficiency in Guangdong. To mitigate potential bias from extreme one-year values, the average yield data from 2014 to 2023 were used. Additionally, the spatial distribution of grain self-sufficiency was analyzed to reveal the patterns of grain production and consumption across Guangdong.

Third, the potential for improving grain yields in Guangdong was assessed. As one of China's leading economic provinces, Guangdong has significant potential to increase grain yield through a "storing grain to technology" strategy, which provides practical solutions. It is projected that Guangdong's grain yield could reach the upper quartile level among grain-producing provinces in China. The equation for calculating the potential improvement in grain yield in Guangdong is as follows:

$$Improvement = \frac{Y_{Q3,y} - Y_{GD,y}}{Y_{GD,y}} \tag{3}$$

where *Improvement* (unit: %) represents the potential grain yield improvement in Guangdong; $Y_{Q3,y}$ (unit: %) denotes the grain yield in the upper quartile of China's grain-producing provinces for year, y; and $Y_{GD,y}$ (unit: %) is the grain yield in Guangdong for year, y.

Finally, based on the potential for grain yield improvement and the control of grain loss, this study employed a scenario analysis to explore pathways for strengthening grain security in Guangdong. The equation to calculate the change in grain production in Guangdong is as follows:

$$Y^*_{i,y} = Y_{i,y} \times (1 + Improvement) \times (1 - Loss) \tag{4}$$

where *Loss* (unit: %) represents the grain loss, using a general figure for China (12%). In the grain loss control and grain yield improvement scenario, area-related parameters are adopted and discussed in Section 3.4. In the integrated scenario, both parameters were combined to generate new results.

2.3. Data

The primary data source for analyzing grain self-sufficiency in China is the *China Statistical Yearbook* [40,46,53–58]. In addition, the *Guangdong Statistical Yearbook* [59] serves as supplementary material [15]. Benefiting from a comprehensive and well-established statistical system, this study utilizes annual data from 108 counties and districts in Guangdong Province, sourced from the *Guangdong Rural Statistical Yearbook* [60]. The data include grain production, population, sown area, yield, and per capita grain consumption for both urban and rural residents. The time range of the data spans from 2014 to 2023, according to the availability of information from the *Guangdong Rural Statistical Yearbook* [60].

3. Results

3.1. Structure and Trend of Grain Production

Clarifying the composition of grain is one of the critical challenges in grain-related research, especially in the context of its implications for food security and human behavior. In China, maize, wheat, rice, and potato are recognized as staple grains that serve as the cornerstone of national and regional food security. However, these grains differ significantly in terms of yield potential, farming systems, and socio-cultural influences. Such differences have inspired derivative studies on how grain production and consumption shape social behaviors. For instance, the "rice theory" suggests that the labor-intensive nature of rice farming—requiring almost twice as many the hours from seeding to harvest as wheat—has contributed to the development of collectivist behaviors in Southern China [61]. Conversely, the less labor-intensive wheat farming in Northern China is associated with more individualistic tendencies.

Given these disparities, this study prioritized analyzing the structural composition of grain production in Guangdong Province to provide a precise understanding and avoid ambiguity. The figure illustrates the proportion of different grain crops (rice, potato, and others) in terms of sowing area and production from 2014 to 2023 (Figure 3). Throughout the period, rice consistently dominated, accounting for over 75% of the total grain production and sowing area each year. This highlights its critical role in maintaining food security in Guangdong Province. Meanwhile, potato and other grains occupied relatively minor shares, collectively accounting for less than 25%. Despite slight annual fluctuations, the structural composition of grain production and sowing area remained relatively stable over the study period, reflecting the limited diversification of crop production at the provincial level. Notably, the consistent dominance of rice highlights its strategic importance in regional agricultural planning and food policy, especially in urban areas where food security concerns are more pressing.

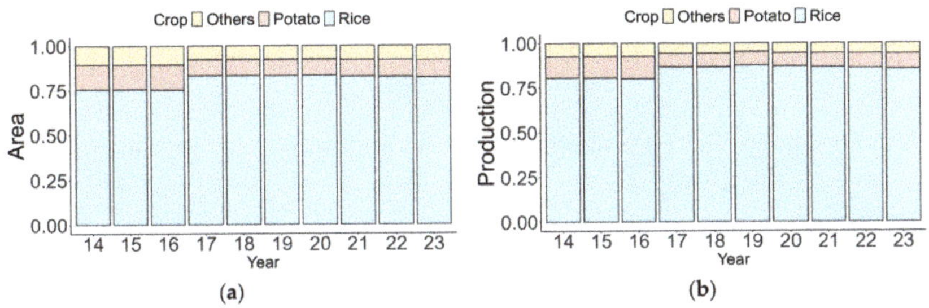

Figure 3. Structural composition and trends of grain-sowing area (**a**) and production (**b**) in Guangdong Province, 2014–2023. Data source: *Guangdong Rural Statistical Yearbook* [60].

3.2. Self-Sufficiency of Grain

Table 1 presents the *GSSG* and *RSSG* for both grain-producing areas and the entire area of Guangdong Province under current yield conditions. The results reveal significant regional disparities in both *RSSG* and *GSSG*. In grain-producing areas, the RSSG ranges from 22.96% in the east to 45.76% in the north, with an average of 33.08% across all regions. In contrast, the *GSSG* shows a deficit across all regions, with the Pearl River Delta exhibiting the largest gap (−8.12 million tons), followed by the east (−5.85 million tons), west (−4.59 million tons), and north (−4.32 million tons). When considering the entire area of Guangdong, which includes non-agricultural districts that rely entirely on grain imports, the *RSSG* decreases slightly, with an average of 28.14%, falling short of the government's target of 30%. The *GSSG* also worsens, particularly in the Pearl River Delta, where the deficit rises to −13.74 million tons.

Table 1. Gap and rate of grain self-sufficiency (*GSSG*, *RSSG*) in grain-producing areas and the entire area of Guangdong under current yield.

Region	Grain-Producing Area		Whole Area	
	GSSG	*RSSG*	*GSSG*	*RSSG*
Delta	−8.12	25.17	−13.74	16.58
East	−5.85	22.96	−5.98	22.57
North	−4.32	45.76	−4.36	45.55
West	−4.59	41.01	−4.8	39.9
All	−22.88	33.08	−28.89	28.14

Units: *GSSG* (million tons); *RSSG* (%).

The map (Figure 4) shows the spatial distribution of grain self-sufficiency (%) across Guangdong Province, highlighting significant regional disparities and distinct agglomeration patterns: The spatial distribution of grain self-sufficiency in Guangdong Province, as detailed in the figures, reveals significant regional disparities. The Delta region, which includes both urban districts and rural counties, shows a wide range of *RSSG*, indicating different levels of grain self-sufficiency. The east region also exhibits variations, with some counties showing high self-sufficiency and others indicating a need for external grain sources. The west region presents a mixed picture, with some districts and counties having high *RSSG*, indicating surplus grain production, while others have lower values, indicating a reliance on external supplies. The north region shows a spread in *RSSG*, with some counties standing out as highly self-sufficient.

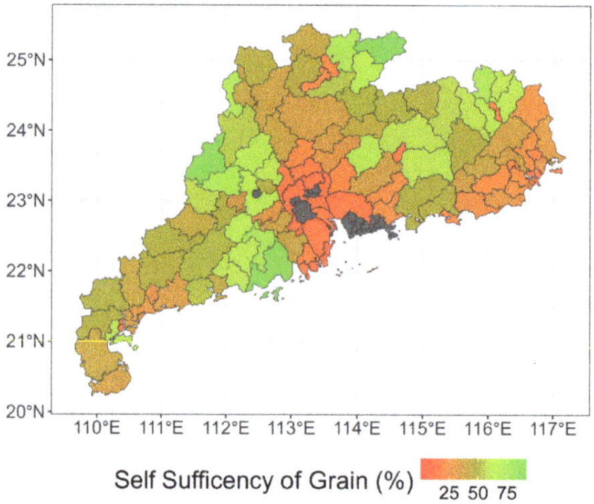

Figure 4. Spatial distribution of grain self-sufficiency in Guangdong Province. Data source: *Guangdong Rural Statistical Yearbook* [60].

The spatial analysis of grain self-sufficiency in Guangdong Province reveals a complex picture of regional disparities in grain production and consumption. The results indicate that while some districts and counties have a high capacity for grain self-sufficiency, others face significant challenges in meeting their grain needs. The yield gap between districts and counties, as indicated by the *RSSG*, is not statistically significant, suggesting that both urban and rural areas play a crucial role in the province's food security. These insights highlight the importance of targeted strategies to enhance grain productivity, particularly in areas with lower *RSSG*, to ensure a more balanced and resilient food supply system across Guangdong Province.

3.3. Comparison of Grain Yield Between Guangdong and Other Provinces in China

Figure 5 presents a comparative analysis of grain yield (measured in kg/ha) between Guangdong Province and other provinces in China from 2015 to 2023. The red dot in the figure highlights Guangdong's grain yield, placing it within the broader context of China's agricultural productivity. The figure reveals that Guangdong's grain yield, while showing modest improvements over the years, remains below the upper quartile of grain-producing provinces in China. Over the nine-year period, Guangdong's grain yield fluctuated between 5636 kg/ha in 2015 and 5911 kg/ha in 2020, with a slight decline to 5764 kg/ha in 2023. In contrast, provinces such as Jilin, Heilongjiang, and Jiangsu consistently achieved higher yields, with Jilin reaching 7494 kg/ha in 2015 and Heilongjiang achieving 6968 kg/ha in 2020. These provinces, which represent the upper quartile of grain-producing regions, demonstrate the potential for significantly higher agricultural productivity. The comparison highlights a notable gap between Guangdong's grain yield and that of the leading provinces. For example, in 2023, Guangdong's yield of 5764 kg/ha was substantially lower than those of Jilin (7186 kg/ha) and Liaoning (7164 kg/ha), despite the latter experiencing a decline in recent years. The figure also emphasizes the importance of closing this yield gap to ensure food security in Guangdong Province, one of China's most economically developed and densely populated regions; it also illustrates the potential for Guangdong to improve its grain yield and align with the leading grain-producing provinces in China.

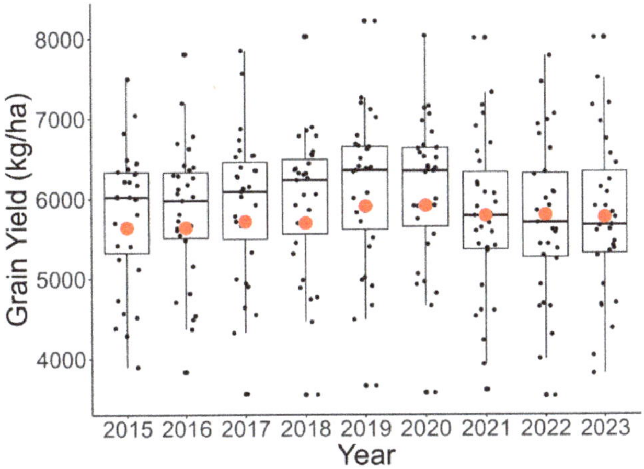

Figure 5. Comparison of grain yield between Guangdong Province and other provinces in China (2015–2023). The red dot in the figure highlights Guangdong's grain yield. Data source: *China Statistical Yearbook* [40].

3.4. Enhancing Grain Self-Sufficiency: Scenarios and Projections in Guangdong Province

Most existing studies focus on identifying the risk of regional grain self-efficiency, and this study makes further exploration by simulating the change in self-sufficiency of grain under different scenarios. Table 2 presents the projected changes in the rate of grain self-sufficiency (RSSG) for Guangdong Province under four distinct scenarios: Scenario A (baseline), Scenario B (grain loss reduction), Scenario C (grain yield improvement), and Scenario D (integrated scenario combining grain loss reduction and yield improvement). The simulations are conducted at both regional and provincial levels, covering the Pearl River Delta, east, north, and west Guangdong.

Table 2. Simulation of change of rate of grain self-sufficiency (*RSSG*) under different scenarios.

Region	Scenario A	Scenario B	Scenario C	Scenario D
Delta	16.58	18.27	18.23	20.1
East	22.57	24.63	25.28	27.58
North	45.55	49.17	49.19	53.1
West	39.9	43.07	43.09	46.52
All	28.14	30.58	30.7	34.14

Scenario A (baseline): This scenario reflects the current conditions of grain production and loss, with an *RSSG* of 28.14% for the entire province. Regional disparities are evident, with the north exhibiting the highest *RSSG* (45.55%) and the Pearl River Delta having the lowest (16.58%). These results align with the findings presented in Table 1, serving as a reference point for assessing the potential impacts of the proposed interventions.

Scenario B (grain loss reduction): In this scenario, grain loss in various regions of Guangdong Province decreases from the national average of 12%, reflecting Guangdong's potential to reduce post-harvest losses. The grain loss rate for different regions is set based on factors such as infrastructure, storage conditions, and transportation networks. The northern region, characterized by generally high elevations and mostly mountainous terrain, faces higher loss rates. In contrast, the southern region has flat terrain, dominated by plains and terraces. The Pearl River Delta is the largest plain in the province, followed by the Chaoshan Plain in the east [62]. Due to well-developed logistics facilities and advanced storage technologies, grain loss is lower in the eastern and Pearl River Delta regions. However, the loss rate is higher in the western and northern regions due to poor terrain conditions, inconvenient transportation, and relatively weak infrastructure [63,64]. Based on this, the Pearl River Delta region is assigned the lowest loss rate (3%), followed by the eastern region (4%), and then the western and northern regions, which have higher loss rates (5%, 5%).

The simulation results show a moderate increase in *RSSG* across all regions, with the provincial average rising to 30.58%. The north experiences the most significant improvement, with its *RSSG* increasing to 49.17%, while the Pearl River Delta experiences a more modest rise to 18.27%. This scenario highlights the importance of reducing grain loss as a viable strategy to enhance grain self-sufficiency.

Scenario C (grain yield improvement): This scenario assumes that Guangdong's grain yield can be increased. Different grain yield coefficients are set for each region based on factors such as terrain conditions, arable land resources, and agricultural technology levels. For example, in terms of arable land pressure, northern Guangdong has the highest pressure, followed by the Pearl River Delta and eastern Guangdong, with the lowest pressure in western Guangdong [36]. Additionally, the overall spatial distribution of arable land loss at the county level in Guangdong shows a trend of gradual increase from south to north and from west to east. The northern region has less arable land due to a higher proportion of mountainous and hilly areas. The Pearl River Delta plain is continuously losing arable land due to the influence of the economic zone, resulting in insufficient reserve land. In contrast, the coastal areas of eastern Guangdong, such as the Chaoshan Plain and the Leizhou Peninsula–Dianbai–Yangjiang area in the western coastal region of Guangdong Province, have relatively larger amounts of arable land [65,66]. Based on these differences, we have set different grain yield coefficients for each region. Eastern Guangdong has the highest yield (12%), aligning with the upper quartile of grain-producing provinces in China (as shown in Figure 4), followed by the Pearl River Delta (10%), and then western and northern Guangdong with lower yields (8%, 8%).

The results indicate a further improvement in *RSSG*, with the provincial average reaching 30.7%. The north again benefits the most, achieving an *RSSG* of 49.19%, while the Pearl River Delta improves to 18.23%. This scenario highlights the potential for yield improvement through technological progress and improved agricultural practices.

Scenario D (integrated scenario): Combining the effects of grain loss reduction and yield improvement, Scenario D represents the most optimistic projection. The provincial *RSSG* rises to 34.14%, exceeding the government's target of 30%. Regionally, the north achieves the highest *RSSG* at 53.1%, followed by the west (46.52%), the east (27.58%), and the Pearl River Delta (20.1%). This integrated approach demonstrates the synergistic effects of simultaneously addressing grain loss and yield improvement, providing a comprehensive pathway to enhance grain self-sufficiency in Guangdong.

Hence, Table 2 illustrates the potential for Guangdong to significantly improve its grain self-sufficiency through targeted interventions. The results suggest that a combination of grain loss reduction and yield improvement can elevate the provincial *RSSG* to 34.14%, thereby contributing to food security and aligning with the national "dual circulation" strategy. These results provide a clear roadmap for policymakers and stakeholders to prioritize investments in agricultural technology and infrastructure, ensuring a stable and resilient grain supply for Guangdong's growing population.

4. Discussion and Policy Recommendation

4.1. Storing Grain in the Land: Farmland Reservation for Grain Supply

China has experienced rapid and massive urban expansion in recent decades, and several studies have assessed the cropland loss due to settlement. Wang et al. (2019) [67] estimated that 5.92 million hectares (ha) or 3.31% of cropland were lost from 2000 to 2010. Huang et al. (2019) [68] found that the annual expansion and annual growth rate of farmland occupation were nearly 1900 ha and 5% in 291 Chinese cities during 1990–2015. Wu et al. (2024) [69] calculated that, from 1990 to 2020, approximately 4000 hectares (ha) of farmland were occupied annually. In order to mitigate the continuous farmland occupation from urbanization and industrialization, Chinese government stipulates that the limit of 1.8 billion mu (120 million ha) of farmland should be maintained to ensure national food security [57,70,71], which is called "Red Line of Farmland Protection Policy". In 2015, this policy has been emphasized by the "storing grain in the land" strategy [72]. Apart from urban expansion, a significant proportion of farmland has been converted for ecological restoration, through initiatives such as the grain-for-green programs [73–75]. Auch approaches can protect environmental and ecological systems, thereby ultimately enhancing the resilience of agricultural production. The trade-off between worries of short-term grain [73,75] and long-term agricultural sustainability suggests that these programs could bring substantial social and ecological benefits [74].

According to the trend in the sown area of grain in Guangdong, it has remained stable for a long time between district and county. This is because the Guangdong government has demonstrated a heightened commitment to farmland preservation by implementing stringent policies, which emphasize the maintenance of the red line of farmland protection, thereby reinforcing the foundation for sustainable food security [76].

4.2. Storing Grain in Technology: Improvement of Grain Yield

The "storing grain in the technology" strategy is a critical component in China's efforts to enhance agricultural productivity and ensure food security [77–79]. While the "storing grain in the land" policy focuses on preserving arable land, the technological approach emphasizes innovation to boost grain yields. However, as the comparative analysis of grain yields between Guangdong Province and other provinces reveals, a significant yield gap persists, highlighting both the challenges and opportunities for Guangdong Province to achieve higher agricultural productivity.

Although Guangdong's grain yield has shown some improvements in recent years, it remains below the upper quartile of grain-producing provinces such as Jilin and Heilongjiang. This gap highlights the urgent need to improve grain yield, which requires the adoption of advanced agricultural technologies [80], improved resource management [81], and stronger policy support [82]. Specific strategies include investing in precision farming technologies such as remote sensing, data analysis, and automated machinery to optimize resource use and increase yields. Additionally, improving soil quality through sustainable practices, optimizing irrigation systems, and promoting crop diversification can contribute to more stable and higher grain yields. Strengthening support for farmers through policy measures, financial incentives, and education on best agricultural practices is also essential. By improving its grain yield, Guangdong could enhance its food supply, contribute to the national "dual circulation" strategy, and strengthen its resilience against food supply disruptions, ultimately achieving food self-sufficiency.

The recent "New Round of Action Plan for Increasing Grain Production Capacity by 100 Billion Jin (2024–2030)", issued by the State Council in 2023, provides a timely framework for addressing these challenges [83]. The plan sets an ambitious target of increasing grain production capacity by 50 million tons by 2030, emphasizing the need for breakthroughs in several key areas: efficient utilization of arable land, optimization of planting structures, prevention of natural disasters, and reduction in grain processing loss. These priorities are closely aligned with Guangdong's current agricultural challenges, particularly in the context of its rapid urbanization and industrialization, which have placed significant strain on its limited agricultural resources.

One of the most promising aspects of the "grain storage through technology" strategy is the development and promotion of high-yield and high-quality crop varieties. In recent years, China has made significant progress in this area; Guangdong Province has also strengthened farmland protection and land use management, implementing multiple measures to enhance farmers' enthusiasm for grain production. The promotion of high-yield and high-quality varieties, the implementation of site-specific practices such as reasonable planting density, and the application of technologies like "one spray, multiple promotions" have all contributed to increasing grain yields and continuously enhancing food security [84]. However, compared to other provinces, Guangdong Province still faces some challenges in the area of food security technology. The growing constraints of resources and the environment, such as soil degradation and water scarcity, are particularly severe in Guangdong due to its high population density and intensive land use [85]. Furthermore, climate change has intensified resource–environmental pressures in the region, while the frequent occurrence of natural disasters poses a significant threat to grain production. To address these challenges, China's Grain Production Capacity Enhancement Action Plan proposes several reasonable solutions, including the improvement of soil quality, advancement in the construction of high-quality farmland, and optimization of planting patterns, with an emphasis on concentrating high-yield crops in favorable production areas [86]. These strategies can not only increase productivity, but also integrate environmental and ecological protection to achieve dual goals of food security and sustainable development. Specifically, by combining technological innovations with ecological measures—such as promoting water-saving irrigation technologies and constructing high-quality farmland—it is possible to increase grain yields while restoring and improving soil quality, alleviating water shortages, and enhancing the resilience of agricultural systems to climate change and natural disasters. This integrated approach, which prioritizes both "grain storage through technology" and ecological conservation, provides a sustainable pathway toward food sovereignty, ensuring the stability and resilience of grain production in Guangdong Province despite the pressures of climate change and resource–environmental constraints.

For Guangdong Province, these strategies provide a clear pathway for improving grain yield and achieving self-sufficiency. However, strategies must also address the broader socioeconomic factors that have hindered agricultural innovation. For instance, the rapid urbanization and industrialization of Guangdong have led to a shift in labor and investment away from agriculture, creating a need for policies that incentivize technological adoption and promote sustainable farming practices. Moreover, the province's reliance on grain imports from other regions highlights the urgency of enhancing local production capacity to reduce vulnerability to external supply disruptions.

In conclusion, while the "storing grain in the technology" strategy provides a robust framework for improving grain yields, its success in Guangdong Province will depend on the province's ability to address both technological and structural challenges. By leveraging national policies and investing in agricultural innovation, Guangdong Province can close the yield gap with leading grain-producing provinces, enhance its food security, and contribute to the broader national agenda of sustainable agricultural development. This dual focus on technology and policy will be essential for ensuring stable and resilient food supply in the face of growing internal and external pressures.

4.3. Grain Loss Reduction

Grain loss occurs during the stages of sowing, cultivation, harvesting, post-harvest handling, storage, and distribution [87]. Grain loss reduction is a critical component of enhancing food security, particularly in a high-density and economically developed region like Guangdong Province. The province faces significant challenges in minimizing post-harvest losses. Reducing these losses is not only essential for improving grain self-sufficiency but also aligns with the national "dual circulation" strategy, which emphasizes domestic production and resilience against external disruptions.

One of the major challenges in reducing grain loss is the impact of natural disasters. China loses millions of tons of grain annually due to extreme weather events such as droughts, floods, and typhoons [88]. In Guangdong Province, these events occur frequently due to its unique geography and climate, making it crucial for local governments to refine flood prevention and preparedness measures to enhance overall disaster response capabilities [89]. Recently, several regions in Guangdong Province have proactively planned for the 2025 flood prevention work, including hazard investigation and remediation, improving emergency disaster relief supply systems, and bolstering emergency material support during extreme events [90]. Although Guangdong Province has made significant progress in agricultural disaster prevention, it is recommended that the province strengthen meteorological monitoring and early warning systems to improve the accuracy and timeliness of disaster alerts [91]. In addition to preventive measures, effective crisis management strategies need to be implemented to minimize the impact of disasters on grain production. This includes flood control infrastructure such as reservoirs and barriers and ensuring that irrigation systems are capable of managing both drought and excessive rainfall. Furthermore, disaster-resistant agricultural technologies, such as drought-tolerant crop varieties and improved drainage systems, should also be promoted. Rapid post-crisis recovery plans should be in place, providing financial support, agricultural subsidies, and technical assistance for replanting and soil recovery. By strengthening both disaster prevention and recovery strategies, Guangdong can reduce the vulnerability of its grain production to climate-related disruptions and ensure a more resilient food supply [92].

In addition to natural disaster prevention, it is equally important to focus on reducing loss during grain processing, storage, and transportation. These stages of the supply chain are often overlooked but can result in substantial grain loss. For example, inefficient storage facilities, inadequate infrastructure, and poor transportation networks can lead to spoilage, contamination, and waste [93]. To address these challenges, the People's Government of Guangdong Province issued the 2022 Implementation Plan for Grain Saving and Loss Reduction. The plan focuses on strengthening grain saving and loss reduction across various stages of the supply chain, including agricultural production, grain storage, transportation, and processing. It also emphasizes enhancing public education and awareness on grain saving and loss reduction [94]. At present, measures for grain saving and loss reduction should be further strengthened and detailed across all stages of the supply chain. Technological innovation plays a key role in minimizing grain loss. The development of high-efficiency drying and storage technologies can help preserve grain quality and reduce post-harvest losses. Additionally, integrating digital tools and data analytics into the supply chain can improve monitoring and decision-making processes. Real-time data tracking can detect problems early and adjust operations accordingly, thus reducing losses [95].

Strategic policy interventions, such as financial support for infrastructure upgrades, training for better management practices, and the development of digital supply chain platforms, are critical in enhancing the effectiveness of these strategies [96]. By combining technological advancements with targeted policies, Guangdong Province can significantly reduce grain loss and improve overall food security. This integrated approach will not only improve the efficiency of grain storage and transportation, but also contribute to more sustainable food systems.

4.4. Interdepartmental Coordination

The successful implementation of food production and technological innovation policies, as discussed in Sections 4.1–4.3, depends on Guangdong's ability to address broader socioeconomic and infrastructure challenges contributing to grain loss. Additionally, Guangdong Province needs to focus on public policy areas, particularly urban planning and land protection. Rapid urbanization and industrialization have led to the occupation of large areas of farmland, reducing the amount of land available for grain production. A balance between urbanization and land protection is needed to ensure food security [97].

As China's largest province in terms of economy and population, Guangdong Province has made the strict protection of arable land a key political objective. The "Special Plan for Arable Land Protection in Guangdong Province (2021–2035)" was developed to guide the remediation of arable land, improve land quality, and support high-quality development in counties, towns, and villages [98]. It is also important to ensure the long-term stability of land for food production, prevent illegal or excessive land development, and promote efficient land use to avoid unregulated development that could affect grain production [99].

Furthermore, it is essential to enhance land reclamation and transfer policies. Some agricultural land may be left unused during urbanization, and an appropriate land transfer mechanism can improve land use [100]. The government should also encourage farmers to transfer land to support modern agricultural development and enhance land productivity.

Infrastructure development is also crucial for ensuring food supply. With urbanization, the construction of logistics and storage facilities becomes increasingly important [101]. Guangdong Province has held meetings to strengthen coordination on food security [102]. However, inter-departmental cooperation remains a challenge. The government needs to integrate policies on agriculture, urban planning, environmental protection, and economic development for unified action. This collaboration will promote land protection, sustainable agriculture, and food security. By raising public awareness of land resource protection, society can foster harmonious urban–rural development and ensure the sustainability of food production. As shown in Table 3, based on the analysis results of this study, the policy recommendations for the above aspects are summarized.

Table 3. Summary of policy recommendations for promoting grain self-sufficiency in Guangdong Province.

Policy Name	Policy Goal	Policy Measures
Storing Grain in the Land: Farmland Reservation for Grain Supply	Protect farmland resources and ensure food supply.	• Strengthen farmland protection policies. • Establish the "Red Line of Farmland Protection". • Limit the conversion of arable land.
Storing Grain in Technology: Improvement of Grain Yield	Increase grain yield and enhance agricultural productivity.	• Improve resource management. • Adopt advanced agricultural technologies. • Strengthen policy support and enhance farmer training.
Grain Loss Reduction	Reduce grain loss from production to consumption.	• Prevent natural disasters and ensuring post-disaster recovery. • Improve storage and transportation networks within the supply chain. • Develop and implement efficient drying and storage technologies.

Table 3. *Cont.*

Policy Name	Policy Goal	Policy Measures
Interdepartmental Coordination	Strengthen coordination between government departments and promote policy implementation.	• Optimize urban planning policies. • Strengthen the implementation of land reclamation and land transfer policies. • Enhance infrastructure construction and raise public awareness.

5. Conclusions

This study has comprehensively examined the grain self-sufficiency in Guangdong Province, focusing on the spatial and temporal patterns of the *RSSG* from 2014 to 2022. The findings reveal that Guangdong Province faces significant challenges in achieving its grain self-sufficiency target, with an average *RSSG* of 28.14%, falling short of the government's goal of 30%. The study also highlights significant regional disparities, with the Pearl River Delta region exhibiting the lowest *RSSG* due to its high population density and limited agricultural land. In contrast, the north region demonstrates a higher RSSG, primarily due to its relatively lower population density and greater agricultural capacity.

The analysis of grain production structure indicates that rice remains the predominant grain crop in Guangdong Province, accounting for over 75% of the total sowing area and production. This highlights the critical role of rice in maintaining food security in the province. However, the yield gap between Guangdong Province and other leading grain-producing provinces in China suggests that there is significant potential for improvement through technological advancements and policy support.

Scenario simulations show that improving grain yield and reducing grain loss could significantly increase the grain self-sufficiency rate, potentially raising it to 34.14%. This finding highlights the importance of implementing a dual strategy of "storing grain in the land" and "storing grain in the technology" to address challenges in farmland preservation and agricultural productivity. Specifically, this includes strengthening farmland protection policies, promoting sustainable agricultural practices, and investing in advanced agricultural technologies. Additionally, the study proposes several other policy recommendations, such as improving post-harvest management, enhancing storage facilities, and strengthening disaster prevention measures to reduce grain loss. It also emphasizes the need for strengthening cross-sectoral coordination, fostering collaboration between agriculture, environmental, and urban planning departments, and collectively promoting food security and higher self-sufficiency in grain production.

However, in implementing these measures, we may face some challenges. For example, in strengthening farmland protection, finding approaches for balancing urbanization with farmland preservation, encouraging farmers to adopt high-yield technologies, and improving infrastructure with limited resources are all difficulties that need to be overcome. Additionally, Guangdong's grain self-sufficiency rate is influenced by factors such as natural disasters, climate change, and fluctuations in the global supply chain, all of which require the government to adopt flexible responses.

The research methods and results are not only applicable to Guangdong Province but also provide valuable insights for other countries or regions, particularly those facing similar urbanization processes, agricultural production issues, and food security challenges. Future research could further explore the applicability of grain self-sufficiency policies in other countries, especially how to adjust policies based on different countries' economic conditions, social structures, and climatic characteristics. This would help provide more practical experiences and references for global food security and sustainable agricultural development.

In the context of increasing global uncertainties, such as natural disasters, epidemics, and international conflicts, the findings of this study highlight the urgency of enhancing Guangdong's grain self-sufficiency. By improving the level of grain self-sufficiency, Guangdong can contribute to China's overall food security strategy and reduce its vulnerability to external supply disruptions. The insights provided in this study offer a clear roadmap for policymakers and stakeholders, guiding them to prioritize investments in agricultural technology and infrastructure to ensure the stable and resilient food supply for Guangdong's growing population. Future research should focus on exploring the socioeconomic and environmental impacts of grain self-sufficiency policies and finding innovative solutions to address the challenges of sustainable agricultural development in the context of rapid urbanization and climate change.

Author Contributions: Conceptualization, S.W.; methodology, S.W.; software, S.W.; validation, Y.X.; formal analysis, S.W. and Y.X.; investigation, S.W. and Y.Z.; resources, S.W. and J.Z.; data curation, Y.Z.; writing—original draft preparation, S.W. and Y.X.; writing—review and editing, Y.Z. and J.Z.; visualization, S.W.; supervision, Y.X.; project administration, Y.X.; funding acquisition, S.W. and Y.X. All authors have read and agreed to the published version of the manuscript.

Funding: Guangdong Rural Science and Technology Special Commissioner Project: Sustainable Utilization of Forest Resource Advantages in Luofu Town (2024A0301036); Jinan University Enterprise Development Research Institute 2023 Annual Research Project (QF20230902); Guangdong Province College Students' Innovation and Entrepreneurship Training Program Supported Project (S202410559033); Special Funds for Cultivation of Guangdong College Students' Scientific and Technological Innovation ("Climbing Program" Special Funds).

Institutional Review Board Statement: Not applicable.

Informed Consent Statement: Not applicable.

Data Availability Statement: The raw data supporting the conclusions of this article will be made available by the authors on request due to specific reason or restrictions.

Conflicts of Interest: The authors declare no conflicts of interest.

Abbreviations

The following abbreviations are used in this manuscript:

RSSG	rate of grain self-sufficiency
GSSG	gap of grain self-sufficiency

References

1. Mulders, P.J.; van den Heuvel, E.R.; van de Molengraft, M.J.G.; Heemels, W.P.M.H.; Reidsma, P. Extreme drought and rainfall had a large impact on potato production in the Netherlands between 2015 and 2020. *Commun. Earth Environ.* **2024**, *5*, 496. [CrossRef]
2. van der Veer, S.; Hamed, R.; Karabiyik, H.; Roskam, J.L. Mitigating the effects of extreme weather on crop yields: Insights from farm management strategies in the Netherlands. *Environ. Res. Lett.* **2024**, *19*, 104042.
3. Yıldırım, C.; Önen, H.G. Vulnerabilities of the neoliberal global food system: The Russia–Ukraine War and COVID-19. *J. Agrar. Change* **2024**, *24*, e12601.
4. Singh, S.; Kumar, R.; Panchal, R.; Tiwari, M.K. Impact of COVID-19 on logistics systems and disruptions in food supply chain. *Int. J. Prod. Res.* **2021**, *59*, 1993–2008.
5. Deconinck, K.; Avery, E.; Jackson, L.A. Food Supply Chains and COVID-19: Impacts and Policy Lessons. *Eurochoices* **2020**, *19*, 34–39.
6. Godrich, S.L.; Lo, J.; Kent, K.; Macau, F.; Devine, A. A mixed-methods study to determine the impact of COVID-19 on food security, food access and supply in regional Australia for consumers and food supply stakeholders. *Nutr. J.* **2022**, *21*, 17. [PubMed]
7. Huss, M.; Brander, M.; Kassie, M.; Ehlert, U.; Bernauer, T. Improved storage mitigates vulnerability to food-supply shocks in smallholder agriculture during the COVID-19 pandemic. *Glob. Food Secur. Agric. Policy Econ. Environ.* **2021**, *28*, 100468.

8. Zhong, T.Y.; Crush, J.; Si, Z.Z.; Scott, S. Emergency food supplies and food security in Wuhan and Nanjing, China, during the COVID-19 pandemic: Evidence from a field survey. *Dev. Policy Rev.* **2022**, *40*, e12575.
9. Hellegers, P. Food security vulnerability due to trade dependencies on Russia and Ukraine. *Food Secur.* **2022**, *14*, 1503–1510.
10. Lin, F.; Li, X.; Jia, N.; Feng, F.; Huang, H.; Huang, J.; Fan, S.; Ciais, P.; Song, X.-P. The impact of Russia-Ukraine conflict on global food security. *Glob. Food Secur.* **2023**, *36*, 100661.
11. Veeck, G.; Veeck, A.; Yu, H. Challenges of agriculture and food systems issues in China and the United States. *Geogr. Sustain.* **2020**, *1*, 109–117.
12. Marchant, M.A.; Wang, H.H. Theme Overview: U.S.—China Trade Dispute and Potential Impacts on Agriculture. *Choices* **2018**, *33*, 1–3.
13. Genevieve Donnellon-May. China's 'Food Silk Road' Holds Key to Thwarting Trump Tariffs. *South China Morning Post.* **2024**. Available online: https://www.scmp.com/opinion/china-opinion/article/3291017/chinas-food-silk-road-holds-key-thwarting-trump-tariffs (accessed on 10 March 2025).
14. Zhu, X.; Zhang, Y.; Zhu, Y.; Li, Y.; Cui, J.; Yu, B. Multidimensional deconstruction and workable solutions for addressing China's food security issues: From the perspective of sustainable diets. *Land Use Policy* **2025**, *148*, 107401.
15. Ito, J.; Li, X. Interplay between China's grain self-sufficiency policy shifts and interregional, intertemporal productivity differences. *Food Policy* **2023**, *117*, 102446. [CrossRef]
16. Chen, H.; He, Y.; Wang, X.; Luo, S.; Jiao, Y.; Han, S. Mapping China's reserved arable lands and its sustainable cropping layout. *J. Sci. Food Agric.* **2024**, *104*, 9180–9190. [PubMed]
17. Zhang, L.; Cai, C.; Kripal, S.; Zhong, K. Green technology innovation, trade deficit and carbon emission transfer in agriculture under the new "dual circulation" development pattern of China. *Ecol. Indic.* **2024**, *159*, 111757.
18. Schillerberg, T.; Tian, D. Changes in crop failures and their predictions with agroclimatic conditions: Analysis based on earth observations and machine learning over global croplands. *Agric. For. Meteorol.* **2023**, *340*, 109620.
19. Liu, J.; Ma, X.; Zhao, B.; Dang, Z. Steering toward sustainability: Can dual circulation development mitigate CO_2 emissions?—Evidence from China. *J. Clean. Prod.* **2024**, *474*, 143548.
20. Chang, Y. Accelerating domestic and international dual circulation in the Chinese shipping industry—An assessment from the aspect of participating RCEP. *Mar. Policy* **2022**, *135*, 104862. [CrossRef]
21. Kazancoglu, Y.; Sezer, M.D.; Ozbiltekin-Pala, M.; Lafçi, C.; Sarma, P.R.S. Evaluating resilience in food supply chains during COVID-19. *Int. J. Logist. Res. Appl.* **2024**, *27*, 688–704.
22. Liu, Y.; Zhou, Y. Reflections on China's food security and land use policy under rapid urbanization. *Land Use Policy* **2021**, *109*, 105699. [CrossRef]
23. Zhu, Y.; Wang, Z.; Zhu, X. New reflections on food security and land use strategies based on the evolution of Chinese dietary patterns. *Land Use Policy* **2023**, *126*, 106520. [CrossRef]
24. Yu, W.; Elleby, C.; Zobbe, H. Food security policies in India and China: Implications for national and global food security. *Food Secur.* **2015**, *7*, 405–414. [CrossRef]
25. Zhang, Y.; Lu, S. Food politics in China: How strengthened accountability enhances food security. *Food Policy* **2024**, *128*, 102692. [CrossRef]
26. Zhou, Z.Y. *Achieving Food Security in China: The Challenges Ahead*; Routledge: London, UK, 2017.
27. Ghose, B. Food security and food self-sufficiency in China: From past to 2050. *Food Energy Secur.* **2014**, *3*, 86–95. [CrossRef]
28. Chen, Z.; Zhang, M.; Dong, R.K.; Wang, S. Building Resilient Food Security Against Global Crisis: New Evidence from China. *Food Energy Secur.* **2024**, *13*, e70008. [CrossRef]
29. Zeng, Q.; Chen, X. Identification of urban-rural integration types in China–an unsupervised machine learning approach. *China Agric. Econ. Rev.* **2023**, *15*, 400–415. [CrossRef]
30. Zhang, Q.; Bista, R.; Bilsborrow, R.E.; Zhang, Z.; Huang, Q.; Song, C. Understanding the mediating role of labor out-migration in household income generation and distribution under a reforestation policy in rural China. *J. Environ. Manag.* **2023**, *345*, 118539. [CrossRef]
31. Sun, Y.; Xu, C.; Zhang, H.; Wang, Z. Migration in response to climate change and its impact in China. *Int. J. Clim. Change Strateg. Manag.* **2017**, *9*, 352–373. [CrossRef]
32. Chen, A.; Groenewold, N. Reducing regional disparities in China: An evaluation of alternative policies. *J. Comp. Econ.* **2010**, *38*, 189–198. [CrossRef]
33. Cao, Q.; Song, J.; Liu, C.; Yang, W. Evolving water, energy and carbon footprints in China's food supply chain. *J. Clean. Prod.* **2023**, *423*, 138716.
34. Fan, S.; Brzeska, J. Feeding more people on an increasingly fragile planet: China's food and nutrition security in a national and global context. *J. Integr. Agric.* **2014**, *13*, 1193–1205.
35. Yu, X.; Schweikert, K.; Li, Y.; Ma, J.; Doluschitz, R. Farm size, farmers' perceptions and chemical fertilizer overuse in grain production: Evidence from maize farmers in northern China. *J. Environ. Manag.* **2023**, *325*, 116347.

36. Fang, W.; Yang, Z.; Liu, Z.; Hatab, A.A. Green recovery of cropland carrying capacity in developed regions: Empirical evidence from Guangdong, China. *Econ. Change Restruct.* **2023**, *56*, 2405–2436.
37. Liu, G.; Wang, H.; Cheng, Y.; Zheng, B.; Lu, Z. The impact of rural out-migration on arable land use intensity: Evidence from mountain areas in Guangdong, China. *Land Use Policy* **2016**, *59*, 569–579.
38. Zhang, P.; Zhou, Y.; Xie, Y.; Wang, Y.; Yang, Z.; Cai, Y. Spatial transmission mechanism of the water, energy and food nexus risks for the Guangdong-Hong Kong-Macao region of China. *J. Clean. Prod.* **2023**, *405*, 136906. [CrossRef]
39. Yang, H.; Zou, R.; Hu, Y.; Wang, L.; Xie, Y.; Tan, Z.; Zhu, Z.; Zhu, A.-X.; Gong, J.; Mao, X. Sustainable utilization of cultivated land resources based on element coupling-function synergy analytical framework: A case study of Guangdong, China. *Land Use Policy* **2024**, *146*, 107316.
40. National Bureau of Statistics of China. *China Statistical Yearbook*; China Statistics Press: Beijing, China, 2023.
41. Mutelo, A.; Zhen, C.; Wei, S.; Hamukwaya, S. High-standard farmland destruction monitoring by high-resolution remote sensing methods: A 2017–2018 case study of Hebei and Guangdong, China. *Environ. Monit. Assess.* **2023**, *195*, 742. [CrossRef]
42. Chen, Y.; Chen, X.; Zhang, P.; Tan, K.; Liu, S.; Chen, S.; Yang, Z.; Wang, X. Value compensation of net carbon sequestration alleviates the trend of abandoned farmland: A quantification of paddy field system in China based on perspectives of grain security and carbon neutrality. *Ecol. Indic.* **2022**, *138*, 108815.
43. Guangdong Government. Plan of Rural Revitalization Strategy in Guangdong Province. 2018. Available online: https://drc.gd.gov.cn/attachment/0/371/371110/2575321.pdf (accessed on 10 March 2025).
44. Clapp, J. Food self-sufficiency: Making sense of it, and when it makes sense. *Food Policy* **2017**, *66*, 88–96. [CrossRef]
45. Niu, Y.; Xie, G.; Xiao, Y.; Liu, J.; Zou, H.; Qin, K.; Wang, Y.; Huang, D. The story of grain self-sufficiency: China's food security and food for thought. *Food Energy Secur.* **2022**, *11*, e344. [CrossRef]
46. Li, Y.; Zhang, W.; Ma, L.; Wu, L.; Shen, J.; Davies, W.J.; Oenema, O.; Zhang, F.; Dou, Z. An analysis of China's grain production:looking back and looking forward. *Food Energy Secur.* **2014**, *3*, 19–32.
47. Wu, G.; Xia, B.; Dong, S.; Zhang, J.; Li, Z.; Yang, G. Spatio-Temporal Differentiation and Driving Factors of County-Level Food Security in the Yellow River Basin: A Case Study of Ningxia, China. *Land* **2024**, *13*, 1885. [CrossRef]
48. Gareau, S.E. The development of guidelines for implementing information technology to promote food security. *Agric. Hum. Values* **2004**, *21*, 273–285. [CrossRef]
49. Pan, J.; Chen, Y.; Zhang, Y.; Chen, M.; Fennell, S.; Luan, B.; Wang, F.; Meng, D.; Liu, Y.; Jiao, L. Spatial-temporal dynamics of grain yield and the potential driving factors at the county level in China. *J. Clean. Prod.* **2020**, *255*, 120312.
50. Xiao, Y.; Cheng, S.; Xie, G.; Liu, A.M.; Lu, C.X.; Wang, Y.Y. The balance between supply and consumption of the main types of grain in China. *J. Nat. Resour.* **2017**, *32*, 927–936. (In Chinese)
51. Wu, W.H. Research Report on Cherishing Food and Opposing Waste by the Special Investigation Team of the Standing Committee of the National People's Congress. In Proceedings of the 24th Meeting of the 13th Standing Committee of the National People's Congress, Beijing, China, 23 December 2020.
52. Li, C. China Must Have Control over Its Own Food Supply. 2022. Available online: http://en.cppcc.gov.cn/2022-07/08/c_788500.htm (accessed on 10 March 2025).
53. Jin, T.; Xia, Q.; Yue, M.M.; Pan, Q.Q.; Nai, J. Analysis of Grain Production and Demand Pattern and Optimization Strategy in Jiangsu Province from the Perspective of Grain and Feed. *Econ. Geogr.* **2016**, *36*, 136–141.
54. Yang, D.Q.; Jiang, H.P. Study on Food Security Issues in China's Major Grain Marketing Areas-From the Perspective of the Gap in Grain Production and Demand Balance. *China Agric. Sci. Technol. Bull.* **2017**, *19*, 1–9.
55. Xin, X.F.; Wang, J.M. It is Necessary to Attach Great Importance to the Over-concentration of Grain Net Export Provinces in China. *Macroecon. Manag.* **2016**, *32*, 53–55+60.
56. Meng, Z.D.; Li, G.X. Analysis of the Evolution of the Spatiotemporal Pattern of China's Grain Production and Demand Balance-From the Perspective of Grain Use and Provincial Level. *Agric. Mod. Res.* **2020**, *41*, 928–936.
57. Qi, X.; Vitousek, P.M.; Liu, L. Provincial food security in China:a quantitative risk assessment based on local food supply and demand trends. *Food Secur.* **2015**, *7*, 621–632.
58. Li, J.; Li, Z. Physical limitations and challenges to Grain Security in China. *Food Secur.* **2014**, *6*, 159–167.
59. Bureau of Statistics of Guangdong. Guangdong Statistical Yearbook. China Statistics Press: Beijing, China, 2015–2024. Available online: https://stats.gd.gov.cn/gdtjnj/ (accessed on 10 March 2025).
60. Bureau of Statistics of Guangdong. Guangdong Rural Statistical Yearbook. China Statistics Press: Beijing, China, 2015–2024. Available online: https://stats.gd.gov.cn/gdnctjnj/ (accessed on 10 March 2025).
61. Talhelm, T.; Zhang, X.; Oishi, S.; Shimin, C.; Duan, D.; Lan, X.; Kitayama, S. Large-scale psychological differences within China explained by rice versus wheat agriculture. *Science* **2014**, *344*, 603–608.
62. Yang, R.; Zhang, X.; Xu, Q. Spatial distribution characteristics and influencing factors of agricultural specialized villages in Guangdong Province, China. *Chin. Geogr. Sci.* **2022**, *32*, 1013–1034.

63. Yang, R.; Xu, Q.; Xu, X.; Chen, Y. Rural settlement spatial patterns and effects: Road traffic accessibility and geographic factors in Guangdong Province, China. *J. Geogr. Sci.* **2019**, *29*, 213–230.
64. Zhang, S.; Ma, W.; Wu, F.; Zhao, K. Unveiling the Impact of Transportation Infrastructure Construction on Rurality: A Case Study from Guangdong, China. *Buildings* **2024**, *14*, 2288. [CrossRef]
65. Wang, D.; Yang, H.; Hu, Y.; Zhu, A.X.; Mao, X. Analyzing Spatio-Temporal Characteristics of Cultivated Land Fragmentation and Their Influencing Factors in a Rapidly Developing Region: A Case Study in Guangdong Province, China. *Land* **2022**, *11*, 1750. [CrossRef]
66. Lu, Y.F.; Lin, J.Y. Predicting the impact of urban agglomeration development on grain production regions: A case study of Pearl River Delta. *J. Nat. Resour.* **2023**, *38*, 1532–1549.
67. Wang, L.Y.; Anna, H.; Zhang, L.Y.; Xiao, Y.; Wang, Y.Q.; Xiao, Y.; Liu, J.G.; Ouyang, Z.Y. Spatial and Temporal Changes of Arable Land Driven by Urbanization and Ecological Restoration in China. *Chin. Geogr. Sci.* **2019**, *29*, 809–819.
68. Huang, X.; Xia, J.H.; Xiao, R.; He, T. Urban expansion patterns of 291 Chinese cities, 1990–2015. *Int. J. Digit. Earth* **2019**, *12*, 62–77.
69. Wu, X.R.; Zhao, N.; Wang, Y.W.; Zhang, L.Q.; Wang, W.; Liu, Y.S. Cropland non-agriculturalization caused by the expansion of built-up areas in China during 1990–2020. *Land Use Policy* **2024**, *146*, 107312. [CrossRef]
70. Cao, Y.; Zhang, X.L.; Zhang, X.K.; Li, H. The incremental construction land differentiated management framework: The perspective of land quota trading in China. *Land Use Policy* **2020**, *96*, 104675.
71. Chen, G.Q.; Han, M.Y. Virtual land use change in China 2002–2010: Internal transition and trade imbalance. *Land Use Policy* **2015**, *47*, 55–65.
72. Wu, S.; Wang, D. Storing Grain in the Land: The Gestation, Delineation Framework, and Case of the Two Zones Policy in China. *Land* **2023**, *12*, 806. [CrossRef]
73. Shi, P.; Feng, Z.H.; Gao, H.D.; Li, P.; Zhang, X.M.; Zhu, T.T.; Li, Z.B.; Xu, G.C.; Ren, Z.P.; Xiao, L. Has "Grain for Green" threaten food security on the Loess Plateau of China? *Ecosyst. Health Sustain.* **2020**, *6*, 1709560.
74. Yibin, W.; Fei, W.; Jian, W.; Hongyu, C.; Mengfei, L. The social-ecological benefits of grain for green program based on coupled coordination network:Taking the China's Loess Plateau as an example. *Land Use Policy* **2024**, *143*, 107211.
75. Dong, Y.; Ren, Z.; Fu, Y.; Yang, R.; Sun, H.; He, X. Land use/cover change and its policy implications in typical agriculture-forest ecotone of central Jilin Province, China. *Chin. Geogr. Sci.* **2021**, *31*, 261–275.
76. Department of Natural Resources of Guangdong Province. Preserving Cultivated Land and Safeguarding the Red Line to Strengthen the Foundation of Grain Production. Available online: https://nr.gd.gov.cn/ztzlnew/dzhdflzjs/content/post_4382011.html (accessed on 27 February 2024).
77. Sun, C.Z.; Hao, S. Research on the competitive and synergistic evolution of the water-energy-food system in China. *J. Clean. Prod.* **2022**, *365*, 132743.
78. Li, X.L.; Wu, K.N.; Yang, Q.J.; Hao, S.H.; Feng, Z.; Ma, J.L. Quantitative assessment of cultivated land use intensity in Heilongjiang Province, China, 2001–2015. *Land Use Policy* **2023**, *125*, 106505.
79. Xu, D.D.; Liu, Y.; Li, Y.C.; Liu, S.Q.; Liu, G.H. Effect of farmland scale on agricultural green production technology adoption: Evidence from rice farmers in Jiangsu Province, China. *Land Use Policy* **2024**, *147*, 107381.
80. Fan, M.; Shen, J.; Yuan, L.; Jiang, R.; Chen, X.; Davies, W.J.; Zhang, F. Improving crop productivity and resource use efficiency to ensure food security and environmental quality in China. *J. Exp. Bot.* **2012**, *63*, 13–24. [CrossRef]
81. McLaughlin, D.; Kinzelbach, W. Food security and sustainable resource management. *Water Resour. Res.* **2015**, *51*, 4966–4985. [CrossRef]
82. Ingram, J.; Gregory, P.; Izac, A.-M. The role of agronomic research in climate change and food security policy. *Agric. Ecosyst. Environ.* **2008**, *126*, 4–12. [CrossRef]
83. Ministry of Agriculture and Rural Affairs (2024). New Round of Action Plan for Increasing Grain Production Capacity by 100 Billion Jin (2024–2030). Available online: http://www.jhs.moa.gov.cn/zcjd/202406/t20240620_6457551.htm (accessed on 20 June 2024).
84. Guangdong Provincial Development and Reform Commission. Guangdong Provincial Modern Industrial System Development Report (2023–2024). Available online: https://www.nfncb.cn/yaowen/47983.html (accessed on 5 February 2025).
85. Lan, Y.; Xu, B.; Huan, Y.; Guo, J.; Liu, X.; Han, J.; Li, K. Food security and land use under sustainable development goals: Insights from food supply to demand side and limited arable land in China. *Foods* **2023**, *12*, 4168. [CrossRef] [PubMed]
86. Bommarco, R.; Kleijn, D.; Potts, S.G. Ecological intensification: Harnessing ecosystem services for food security. *Trends Ecol. Evol.* **2013**, *28*, 230–238. [CrossRef] [PubMed]
87. Yan, D.; Liu, X.J.; Hao, X.P.; Li, J.K. Tracing environmental impacts of grain losses along the supply chain in the North China Plain:An integrated framework. *Resour. Conserv. Recycl.* **2023**, *189*, 106771. [CrossRef]
88. Chavez, E.; Conway, G.; Ghil, M.; Sadler, M. An end-to-end assessment of extreme weather impacts on food security. *Nat. Clim. Change* **2015**, *5*, 997–1001. [CrossRef]

89. Cheng, J.; Bambrick, H.; Yakob, L.; Devine, G.; Frentiu, F.D.; Williams, G.; Li, Z.; Yang, W.; Hu, W. Extreme weather conditions and dengue outbreak in Guangdong, China: Spatial heterogeneity based on climate variability. *Environ. Res.* **2021**, *196*, 110900. [CrossRef]
90. Department of Emergency Management of Guangdong Province. Better Safe Than Sorry! Several Regions in Guangdong Have Proactively Planned and Effectively Prepared for the 2025 Flood Prevention and Preparedness Work. Available online: http://yjgl.gd.gov.cn/xw/dsgzdt/content/post_4674126.html (accessed on 28 February 2025).
91. Wang, Y.; Liu, C.; Wang, Y.; Liu, Y.; Liu, T. Climate risk assessment and adaption ability in China's coastal urban agglomerations-A case study of Guangdong-Hong Kong-Macao greater bay area. *J. Clean. Prod.* **2024**, *452*, 142036. [CrossRef]
92. Wei, T.; Glomsrød, S.; Zhang, T. Extreme weather, food security and the capacity to adapt–the case of crops in China. *Food Secur.* **2017**, *9*, 523–535. [CrossRef]
93. Maiyar, L.M.; Thakkar, J.J. Modelling and analysis of intermodal food grain transportation under hub disruption towards sustainability. *Int. J. Prod. Econ.* **2019**, *217*, 281–297.
94. People's Government of Guangdong Province. Notice of the General People's Government of Guangdong Province on Issuing the Implementation Plan for Grain Saving and Loss Reduction in Guangdong Province. Available online: https://www.gd.gov.cn/zwgk/gongbao/2022/24/content/post_4007405.html (accessed on 10 March 2025).
95. Bomoi, M.I.; Nawi, N.M.; Aziz, S.A.; Kassim, M.S.M. Sensing technologies for measuring grain loss during harvest in paddy field: A review. *AgriEngineering* **2022**, *4*, 292–310. [CrossRef]
96. Tian, S.; Wu, L.; Pawar, K.S. Achieving food supply chain sustainability through digital sharing platforms: Prospects in China. *Ind. Manag. Data Syst.* **2025**, *125*, 1023–1051.
97. Gren, Å.; Andersson, E. Being efficient and green by rethinking the urban-rural divide—Combining urban expansion and food production by integrating an ecosystem service perspective into urban planning. *Sustain. Cities Soc.* **2018**, *40*, 75–82.
98. People's Government of Guangdong Province. Public Announcement of the Special Plan for Arable Land Protection in Guangdong Province: Building a New Pattern of Arable Land Protection. Available online: https://www.gd.gov.cn/zwgk/zcjd/snzcsd/content/post_4268206.html (accessed on 10 March 2025).
99. Zhou, Y.; Li, X.; Liu, Y. Cultivated land protection and rational use in China. *Land Use Policy* **2021**, *106*, 105454. [CrossRef]
100. Petrescu-Mag, R.M.; Petrescu, D.C.; Reti, K.O. My land is my food: Exploring social function of large land deals using food security–land deals relation in five Eastern European countries. *Land Use Policy* **2019**, *82*, 729–741.
101. Hammond, S.T.; Brown, J.H.; Burger, J.R.; Flanagan, T.P.; Fristoe, T.S.; Mercado-Silva, N.; Okie, J.G. Food spoilage, storage, and transport: Implications for a sustainable future. *BioScience* **2015**, *65*, 758–768. [CrossRef]
102. National Food and Strategic Reserves Administration. The First Plenary Meeting of the Inter-Departmental Joint Conference on Food Security in Guangdong Province Was Held in Guangzhou. Available online: https://www.lswz.gov.cn/html/hybd2022/2022-01/25/content_269201.shtml (accessed on 10 March 2025).

foods

MDPI

Article

Getting Insights in Stakeholder Collaboration in the Transition Toward Safe and Sustainable Food Production: Net-Map Analysis of the Italian Wheat Supply Chain

Biancamaria Ciasca *, Nunzia M. Cito and Veronica M. T. Lattanzio

Institute of Sciences of Food Production, National Research Council of Italy, Via Amendola 122/O, 70126 Bari, Italy; nunzia.cito@ispa.cnr.it (N.M.C.); veronicamariateresa.lattanzio@cnr.it (V.M.T.L.)
* Correspondence: biancamaria.ciasca@ispa.cnr.it

Abstract: The transition toward sustainable food systems is a critical challenge for the European Union (EU), requiring integrated policies and collaborative action. However, the complexity of food systems and the diverse interests of stakeholders often hinder effective progress. To apply a systemic approach and deliver context-specific solutions, it is necessary to establish key commodity/supply chain groups that include diverse perspectives from multiple stakeholders. Focusing on the Italian wheat supply chain, this research addresses the gap in understanding stakeholder relationships and their influence on the definition and implementation of sustainability measures. It introduces Net-Map analysis as a fit-for-purpose, multistakeholder, participatory approach to map and analyze these relationships, identifying key actors, power dynamics, and opportunities for fostering collaboration. Through a participatory workshop and in-depth interviews, the study gathered data providing insights into stakeholder roles and leverage points. The findings contribute to improved stakeholder involvement in decision-making processes, ultimately supporting the development of a more resilient and sustainable wheat supply chain in Italy.

Keywords: food chain sustainability; multi-actor approach; wheat supply chain

1. Introduction

Reaching sustainable food systems requires building sustainability culture, setting an enabling policy framework, and taking collective actions. A harmonized legislative framework would allow food systems' stakeholders across EU countries to jointly respond to sustainability demands set in a number of policies, seeking synergies while respecting local and context-specific situations. In this respect, the EU policy initiative for a Legislative Framework on Sustainable Food Systems [1] is expected to play a pivotal role in accelerating the transition toward sustainable food production. The proposal for regulation should lay down general principles and objectives, together with the requirements and responsibilities of all actors in the EU food system. It should ensure coherence with all EU food-related policies (e.g., agriculture, fisheries, and aquaculture) in terms of sustainability objectives, including biodiversity and climate objectives.

The proposal was expected to be tabled in the fourth quarter of 2023. To date, it is not mentioned on the tentative agenda for forthcoming Commission meetings or in the Commission work program for 2024. The timeline for presenting this proposal is still undetermined, likely due to the complexity of the issue given its requirements for a multidimensional and integrative approach that considers all relevant actors.

Shifting to sustainable food production requires food safety attention, even more, due to new food system transformation practices targeting circularity, diversity, etc. [2]. The Farm to Fork Strategy, as a key element of the European Green Deal, recognizes that food safety and security are at the core of sustainable food systems [3]. Bené et al. [4] highlighted this aspect in their rigorous analysis of the metrics/indicators for food system sustainability. Four dimensions are widely considered when exploring sustainable food systems: environmental, economic, social, and food/nutritional. The food/nutritional dimension may be further categorized to include the food safe subdimension, along with food security, waste/losses, and nutrition.

Due to the multidimensional nature of food systems sustainability and the related indicators, coordinated action and collective efforts by multiple stakeholders along the supply chain are essential to unravel and address challenges in an integrative manner [5] One of the primary gaps to be addressed for effective implementation of sustainability measures is the alignment of stakeholders around a common vision and goals, transcending their specific sectors.

Recent studies have shown a growing focus on the implications and challenges of establishing meaningful collaborations among the diverse range of stakeholders in sustainable food supply chains. This topic has been extensively discussed by Dania et al. (2018) [6], unravelling both vertical and horizontal collaboration approaches. Vertical collaboration, which is the focus of the present study, occurs between actors at different levels of the supply chain (e.g., farmers, processors, retailers), while horizontal collaboration involves actors at the same level (e.g., farmer cooperatives) [7]. The review by Dania et al. (2018) [6] highlights how vertical collaboration is crucial in food supply chains since it improves traceability systems, quality control mechanisms, and the efficiency of resource allocation and enhances the information flow from farm to fork. A quite comprehensive bibliometric analysis of the topic "Collaboration for the sustainable food supply chain", built on a sample of 528 articles, has been reported by Duong et al. (2023) [8]. This analysis reveals a consensus on the need for collaboration due to the complexity of the food industry and the nature of the food products. Some gaps to be covered by future research have been identified, including the need for more studies to explore ways to engage stakeholders in the sustainability transition, with a stronger focus on social sustainability, and in the emerging economies context. A promising approach to foster stakeholder engagement has been proposed by Ammirato et al. (2021) [9] who introduced the concept of network-based collaboration, combining both vertical and horizontal elements. The authors see this approach as an effective strategy to help supply chain actors in reaching sustainability goals, since it creates multi-stakeholder spaces for dialogue, facilitates knowledge exchange across the entire supply chain, enables collective problem-solving, and promotes innovation through diverse perspectives.

A specific focus on collaboration schemes for sustainability practices' implementation in Italian supply chains can be found in León-Bravo et al. (2017) [10]. Their findings suggest that sustainability performance, in several areas including sustainable sourcing and green processing, packaging, transportation, and social sustainability can be associated with some collaborative practices implemented in the upstream and/or downstream. With respect to vertical collaboration, the study also clearly shows the heterogeneity of relationships established among actors for sustainability purposes, concluding that no single standard frame can be used to manage them.

Establishing key commodity/supply chain groups that include diverse perspectives from multiple stakeholders is necessary for delivering more innovative and context-specific solutions [11]. The wheat supply chain in Italy comprises numerous stakeholders, including farmers, processors, and retailers, but it remains fragmented. Different stakeholders have varying priorities and levels of commitment to sustainability, which complicates collaborative efforts. For example, while some companies may prioritize sustainability for brand enhancement, farmers often focus more on immediate economic concerns. Effective coordination among these parties is crucial for implementing sustainability measures; however, existing communication channels are frequently inadequate, leading to inefficiencies and misunderstandings about sustainability goals [12]. This situation can result in limited participation in the co-creation of knowledge and practices related to sustainability.

To cope with this fragmentation, a first step is represented by the characterization and assessment of multi-stakeholder roles and connections. In this context, Net-Map analysis has been proven to be a fit-for-purpose tool. The Net-Map methodology is an interview-based mapping tool that involves many categories of stakeholders (e.g., policy, food business, society, and science) within a given system in a structured participatory process. It allows for (1) visualizing formal and informal networks within the defined system, unravelling relationships, spheres of influence, and objectives; (2) discovering sources of conflicts and possibilities for cooperation; (3) facilitating knowledge exchange and mutual learning processes to ultimately foster the development of visions and strategies to achieve common goals. It was originally developed to map multiple social relationships for a diversity of purposes, such as understanding knowledge flows in governance networks or detecting actors that have influence in certain governance processes based on their position in the network [13,14]. This analysis can facilitate the understanding of some national (or regional) levels of connections between stakeholders, which sometimes may not be fully integrated, due to different roles and rigid institutional functions. The Net-Map analysis has been also implemented to support the evaluation of the City Region Food System to identify the key actors involved in the transformation of the local food system [15].

The Net-Map methodology, adapted in a digital version, has been implemented in a recent study to analyze multi-stakeholder networks in risk analysis of emerging food safety issues [16]. The study resulted in complex Net-Maps, visualizing how the science–policy–society collaboration system regarding risk analysis of the selected food safety issues (food contact materials) is set up for four target EU countries. The maps allowed us to identify the most prominent stakeholders who take up a central role in the risk analysis process as well as linkages between them. A further application of Net-Map analysis to food safety systems has been reported by Pilamala Rosales et al. [17] to gain insights on stakeholders' roles and influence in the safety of the street food system at local markets in Ecuador.

In the present study, Net-Mapping was used to analyze stakeholder relationships within the Italian wheat supply chain. The overall mapping process was based on a participatory workshop complemented by in-depth interviews to validate and integrate the collected data. This methodology helped identify and visualize these relationships and how they influence stakeholder positioning and their role in defining and implementing sustainability indicators. The analysis of these stakeholder maps allowed us to envisage potential barriers and opportunities related to the visualized connections. Building on each participating stakeholder's unique knowledge and perspective, this method supports more robust multi-stakeholder collaborations aimed at orienting the national wheat supply chain toward long-term sustainability.

2. Materials and Methods

2.1. Study Participants

The collaborative exercise (online workshop and in-depth interviews) involved 14 experts covering the following stakeholder categories: science (academia and research), policy (policy officers), food business (small and medium enterprises), and society (associations).

The anonymized list of the Net-Map analysis (participatory workshop and in-depth interviews), their role in the wheat supply chain and relevant expertise is reported in Table 1.

Table 1. Anonymized list of the Net-Map participants (participatory workshop and in-depth interviews), their role in the wheat supply chain, and relevant expertise.

Stakeholder Category	Role/Expertise
NGO—European Community of Consumer Cooperatives (COOP Europe)	Sustainability Policy Manager
NGO—European Community of Consumer Cooperatives (COOP Europe)	Communications Policy Advisor
NGO—National Food Bank (Italy)	Secretary General
Academia—University of Siena	Scientific expert—development and implementation of indicators for the sustainable development goals
Academia—University of Torino	Scientific expert—City region food system
Academia—University of Torino	Scientific expert—City region food system
Research—National Research Council of Italy	Scientific expert—Food safety and communication
Academia—University of Bari	Scientific expert—economic and environmental sustainability of agricultural practices
Academia—University of Bologna	Scientific expert—Political geography, political science, and food security
Secondary Production—Valle Fiorita, SME (bakery products)	R&D Responsiblities—quality, authenticity and sustainability of bakery products
Primary production trade association—CIA (Italian Confederation of Agriculture)	Director—support to wheat farmers for sustainability and efficiency of the wheat sector; communication to consumers
Primary production trade association—Liberi agricoltori	Farmer and Manager of Liberi Agricoltori—policy representation, financial support, and technical assistance to small and medium wheat farmers
Secondary production trade association—COLDIRETTI	President of Coldiretti Bari—promotion of sustainable farming practices, policy advocacy
Local Authority—Metropolitan city of Bari (Italy)	Manager of the Service for Protection and Valorization of the Environment, Promotion, and Coordination of Economic Development

2.2. Net-Map Workshop

The applied Net-Map methodology is based on recent adaptations developed and tested by Pilamala Rosales et al. [17] and van der Linden et al. [16]. Before conducting the actual Net-Map analysis, as recommended by Schiffer (2007) [13], some background information on stakeholders' categories and goals, here called preliminary analysis, were collected.

The resulting methodology, tailored to the selected case study, is described in Figure 1.

Briefly, after the preliminary analysis, the participatory workshop complemented by in depth interviews allowed for (i) building the Net-Maps by identifying and categorizing the key stakeholders who play a role in the network, identifying relationships that connect them and consider domains of influence (e.g., information, data, cost flows); (ii) envisaging barriers and enabling factors in relation to the visualized connections in the Net-Maps.

To guide the participants in providing their contributions in a structured pathway, a digital board was designed using the Miro tool (https://miro.com/it/, accessed on 12 November 2024). The Net-Map protocol was used to create a whiteboard with all required frames for each step, including the task description (see Supplementary Material S1).

A training session was planned the week before the workshop to ensure smooth participation and make participants confident with the use of the digital board.

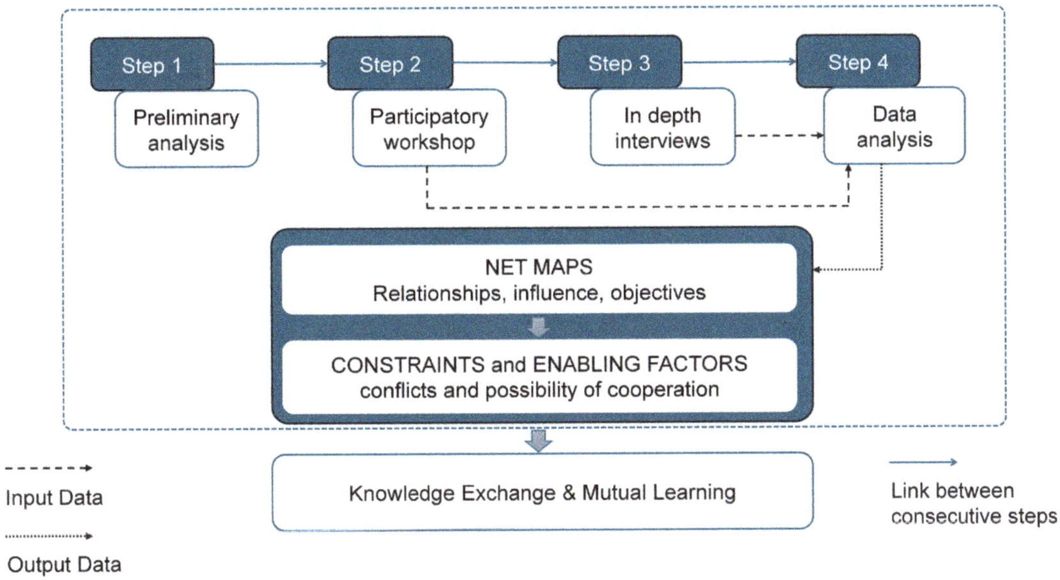

Figure 1. Steps of the Net-Map methodology.

Details of each step are given in the following.

2.2.1. Step 1—Preliminary Analysis

A preliminary analysis that involved scanning available literature and sustainability guidelines/statements available from food business operators was carried out to define terminology, to draft an initial list of sustainability indicators, and to identify the network members and the linkage types [4,18–21].

2.2.2. Step 2—Collaborative Exercise on the Digital Board

The participatory workshop comprised four phases, as described below.

A1. Stakeholders' identification: A provisional list of the stakeholders and their role was provided to the participants (see Supplementary Material S1—board 5). Participants were subsequently asked to revise and integrate it.

A2. Characterization of the relations between the stakeholders and their character: Participants were asked to connect the identified stakeholders with arrows of three different colors based on the three different linkage types: (a) red arrows for legally required information and data sharing, (b) green arrow for voluntary information and data sharing, and (c) blue arrow for costs flow. The term "data" included metric measures of the sustainability indicator, whereas "information" meant origin and social indicators.

This step led to the construction of Net-Maps.

A3. Identification and prioritization of actual or potential constraints: In relation to the visualized connections in the Net-Maps and taking into account four categories of constraints, namely (1) data: lack of data sources, lack of detail at the supply chain/regional level, and the shortage/absence of methodologies for data processing; (2) capabilities: shortage/absence of obligations (legislative or disciplinary); (3) resources: shortage/absence of funds, human resources, time, knowledge, or experience; (4) relations: shortage/absence of contacts, non-functional/non-operational relationships, and participants were asked to indicate a maximum of two main constraints for each stakeholder category, where they thought it relevant.

A4. Identification and prioritization of actual or potential enabling factors: In relation to the visualized connections in the Net-Maps, participants were asked to envisage potential enabling factors. Once a list of enabling factors was drafted collectively, the participants were asked to individually rank each listed enabling factor by assigning the following scores: 1 = low impact enabling factor, 2 = medium impact enabling factor, and 3 = high impact enabling factor.

2.2.3. Step 3—In-Depth Interviews

To validate and integrate the collected data, five further experts were involved in bilateral in-depth interviews. For this purpose, step 2 exercises were reported in the form of questions and tables to be filled in by participants (see Supplementary Material S2).

2.2.4. Step 4—Data Analysis

Data from the activity A2 of step 2 (participatory workshop) and relevant questions/tables in step 3 (in depth interviews) were entered into Microsoft Excel worksheets. Four separate worksheets corresponding to four symmetrical square matrices were generated for each link type, namely (1) information required by law and data sharing, (2) voluntary information and data sharing, (3) cost flow, and (4) combined links including all the previous link types between the stakeholders. In these matrices, all stakeholders were arranged in the same order in columns and rows. A score equal to "1" indicated a link, while a score of "0" indicated no link. The score equal to 1 indicated either a mono directional or a bidirectional link. In the matrix of combined links, when two actors are connected to each other by links of a different type, the link is counted only once.

The three worksheets were imported into yEd (version 3.21.1, yWorks GmbH, https://www.yworks.com/products/yed/download, accessed on 2 January 2025) to create the corresponding Net-Maps. The organic layout algorithm was selected to represent the Net-Map graphs. Specific centrality values (indegree, outdegree, degree, and eigenvector centrality) were calculated using Gephi version 0.10. The results were four Net-Maps in the domain of the following linkages: (1) legally required information and data sharing, (2) voluntary information and data sharing, (3) cost flow, and (4) combined links where lines and arrows reflect the relationship between the stakeholders. In each map, the node size indicates a stakeholder's influence.

3. Results

3.1. Participatory Workshop

3.1.1. Setting the Scene

Introduction to the EU Sustainable Food System Initiative

Understanding the legislative framework is crucial to determining the involvement and impact or role of various stakeholders in the implementation of sustainability indicators. Therefore, to provide all participants with a general understanding of current laws and future possibilities related to food system sustainability, in the first session of the workshop, an overview of the legislative framework on sustainability was given, also considering the ongoing EC initiative on sustainable food systems [22]. The main insights are summarized below.

The ambition of this initiative is to provide the EU with a wide sustainability approach to reduce not only the social and environmental impacts of food systems but also to make them more resilient. The new framework should ensure coherence with all EU food-related policies (including food safety policies). The envisaged possible policy options are based on different scenarios:

1. Baseline: This involves pursuing sustainable food systems in the context of the implementation of existing legislation and consistently with the objectives of the Farm to Fork Strategy, while taking into account the possible multitude of approaches to sustainability of food systems by Member States and stakeholders.
2. Voluntary approaches: This option foresees the assessment of whether and to which extent voluntary approaches through soft law instruments can contribute in the long-term to the transition toward sustainable food systems
3. Reinforcement of existing legislation: This involves pursuing a comprehensive transition toward an Union sustainable food system, considering the objectives of the Farm to Fork Strategy, by having a number of targeted sectoral EU level interventions through existing Union Acquis (i.e., the collection of common rights and obligations that constitute the body of EU law).
4. Building a new comprehensive legislative framework on sustainability that could serve as a *lex generalis*: This would apply to all actors of the food system. It would set out the common basis composed of general objectives, definitions, principles, and requirements for ensuring sustainability considerations, beyond the already applicable safety-based requirements set in the General Food Law (Regulation (EC) No 178/2002). This common basis will serve as an integrated general approach for *lex specialis*, when addressing specific contexts.

Referring to the existing and upcoming EU regulatory framework, experts have pointed out that numerous programs are being implemented at the national level, with each member state advancing independently in its own way. Therefore, among the key elements for the evaluation and national implementation of the possible envisaged policy options, the experts indicated the following:

- Setting sustainability principles and objectives to provide a common understanding of the goals to be achieved;
- Setting general minimum standards to be met for foods produced or placed on the EU market and related food operations;
- Establishing clear responsibilities of food system actors;
- Including horizontal elements for sustainability analysis in relation to regulated products in the food chain, complementing the existing 'risk analysis' principle (food safety and general food law).

Introduction to the Case Study

The wheat supply chain was proposed as a case study due to its crucial role in Italy's agro-food landscape, national economy, and cultural heritage. Considering both durum and soft wheat, the total wheat production is estimated at around 4 million tons, which would account for roughly 25.9% of total cereal production [23,24]. Wheat is integral to Italian cuisine, especially in pasta and bread production. Italians consume over 100 kg per capita per year of durum and soft wheat products, which includes bread, cakes, and pizzas [25]. The supply chain of durum wheat, semolina, and pasta plays a vital role in the global image of 'Made in Italy' and the Mediterranean diet, particularly as a unique resource for some agricultural districts in Southern Italy [26]. Some Italian pasta producers are strongly committed enhancing the wheat value chain through collaborations to create tools that support farmers in sustainable cultivation and to address consumers' demand for higher quality and environmentally friendly products [26,27].

Introduction to the Methodology

A brief introduction to the Net-Map analysis methodology is given, explaining the goals and non-goals. Specifically, the goal of the participatory exercise was to identify stakeholders of the Italian supply chain that contribute to the definition/implementation of sustainability indicators, visualize and characterize their connections, and finally reflect on constraints and opportunities in relation to the visualized connections. Identifying or proposing sustainability indicators was out of the scope of the collaborative exercise (non-goal).

3.1.2. Background Information

To bring together all participants (either in the workshop or in interviews) at the same background level and refer to a real/concrete situation when performing the collaborative exercise, a non-exhaustive list of sustainability indicators for the cereal supply chain was presented (Table 2) as well as a scheme of the Italian wheat supply chain depicting the main stakeholders (Figure 2).

Table 2. Example list of sustainability indicators for the cereal supply chain. The list was drafted from data collected in Step 1—preliminary analysis [4,19–22].

Dimension	Subdimension	Indicator Example
Enviromental	Air	Quality: gas emissions in total agriculture)
	Water	Quality: Water pH
		Use: Agricultural water withdrawal as percentage of total renewable water (%)
	Soil and land	Quality: Soil carbon content (as percentage in weight)
		Use: Agricultural land as % of arable land
	Biodiversity	Benefits of biodiversity index (0 = no diversity potential to 100 = maximum)
		Crop diversity (Shannon index)
		Absence of genetically modified organisms
	Use of energy	Agriculture and forestry energy use as % of total
		Primary energy resources—renewable (MJ)
		Primary energy resources—non renewable (MJ)
	Integrated pest management	Use of integrated pest management
Food and Nutrition	Food security	Availability: Per capita food available for human consumption (kcal/capita/day)
		Access: Per capita food available for human consumption (kcal/capita/day)
		Utilization: Access to improved water resource (% of total population)
		Stability: Price volatility index; per capita food supply variability (kcal/capita/day)
	Food safety	Traciability
		Burden of foodborne illness (number of cases)
		Deoxinivalenol content
	Food waste and use	Food loss as % of total food produced
		Hazardous waste disposed (g on 1 kg of product)
		Non-hazardous waste disposed (g on 1 kg of product)
		Radioactive waste disposed (g on 1 kg of product)
	Nutrition	Diet diversification
		Undernutrition: Stunting, children aged < 5 years stunted (%)
		Obesity: Prevalence of obesity (% of the population, over 18 y of age)

Table 2. *Cont.*

Dimension	Subdimension	Indicator Example
Economic		Financial performance: Agriculture value-added per worker (constant 2010 USD)
		Employment rate: Agriculture under-employment (%)
		Economic distribution: Gini index for land distribution and tendency
		Use of technological innovation
Social		Gender equality: labour force participation rate, female (% of female population ages 15+)
		Inclusion: predominant fair-trade organizations and producers
		Recognition of fair prices to farmers
		Labor condition: Management performance, parental leave, average hours of training per years per employment

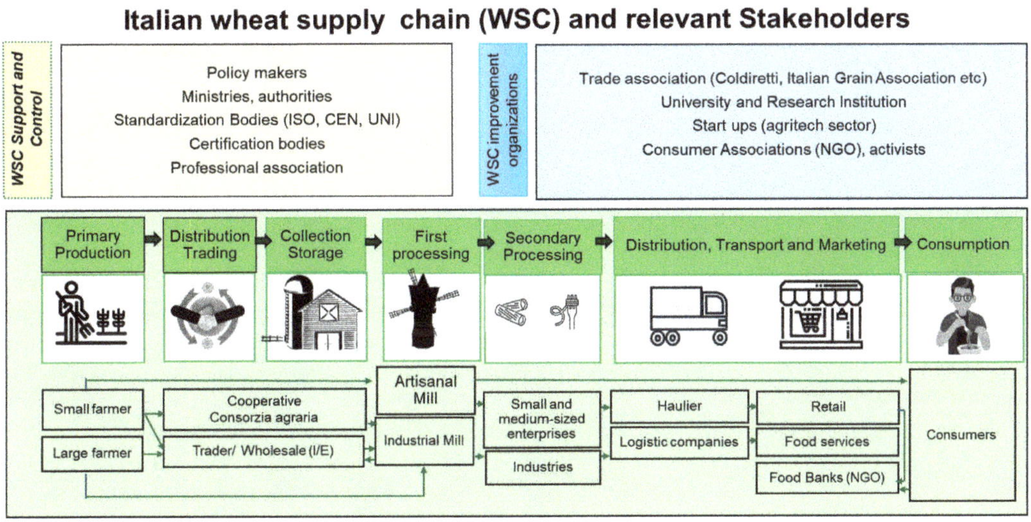

Figure 2. Simplified scheme of the Italian wheat supply chain and relevant stakeholders.

3.1.3. Identification of Stakeholder Roles and Linkages

After providing the above information, participants were guided through the collaborative activities.

The first activity consisted of two parallel tasks: (1) identification of the stakeholders that in Italy contribute to the definition/implementation of sustainability indicators and (2) identification of their potential contribution or actual role relevant to the definition or implementation of sustainability indicators. Starting from a draft list of stakeholders and roles provided as background material (see Section 2.2.1—Step 1), with respect to the case study "sustainable wheat supply chain", participants identified a wide variety of stakeholders, belonging to several categories and with different roles, contributing directly or indirectly to the definition/implementation of sustainability indicators.

The final stakeholder list (Table 3), drafted during the participatory workshop and fine-tuned through bilateral interviews, showed the diversity of contributions and roles that are necessary to achieve the common overall objective or increasing the sustainability of the Italian wheat supply chain.

Table 3. Identified stakeholders of the wheat Italian supply chain and their potential contribution or actual role relevant to the definition or implementation of sustainability indicators.

Stakeholder	Role/Contribution
Policy makers/Law makers	(EU) Establishing principles, definitions, responsibilities, standards, and metrics that also allow ensuring the traceability of raw materials (National) Transpose European directives and implement regulations and programming
Ministries (e.g., Ministry of Agricultural, Food and Forestry Policies, Ministry of Enterprises and Made in Italy)	Resource allocation, define strategic plans
Authorities (national, regional, and local)	May be involved in the definition of guidelines for the implementation of indicators and in monitoring sustainability performance Setting taxation, incentives, local policy-making
Standardization Bodies (ISO, NI)	Definition and development of voluntary standards, may contribute to the definition of sustainability standards
Certification bodies	Evaluate and certify compliance with sustainability standards, boost sharing of good practices and methodologies on sustainable production
Professional associations	Represent and safeguard the interests of the specific categories (farmers, enterprise) in the context of economic policy, contribute to the identification and sharing of good practices on inclusive and responsible business models in agriculture
Research	Development of innovations for sustainable production (variety selections, waste recovery strategies, packaging, logistic, food distribution) Contribution to the definition of decision support systems Development of tools, reports, surveys, etc. to support political decisions (collect measures for the definition of sustainability indicators) and market analysis
Start-ups	Development of innovations for sustainable production
Primary production (including trade associations, cooperatives, agricultural consortia)	Implementation of good practices/specific decision support systems
Primary production (organic seeds)	
First processing (semi-finished products—mills, etc.)	Assessment and selection of primary production based on sustainability performance
	Implementation of good practices/specific decision support systems
Secondary processing (industries)	Development of innovations, supply chain specifications, investment of resources Define and implement decision support systems Communication campaigns Assessment of consumer or supply chain perception (focus groups, surveys, data collection) Assessment and choice of primary production on the basis of sustainability performance Ensure transparency on conditions of use of raw materials
Secondary processing (small and medium-sized enterprises, including trade associations)	As above
Food services (ready-to-eat-meals, catering)	Assessment and choice of suppliers based on sustainability performance (short supply chains, certifications)
Retail	
Transport and ancillary activities (loading/unloading)	Optimization of logistics by intelligent transportation systems, implementation of sustainable transportation practices
Consumers	Receiving information, conscious consumption, paying for sustainable supply chains

Table 3. *Cont.*

Stakeholder	Role/Contribution
Food banks, social cooperatives operating in the agricultural sector	Implementation of good practices/specific decision support systems Commitment for the affordability of quality and sustainable foodstuffs Donation management, good (hygienic) practices for safe handling of donations
Associations, activists	Stimulate thinking about sustainability at various levels (environmentalists, food sovereignty movements, commitment to quality food affordability, etc.)

With a view to the incoming sustainability regulation and the possible scenarios for national implementation, the role of the "standardization bodies" and "certification bodies" was discussed. In principle, the two stakeholder categories have different although linked roles: the "standardization bodies" draft standards, while the certification bodies provide services to companies to register the sustainability standards. However, the frontier between standard setting and standard certification is increasingly blurred: many certification bodies are involved both in standard setting and certification (albeit in formally distinct guises), while some non-governmental organization traditionally engaged in defining sustainability standards now also take part in certification [28].

The second collaborative activity consisted of identifying and visualizing the connections between the stakeholders and defining the linkage type, choosing from the following: (1) legally required information and data sharing, (2) voluntary information and data sharing, and (3) cost flow. This activity resulted in the Net-Maps depicted in Figure 3A, B, and C, respectively. To build the Net-Maps, the connections visualized by colored arrows on the digital board, during the participatory workshop, were integrated with connections indicated by the experts in the in-depth interviews (see Methods Section 2.2.4—data analysis).

For the interpretation of the visual data (Net-Maps), centrality measures (Table 4) were also taken into account, since they provided quantitative information on the number of inputs that a stakeholder receives from the other stakeholders (in DG), provides to the other stakeholders (out DG), and exchanges with the other stakeholders (DG). The normalized eigenvector centrality (EV) provided information on the extent to which a stakeholder is connected to other well-connected stakeholders.

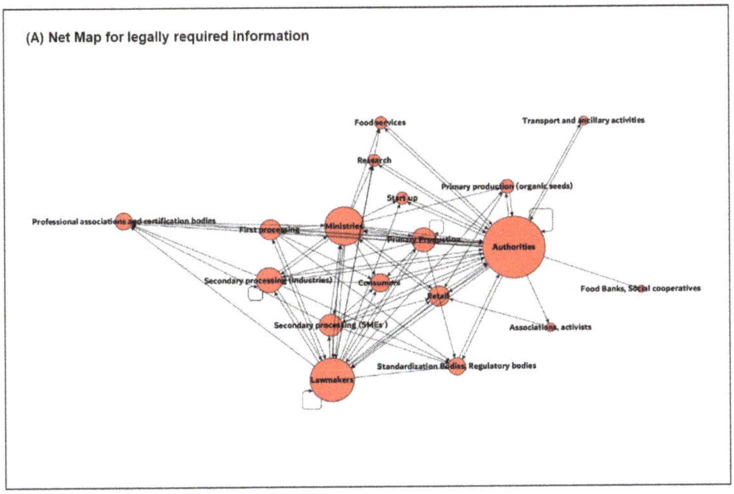

Figure 3. *Cont.*

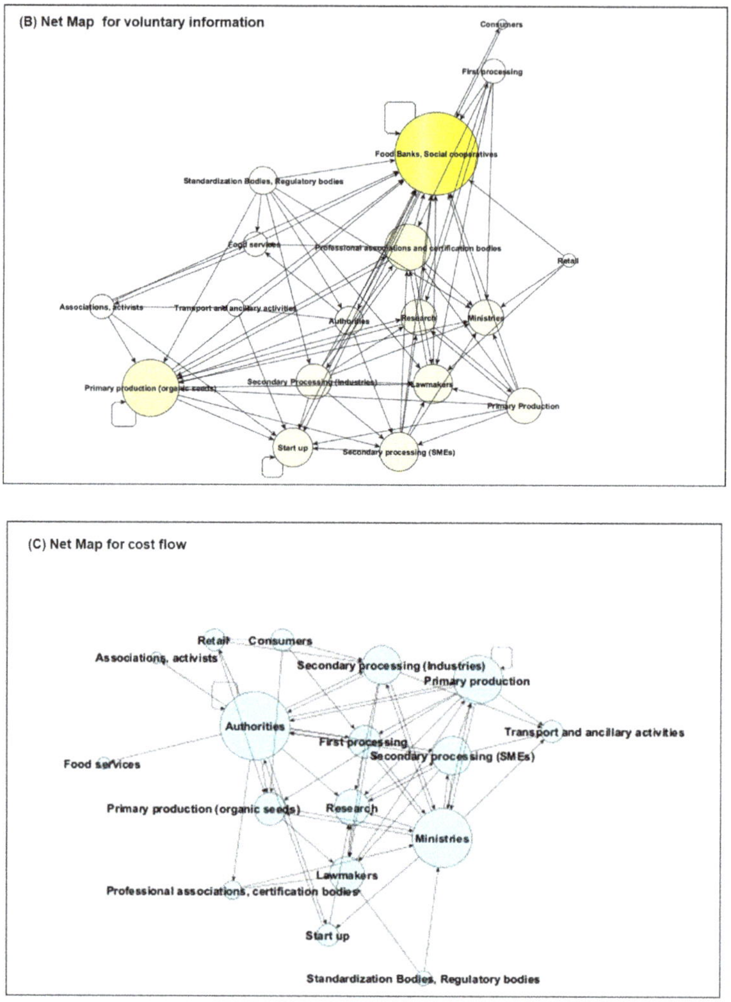

Figure 3. Net-Maps depicting the relationships among stakeholders of the Italian wheat supply chain contributing directly or indirectly to the definition/implementation of sustainability indicators. Arrows indicate three linkage types: (**A**) exchange of legally required information, (**B**) exchange of voluntary information, (**C**) cost flow. Node size indicates the stakeholder's influence (total number of connections—degree centrality measure).

Table 4. Centrality measure of Net-Maps (indegree centrality (in DG): number of inputs stakeholder receive from the other stakeholders, out DG: number of inputs stakeholder provide to the other stakeholders, normalized degree centrality (DG): number of inputs stakeholder exchange with the other stakeholders, normalized eigenvector centrality (EV): the extent to which a stakeholder is connected to other well-connected stakeholders).

Stakeholders	Legally Required Information				Voluntary Information				Flow Cost				Combined			
	In DG	Out DG	DG	EV	In DG	Out DG	DG	EV	In DG	Out DG	DG	EV	In DG	Out DG	DG	EV
Lawmakers	8	15	0.68	0.58	8	2	0.45	0.24	6	2	0.44	0.55	11	16	0.77	0.62
Ministries	7	13	0.59	0.51	8	1	0.41	0.24	7	8	0.83	0.74	10	15	0.71	0.56
Authorities	16	18	1.00	1.00	2	5	0.32	0.25	6	12	1.00	0.82	17	18	1.00	0.98
Standardization bodies	6	1	0.21	0.44	0	7	0.32	0.00	0	2	0.11	0.00	6	7	0.37	0.36

Table 4. *Cont.*

Stakeholders	Legally Required Information				Voluntary Information				Flow Cost				Combined			
	In DG	Out DG	DG	EV	In DG	Out DG	DG	EV	In DG	Out DG	DG	EV	In DG	Out DG	DG	EV
Professional associations certification bodies	6	1	0.21	0.46	6	6	0.55	0.24	1	2	0.17	0.19	8	7	0.43	0.48
Research	3	1	0.12	0.31	5	4	0.41	0.17	7	1	0.44	1.00	9	6	0.43	0.55
Start-up	3	1	0.12	0.31	8	2	0.45	0.57	2	2	0.22	0.36	11	4	0.43	0.67
Primary production	4	7	0.32	0.36	1	8	0.41	0.05	5	7	0.67	0.82	6	15	0.60	0.38
Primary production (organic seeds)	3	2	0.15	0.31	7	8	0.68	0.28	5	2	0.39	0.60	11	10	0.60	0.60
First processing	3	6	0.26	0.31	1	5	0.27	0.05	4	3	0.39	0.55	6	8	0.40	0.39
Secondary processing (Industries)	5	7	0.35	0.43	2	7	0.41	0.07	4	5	0.50	0.41	9	12	0.60	0.52
Secondary processing (SMEs)	3	7	0.29	0.31	4	6	0.45	0.17	3	6	0.50	0.59	7	12	0.54	0.45
Food services	3	1	0.12	0.31	4	2	0.27	0.14	1	0	0.06	0.19	6	3	0.26	0.34
Retail	6	3	0.26	0.42	0	3	0.14	0.00	2	2	0.22	0.19	7	6	0.37	0.42
Transport and ancillary activities	1	1	0.06	0.15	2	2	0.18	0.07	4	0	0.22	0.60	6	3	0.26	0.33
Consumers	7	1	0.24	0.52	1	1	0.09	0.25	0	4	0.22	0.00	8	6	0.40	0.52
Food banks, social cooperatives	1	0	0.03	0.15	18	4	1.00	1.00	0	0	0.00	0.00	18	4	0.63	1.00
Associations, activists	1	1	0.06	0.15	1	5	0.27	0.04	1	0	0.06	0.19	2	6	0.23	0.14

The network for legally required information and data sharing (Figure 3A) sees authorities, ministries, and law makers as the most influential actors based on the input they provide to other stakeholders (highest out DG values) and the highest number of connections they have with the most influential stakeholders in the system (highest EG values). Other influential stakeholders in this Net-Map, in terms of input provided to the system (high out DG values), are those normally interlinked by contractual/legal obligations, such as primary producers, first and secondary processors, retailers, and, indeed, authorities. On the other side, the least influential actors, showing a lower number of connections and exchanges of obligatory information within the system, are associations (DG: 0.06, EG: 0.16), transport (DG: 0.06, EG: 0.11), and food banks/social cooperative stakeholders (DG: 0.03, EG: 0.11). Consumers have a central positioning in the map but a small node size (influence), being targeted by the large amount of information received, since legally required, from policy actors and food business operators. The network for voluntary information and data sharing (Figure 3B) is quite different. As it turns out, the food bank is the most prominent stakeholder collecting information through connections with a wide diversity of actors along the supply chain (inDG value: 18, EG value: 1.00). The central role of food bank/social cooperative group is mainly due to the high number of voluntary incoming information (18 out of 22 total linkages identified), since collection of data and information is within their tasks either when operating at a national level or as an international federation. Food banks act as intermediaries by collecting surplus and redistributing grain products to vulnerable populations. For this purpose, they establish and feed a relevant information flow through cooperation with growers, millers, and food producers to source food products especially during times of surplus, reducing food waste. For the remaining stakeholders, an equal distribution between incoming and outcoming voluntary information is observed. Stakeholders from the policy side (ministries, authorities, and certification bodies) appear also quite influential in the map, given the high number of connections visualized also in the domain of exchange of voluntary information (DG values from 0.32 to 0.45). Similar DG values, i.e., overall number of connections, have been obtained for primary and secondary (processing) producers, including the small and medium enterprises (SMEs). As associations/federations, these stakeholders either support individual wheat farmers and wheat processors or act at an advocacy level to promote the overall sustainability and efficiency of the wheat sector (link with policy stakeholders). It is worth mentioning that, although not central in the system, start-ups are well connected with primary and secondary producers, indicating some commitment for innovation along the Italian wheat supply chain.

Finally, the Net-Maps for cost flow shows the prominence of the authorities (DG value of 1.00, EG value of 1.00) and the ministries (DG value of 0.83, EG value of 0.85) either investing money or receiving incomes from the implementation of sustainability measures. With one exception, this map shows mostly unidirectional flows rather than bi-directional ones. The cost flow is visualized mainly from food production actors (primary production, processing, logistics, and retail) to policy/law actors. The actors belonging to the primary production group are connected to the most influential stakeholders (EG value of 0.98), although the cost flow is direct to a limited part of the network (DG value of 0.28). Researchers show the highest EG value (EG value of 1.00) because they are well connected to ministries, law makers, and authorities, reflecting current resources allocation in research that is supposed to provide innovations to increase the sustainability of food production as well as to support technology transfer. Consumers are characterized by low DG and EG values as compared with primary/secondary producers, suggesting that the latter is the most impacted by additional production costs arising from the request to comply with sustainability standards.

To sum up, combining all three linkages in a single Net-Map (Figure 4), stakeholders from the policy side, i.e., authorities (DG value of 1.00, EV value of 1.00), lawmakers (DG value of 0.65, EV value of 0.93), and the ministries (DG value of 0.69, EV value of 0.90), reassert themselves as the most influential stakeholders in the definition and implementation of sustainability measures. They are surrounded by primary and secondary producers (EG values from 0.89 to 0.83) setting connections and influencing the system mainly through corporate entities.

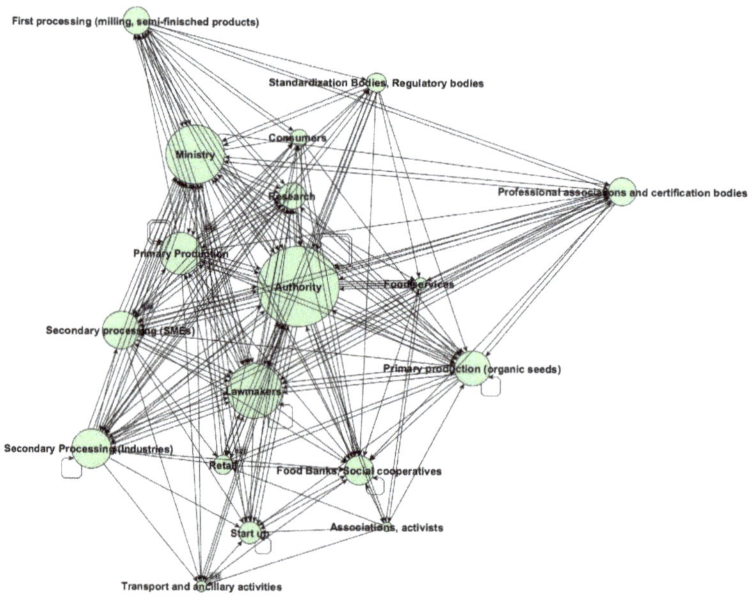

Figure 4. Combined Net-Map of the relationships among stakeholders of the Italian wheat supply chain contributing directly or indirectly to the definition/implementation of sustainability indicators. The map combines data obtained for the three analyzed linkage types: legally required information, voluntary information, and cost flow. Node size indicates the stakeholder's influence (total number of connections—degree centrality measure).

In the overall picture, consumers are a much less influential stakeholder category. Consumers show a DG value of 0.40 and EV of 0.52 mainly due to their connections in the network of legally required information and data sharing, whereas they are not influential in the network of voluntary information data sharing and flow costs.

3.1.4. Identification and Prioritization of (Potential) Constraints Hampering the Definition and/or Implementation of Sustainability Indicators

Starting from the Net-Maps reported in Figures 3 and 4 and relying on their personal experience and knowledge in the field (Table 1), the experts were asked to further elaborate on the visualized connections and node sizes by envisaging constraints encountered by different stakeholders in the definition and implementation of sustainability indicators.

Figure 5 summarizes the participants' opinions. The radar graph combines data obtained from the participatory workshop and the interviews. In both cases, participants considered four categories of constraints: (1) data: lack of data sources, lack of detail at the supply chain/regional level, and shortage/absence of methodologies for data processing; (2) capabilities: shortage/absence of obligations (legislative or disciplinary); (3) resources: shortage/absence of funds, human resources, time, knowledge, or experience; (4) relations: shortage/absence of contacts and non-functional/non-operational relationships. They were asked to identify and indicate up to two main constraints per stakeholder category where relevant. Each radial axis in the figure represents the number of times a given constraint was identified (as sum of contributions from individual participants) for each stakeholder category listed around the chart.

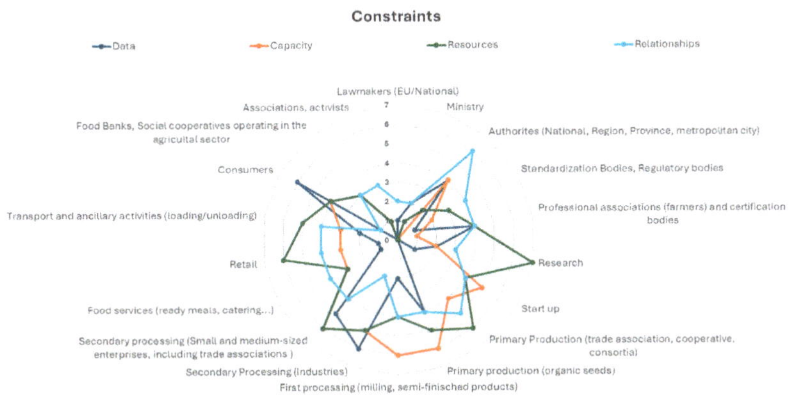

Figure 5. Radar graph of constraints indicated by experts hampering the definition and/or implementation of sustainability indicators by the Italian wheat supply chain stakeholders. The graph combines data obtained from the participatory workshop or the interviews. Four categories of constraints are considered: data, capabilities, resources, and relations. Each radial axis in the figure represents the number of times a given constraint was indicated (as the sum of contributions from individual participants) for each stakeholder category listed around the chart.

In general, relationships (lack of contacts and links, dysfunctional relationships) were mentioned as the areas where most improvement was needed. This constraint was reported to be particularly relevant for authorities at the national, regional, and local (municipality) level. Even though Net-Maps showed these stakeholders to be well connected and to have a high influence in the studied system, experts pointed out that relationships are often monodirectional, and there is a lack of functional/structured channels enabling the other stakeholder categories to reach them. The same consideration applies to relationships within the different geographical levels (i.e., from local to national authorities).

For industries, rather than relationships, data availability was seen as the major limitation in the definition and implementation of sustainability indicators. As shown in the Net-Maps for legally required and voluntary information exchange, the flow of data/information is mainly unidirectional, i.e., from business operators to actors responsible for drafting, implementing, or enforcing regulations, while suitable tools and channels for data exchange and integration, e.g., among business operators of the wheat supply chain, are still missing or inadequate.

Resources (finances, manpower, equipment) were identified as a major limitation mainly for researchers, followed by primary and secondary producers and retailers. All these actors are increasingly requested to allocate more effort to deliver and implement innovative solutions to make the Italian wheat supply chain more sustainable, in addition to other mandatory requirements for safety, quality, traceability, and nutrition.

Relationships were identified as limitations for authorities but not for policy makers, who Net-Maps identified as the most influential actors in the network. However, during the discussion, two further limitations attributable to policy makers emerged. These are the concrete commitment of policy makers to put into practice new sustainability measures and the difficulty of implementing them in a specific territorial context. This is particularly true when there is little propensity for change or when the implementation of indicators seems to conflict with campaign promises.

3.1.5. Identification and Prioritization of (Potential) Enabling Factors Supporting the Definition and/or Implementation of Sustainability Indicators

This final step served as a basis to formulate recommendations, since it aimed at identifying and prioritizing factors enabling stakeholders to contribute to the definition and/or implementation of sustainability indicators and measures, as well as exploring opportunities and strategies for improved cooperation. Similar to the previous exercise, taking into account connections and spheres of influence visualized through the Net-Maps (Figures 3 and 4) and referring to their own specific experience and knowledge in the field (Table 1), the experts were asked to identify then prioritize actual or potential enabling factors. Once a list of enabling factors was drafted collectively in the participatory workshop, the experts were asked to individually rank each listed enabling factor by assigning the following scores: 1 = low impact enabling factor, 2 = medium impact enabling factor, and 3 = high impact enabling factor. Experts contributing to bilateral interviews were asked to agree on the enabling factors list then to rank them.

The overall results of the ranking are reported in Figure 6. A total score < 26 indicated a low impact (positions from 6 to 8 in Figure 6), a score between 26 and 28 indicated a medium impact (positions from 3 to 2), and a score $\geq$ 28 indicated a high impact enabling factor (position 1 in Figure 6).

The highest impact enabling factor was identified as mutual trust between supply chain operators. It was noted that this ultimate goal can be achieved only if most of the other envisaged enabling factors are properly leveraged.

Among these factors, experts prioritized the availability of practical, implementable tools such as best practices, guidelines, or decision support systems. All these tools rely on existing knowledge and data, which should be accessible to all stakeholders through dedicated cooperative spaces like digital platforms, forums, or events.

Regular participation in training and educational initiatives was highlighted as a prerequisite for efficiently accessing and utilizing available knowledge and data, as well as for generating and sharing updated resources.

Additionally, reflecting on the cost flows visualized in the Net-Map, participants emphasized that adequate incentives were essential to support and accelerate the implementation of these measures. Such incentives should be applied across various stakeholders, including supply chain operators adopting sustainability measures, innovators, and researchers developing solutions or proposing new strategies.

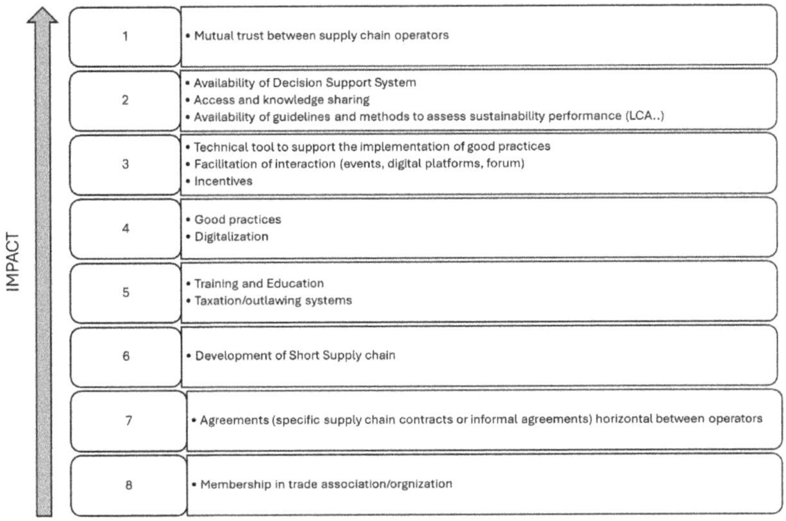

Figure 6. Prioritization of the identified enabling factors supporting the definition and/or implementation of sustainability indicators by the Italian wheat supply chain stakeholders.

Regarding the different types of stakeholder connections analyzed (Figures 3 and 4), formal or informal agreements between operators within a specific supply chain or segment were considered more effective than membership in trade associations or organizations. However, acting as a federation, association, or cooperative allows stakeholders to exert greater influence in the decision-making process.

4. Discussion

Mapping actors, roles, and relationships within a defined formal or informal network is a key step in building approaches to foster cooperation. The importance of identifying the key actors and roles for efficient engagement has been highlighted in the analysis by Ammirato et al. [9]. The reviewed case studies showed that, although food chain actors can be heterogeneous in terms of their operating environment and tasks, they can be guided to share values and culture and to re-orient themselves toward the achievement of common long-term sustainability objectives. With this in mind, the present study intentionally focused on a specific culturally and geographically relevant key commodity group, where the mapping exercise revealed a diversity of contributions and roles that are necessary to achieve the common overall objective of a sustainability-oriented supply chain.

Net-Maps of the relationships within the Italian wheat supply chain networks allowed for the visualization of the positioning and influence of each stakeholder group, showing existing asymmetries. As an example, stakeholders from the governance and policy side turned out to be the most influential group, connected with all other stakeholders mostly through unidirectional relationships. According to the analysis by Dania et al. (2018) [6], this stakeholder group can be identified as the most "powerful" in the analyzed network, having a prevalent influence on the types of information shared, recipients, and sharing mechanism in collaboration activities. The value of horizontal cooperation through associations/cooperatives emerged in leveraging voices of less influential stakeholders, such as individual wheat farmers and wheat processors. On the other hand, even though increasing consumer attention to sustainability issues was acknowledged by the experts as a driver for the implementation of sustainability measures in food production, a passive rather than an active involvement of consumers resulted in the mapped networks.

In addition to asymmetries in stakeholders' size and positioning, the expert group also identified current limitations in cooperation. Relationships (lack of links, dysfunctional relationships) were mentioned as the areas where most improvement was needed, especially with reference to monodirectional (top-down) relationships with policy actors. As previously underlined by Stanco et al. [12] in their exploration of sustainable collective innovation along the Italian wheat supply chain, cooperation processes mostly do not take place autonomously but are supported by policies and governance mechanisms established between the players in the food supply chain. However, representative stakeholders in the Net-Map analysis advocated for bidirectional (bottom-up) communication channels to enable food supply chain actors to shape policies and governance mechanisms.

Indeed, cooperation implies the allocation of human, technological, physical, and financial resources. Therefore, in agreement with previous findings [6], overcoming the heterogeneity of resources and capabilities among stakeholders was identified as a critical challenge in establishing collaboration systems.

Trust and relationship quality emerged as the top priority among enabling factors. Fischer [29] demonstrate that long-term relationships built on trust are fundamental for successful collaboration in cereal supply chains. His analysis also indicated that trust can be significantly improved by effective communication among stakeholders. Enhanced communication and information sharing can also minimize uncertainty in production, distribution, and marketing processes [30].

In this respect, the present study identified digital platforms as the most efficient tool to foster multi-stakeholder dialogue and facilitate data and knowledge sharing, thus fostering transparency and trust. In line with the analysis by Duong et al. [8], the implementation of digital technologies to share information in a timely and accurate manner has been recognized as a key enabling factor to support collaboration toward sustainability objectives is a key enabling factor. Currently, the data shared in the food supply chain are often incomplete or inaccurate [8]. Our mapping confirms this, showing that particularly, farmers and SMEs are less involved in data sharing due to limited financial resources and necessary skills. This issue is being addressed in Italy by the ongoing research devoted to delivering an open access and integrated information system of statistical indicators regarding new technologies and methods for traceability, quality, and safety of agrifoods, including the wheat supply chain [31].

5. Conclusions

In this study, the Net-Map methodology was applied to analyze stakeholder relationships within the Italian wheat supply chain. The mapping process revealed a diversity of stakeholder roles and contributions. The obtained Net-Maps enabled us to visualize the interrelationships (formal and informal networks) among various stakeholders, including policymakers, food business operators, and consumers, highlighting their distinct roles and influences in driving the transition toward sustainable food systems. Starting from the analysis of the state-of-the-art of the EU sustainability framework, the participatory workshop and subsequent interviews provided valuable insights into existing constraints and enabling factors within the stakeholder network.

Based on these insights, several recommendations for actions to improve stakeholder relationships and their involvement in the decision-making process can be derived.

Considering the current state and the existing connections visualized in the Net- Maps, goals to be pursued in the short- to medium-term (0–3 years) and relevant actions include the following:

- Establishing a common language and understanding of sustainability: Participatory workshops and training sessions would address the current misalignment among stakeholders and varying priorities regarding sustainability.
- Providing spaces for building connections: This involves creating dedicated digital platforms or fora for stakeholders to share information, best practices, and discuss challenges related to sustainability that would foster and feed dialogue and collaboration.
- Facilitating knowledge exchange for innovation: This involves establishing or reinforcing networks with research institutions, agricultural extension services, and industry experts to provide technical support and promote the co-creation of knowledge and sustainable practices.
- Fostering the creation of data sources and methodologies: This includes establishing data collection, sharing, and reporting protocols for key sustainability indicators across the wheat supply chain.
- Launching small-scale pilot projects: These projects should address specific sustainability challenges (e.g., reducing water usage, improving soil health, minimizing food waste) while foreseeing multi-actor participation by farmers, processors, and retailers. This call for action, specifically addressed to funding organizations, would provide tangible examples of successful collaboration and demonstrate the benefits of sustainable practices.
- Developing and implementing incentive programs: This initiative, to be undertaken by governmental actors, should encourage farmers and processors to adopt sustainable practices, addressing their economic concerns. The development of certification schemes or labels for sustainably produced wheat products would complement this, acknowledging the producers' commitment and addressing consumer demand for transparent information.

A long-term impact (3–5 years and beyond) is expected to be achieved by implementing the following actions:

- Advocating for supportive policies and regulations: This involves acting through federations and associations and working with policymakers to develop and implement policies and regulations that promote long-term investment in sustainability across the entire wheat supply chain. Promoting long-term multi-stakeholder partnerships would make these actions more impactful at the national and EU levels.
- Monitoring and evaluating progress: This involves developing systems for monitoring and evaluating the impact of sustainability initiatives on environmental, economic, and social outcomes based on harmonized indicators and metrics. Progress toward sustainability goals should be communicated to stakeholders and the public.

These proposed recommendations are intended to address the fragmentation and varying priorities within the Italian wheat supply chain, fostering collaboration among stakeholders and ultimately contributing to the broader goal of sustainable food systems in the EU.

Study limitations. The data obtained reflect the participants' perception; this might represent a limitation of this study. Although the deep knowledge of participants and the modified Net-Map approach (participatory workshop complemented by in-depth interviews) ensure that the connections depicted in the Net-Maps reflect reality, the small size of the multi-actor group might limit the generalizability of the results. When elaborating on constraints and enabling factors, some participants did not feel fully confident in voting for other categories of stakeholders than their own, which explains the absence of data for the less represented categories (e.g., law makers). The identified barriers and opportunities are consistent with previously reported analyses focused on the Italian wheat supply chain [11,12]; however, it is worth remembering that this analysis was conducted for a specific case study within a specific national context.

Supplementary Materials: The following supporting information can be downloaded at: https://www.mdpi.com/article/10.3390/foods14050786/s1, Supplementary Material S1: Boards of the participatory workshop. Supplementary Material S2: In-depth interview questions.

Author Contributions: Conceptualization, B.C. and V.M.T.L.; methodology, B.C. and V.M.T.L.; formal analysis, B.C.; investigation B.C.; resources, V.M.T.L.; writing—original draft preparation, B.C. and V.M.T.L.; writing—review and editing, B.C., V.M.T.L. and N.M.C.; supervision, V.M.T.L.; project administration, N.M.C.; funding acquisition, V.M.T.L. All authors have read and agreed to the published version of the manuscript.

Funding: This study was carried out within the Agritech National Research Center and received funding from the European Union Next-Generation EU (PIANO NAZIONALE DI RIPRESA E RESILIENZA (PNRR)—MISSIONE 4 COMPONENTE 2, INVESTIMENTO 1.4—D.D. 1032 17 June 2022, CN00000022). This manuscript reflects only the authors' views and opinions, neither the European Union nor the European Commission can be considered responsible for them.

Institutional Review Board Statement: Not applicable.

Informed Consent Statement: Not applicable.

Data Availability Statement: The original contributions presented in the study are included in the article, further inquiries can be directed to the corresponding author.

Conflicts of Interest: The authors declare no conflicts of interest.

Abbreviations

The following abbreviations are used in this manuscript:

SMEs	Small and medium enterprises
ISO	International Organization for Standardization
CEN	European Committee for Standardization

References

1. European Commision (EC) Legislative Framework for Sustainable Food System. Available online: https://food.ec.europa.eu/horizontal-topics/farm-fork-strategy/legislative-framework_en (accessed on 11 February 2025).
2. Food and Agriculture Organization of the United Nations (FAO). *Thinking About the Future of Food Safety—A Foresight Report*; FAO: Rome, Italy, 2022.
3. The EU Pathway Towards Sustainable Food Systems Transformation. Available online: https://www.unfoodsystemshub.org/docs/unfoodsystemslibraries/national-pathways/european-union/european-union-pathway.pdf?sfvrsn=48850b55_1 (accessed on 12 November 2024).
4. Béné, C.; Prager, S.D.; Achicanoy, H.A.; Toro, P.A.; Lamotte, L.; Bonilla, C.; Mapes, B.R. Global map and indicators of food system sustainability. *Sci. Data* **2019**, *6*, 279. [CrossRef] [PubMed] [PubMed Central]
5. United Nation Enviroment Programme (UNEP); Food and Agriculture Organization of the United Nations (FAO); United Nations Development Programme (UNDP). Rethinking Our Food Systems: A Guide for Multi-Stakeholder Collaboration; Nairobi. Available online: https://www.undp.org/sites/g/files/zskgke326/files/2023-06/rethinking-food-systems_a_guide_for_multi-stakeholder_collaboration.pdf (accessed on 22 February 2025).
6. Dania, W.A.P.; Xing, K.; Amer, Y. Collaboration behavioural factors for sustainable agri-food supply chains: A systematic review. *J. Clean. Prod.* **2018**, *202*, 193–208. [CrossRef]

7. Barrat, M. Understanding the meaning of collaboration in the supply chain. *Int. J. Phys. Distrib. Logist. Manag.* **2004**, *34*, 66–92. [CrossRef]
8. Nguyen Khanh Duong, L.; Kumar, V.; He, Q. Collaboration for the sustainable food supply chain: A bibliometric analysis. *Bus. Strateg. Environ.* **2024**, *34*, 1287–1301. [CrossRef]
9. Ammirato, S.; Felicetti, A.M.; Ferrara, M.; Raso, C.; Violi, A. Collaborative Organization Models for Sustainable Development in the Agri-Food Sector. *Sustainability* **2021**, *13*, 2301. [CrossRef]
10. León-Bravo, V.; Caniato, F.; Caridi, M.; Johnsen, T. Collaboration for Sustainability in the Food Supply Chain: A Multi-Stage Study in Italy. *Sustainability* **2017**, *9*, 1253. [CrossRef]
11. Ciliberti, S.; Stanco, M.; Frascarelli, A.; Marotta, G.; Martino, G.; Nazzaro, C. Sustainability Strategies and Contractual Arrangements in the Italian Pasta Supply Chain: An Analysis under the Neo Institutional Economics Lens. *Sustainability* **2022**, *14*, 8542. [CrossRef]
12. Stanco, M.; Nazzaro, C.; Lerro, M.; Marotta, G. Sustainable collective innovation in the agri-food value chain: The case of the "Aureo" wheat supply chain. *Sustainability* **2020**, *12*, 5642. [CrossRef]
13. Schiffer, E.; Hauck, J. Net-Map: Collecting social network data and facilitating network learning through participatory influence network mapping. *Field Methods* **2010**, *22*, 231–249. [CrossRef]
14. Hauck, J.; Stein, C.; Schiffer, E.; Vandewalle, M. Seeing the forest and the trees: Facilitating participatory network planning in environmental governance. *Glob. Environ. Change* **2015**, *35*, 400–410. [CrossRef]
15. Leimkühler, M.; Specht, K.; Schröter, B.; Wissmann, A.; Iodice, C.; Fox-Kämper, R.; Steines, A.K.; Monticone, F.; Samoggia, A.; Modarelli, G.C.; et al. Using the Net-Map tool to analyze stakeholder networks in the city region food systems of seven European cities. *Front. Sustain. Food Syst.* **2024**, *8*, 1315399. [CrossRef]
16. Van der Linden, N.; Meerpoel, C.; Schebesta, H.; Luning, P. Development of a digital Net-Map tool to analyse multi-stakeholder networks in risk analysis of emerging food safety issues. *Food Control* **2024**, *169*, 110975. [CrossRef]
17. Pilamala Rosales, A.; Linnemann, A.R.; Luning, P.A. A Net-Map analysis to understand the roles and influence of stakeholders in street food safety—A study in Ecuador. *Food Control* **2023**, *154*, 109966. [CrossRef]
18. Moreno-Miranda, C.; Dries, L. Integrating coordination mechanisms in the sustainability assessment of agri-food chains: From a structured literature review to a comprehensive framework. *Ecol. Econ.* **2022**, *192*, 107265. [CrossRef]
19. Barilla. Barilla Sustainability Report 2021. Available online: https://www.barillagroup.com/media/filer_public/b9/90/b990c3ff-9133-4bb7-bdcb-1b01a78e03b7/eng_barilla_group_2021_sustainability_report_interactive.pdf (accessed on 22 February 2025).
20. Galli, F.; Bartolini, F.; Brunori, G.; Colombo, L.; Gava, O.; Grando, S.; Marescotti, A. Sustainability assessment of food supply chains: An application to local and global bread in Italy. *Agric. Food Econ.* **2015**, *3*, 1–17. [CrossRef]
21. CSQA. Certificate for Sustainable Cereal Productions. Available online: https://www.csqa.it/it-it/certificazioni/agroalimentare/cereali-e-semi-oleosi-sostenibili-dtp-112 (accessed on 22 February 2025).
22. Ad-Hoc Meetings Advisory Group on Sustainability of Food Systems (AGSFS). Available online: https://food.ec.europa.eu/horizontal-topics/expert-groups/advisory-groups-action-platforms/advisory-group-agsfs/ad-hoc-meetings_en (accessed on 12 November 2024).
23. Tridge. Italian Wheat 2024 Production Will Be the Lowest in Recent Years. *Tridge*. Available online: https://www.tridge.com/news/italian-wheat-2024-production-will-be-the-lo-exqbop (accessed on 12 February 2025).
24. Statista. Monthly Producer Price Index of Durum Wheat in Italy. *Statista*. Available online: https://www.statista.com/statistics/1383684/monthly-producer-price-index-of-durum-wheat-italy/ (accessed on 12 February 2025).
25. Italia Regina. The Numbers of Pasta and Bread in Italy. *Italia Regina*. Available online: https://www.italiaregina.it/the-numbers-of-pasta-and-bread-in-italy/ (accessed on 12 February 2025).
26. Dongo, D. Crisis of Italian Durum Wheat. Food Times. Available online: https://www.foodtimes.eu/non-categorizzato/crisis-italian-durum-wheat/ (accessed on 12 February 2025).
27. Barilla Group. 2023 Sustainability Report: Sustainable Supply Chains. *Barilla Group*. Available online: https://www.barillagroup.com/media/filer_public/8a/38/8a38cac7-1207-4fb9-bf58-a6dac6956e80/eng_barilla_group_2023_sustainability_report_-_2_sustainable_supply_chains.pdf (accessed on 12 February 2025).
28. Djama, M.; Fouilleux, E.; Vagneron, I. Standard-setting, certifying and benchmarking: A governmentality approach to sustainability standards in the agro-food sector. In *Governing Through Standards: Origins, Drivers and Limitations*; Ponte, S., Gibbon, P., Vestergaard, J., Eds.; Palgrave Macmillan: New York, NY, USA, 2011; pp. 184–209.
29. Fischer, C. Trust and communication in European agri-food chains. *Supply Chain. Manag.* **2023**, *18*, 208–218. [CrossRef]

30. Jraisat, L.; Upadhyay, A.; Ghalia, T.; Jresseit, M.; Kumar, V.; Sarpong, D. Triads in sustainable supply-chain perspective: Why is a collaboration mechanism needed? *Int. J. Prod. Res.* **2021**, *61*, 4725–4741. [CrossRef]
31. Betti, G.; Evangelista, D.; Gagliardi, F.; Giordano, E.; Riccaboni, A. Towards Integrating Information Systems of Statistical Indicators on Traceability, Quality and Safety of Italian Agrifood Systems for Citizens, Institutions and Policy-Makers. *Sustainability* **2024**, *16*, 6330. [CrossRef]

foods

MDPI

Article

Broccoli, Artichoke, Carob and Apple By-Products as a Source of Soluble Fiber: How It Can Be Affected by Enzymatic Treatment with Pectinex® Ultra SP-L, Viscozyme® L and Celluclast® 1.5 L

Pablo Ayuso , Rocío Peñalver , Jhazmin Quizhpe , María de los Ángeles Rosell and Gema Nieto *

Department of Food Technology, Nutrition and Food Science, Veterinary Faculty, University of Murcia, Regional Campus of International Excellence "Campus Mare Nostrum", 30100 Murcia, Spain; pablo.ayuson@um.es (P.A.); rocio.penalver@um.es (R.P.); jhazminedith.quizhper@um.es (J.Q.); mariaangeles.rosellp@um.es (M.d.l.Á.R.)

* Correspondence: gnieto@um.es; Tel.: +34-86-888-9624; Fax: +34-86-888-4147

Abstract: Dietary fiber (DF), and especially soluble dietary fiber (SDF), is a nutrient of particular interest today because of its anti-inflammatory role and its ability to reduce cardiovascular risk. Therefore, the enhancement of SDF in foods using different techniques has become a promising field of research. In order to prove the possibility of increasing this SDF content, the effects of different commercial enzymes (Pectinex® Ultra SP-L, Viscozyme® L and Celluclast® 1.5 L) were tested on a variety of carob (CE), artichoke (ARE), apple (APE) and broccoli (BE) by-product extracts. Enzymatic treatment significantly affected SDF content in all by-products, showing the greatest increases for CE, ARE and APE using Celluclast® 1.5 L, while Viscozyme® L obtained the best results after application in BE. On the other hand, positive results were reported in the solubility, WHC and FAC of the by-products due to the enzymatic treatment, being increased in all extracts analyzed. Moreover, a general increase in antioxidant capacity (FRAP, ABTS and DPPH) was observed after enzymatic treatment. Finally, high yields were obtained after the application of the enzymatic processes, reaching values of 80–85% for each food by-product. These results evidenced a potential revalorization of carob, artichoke, apple and broccoli by-products after enzymatic treatment, improving its nutritional and physicochemical properties, revealing a possible application as a higher value-added ingredient.

Keywords: prebiotic; soluble fiber; natural extract; food additives; enzymatic treatment

1. Introduction

Consumers are increasingly looking for products with a high content of dietary fiber (DF), which is becoming a key nutrient in the human diet. Dietary fiber is a heterogeneous group of non-carbohydrate plant cell wall compounds (chitins, lignins and waxes) and carbohydrate polymers that are neither digested nor absorbed by the human intestine (non-starch polysaccharides, resistant oligosaccharides and resistant starch) [1]. Its recommended dietary intake by the European Food Safety Authority (EFSA) is 25 g per day and it is naturally present in whole foods, fruits, vegetables, tuberous roots and legumes [2].

DF is considered an essential component of a healthy diet and has been associated with several health benefits. A high dietary fiber intake has been related to a significant reduction in mortality from cardiovascular events [3,4], a lower incidence of diverticular disease [5] and a reduced risk of malignancies such as colorectal cancer [6]. However, the main functional property of DF is its prebiotic activity due to fermentation of intestinal microbiota of the colon. Saccharolytic bacteria are capable of producing short-chain fatty

https://doi.org/10.3390/foods14010010

acids (SCFAs) from DF, such as butyric, propionic and acetic acids [7], which are associated with a wide range of health properties. Different authors have evidenced the anticancer [8], anti-inflammatory [9], immunoregulatory [10], anti-obesity [11] and hepatoprotective [12] activity of these compounds.

According to their physical characteristics, there are two types of dietary fiber, insoluble dietary fiber (IDF) and soluble dietary fiber (SDF). SDF presents better bioactive properties than IDF due to its fermentability and viscosity [13]. In addition, it has the most desirable health effects, contributing to reducing the risk of several systemic and chronic diseases such as type 2 diabetes, obesity, hypercholesterolemia or colorectal cancer [14]. It has been shown that an intake of about 6 g/day of SDF can be associated with several health benefits, such as reduction of serum LDL-cholesterol and estimated risk of coronary heart disease [15]. For this reason, DF composition and the ratio of soluble dietary fiber (SDF)/insoluble dietary fiber (IDF) are important for nutrition, with a recommended SDF/IDF proportion of 1:2 [16].

On the other hand, fiber from plant by-products is mainly composed of IDF [17]. Therefore, the modification and degradation of IDF is considered of special interest to improve its activities and functional properties. In order to transform IDF into SDF content of food products, chemical, biological or physical techniques have been developed by several researchers. Ultrasound treatment [18], microfluidization [19], use of supercritical fluids [20], or microwave degradation [21] are techniques in increasing development. However, the most effective and most widely used is enzymatic treatment.

In fact, enzymatic hydrolysis is the most promising technology due to higher product yield and low energy requirements. Although its cost is usually higher than other physical methods due to enzyme purification [22], this could be considered an environmentally friendly process because it does not use solvents or chemical reagents [23]. Enzymatic modification consists of using enzymes capable of cleaving insoluble polysaccharides present in food matrices, decomposing them into small molecules. Enzymes have the advantage that they are non-toxic, easy to inactivate and can be highly effective at relatively low concentrations [24]. Moreover, their activity can be regulated by controlling the temperature and pH of the medium. Enzymatic treatment has been used extensively by several authors with the aim of treating different foods, improving their nutritional characteristics. Cellulases, pectinases or xylanases have currently been used in different by-products of potato [25], tomato [26], purple turnip [27], rice bran [28], ginger [29] or date [30].

On the other hand, enzymatic treatment of fibers provides other benefits. Numerous studies have reported that enzyme-assisted extraction is a useful technique for the mobilization of phenolic compounds from glycosides [31]. The extracted phenolic compounds can improve the stability and shelf life of food products, increase their antioxidant capacity and inhibit the growth of a number of microorganisms. Napolitano et al. [32] demonstrated that the conversion of IDF to SDF from cereal fiber by enzymatic treatment achieved an increase in free phenolic concentration, water-soluble antioxidant activity as well as bioavailability of phenolic compounds.

Enzymatic treatment has also been shown to modify the physicochemical and structural characteristics of dietary fibers. The modification of IDF by SDF confers to the food matrices a higher gelling capacity and viscosity as well as better fermentability. Numerous studies consider enzyme-treated fibers as potential thickener, emulsifier, stabilizer and fat substitute [33,34]. These aspects seem to validate the possibility of the application of enzyme-treated food by-products as an innovative form of revalorization in the food industry.

The objective of the current study is to test the possible application of different commercial enzymes (Pectinex® Ultra SP-L, Viscozyme® L and Celluclast® 1.5 L) on different food by-products obtained during the processing of artichoke, broccoli, apple and carob. This will be carried out by validating the conversion of IDF to SDF and the change in the physicochemical characteristics and antioxidant activity of the extracts after treatment, thereby verifying the potential revalorization of the extracts as a possible food additive.

2. Materials and Methods

2.1. Materials

Extracts of broccoli (BE) (stems), artichoke (ARE) (external bracts), apple (APE) (by-product generated after the production of apple jam) and carob (CE) (fiber obtained after the production of carob syrup) were supplied by Cricket (Lorca, Spain) and local grocery stores. By-products were oven-dried at 50 °C for 18 h. Then, dried by-products were grounded with a coffee mill and sieved until obtaining a fine flour of 500 µm particle size.

Pectinex® Ultra SP-L (blend of pectinases, hemicellulases and beta-glucanases from *Aspergillus aculeatus*; enzyme activity: ≥3300 EGU/g), Viscozyme® L (blend of beta-glucanases, pectinases, hemicellulases and xylanasee from *Aspergillus aculeatus*; enzyme activity: ≥ 100 FBG/g), and Celluclast®1.5 L (cellulase from *Trichoderma reesei*; enzyme activity: ≥ 700 PGNU/g) was purchased from Novozymes (Bagsværd, Denmark).

2.2. Enzymatic Treatment

By-product extracts were subjected to enzymatic treatment using three different commercial enzymes. For this purpose, diluted extracts (1:5 w/v) were incubated with Pectinex® Ultra SP-L (65 °C, pH 5.5), Viscozyme® L (45 °C, pH 4.5) and Celluclast®1.5 L (50 °C, pH 6) at a concentration of 0.45% (v/w) for 24 h in an orbital shaker, attending to the most optimal conditions indicated by the provider. Each treatment was done in triplicate for every extract.

After hydrolysis, the mixture was oven-dried at 60 °C and the dried by-product extracts were grounded with a coffee mill and sieved until obtaining a fine flour of 500 µm particle size. The drying process was carried out until the final moisture content of the extracts was approximately 4%. The enzymatic process yield of the extracts was calculated as follows:

$$\text{Yield}(\%) = \frac{\textit{Initial weight }(g)}{\textit{Final weight }(g)} \times 100 \tag{1}$$

2.3. Determination of Physicochemical Characteristics

Solubility measurement was performed following the method described by Cano-Chauca et al. [35] with some modifications. For 5 min, 0.5 g of sample was mixed with 50 mL of water. The solution was placed in a tube and centrifuged at 3000× g for 5 min. A 12.5 mL aliquot of the supernatant was transferred to previously weighed Petri dishes and immediately oven-dried at 100 °C for 5 h. Each treatment was done in triplicate for every extract. The solubility (%) was calculated by weight difference according to the following formula:

$$\text{Solubility}(\%) = \frac{\textit{Dried Supernatnt weight }(g) \times 4}{\textit{Sample weight }(g)} \times 100 \tag{2}$$

Water-holding capacity (WHC) was determined following the protocol described by Luo et al. [36]. In a centrifuge tube, 1 g of dry sample was mixed with 20 mL of distilled water for 24 h at room temperature. Samples were centrifuged at 3000 rpm for 20 min, the supernatant was carefully discarded and the pulp residue was collected and weighed. WHC was expressed as grams of water per gram of dry sample.

Fat-adsorption capacity (FAC) was calculated with the method described by Sosulsky and Cadden [37]. For this purpose, 1 g of sample was mixed with sunflower oil for 30 min. Afterwards, the samples were centrifuged at 1600× *g* for 25 min. FAC was calculated as the amount of sunflower oil retained by each sample expressed as g/g.

Color was measured using a Konica Minolta CR 400 chromameter (Minolta Camera Co., Osaka, Japan), standardized using a calibration plate. Lightness (L*), a* chromaticity (red-green) and b* chromaticity (blue-yellow) were measured according to the CIELab system. Measurements were taken in triplicate on untreated and enzymatically treated extracts.

The total color differences (ΔE) of each sample with respect to the control formulation and in relation to day 0 of storage were calculated as follows:

$$\Delta E = \sqrt{(L_s^* - L_c^*)^2 + (a_s^* - a_c^*)^2 + (b_s^* - b_c^*)^2} \quad (3)$$

2.4. Dietary Fiber Quantification

Total dietary fiber (TDF), soluble dietary fiber (SDF) and insoluble dietary fiber (IDF) were determined according to the 985.29:1985 method of the Association of Official Analytical Chemists (AOAC), described by Prosky et al. [38].

For the determination, 0.5 g of each sample extract was weighed in triplicate and diluted with 25 mL of phosphate buffer (50 mM, pH 6). To facilitate starch hydrolysis, samples were incubated with 50 μL of α-amylase enzyme from *Bacillus licheniformis* (500–1500 units/mg protein) (Sigma-Aldrich, Saint Louis, MO, USA) for 30 min at 95 °C in a shaking water bath. Subsequently, the samples were subjected to a second digestion (pH 7.5, 60 °C, 30 min) by adding 100 μL of 0.5 mg/mL protease from *Bacillus licheniformis* (7–15 units/mg solid) (Sigma-Aldrich, Saint Louis, MO, USA). Finally, 150 μL of amyloglucosidase from *Aspergillus niger* (≥260 units/mL) (Sigma-Aldrich, Saint Louis, MO, USA) was added to the mixture, and incubated for 30 min at 60 °C.

After digestion, filtration of the dietary fiber was carried out using a Fibertec System E 1023 (Foss, Högänas, Sweden). An initial filtration was performed to isolate the IDF. The filtrate from the previous step was precipitated after the addition of 140 mL of 95% ethanol at 60 °C and allowed to stand for one hour. The filtration procedure was then repeated and the SDF residue was collected. Once the residues were obtained, the protein and ash content of the residues were analyzed through standard methods of AOAC [39] (955.04:1995 and 923.03:1995, respectively). TDF content was calculated as the sum of IDF and SDF. The Dietary Fiber Conversion rate (DFC) from IDF to SDF was calculated as follows:

$$DFC\ (\%) = \frac{SDF_{AfterTreatment} - SDF_{Before\ Treatment}}{IDF_{BeforeTreatment}} \times 100 \quad (4)$$

2.5. Total Phenolic Content (TPC) and Antioxidant Capacity

Before the analysis, the different formulations of artichoke, broccoli, carob and apple extracts were homogenized in a methanol/water solution (80%, *v*/*v*). The mixture was incubated for 24 h at 4 °C and centrifuged (4500 rpm, 25 min). Supernatants were filtered (0.45 μm) and kept at −18 °C until analysis.

The total phenolic content (TPC) was determined following the method described by Singleton et al. [40] using Folin–Ciocalteau reagent and gallic acid as the standard. The analysis was carried out in triplicate and the TPC was expressed as mg of gallic acid equivalents (GAEs) per g of sample.

The ferric-reducing antioxidant power (FRAP) assay was determined following the protocol described by Benzie and Strain [41]. The FRAP reagent was prepared with a solution of 300 mmol/L acetate buffer, pH = 3.6, 20 mmol/L $FeCl_3 \cdot 6\ H_2O$ and 10 mmol/L TPTZ (2,4,6-tripyridyl-s-triazine) in 40 mmol/L HCl (10:1:1, *v*/*v*/*v*). A 100 μL amount of sample was mixed with 1 mL of FRAP reagent in plastic cuvettes and absorbance was measured at 593 nm after 4 min of incubation. The same procedure was carried out with a standard solution of Trolox. Analyses were carried out in triplicate and the antioxidant capacity was expressed as μM Trolox equivalents (TEs) per g of sample.

Antioxidant activity related to chelating capacity was measured using the 2,2-diphenyl-1-picrylhydrazyl (DPPH) free radical scavenging method described by Brand-Williams et al. [42]. A 100 μL amount of sample was mixed with 3.9 mL DPPH reagent in plastic cuvettes and absorbance was measured at 515 nm after 30 min of incubation in the dark. The same procedure was carried out with a standard solution of Trolox. Analyses were carried out in triplicate and antioxidant activity related to chelating capacity was expressed as μM Trolox equivalents (TEs) per g of sample.

The radical cation scavenging capacity against 2,2-azinobis (3-ethylbenzothiazolin)-6-sulphonic acid (ABTS) radical scavenging was carried out following the method described by Re et al. [43]. The ABTS reagent (7 mM) was prepared in 2.45 mM potassium persulphate (1:1, *v*/*v*), pH = 7.4, and adjusted to an absorbance of 0.7000 at 734 nm. A 1 mL amount of ABTS reagent was mixed with 100 μL of sample and absorbance was measured at 734 nm after 2 min of incubation. Analyses were carried out in triplicate and antioxidant activity related to radical scavenging capacity was expressed as μM Trolox equivalents (TEs) per g of sample.

2.6. Statistical Analysis

Differences between extracts were determined by one-way analysis of variance (ANOVA). A value of $p < 0.05$ was considered statistically significant, using the Tukey post-hoc test. Pearson's correlation was applied to test correlations between all studied markers. The results were expressed as mean ± standard deviation (SD). All statistical analyses were performed using IBM SPSS Statistics 28.0 software (IBM Corporation, Armonk, NY, USA).

3. Results and Discussion

3.1. Dietary Fiber Analysis

The IDF, SDF and TDF content of carob, artichoke, apple and broccoli extracts, as well as after application of enzymatic treatment using Pectinex® Ultra SP-L, Viscozyme® L and Celluclast 1.5 L, all purchased from Novozymes (Bagsværd, Denmark), are represented in Figure 1.

The results obtained show a high fiber content for the four extracts, with TDF representing 67.5, 59.5, 51.1 and 32.5% for carob, apple, artichoke and broccoli, respectively. SDF and IDF results were in accordance with the established literature for carob (65.69% IDF and 1.81% SDF) [44]. The enzymatic treatment applied to this extract significantly increased the SDF values and reduced the IDF values (Figure 1A). Particularly, the treatment with Celluclast 1.5 L achieved the best results, being able to convert up to 12.59% (Table 1) of the insoluble to soluble fiber content of carob by-product extract. Similar results were obtained for APE, in which the enzymatic treatment significantly reduced the levels of IDF, highlighting the results obtained with Viscozyme® L, where a more pronounced decrease was produced. However, the most remarkable increase for soluble fiber content were obtained with the enzyme Celluclast 1.5 L, resulting in a DFC value of 17.01%. Celluclast 1.5 L also obtained the best results after the enzymatic treatment of artichoke, being able to increase SDF content from 1.68 to 8.78% (Figure 1C). However, Viscozyme® L treatment was able to reduce the IDF content to a greater extent. Finally, the treatment also exerted a positive effect on broccoli extract. In this case, Viscozyme® L and Celluclast® 1.5 L enzymes exerted a similar effect after treatment, achieving an IDF to SDF turnover of 13.39 and 11.82%, respectively.

Table 1. Dietary Fiber Conversion rate (DFC) of IDF to SDF (%).

Treatment	Extracts			
	CE	APE	ARE	BE
Without Treatment	0.00 ± 0.00^{c}	0.00 ± 0.00^{c}	0.00 ± 0.00^{c}	0.00 ± 0.00^{c}
Pectinex® Ultra SP-L	2.73 ± 0.49^{b}	7.22 ± 1.54^{b}	3.27 ± 1.00^{b}	7.37 ± 0.57^{b}
Viscozyme® L	1.85 ± 0.32^{bc}	4.34 ± 2.08^{b}	3.82 ± 0.48^{b}	13.39 ± 2.07^{a}
Celluclast® 1.5 L	12.59 ± 1.36^{a}	17.01 ± 1.55^{a}	14.38 ± 1.76^{a}	11.82 ± 0.91^{a}

a–c: Different letters within in the same column indicate significant differences between enzymatic treatments ($p < 0.05$). APE: apple extract; ARE: artichoke extract; BE: broccoli extract; CE: carob extract.

Overall, Celluclast® 1.5 L was the enzyme that caused the greatest increase in SDF in carob, apple and artichoke extracts, with Viscozyme® L being the most effective after the enzymatic treatment of broccoli.

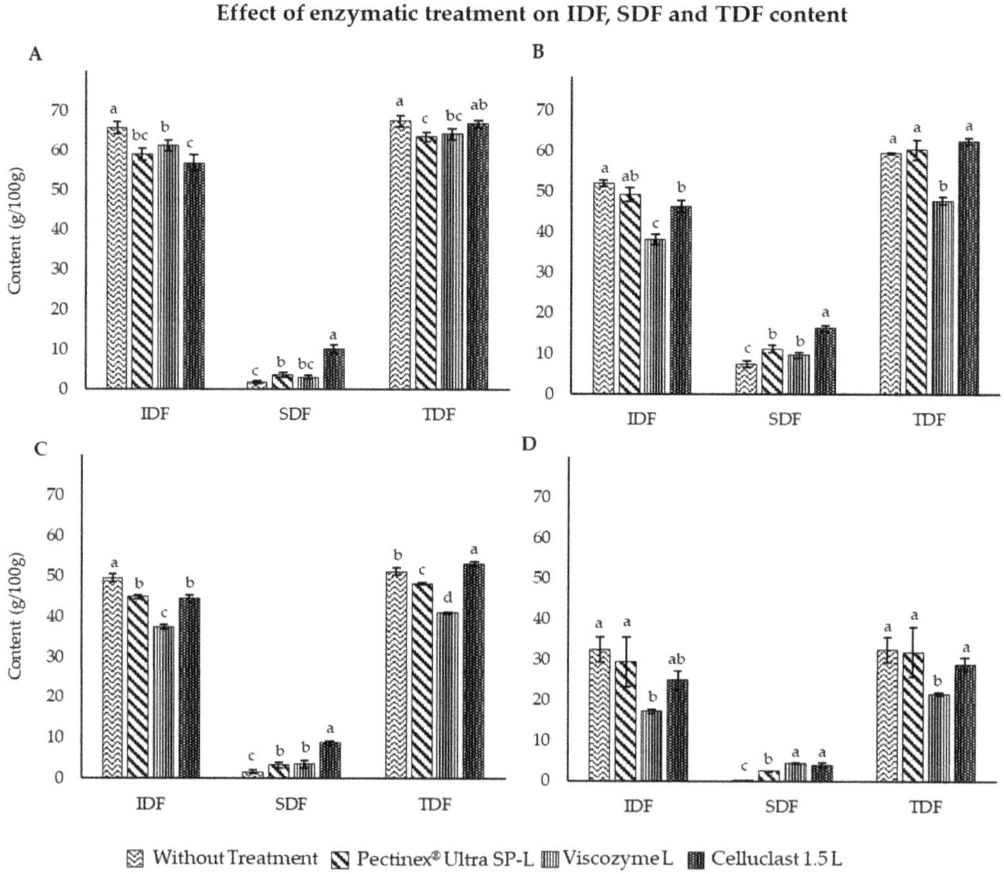

Figure 1. Effect of enzymatic treatment with different enzymes on IDF, SDF and TDF content of by-product extracts of CE (**A**), APE (**B**), ARE (**C**) and BE (**D**). a–d: Different letters for each parameter indicate significant differences between enzyme treatments ($p < 0.05$).

These results may be due to the fact that Celluclast® 1.5 L is an enzyme that catalyzes the hydrolysis of the β-1,4-glycosidic bond of cellulose and hemicellulose, developing cellulo-, xylan- and mannolytic activities [45], giving place to smaller monosaccharides and polysaccharides (cellobiose, cello-oligosaccharides or glucose). Cellulose and hemicellulose are the main components of IDF in artichoke [46], apple [45] and carob [47], so an increase in SDF and a decrease in IDF is probably the result of enzymatic cleavage of cellulose and hemicellulose by this enzyme. On the other hand, broccoli stems present a high content of cellulose, hemicellulose and pectins [48], so both Celluclast® 1.5 L and Viscozyme® L (enzyme cocktail containing xylanase, arabanase, β-glucanase, hemicellulose and cellulase) are very effective in this extract. Finally, Pectinex® Ultra SP-L is a polygalacturonase, an enzyme only able to hydrolyze pectins (soluble fiber), and unable to break cellulose bonds. For this reason, this commercial enzyme is more specific than Celluclast® 1.5 L and Viscozyme® L, showing the lowest results in the conversion of IDF to SDF.

Several authors have studied the beneficial effect of enzymatic treatment on extracts from by-products. Sabater et al. [49] examined the effect of Celluclast® 1.5 L on external bracts, leaves and stems of artichoke, demonstrating a high capacity for fiber hydrolysis and improving the extraction of pectins. Fissore et al. [50] reported similar results in enzymatically treated artichoke by-products, increasing the content of inulin, the main compound present in the SDF of artichoke. In addition, the effect of cellulase and Viscozyme® L has also been tested in extracts of leaves, stems and inflorescences of broccoli, with both being enzymes able to increase the content of soluble fibers such as pectins.

The application of pectinases as enzymatic treatment in apple by-products is the most widely used [24,51,52]. Our results, as well as recent studies conclude that cellulases can induce a higher hydrolytic effect of apple fiber, enhancing the release of oligosaccharides [53] and soluble fibers [53,54]. Jagelaviciute et al. [52] found that hydrolysis with Celluclast® 1.5 L increased content of SDF and SDF/IDF ratio in apple pomace. However, enzymatic treatment with Viscozyme® L and Pectinex® Ultra Tropical resulted in decreased SDF/IDF ratio.

Moreover, Table 2 summarizes the yield of the extracts obtained after enzymatic treatment. The yield was calculated according to formula (2) described in Section 2.2 of the "Materials and Methods" section.

Table 2. Yield (%) after enzymatic treatment of the by-product extracts.

Treatment	Extracts			
	CE	APE	ARE	BE
Pectinex® Ultra SP-L	87.13 ± 0.83 [a]	81.44 ± 0.39 [a]	86.97 ± 0.67 [a]	84.13 ± 0.32 [a]
Viscozyme® L	78.42 ± 0.67 [b]	83.72 ± 1.12 [a]	81.55 ± 0.48 [b]	78.32 ± 0.21 [b]
Celluclast® 1.5 L	86.70 ± 0.87 [a]	79.15 ± 1.23 [a]	80.24 ± 0.57 [b]	82.17± 0.67 [a]

[a,b]: Different letters in the same column indicate significant differences between enzymatic treatments ($p < 0.05$). APE: apple extract; ARE: artichoke extract; BE: broccoli extract; CE: carob extract.

Despite being the enzyme with the lowest soluble fiber-conversion potential, the best yield results for CE, ARE and BE were achieved by the Pectinex® Ultra SP-L treatment (84.13–87.13%). The high yield of pectinase application compared to other enzymatic treatments has been described by other authors [51,52]. The degradation of pectin in the primary matrix of the cell wall and in the middle lamella is necessary to break the water-binding capacity of pectin, increasing the hydration properties of the samples due to the hydroxyl groups that allow water associations through hydrogen bonds [51], and therefore increasing the yield obtained.

On the other hand, similar results were obtained for the extracts treated with Viscozyme® L and Celluclast® 1.5 L. Both enzymatic treatments developed an acceptable yield for all extracts, enabling their potential use for dietary fiber modification.

3.2. Physicochemical Characteristics

Solubility properties constitute an important parameter for functional ingredients, as they are related to the structure of polysaccharides; they can be established as regular (insoluble) or irregular (soluble) in the skeleton or as side chains [55]. The solubility values of the different extracts after enzymatic treatment are shown in Table 3.

Table 3. Solubility, WHC and FAC measurements after enzymatic treatment.

Treatment	Extracts			
	CE	APE	ARE	BE
Solubility (%)				
Without Treatment	17.41 ± 2.91 ^{b}	29.25 ± 0.34 ^{c}	24.88 ± 0.29 ^{c}	38.69 ± 0.51 ^{c}
Pectinex® Ultra SP-L	21.55 ± 3.51 ^{b}	42.38 ± 0.62 ^{a}	36.21 ± 0.41 ^{b}	29.20 ± 0.67 ^{d}
Viscozyme® L	31.17 ± 0.88 ^{a}	36.83 ± 0.27 ^{b}	37.49 ± 0.89 ^{b}	62.97 ± 0.62 ^{a}
Celluclast® 1.5 L	22.04 ± 4.03 ^{b}	42.37 ± 0.26 ^{a}	44.49 ± 0.21 ^{a}	58.04 ± 0.08 ^{b}
WHC (g/g)				
Without Treatment	2.32 ± 0.16 ^{b}	1.45 ± 0.06 ^{c}	2.89 ± 0.07 ^{c}	1.91 ± 0.09 ^{a}
Pectinex® Ultra SP-L	3.79 ± 0.63 ^{ab}	1.72 ± 0.04 ^{bc}	3.77 ± 0.04 ^{a}	2.11 ± 0.00 ^{a}
Viscozyme® L	2.63 ± 0.03 ^{ab}	1.84 ± 0.12 ^{b}	3.33 ± 0.05 ^{b}	1.79 ± 0.18 ^{a}
Celluclast® 1.5 L	3.11 ± 0.30 ^{b}	2.37 ± 0.07 ^{a}	3.63 ± 0.05 ^{a}	2.13 ± 0.14 ^{a}
PAC (g/g)				
Without Treatment	1.03 ± 0.02 ^{c}	0.77 ± 0.02 ^{a}	1.69 ± 0.00 ^{c}	0.56 ± 0.01 ^{c}
Pectinex® Ultra SP-L	3.64 ± 0.13 ^{a}	1.08 ± 0.03 ^{a}	2.69 ± 0.03 ^{ab}	0.97 ± 0.02 ^{b}
Viscozyme® L	1.40 ± 0.06 ^{ab}	0.71 ± 0.20 ^{a}	2.38 ± 0.15 ^{b}	0.92 ± 0.03 ^{bc}
Celluclast® 1.5 L	1.32 ± 0.02 ^{b}	1.13 ± 0.06 ^{a}	2.90 ± 0.18 ^{a}	1.07 ± 0.03 ^{a}

$^{a–d}$: Different letters in the same column indicate significant differences between enzymatic treatments ($p < 0.05$). APE: apple extract; ARE: artichoke extract; BE: broccoli extract; CE: carob extract.

Solubility values obtained showed a clear relationship between its increase and the enzymatic treatment. The untreated extracts presented the lowest solubility results, showing differences between extracts, although BE showed the highest values and CE the lowest values. The increase in solubility for each extract relied on the enzyme used during treatment. Viscozyme® L was the most effective for CE and BE, achieving an increase in solubility of 13.76 and 24.28%, respectively. On the other hand, the treatment with Pectinex® Ultra SP-L was the best performer for APE, increasing by 13.13%. Finally, Celluclast® 1.5 L was the most effective for ARE, increasing solubility values from 24.88% to 44.49%.

Similar results were obtained for water-holding capacity (WHC) and fat-adsorption capacity (FAC). A general increase in fluid retention was observed after the application of enzymatic treatment in CE, APE, ARE and BE. In particular, a greater increase was found in the extracts treated with Pectinex® Ultra SP-L, reaching increases in WHC of 163, 118, 130 and 109% for CE, APE, ARE and BE, respectively. The possible reason for the improvement in solubility and liquid retention may be certain changes in the microstructure increasing the surface area of the modified dietary fiber, exposing in turn more water-binding sites, which favored the absorption and penetration of water molecules [33,56]. Other authors have demonstrated this improvement in water solubility after enzymatic treatment. Similar results for extracts from broccoli by-products were obtained by Rivas et al. [57], reaching water solubility values of 53.7% after treatment with Viscozyme® L, as well as an increase in WHC and FAC in leaves, stems and inflorescences of broccoli. In addition, the effect on solubility that pectinases can induce in apple extracts has also been demonstrated [24], with some authors reporting the highest solubility index (43.80 ± 2.10%) obtained using Pectinex® Ultra SP-L [52]. Pectin can increase the hydration properties of the samples due to its hydroxyl groups that allow water associations through hydrogen bonds, which could be the reason it is the enzyme that gives the highest results for WHC and FAC.

Moreover, it has been shown that the effect of cellulase in carob juice can increase the water-retention capacity and, consequently, a reduction in the viscosity of the juice and facilitating filtration [58]. On the other hand, it has been possible to improve the water-holding capacity of extracts from other food matrices applying different enzymes, such as potato pulp using Viscozyme® L [25], purple turnip using cellulases [27] or corn bran using xylanases and cellulases [59]. As expected, these improvements in solubility correspond with the results obtained for SDF content, reaffirming Celluclast® 1.5 L as the most effective enzyme for APE and ARE, and Viscozyme® L as the most appropriate for BE treatment.

It has been shown that dietary fiber with high solubility and WHC could be used as a functional ingredient to modify the viscosity and texture of some foods (bakery products, dairy products, jams, meats, soups), as well as to stabilize fat in emulsion-based products [60], and to improve shelf life [55]. Some studies also evidenced that foods with high water-holding capacity and solubility prevent and treat intestinal disorders by increasing fecal bulk and reducing gastrointestinal transit time [57], and may have a positive impact on reducing the risk of chronic disorders, coronary heart disease, diabetes, obesity and some forms of cancer [55].

On the other hand, Table 4 summarizes the color (CIElab) measurements of the untreated and enzymatically treated extracts.

Table 4. Color evolution (CIELab) of untreated and treated extracts.

Treatment	CE	APE	ARE	BE
L*				
Without Treatment	44.20 ± 1.93 ^{a}	50.83 ± 2.17 ^{a}	54.39 ± 0.42 ^{a}	54.8 ± 2.02 ^{a}
Pectinex® Ultra SP-L	46.94 ± 1.21 ^{a}	39.77 ± 0.48 ^{c}	49.41 ± 0.13 ^{b}	46.56 ± 0.11 ^{b}
Viscozyme® L	41.19 ± 0.01 ^{b}	36.76 ± 0.76 ^{c}	47.73 ± 0.46 ^{ca}	43.64 ± 0.72 ^{c}
Celluclast® 1.5 L	39.80 ± 0.31 ^{b}	44.30 ± 2.10 ^{b}	49.20 ± 0.09 ^{b}	48.97 ± 0.29 ^{b}
a*				
Without Treatment	11.17 ± 0.12 ^{a}	12.19 ± 0.32 ^{a}	8.74 ± 0.07 ^{a}	4.94 ± 0.27 ^{a}
Pectinex® Ultra SP-L	4.62 ± 0.04 ^{c}	9.12 ± 0.02 ^{b}	4.84 ± 0.03 ^{c}	3.64 ± 0.07 ^{b}
Viscozyme® L	11.07 ± 0.03 ^{a}	8.51 ± 0.03 ^{c}	6.50 ± 0.03 ^{b}	3.74 ± 0.26 ^{b}
Celluclast® 1.5 L	10.50 ± 0.08 ^{a}	8.74 ± 0.28 ^{bc}	4.68 ± 0.14 ^{c}	3.83 ± 0.04 ^{b}
b*				
Without Treatment	19.03 ± 0.39 ^{a}	28.47 ± 0.31 ^{a}	26.27 ± 0.45 ^{a}	27.62 ± 0.26 ^{a}
Pectinex® Ultra SP-L	19.35 ± 0.20 ^{a}	19.51 ± 0.23 ^{b}	21.51 ± 0.07 ^{ab}	19.67 ± 0.11 ^{b}
Viscozyme® L	18.24 ± 0.01 ^{b}	16.83 ± 1.29 ^{c}	21.93 ± 0.08 ^{b}	17.82 ± 0.75 ^{c}
Celluclast® 1.5 L	16.46 ± 0.10 ^{c}	17.90 ± 0.37 ^{bc}	21.29 ± 0.14 ^{c}	20.51 ± 0.19 ^{b}
ΔE				
Without Treatment	0.00 ± 0.00 ^{c}	0.00 ± 0.00 ^{c}	0.00 ± 0.00 ^{b}	0.00 ± 0.00 ^{c}
Pectinex® Ultra SP-L	7.13 ± 0.30 ^{a}	14.61 ± 2.08 ^{ab}	7.92 ± 0.28 ^{a}	11.59 ± 1.57 ^{ab}
Viscozyme® L	3.20 ± 1.95 ^{bc}	18.79 ± 0.45 ^{a}	8.27 ± 0.09 ^{a}	14.95 ± 0.79 ^{a}
Celluclast® 1.5 L	5.30 ± 1.58 ^{ab}	10.08 ± 2.87 ^{b}	8.27 ± 0.33 ^{a}	9.33 ± 1.21 ^{bc}

$^{a–c}$: Different letters in the same column indicate significant differences between enzymatic treatments ($p < 0.05$). APE: apple extract; ARE: artichoke extract; BE: broccoli extract; CE: carob extract.

Overall, significant differences ($p < 0.05$) were observed between the results of the L*, a* and b* coordinates between the extracts before and after enzymatic treatment, showing that this has an impact on the color. A loss of reddish color (a*) was found after treatment with all enzymes in APE, ARE and BE, while only Pectinex® Ultra SP-L changed the color in CE. In addition, changes in the blue-yellow chromaticity (b*) and Lightness (L*) were also observed, being reduced during all treatments of APE, ARE and BE. A parameter that was taken into account when evaluating color modifications during treatment is the total color difference (ΔE) relative to the original sample. Several studies have mentioned that ΔE values greater than 3 units correspond to changes perceptible to the human eye [61]. After analysis, a significant change greater than 3 was observed for all treatments in CE, APE, ARE and BE, denoting a clear relationship between the enzymatic treatment and a visually appreciable color change. In addition, a greater change was seen in the APE and BE extracts, reaching values above 10 after the use of the three different enzymes. On the other hand, CE was the extract that underwent the least color change, especially after treatment with Viscozyme® L.

Several studies have examined the impact that enzymatic treatment can have on the color of extracts. Pectinases and cellulases have been used to change the color of saffron tepals [62] or black tea [63] extracts, improving their acceptability. Konopacka et al. [64] observed changes in L*, a* and b* after enzymatic treatment of pumpkin and carrot, increasing these values after application of commercial pectinases. The reduction of L*, a * and b* values obtained after treatment may be due to a browning effect of the extracts. The application of temperatures above 50 °C during processing may accelerate a chemical reaction of the polyphenol oxidase enzyme, catalyzing the production of melanins and benzoquinone from polyphenols present in the extracts [65].

3.3. Antioxidant Activity and Total Phenolic Content (TPC) Determination

The antioxidant capacity of the extracts (FRAP, ABTS and DPPH) before and after enzymatic treatment as well as the total phenolic content (TPC) are summarized in Table 5. Treatment with different enzymes caused significant changes ($p < 0.05$) in antioxidant activity. A significant increase in DPPH was observed for CE, after the use of Pectinex® Ultra SP-L, Viscozyme® L and Celluclast® 1.5 L, while positive trends were observed in FRAP and ABTS assays. Concerning ARE, Pectinex® Ultra SP-L and Celluclast® 1.5 L were able to significantly increase the ABTS values with respect to the untreated extract. The results for APE were also positive, as the Viscozyme® L treatment significantly increased the antioxidant capacity in the FRAP, ABTS and DPPH assays, increasing by 149, 156 and 171%, respectively. Finally, significant changes were also observed in BE, with Pectinex® Ultra SP-L and Viscozyme® L treatments being beneficial in ABTS and DPPH tests.

Similar results have been found in the literature, Napolitano et al. [32] showed that the conversion of IDF to SDF from barley spent grain and durum wheat fiber produced an increase in antioxidant activity measured by ABTS assay after enzymatic treatment. Another study performed on black currant extracts showed an increase in the inhibition of human LDL Oxidation after treatment with commercial enzymes (Pectinex BE, Macer8 FJ and Novozyme) [66]. Mrabet et al. [30] also demonstrated that treatment with Viscozyme® L produced an increase in the DPPH assay in date fruit. On the other hand, the results for total phenolic content were more diverse. No significant differences were observed after CE enzyme treatment; however, an increase in TPC was observed in Celluclast® 1.5 L-treated ARE and a decrease after Viscozyme® L treatment. The results with ARE showed a clear relationship between enzyme treatment and higher polyphenol content, as both Viscozyme® L and Celluclast® 1.5 L treatments increased TPC levels. Finally, the results showed an increase after treatment with Pectinex® Ultra SP-L and Viscozyme® L.

Table 5. Antioxidant activity (μmol TE/g) and Total Phenolic Content (TPC) (mg GAE/g) after enzymatic treatment of the by-product extracts.

Assay	Treatment	CE	APE	ARE	BE
FRAP	Without Treatment	140.81 ± 37.74 [a]	94.70 ± 11.03 [a]	123.22 ± 2.45 [b]	74.10 ± 2.17 [a]
	Pectinex® Ultra SP-L	163.32 ± 0.54 [a]	98.94 ± 16.92 [a]	130.47 ± 4.32 [b]	84.67 ± 2.98 [a]
	Viscozyme® L	179.07 ± 4.38 [a]	63.69 ± 0.27 [b]	182.12 ± 4.90 [a]	85.56 ± 7.10 [a]
	Celluclast® 1.5 L	181.60 ± 18.6 [a]	113.79 ± 4.09 [a]	130.61 ± 2.73 [b]	65.22 ± 7.37 [a]
ABTS	Without Treatment	124.40 ± 20.49 [b]	58.22 ± 0.39 [ab]	80.92 ± 12.35 [b]	55.83 ± 1.19 [b]
	Pectinex® Ultra SP-L	171.90 ± 3.97 [ab]	82.47 ± 0.00 [ab]	83.58 ± 6.00 [b]	68.51 ± 2.60 [a]
	Viscozyme® L	179.63 ± 14.41 [ab]	41.02 ± 4.39 [b]	126.10 ± 1.20 [a]	76.66 ± 0.40 [a]
	Celluclast® 1.5 L	198.32 ± 18.00 [a]	85.55 ± 11.47 [a]	85.86 ± 4.34 [b]	54.01 ± 2.79 [b]
DPPH	Without Treatment	199.00 ± 44.07 [a]	129.75 ± 31.41 [a]	126.85 ± 19.08 [b]	106.06 ± 25.31 [ab]
	Pectinex® Ultra SP-L	269.93 ± 31.71 [a]	129.33 ± 22.31 [a]	129.20 ± 31.41 [b]	137.90 ± 3.71 [a]
	Viscozyme® L	276.81 ± 0.00 [a]	68.56 ± 3.18 [a]	217.02 ± 25.47 [a]	143.06 ± 3.91 [a]
	Celluclast® 1.5 L	280.93 ± 28.73 [a]	137.21 ± 22.12 [a]	174.96 ± 3.20 [ab]	69.67 ± 7.97 [b]
TPC	Without Treatment	59.22 ± 24.82 [a]	74.20 ± 9.32 [ab]	84.78 ± 0.74 [b]	63.72 ± 0.74 [bc]
	Pectinex® Ultra SP-L	71.11 ± 3.75 [a]	81.44 ± 0.39 [ab]	85.85 ± 1.72 [ab]	74.45 ± 2.62 [ab]
	Viscozyme® L	47.57 ± 0.99 [a]	53.48 ± 0.50 [b]	127.91 ± 0.00 [a]	88.75 ± 5.48 [a]
	Celluclast® 1.5 L	55.32 ± 1.00 [a]	83.36 ± 1.97 [a]	109.2 ± 23.71 [ab]	51.01 ± 3.49 [c]

[a–c]: Different letters in the same column indicate significant differences between enzymatic treatments ($p < 0.05$). APE: apple extract; ARE: artichoke extract; BE: broccoli extract; CE: carob extract.

These results show that there is a relationship between the application of different enzymes and a higher amount of polyphenols depending on the extract used. This may be because phenols that were trapped or structurally bound to the fiber could be partially released after digestion and are observed in a higher quantity [67].

Similar results were found in the research of Jagelaviciute et al. [52] where an increase in TPC of apple pomace was found after treatment with Viscozyme® L, Pectinex® Ultra Tropical and Celluclast® 1.5 L. In fact, these results also agree with those obtained by Landbo and Meyer [66] where the use of different pectinases increased the total phenolic content of black currant. On the other hand, it has also been shown that the conversion of IDF to SDF from coffee silver skin by enzymatic treatment can also lead to the release of hydroxycinnamic acids, mainly ferulic acid, which are bound to polysaccharide chains [32].

3.4. Pearson Correlations

A Pearson analysis was performed to determine the correlation of the different parameters studied; the results are presented in Table 6.

The results obtained showed significant differences ($p < 0.05$) between the different parameters studied. A relationship was observed between the different color parameters (a^*, b^*, L^* and ΔE) and soluble fiber content, suggesting that enzymatic treatment influences the color of the extracts, making them darker. As mentioned above, a possible hypothesis for this phenomenon is enzymatic browning due to temperatures above 50 °C during the enzymatic treatment process, thereby accelerating a chemical reaction of the polyphenol oxidase enzyme and catalyzing the production of melanins from the polyphenols present in the extracts [65].

Another aspect to highlight is the significant relationship between SDF content and solubility; this may be due to the fact that certain changes in the structure increased the surface area of the modified dietary fiber, in turn exposing more water-binding sites, which favored the absorption of water molecules [56]. However, the relationship between SDF and WHC/FAC was not significant, indicating that there are more factors to consider between these parameters.

Table 6. Pearson correlations between different measured parameters.

	SDF	DFC	FRAP	TPC	DPPH	ABTS	a*	b*	L*	ΔE	WHC	FAC	Solubility
SDF		0.729 **	−0.188	0.067	−0.112	−0.105	0.208	−0.447 **	−0.528 **	0.524 **	−0.216	−0.186	0.342 *
DFC	0.729 **		−0.152	0.257	−0.034	−0.029	−0.299 *	−0.533 **	−0.321 *	0.531 **	0.006	−0.028	0.627 **
FRAP	−0.188	−0.152		0.26	0.840 **	0.859 **	0.318	−0.213	−0.154	−0.33	0.663 **	0.574 **	−0.518 **
TPC	0.067	0.257	0.26		0.044	−0.09	−0.189	0.157	0.222	0.044	0.291	0.294	0.29
DPPH	−0.112	−0.034	0.840 **	0.044		0.944 **	0.276	−0.322	−0.235	−0.282	0.517 **	0.459 **	−0.532 **
ABTS	−0.105	−0.029	0.859 **	−0.09	0.944 **		0.311	−0.415 *	−0.322	−0.26	0.531 **	0.421 *	−0.541 **
a*	0.208	−0.299 *	0.318	−0.189	0.276	0.311		0.025	−0.335 *	−0.380 **	−0.242	−0.329	−0.526 **
b*	−0.447 **	−0.533 **	−0.213	0.157	−0.322	−0.415 *	0.025		0.843 **	−0.623 **	−0.142	−0.059	−0.106
L*	−0.528 **	−0.321 *	−0.154	0.222	−0.235	−0.322	−0.335 *	0.843 **		−0.592 **	0.134	0.176	−0.001
ΔE	0.524 **	0.532 **	−0.33	0.044	−0.282	−0.26	−0.380 **	−0.623 **	−0.592 **		−0.169	−0.059	0.531 **
WHC	−0.216	0.006	0.663 **	0.291	0.517 **	0.531 **	−0.242	−0.142	0.134	−0.169		0.900 **	−0.297
FAC	−0.186	−0.028	0.574 **	0.294	0.459 **	0.421 *	−0.329	−0.059	0.176	−0.059	0.900 **		−0.208
Solubility	0.342 *	0.627 **	−0.518 **	0.290	−0.532 **	−0.541 **	−0.526 **	−0.106	−0.001	0.532 **	−0.297	−0.208	

Significance levels: *: $p < 0.05$; **: $p < 0.01$.

4. Conclusions

Enzymatic treatment of food by-products with different commercial enzymes achieved considerable nutritional and physicochemical benefits. Increases in SDF levels were demonstrated for all by-products, with the treatment being especially positive with Viscozyme® L for BE and Cellucalst® 1.5 L for CE, ARE and APE. Moreover, this increase in the SDF content was related to an improvement in the solubility, WHC and FAC of the extracts, which showed an increase after almost all possible enzymatic treatments. Additionally, treatment with the different enzymes had an impact on the antioxidant capacity of all the extracts, conferring them health-promoting benefits. On the other hand, the processes maintained a high yield after enzyme application, indicating a good potential application of this technology on a larger scale. These results evidence potential benefits of different extracts after the application of an enzymatic treatment, which could represent a key opportunity for the food industry and an attractive revaluation for food by-products.

Author Contributions: Conceptualization, R.P., P.A. and G.N.; methodology, R.P. and P.A.; validation, P.A., J.Q., M.d.l.Á.R., R.P. and G.N.; investigation, P.A.; writing—original draft preparation, P.A.; writing—review and editing, P.A., J.Q., M.d.l.Á.R., R.P. and G.N.; visualization, G.N.; supervision, G.N.; project administration, G.N.; funding acquisition, G.N. All authors have read and agreed to the published version of the manuscript.

Funding: This study forms part of the Agroalnext program and was supported by MCIU with funding from European Union Next GenerationEU (PRTR-C17.I1) and by Comunidad Autónoma de la Región de Murcia—Fundación Séneca.

Institutional Review Board Statement: Not applicable.

Informed Consent Statement: Not applicable.

Data Availability Statement: The original contributions presented in the study are included in the article, further inquiries can be directed to the corresponding author.

Conflicts of Interest: The authors declare no conflicts of interest.

References

1. Stephen, A.M.; Champ, M.M.-J.; Cloran, S.J.; Fleith, M.; van Lieshout, L.; Mejborn, H.; Burley, V.J. Dietary Fibre in Europe: Current State of Knowledge on Definitions, Sources, Recommendations, Intakes and Relationships to Health. *Nutr. Res. Rev.* **2017**, *30*, 149–190. [CrossRef] [PubMed]
2. Venter, C.; Meyer, R.W.; Greenhawt, M.; Pali-Schöll, I.; Nwaru, B.; Roduit, C.; Untersmayr, E.; Adel-Patient, K.; Agache, I.; Agostoni, C.; et al. Role of Dietary Fiber in Promoting Immune Health—An EAACI Position Paper. *Allergy* **2022**, *77*, 3185–3198. [CrossRef] [PubMed]
3. Threapleton, D.E.; Greenwood, D.C.; Evans, C.E.L.; Cleghorn, C.L.; Nykjaer, C.; Woodhead, C.; Cade, J.E.; Gale, C.P.; Burley, V.J. Dietary Fiber Intake and Risk of First Stroke. *Stroke* **2013**, *44*, 1360–1368. [CrossRef] [PubMed]
4. Threapleton, D.E.; Greenwood, D.C.; Evans, C.E.L.; Cleghorn, C.L.; Nykjaer, C.; Woodhead, C.; Cade, J.E.; Gale, C.P.; Burley, V.J. Dietary Fibre Intake and Risk of Cardiovascular Disease: Systematic Review and Meta-Analysis. *BMJ* **2013**, *347*, f6879. [CrossRef] [PubMed]
5. Kapler, M.; Kotula, J.; Mąsior, M.N.; Zinkow, K.; Gawlik, Ł.M. The Role of Dietary Fiber in Gastrointestinal Health: Focus on Irritable Bowel Syndrome and Diverticular Disease. *Qual. Sport* **2024**, *18*, 53478. [CrossRef]
6. Aune, D.; Chan, D.S.M.; Lau, R.; Vieira, R.; Greenwood, D.C.; Kampman, E.; Norat, T. Dietary Fibre, Whole Grains, and Risk of Colorectal Cancer: Systematic Review and Dose-Response Meta-Analysis of Prospective Studies. *BMJ* **2011**, *343*, d6617. [CrossRef]
7. Xiong, R.-G.; Zhou, D.-D.; Wu, S.-X.; Huang, S.-Y.; Saimaiti, A.; Yang, Z.-J.; Shang, A.; Zhao, C.-N.; Gan, R.-Y.; Li, H.-B. Health Benefits and Side Effects of Short-Chain Fatty Acids. *Foods* **2022**, *11*, 2863. [CrossRef]
8. Yusefi, M.; Shameli, K.; Ali, R.R.; Pang, S.-W.; Teow, S.-Y. Evaluating Anticancer Activity of Plant-Mediated Synthesized Iron Oxide Nanoparticles Using Punica Granatum Fruit Peel Extract. *J. Mol. Struct.* **2020**, *1204*, 127539. [CrossRef]
9. Olsson, A.; Gustavsen, S.; Nguyen, T.D.; Nyman, M.; Langkilde, A.R.; Hansen, T.H.; Sellebjerg, F.; Oturai, A.B.; Bach Søndergaard, H. Serum Short-Chain Fatty Acids and Associations with Inflammation in Newly Diagnosed Patients With Multiple Sclerosis and Healthy Controls. *Front. Immunol.* **2021**, *12*, 661493. [CrossRef]

10. Yang, W.; Yu, T.; Huang, X.; Bilotta, A.J.; Xu, L.; Lu, Y.; Sun, J.; Pan, F.; Zhou, J.; Zhang, W.; et al. Intestinal Microbiota-Derived Short-Chain Fatty Acids Regulation of Immune Cell IL-22 Production and Gut Immunity. *Nat. Commun.* **2020**, *11*, 4457. [CrossRef]
11. Cao, S.-Y.; Zhao, C.-N.; Xu, X.-Y.; Tang, G.-Y.; Corke, H.; Gan, R.-Y.; Li, H.-B. Dietary Plants, Gut Microbiota, and Obesity: Effects and Mechanisms. *Trends Food Sci. Technol.* **2019**, *92*, 194–204. [CrossRef]
12. Shimizu, H.; Masujima, Y.; Ushiroda, C.; Mizushima, R.; Taira, S.; Ohue-Kitano, R.; Kimura, I. Dietary Short-Chain Fatty Acid Intake Improves the Hepatic Metabolic Condition via FFAR3. *Sci. Rep.* **2019**, *9*, 16574. [CrossRef] [PubMed]
13. Ma, M.; Mu, T. Effects of Extraction Methods and Particle Size Distribution on the Structural, Physicochemical, and Functional Properties of Dietary Fiber from Deoiled Cumin. *Food Chem.* **2016**, *194*, 237–246. [CrossRef] [PubMed]
14. Yapo, B.M.; Gnakri, D. PecticPolysaccharides and Their Functional Properties. In *Polysaccharides*; Springer International Publishing: Cham, Switzerland, 2014; pp. 1–18.
15. Schneeman, B.O. Soluble vs Insoluble Fiber: Different Physiological Responses. *Food Technol.* **1987**, *41*, 81–82.
16. Borchani, C.; Besbes, S.; Masmoudi, M.; Bouaziz, M.A.; Blecker, C.; Attia, H. Influence of Oven-Drying Temperature on Physicochemical and Functional Properties of Date Fibre Concentrates. *Food Bioprocess Technol.* **2012**, *5*, 1541–1551. [CrossRef]
17. He, C.; Sampers, I.; Van de Walle, D.; Dewettinck, K.; Raes, K. Encapsulation of Lactobacillus in Low-Methoxyl Pectin-Based Microcapsules Stimulates Biofilm Formation: Enhanced Resistances to Heat Shock and Simulated Gastrointestinal Digestion. *J. Agric. Food Chem.* **2021**, *69*, 6281–6290. [CrossRef]
18. Xu, Y.; Zhang, X.; Yan, X.-H.; Zhang, J.-L.; Wang, L.-Y.; Xue, H.; Jiang, G.-C.; Ma, X.-T.; Liu, X.-J. Characterization, Hypolipidemic and Antioxidant Activities of Degraded Polysaccharides from Ganoderma Lucidum. *Int. J. Biol. Macromol.* **2019**, *135*, 706–716. [CrossRef]
19. Chen, J.; Gao, D.; Yang, L.; Gao, Y. Effect of Microfluidization Process on the Functional Properties of Insoluble Dietary Fiber. *Food Res. Int.* **2013**, *54*, 1821–1827. [CrossRef]
20. Escobar, E.L.N.; da Silva, T.A.; Pirich, C.L.; Corazza, M.L.; Pereira Ramos, L. Supercritical Fluids: A Promising Technique for Biomass Pretreatment and Fractionation. *Front. Bioeng. Biotechnol.* **2020**, *8*, 252. [CrossRef]
21. Hashemifesharaki, R.; Xanthakis, E.; Altintas, Z.; Guo, Y.; Gharibzahedi, S.M.T. Microwave-Assisted Extraction of Polysaccharides from the Marshmallow Roots: Optimization, Purification, Structure, and Bioactivity. *Carbohydr. Polym.* **2020**, *240*, 116301. [CrossRef]
22. Gan, J.; Xie, L.; Peng, G.; Xie, J.; Chen, Y.; Yu, Q. Systematic Review on Modification Methods of Dietary Fiber. *Food Hydrocoll.* **2021**, *119*, 106872. [CrossRef]
23. Meyabadi, T.F.; Dadashian, F. Optimization of Enzymatic Hydrolysis of Waste Cotton Fibers for Nanoparticles Production Using Response Surface Methodology. *Fibers Polym.* **2012**, *13*, 313–321. [CrossRef]
24. Canela-Xandri, A.; Balcells, M.; Villorbina, G.; Cubero, M.Á.; Canela-Garayoa, R. Effect of Enzymatic Treatments on Dietary Fruit Fibre Properties. *Biocatal. Biotransformation* **2018**, *36*, 172–179. [CrossRef]
25. Meyer, A.S.; Dam, B.P.; Lærke, H.N. Enzymatic Solubilization of a Pectinaceous Dietary Fiber Fraction from Potato Pulp: Optimization of the Fiber Extraction Process. *Biochem. Eng. J.* **2009**, *43*, 106–112. [CrossRef]
26. Gu, M.; Fang, H.; Gao, Y.; Su, T.; Niu, Y.; Yu, L. (Lucy) Characterization of Enzymatic Modified Soluble Dietary Fiber from Tomato Peels with High Release of Lycopene. *Food Hydrocoll.* **2020**, *99*, 105321. [CrossRef]
27. Ma, R.; Chen, J.-N.; Zhou, X.; Lin, H.; Gao, Q.; Peng, X.; Tanokura, M.; Xue, Y. Effect of Chemical and Enzymatic Modifications on the Structural and Physicochemical Properties of Dietary Fiber from Purple Turnip (*Brassica Rapa* L.). *LWT* **2021**, *145*, 111313. [CrossRef]
28. Hernandez, N.; Rodriguez-Alegría, M.E.; Gonzalez, F.; Lopez-Munguia, A. Enzymatic Treatment of Rice Bran to Improve Processing. *J. Am. Oil Chem. Soc.* **2000**, *77*, 177–180. [CrossRef]
29. Wang, C.; Song, R.; Wei, S.; Wang, W.; Li, F.; Tang, X.; Li, N. Modification of Insoluble Dietary Fiber from Ginger Residue through Enzymatic Treatments to Improve Its Bioactive Properties. *LWT* **2020**, *125*, 109220. [CrossRef]
30. Mrabet, A.; Rodríguez-Gutiérrez, G.; Rubio-Senent, F.; Hamza, H.; Rodríguez-Arcos, R.; Guillén-Bejarano, R.; Sindic, M.; Jiménez-Araujo, A. Enzymatic Conversion of Date Fruit Fiber Concentrates into a New Product Enriched in Antioxidant Soluble Fiber. *LWT* **2017**, *75*, 727–734. [CrossRef]
31. Takó, M.; Zambrano, C.; Kotogán, A.; Kerekes, E.B.; Papp, T.; Krisch, J.; Vágvölgyi, C. Fermentative and Enzyme-Assisted Production of Phenolic Antioxidants from Plant Residues. In *Microbial Fermentation and Enzyme Technology*; CRC Press: Boca Raton, FL, USA, 2020; pp. 175–193.
32. Napolitano, A.; Lanzuise, S.; Ruocco, M.; Arlotti, G.; Ranieri, R.; Knutsen, S.H.; Lorito, M.; Fogliano, V. Treatment of Cereal Products with a Tailored Preparation of *Trichoderma* Enzymes Increases the Amount of Soluble Dietary Fiber. *J. Agric. Food Chem.* **2006**, *54*, 7863–7869. [CrossRef]
33. Shen, M.; Weihao, W.; Cao, L. Soluble Dietary Fibers from Black Soybean Hulls: Physical and Enzymatic Modification, Structure, Physical Properties, and Cholesterol Binding Capacity. *J. Food Sci.* **2020**, *85*, 1668–1674. [CrossRef] [PubMed]

34. Feng, Z.; Dou, W.; Alaxi, S.; Niu, Y.; Yu, L. (Lucy) Modified Soluble Dietary Fiber from Black Bean Coats with Its Rheological and Bile Acid Binding Properties. *Food Hydrocoll.* **2017**, *62*, 94–101. [CrossRef]
35. Cano-Chauca, M.; Stringheta, P.C.; Ramos, A.M.; Cal-Vidal, J. Effect of the Carriers on the Microstructure of Mango Powder Obtained by Spray Drying and Its Functional Characterization. *Innov. Food Sci. Emerg. Technol.* **2005**, *6*, 420–428. [CrossRef]
36. Luo, X.; Wang, Q.; Zheng, B.; Lin, L.; Chen, B.; Zheng, Y.; Xiao, J. Hydration Properties and Binding Capacities of Dietary Fibers from Bamboo Shoot Shell and Its Hypolipidemic Effects in Mice. *Food Chem. Toxicol.* **2017**, *109*, 1003–1009. [CrossRef]
37. Sosulsky, F.W.; Cadden, A.M. Composition and Physiological Properties of Several Sources of Dietary Fiber. *J. Food Sci.* **1982**, *47*, 1472–1477. [CrossRef]
38. Prosky, L.; Asp, N.G.; Schweizer, T.F.; DeVries, J.W.; Furda, I. Determination of Insoluble, Soluble, and Total Dietary Fiber in Foods and Food Products: Interlaboratory Study. *J. AOAC Int.* **1988**, *71*, 1017–1023. [CrossRef]
39. *AOAC Official Methods of Analysis of AOAC International*, 19th ed.; Association of Official Analytical Chemists: Gaithersburg, MD, USA, 2012.
40. Singleton, V.L.; Orthofer, R.; Lamuela-Ravents, R.M. Analysis of Total Phenols and Other Oxidation Substrates and Antioxidants by Means of Folin-Ciocalteu Reagent. In *Methods in Enzymology*; Academic Press: Cambridge, MA, USA, 1999.
41. Benzie, I.F.F.; Strain, J.J. The Ferric Reducing Ability of Plasma (FRAP) as a Measure of "Antioxidant Power": The FRAP Assay. *Anal. Biochem.* **1996**, *239*, 70–76. [CrossRef]
42. Brand-Williams, W.; Cuvelier, M.E.; Berset, C. Use of a Free Radical Method to Evaluate Antioxidant Activity. *LWT-Food Sci. Technol.* **1995**, *28*, 25–30. [CrossRef]
43. Re, R.; Pellegrini, N.; Proteggente, A.; Pannala, A.; Yang, M.; Rice-Evans, C. Antioxidant Activity Applying an Improved Abts Radical Cation Decolorization Assay. *Free. Radic. Biol. Med.* **1999**, *26*, 1231–1237. [CrossRef]
44. Gruendel, S.; Garcia, A.L.; Otto, B.; Mueller, C.; Steiniger, J.; Weickert, M.O.; Speth, M.; Katz, N.; Koebnick, C. Carob Pulp Preparation Rich in Insoluble Dietary Fiber and Polyphenols Enhances Lipid Oxidation and Lowers Postprandial Acylated Ghrelin in Humans. *J. Nutr.* **2006**, *136*, 1533–1538. [CrossRef]
45. De La Peña-Armada, R.; Villanueva-Suárez, M.J.; Rupérez, P.; Mateos-Aparicio, I. High Hydrostatic Pressure Assisted by Celluclast® Releases Oligosaccharides from Apple By-Product. *Foods* **2020**, *9*, 1058. [CrossRef] [PubMed]
46. Francavilla, M.; Marone, M.; Marasco, P.; Contillo, F.; Monteleone, M. Artichoke Biorefinery: From Food to Advanced Technological Applications. *Foods* **2021**, *10*, 112. [CrossRef] [PubMed]
47. Zhu, B.J.; Zayed, M.Z.; Zhu, H.X.; Zhao, J.; Li, S.P. Functional Polysaccharides of Carob Fruit: A Review. *Chin. Med.* **2019**, *14*, 40. [CrossRef] [PubMed]
48. Houben, K.; Jolie, R.P.; Fraeye, I.; Van Loey, A.M.; Hendrickx, M.E. Comparative Study of the Cell Wall Composition of Broccoli, Carrot, and Tomato: Structural Characterization of the Extractable Pectins and Hemicelluloses. *Carbohydr. Res.* **2011**, *346*, 1105–1111. [CrossRef]
49. Sabater, C.; Corzo, N.; Olano, A.; Montilla, A. Enzymatic Extraction of Pectin from Artichoke (*Cynara scolymus* L.) by-Products Using Celluclast®1.5L. *Carbohydr. Polym.* **2018**, *190*, 43–49. [CrossRef]
50. Fissore, E.N.; Domingo, C.S.; Pujol, C.A.; Damonte, E.B.; Rojas, A.M.; Gerschenson, L.N. Upgrading of Residues of Bracts, Stems and Hearts of *Cynara cardunculus* L. var. *scolymus to Functional Fractions Enriched in Soluble Fiber. Food Funct.* **2014**, *5*, 463–470. [CrossRef]
51. Oszmiański, J.; Wojdyło, A.; Kolniak, J. Effect of Pectinase Treatment on Extraction of Antioxidant Phenols from Pomace, for the Production of Puree-Enriched Cloudy Apple Juices. *Food Chem.* **2011**, *127*, 623–631. [CrossRef]
52. Jagelaviciute, J.; Staniulyte, G.; Cizeikiene, D.; Basinskiene, L. Influence of Enzymatic Hydrolysis on Composition and Technological Properties of Apple Pomace and Its Application for Wheat Bread Making. *Plant Foods Hum. Nutr.* **2023**, *78*, 307–313. [CrossRef]
53. De la Peña-Armada, R.; Villanueva-Suárez, M.J.; Molina-García, A.D.; Rupérez, P.; Mateos-Aparicio, I. Novel Rich-in-Soluble Dietary Fiber Apple Ingredient Obtained from the Synergistic Effect of High Hydrostatic Pressure Aided by Celluclast®. *LWT* **2021**, *146*, 111421. [CrossRef]
54. Zhao, Y.; Yu, K.; Tian, X.; Sui, W.; Wu, T.; Wang, S.; Jin, Y.; Zhu, Q.; Meng, J.; Zhang, M. Combined Modification of Soluble Dietary Fibers from Apple Pomace by Steam Explosion and Enzymatic Hydrolysis to Improve Its Structural, Physicochemical and Functional Properties. *Waste Biomass Valorization* **2022**, *13*, 4869–4879. [CrossRef]
55. Elleuch, M.; Bedigian, D.; Roiseux, O.; Besbes, S.; Blecker, C.; Attia, H. Dietary Fibre and Fibre-Rich by-Products of Food Processing: Characterisation, Technological Functionality and Commercial Applications: A Review. *Food Chem.* **2011**, *124*, 411–421. [CrossRef]
56. Huang, K.; Du, B.; Xu, B. Alterations in Physicochemical Properties and Bile Acid Binding Capacities of Dietary Fibers upon Ultrafine Grinding. *Powder Technol.* **2018**, *326*, 146–150. [CrossRef]

57. Rivas, M.Á.; Benito, M.J.; Martín, A.; de Guía Córdoba, M.; Ruíz-Moyano, S.; Casquete, R. Improve the Functional Properties of Dietary Fibre Isolated from Broccoli By-Products by Using Different Technologies. *Innov. Food Sci. Emerg. Technol.* **2022**, *80*, 103075. [CrossRef]
58. Tounsi, L.; Kechaou, N. Influence of Enzymatic Treatment on Improving Carob Juice Extraction and Quality. *J. Biochem. Int.* **2022**, *9*, 12–21. [CrossRef]
59. Li, S.; Hu, N.; Zhu, J.; Zheng, M.; Liu, H.; Liu, J. Influence of Modification Methods on Physicochemical and Structural Properties of Soluble Dietary Fiber from Corn Bran. *Food Chem. X* **2022**, *14*, 100298. [CrossRef]
60. Grigelmo-Miguel, N.; Martín-Belloso, O. Characterization of Dietary Fiber from Orange Juice Extraction. *Food Res. Int.* **1998**, *31*, 355–361. [CrossRef]
61. Martínez, J.A.; Melgosa, M.; Pérez, M.M.; Hita, E.; Negueruela, A.I. Note. Visual and Instrumental Color Evaluation in Red Wines. *Food Sci. Technol. Int.* **2001**, *7*, 439–444. [CrossRef]
62. Lotfi, L.; Kalbasi-Ashtari, A.; Hamedi, M.; Ghorbani, F. Effects of Enzymatic Extraction on Anthocyanins Yield of Saffron Tepals (*Crocos sativus*) along with Its Color Properties and Structural Stability. *J. Food Drug Anal.* **2015**, *23*, 210–218. [CrossRef]
63. Chandini, S.K.; Rao, L.J.; Gowthaman, M.K.; Haware, D.J.; Subramanian, R. Enzymatic Treatment to Improve the Quality of Black Tea Extracts. *Food Chem.* **2011**, *127*, 1039–1045. [CrossRef]
64. Konopacka, D. The Effect of Enzymatic Treatment on Dried Vegetable Color. *Dry. Technol.* **2006**, *24*, 1173–1178. [CrossRef]
65. Singh, B.; Suri, K.; Shevkani, K.; Kaur, A.; Kaur, A.; Singh, N. Enzymatic Browning of Fruit and Vegetables: A Review. In *Enzymes in Food Technology*; Springer: Singapore, 2018; pp. 63–78.
66. Landbo, A.-K.; Meyer, A.S. Enzyme-Assisted Extraction of Antioxidative Phenols from Black Currant Juice Press Residues (*Ribes nigrum*). *J. Agric. Food Chem.* **2001**, *49*, 3169–3177. [CrossRef]
67. Palafox-Carlos, H.; Ayala-Zavala, J.F.; González-Aguilar, G.A. The Role of Dietary Fiber in the Bioaccessibility and Bioavailability of Fruit and Vegetable Antioxidants. *J. Food Sci.* **2011**, *76*, R6–R15. [CrossRef]

foods

MDPI

Article

A Structural Equation Model for Sustainable Supply Chain Management in the Food Industry †

Theofilos D. Mastos [1,*], Katerina Gotzamani [1], Petros Ieromonachou [2] and George Tsiotras [1]

1 Business Excellence Lab, School of Business Administration, University of Macedonia, 546 36 Thessaloniki, Greece; kgotza@uom.edu.gr (K.G.); tsiotras@uom.edu.gr (G.T.)
2 Connected Cities Research Group, School of Business, Operations and Strategy, University of Greenwich, London SE10 9LS, UK; p.ieromonachou@gre.ac.uk
* Correspondence: tmastos@uom.edu.gr
† This article is part of the Ph.D. thesis of Theofilos D. Mastos.

Abstract: This paper presents a model designed to measure and investigate the relationships between critical factors, practices, and performance of sustainable supply chain management (SSCM) in the food industry. A survey of 423 firms in the Greek food industry was conducted to meet these objectives. The data were analyzed using exploratory factor analysis, followed by confirmatory factor analysis and structural equation modeling, to explore the relationships between the model's constructs. The results indicate that "firm-level critical sustainability factors" and "supply chain critical sustainability factors" significantly enhance "supply chain collaboration" and "supply chain strategic orientation". Additionally, "supply chain strategic orientation" positively influences "social performance" and "environmental performance", while "supply chain collaboration" positively affects "economic performance" and "environmental performance". Furthermore, "social performance" is found to have a significant positive impact on "economic performance". This study provides empirical evidence that helps managers understand the importance of the relationships among SSCM critical factors, SSCM practices, and SSCM performance, and enables them to allocate resources effectively and design SSCM strategies. Finally, the developed constructs offer a measurement tool useful for both practitioners implementing SSCM in their firms and researchers who wish to apply or test the proposed scales in other projects or use them as benchmarks.

Keywords: food industry; Greece; structural equation model; sustainable supply chain management

1. Introduction

Over the past two decades, sustainable supply chain management (SSCM) has become a central focus in the business world, drawing significant attention from both academicians and practitioners [1,2]. The globalization of operations has dispersed processes worldwide, linking all supply chain members, through the exchange of information, materials, and capital flows [2]. Various stakeholders, including regulatory bodies, NGOs, community organizations, suppliers, customers, and global competitors, have pressured firms to balance environmental, social, and economic issues in their supply chains [3] through the adoption of sustainable supply chain management practices.

Like any other business operation, SSCM aims to achieve clear performance goals [4]. However, the complexity of the supply chains makes this challenging, as individual members may have different and sometimes even conflicting goals and performance measures. These differing measures may not always positively impact the entire chain's performance, as one firm's outcomes can be detrimental to others [5]. As a result, the overall performance of the supply chain can only improve if viewed as a whole, beyond the firm level [6].

SSCM is implemented through specific practices such as environmental purchasing, reverse logistics, and sustainable packaging among others. These practices are anticipated to enhance the sustainability performance of supply chains [7]. The development of SSCM practices can be influenced by various critical factors, which may either enable or inhibit them. Different industries address these factors based on their size, culture, location, and supply chain partnerships. Researchers have studied SSCM across various sectors like manufacturing [2,8], the automotive industry [9], oil and gas [10], energy [11], and the food industry [12]. It is widely accepted, that the food industry faces significant environmental, economic, social, and political challenges such as deforestation, climate change, energy consumption, food safety, production practices, fair wages, and population growth coming to the forefront [12,13].

Other critical SSCM issues include measuring supply chain impacts, collaboration and networking, stakeholder engagement, and sustainable development goals [14]. These challenges highlight the unique variability and risk factors in food supply chains due to product-specific characteristics like perishability, seasonality, and storage conditions [15]. Both customers and firms have expressed concerns about product origins, food safety, product quality, sustainable production [16], animal welfare, and environmental protection [15].

Numerous studies have examined the relationship between SSCM practices and performance to better understand these challenges. In various industries, direct and indirect impacts between sustainability performance dimensions have also been observed. For instance, a positive correlation exists between corporate social performance and financial performance [17]. In the wine industry, ref. [18] found that socially sustainable employee practices reduce costs. The authors of [19] demonstrated that environmental practices lead to positive environmental outcomes and indirectly improve cost performance through quality enhancements. Despite substantial theoretical developments in SSCM, there is still a lack of empirical research on industry and location-specific SSCM critical factors, practices, and performance [12,20–23]. The food industry, which is characterized by strong supply chain relationships, aims to achieve a high sustainability performance [12].

Based on the above introduction, the following two research gaps have been identified:

- Different industries encounter a single critical factor from varying perspectives, influenced by their size, culture, location, and the number of supply chain members. Research on investigating industry and location-specific SSCM critical factors is still limited [20,21].
- There is also a scarcity of empirical studies on industry-specific SSCM practices and their relationships with SSCM performance.

In this context, this paper aims to address these two gaps by examining the relationships among SSCM critical factors, SSCM practices, and SSCM performance in the Greek food industry.

The rest of the paper is organized as follows: The next section provides an overview of the literature related to SSCM critical factors, SSCM practices, and SSCM performance measures and develops the relative research hypotheses; Section 3 describes the survey methodology that has been utilized; Section 4 presents the results in conjunction with previous research works and discusses the proposed structural equation model. Section 5 concludes the paper and provides the implications and limitations of this study, as well as future research opportunities.

2. Literature Review and Research Hypotheses

Since the identification and conceptualization of SSCM is still unclear, a literature review was conducted on three important sustainable supply chain management constructs i.e., SSCM critical factors, SSCM practices, and SSCM performance. The search for relevant scientific articles was conducted using keywords and author names in major bibliographic databases and publishers, such as Scopus, Elsevier, Emerald, Springer, Wiley, Taylor & Francis, Sage Publications, and Inderscience, covering a twenty-year span from 2000 onward. Keywords included terms like "sustainable supply chain management", "drivers", "barriers", "enablers", "motivators", "critical factors", "sustainable supply chain management practices", "sustainability performance", and "food industry". The author search focused on researchers such as Seuring S., Beske P., Gualandris J., Govindan K., and Pagell M., who have consistently contributed to SSCM research [1]. A secondary search was also conducted through cited references. Only peer-reviewed articles in English-language scientific journals were considered. This review primarily focused on studies with applications in the food industry but also included articles from other sectors. The measures identified by the comprehensive literature review are named and grouped based on the affinity method, which is utilized to organize into categories common themes from a large amount of information [24].

2.1. SSCM Critical Factors

The critical factors for SSCM are reviewed and categorized into firm-level, supply chain-level, and external critical factors. Firm-level critical factors are defined as organizational factors, while supply chain-level factors encompass factors that are related to the firm's supply chain environment. External critical factors may originate from a plethora of stakeholders, such as government, customers, suppliers, media, non-governmental organizations (NGOs), etc. These SSCM critical factors are found to be linked with the adoption of SSCM practices.

Previous scholarly works have recognized factors such as top management commitment and proficiency (knowledge and expertise) in sustainability, as significant firm-level determinants in the realm of SSCM. For instance, the absence of commitment and support from top management obstructs the advancement of SSCM [1,5,25]. SSCM requires proactive top management that understands that sustainability is an organizational commitment [26]. Fulfilling customer demands, expectations, and requirements is frequently mentioned as a crucial factor for the successful implementation of SSCM [5,27,28]. Ref. [28] recognized that knowledge and expertise in sustainability are key drivers in developing a firm's SSCM strategy. Similarly, ref. [27] emphasized that knowledge is a vital intangible asset for implementing SSCM. Other factors that have been highlighted as critical for SSCM in the food industry include operational cost reduction and market, retailer pressure, brand image, and corporate reputation [29].

Supply chain-level critical factors are closely tied to firm-level critical factors. According to the literature, the alignment between firm-level and supply chain-level factors significantly influences their successful integration [30]. Information sharing has been identified as a crucial enabler for adopting SSCM practices [31–33]. Ref. [20] suggests that information sharing facilitates the development of new sustainability ideas and enhances collaboration throughout the supply chain. In the food industry, information sharing among supply chain members is seen as a novel method for traceability and is associated with improved supply chain performance [26]. The existing body of literature suggests that the act of sharing information fosters the creation of novel sustainable concepts and amplifies cooperation across the supply chain, as pointed out by [20]. In contrast, the lack of information sharing has a negative impact on SSCM implementation [32,34]. Relationships that are built on trust and commitment among supply chain members are reported as another key factor for SSCM adoption. Ref. [29] explored the implications of sustainability in the Italian meat supply chain and found that trustful relationships are a critical element for deploying exceptional supply chain practices, such as collaboration for sustainability.

Regarding the external environment of critical factors, the identification, engagement and communication with customers, legislative pressures, local community, and NGOs are highlighted as essential elements for effectively executing sustainable supply chain management (SSCM) practices [35]. This corresponds with earlier research, affirming that stakeholders play a pivotal role in propelling the incorporation of SSCM practices [21]. Especially in the food retail industry, NGO pressure is found to be critical for the adoption of SSCM [36].

2.2. SSCM Practices

Considering the practices that are adopted in the food industry in order to implement SSCM, five categories are proposed in the literature. The SSCM practices, as proposed by Beske et al. [12] ((1) strategic orientation, (2) supply chain continuity, (3) collaboration, (4) risk management, and (5) pro-activity), are used as a key starting point and as a guiding tool for developing a model of SSCM in the food industry since they cover the aspects of SSCM and focus on strengthening the relationships among supply chain partners, the flow of goods and information, and the sustainability aspects. Strategic orientation is encompassing the commitment of firms to supply chain management and to the triple bottom line (TBL) [12]. Continuity of supply chains involves the development (design and structure) of a supply chain network [12]. Supply chain collaboration, which represents the third SSCM practice, goes beyond the traditional modus operandi between organizations since it is not restricted only to new product development but also to the development and enhancement of business processes [12,37]. Supply chain risk management includes the adoption of risk mitigation practices to avoid exposure to risks, while proactivity refers to the actions taken by a firm in order to control and manage a specific situation regarding sustainability before it happens, rather than responding to it after it happens.

Based on the above, the following research hypotheses have been developed:

Hypothesis 1 (H1). *SSCM critical factors have a significant positive impact on SSCM practices.*

Hypothesis 1a (H1a). *Firm level critical factors have a significant positive impact on SSCM practices.*

Hypothesis 1b (H1b). *Supply chain level critical factors have a significant positive impact on SSCM practices.*

Hypothesis 1c (H1c). *External critical factors have a significant positive impact on SSCM practices.*

2.3. SSCM Performance

SSCM performance refers to how well supply chains achieve their environmental, economic, and social goals. Most studies in the literature focus on the economic and environmental performance aspects, whereas the social dimension and the integration of the three sustainability dimensions are still lagging [2]. Regarding environmental performance, the most frequently used measure is related to either the reduction or avoidance of hazardous/harmful/toxic materials. The second most cited measure is water consumption, followed by energy consumption, recycled materials, life cycle analysis (LCA), and environmental penalties. Energy efficiency, air emissions, and greenhouse gas emissions are reported as some of the most important measures in the literature. The most common measure for assessing economic performance is quality. This can relate to the quality of the purchased products [38] or the quality of the production process itself [39]. Sales, market share, and profit are the next most commonly used metrics, with delivery time and customer satisfaction following. Research indicates that only a few metrics are consistently used, underscoring the limited focus on the social aspect of sustainable supply chain management (SSCM). Among the social metrics, recordable accidents are the most frequent metric, followed by training, education, and labor practices. Several researchers have found that SSCM practices have the ability to enhance environmental and social performance [7,35]. Evidence from the food industry, provided by [29], demonstrates that traceability practices in the meat supply chain are closely associated with social sustainability and food safety.

Based on the above observations, this study developed the following research hypotheses:

Hypothesis 2 (H2). *SSCM practices have a significant positive impact on SSCM performance.*

Hypothesis 2a (H2a). *SSCM practices have a significant positive impact on social performance.*

Hypothesis 2b (H2b). *SSCM practices have a significant positive impact on economic performance.*

Hypothesis 2c (H2c). *SSCM practices have a significant positive impact on environmental performance.*

Ref. [35] suggests that environmental performance improvements such as energy efficiency practices, have visible cost reductions in the short term. In the same line, for the Italian meat supply chain, ref. [29] found that SSCM practices, such as the adoption of cleaner technologies, confer a competitive edge by enhancing both economic and environmental or social performance. Ref. [40] also found a positive correlation between corporate social performance and corporate financial performance.

Based on the above arguments, this study developed the following two hypotheses:

Hypothesis 3 (H3). *Environmental performance has a significant positive impact on economic performance.*

Hypothesis 4 (H4). *Social performance has a significant positive impact on economic performance.*

The theoretical developments that have been conducted in the previous paragraphs have assisted in the development of a conceptual model of relations (Figure 1) consisting of three pivotal SSCM concepts: (a) SSCM critical factors; (b) SSCM practices; and (c) SSCM performance. The relations and interactions among these constructs are determined and the research hypotheses are tested in the following sections.

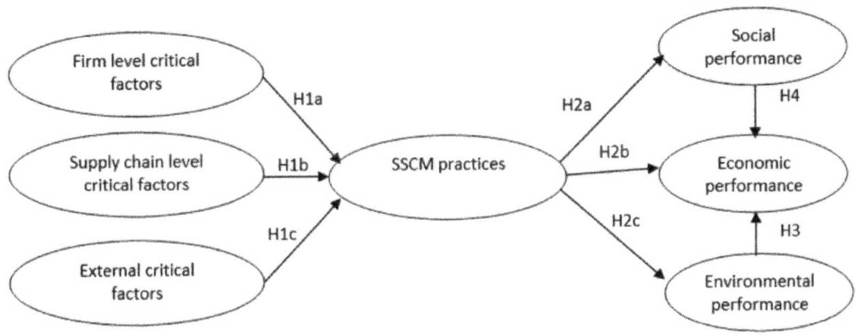

Figure 1. Theoretical model for SSCM in the food industry.

3. Research Methodology

The data for this study were gathered through a structured survey questionnaire, which was developed based on a comprehensive literature review on sustainable supply chain management [12,22,27,28,32,33,41–43]. Similar research in the SSCM field has also utilized surveys, highlighting the relevance of this approach [44]. The questionnaire was divided into four sections: the first section covered the critical factors for effective SSCM implementation; the second examined various SSCM practices; the third section assessed the SSCM performance; and the fourth focused on firm and respondent profiles. Responses were collected within a period of 8 months and the survey length was approximately 15 min. Content validity was ensured by reviewing the extensive literature, leading to an initial list of 80 items. A pilot study involving 10 food industry experts was then conducted and their feedback was incorporated during the pre-testing phase [45,46] to improve question clarity, define terms, outline the research scope, and refine expected outcomes. The purpose of pilot testing was to examine if each question includes the information that needs to be measured. The detailed expertise support helped capture some industry-specific and cultural issues of the questionnaire. In addition, four academic reviewers also evaluated and revised the draft, resulting in a final list of 68 items [47]. Responses were recorded using a seven-point Likert scale (1 = "strongly disagree" to 7 = "strongly agree") and data processing was carried out using SPSS 24 and AMOS 21.

3.1. Research Sample

The survey sample comprised firms from the food and beverage sector listed in Greek sustainability databases, such as "CSR HELLAS NET", "Sustainable Greece 2020", and "CSR Index GR". Additionally, the sample included firms from other business directories, like the "Federation of Hellenic Food Industries" and ICAP (Greece's largest business information and consulting firm), resulting in an initial sample of 904 firms. All firms were contacted via email. The questionnaire was aimed at the personnel responsible for supply chain management. It was sent by the authors and university students, who had undergone training relevant to the study's purpose and content. The email included a cover letter assuring respondents of the confidentiality of their answers and invited them to leave their contact information if they wished to receive the research results. A total of 423 (46.8% response rate) completed questionnaires were received, which was deemed acceptable for data processing, compared to similar studies [22].

3.2. Non-Response Bias and Common Method Bias

To evaluate the dataset for non-response bias, the sample was divided into early and late respondents, with the latter group representing theoretical non-respondents, following the approach of ref. [48]. A Mann–Whitney U test was conducted to compare these two groups, and no statistically significant differences were detected, suggesting that non-response bias does not affect this study.

Additionally, common method bias, a significant validity concern in behavioral research [49], was assessed. To mitigate this issue, Harman's single-factor test was applied to determine if a single factor accounted for more than 50% of the variance in the dataset. The analysis showed that all items loaded in one factor explained only 35.2% of the total variance, hence confirming the absence of common method bias in this study.

4. Data Analysis and Results

The sample consisted of firms from various sub-sectors within the food industry, encompassing the entire food supply chain and ensuring the results apply to a broad range of supply chain participants. A total of 40% of the firms operated in multiple supply chain activities, while 22% operated in the food services sub-sector. The rest of the sample included retail firms, food manufacturers, wholesale, beverage producers and, transportation and storage. In terms of firm size more than 87% were small and medium enterprises and 13% were large enterprises with more than 250 employees.

The construct validity of the latent factors was verified through several measures: convergent validity (with AVE > 0.5), discriminant validity (where AVE > $Corr^2$ [50,51], face-content validity (obtained via feedback from food industry experts on the questionnaire), and nomological validity (demonstrated by significant correlations among the latent factors identified) [52]. Convergent validity was established by examining the factor loadings (>0.6), the average variance extracted (AVE) (>0.4), and construct reliability (CR) (>0.7) for all constructs [53] (see Table 1). It is worth noting that the AVE value for the construct "Supply Chain Critical Sustainability Factors" was below 0.50. However, when AVE falls between 0.4 and 0.5 and the composite reliability (CR) exceeds 0.6, the construct's convergent validity is acceptable [54,55]. Discriminant validity was assessed by comparing the AVE with the highest squared correlation between the factor in question and other latent factors [52]. The results showed that the AVE surpasses $Corr^2$, which confirms discriminant validity [53].

Table 1. Reliability and convergent validity.

Latent Factors	Items	Factor Loadings	AVE	CR	Cronbach's Alpha
Firm-level Critical Sustainability Factors	4	0.741	0.549	0.829	0.817
Supply Chain Critical Sustainability Factors	3	0.656	0.431	0.694	0.706
Supply Chain Collaboration	4	0.745	0.561	0.835	0.824
Supply Chain Strategic Orientation	3	0.710	0.504	0.753	0.776
Economic Performance	4	0.880	0.765	0.929	0.932
Social Performance	6	0.730	0.536	0.873	0.877
Environmental Performance	3	0.729	0.535	0.774	0.770

4.1. Modified Research Model and Hypotheses

In order to develop a robust structural equation model (SEM), an exploratory factor analysis (EFA) was conducted followed by a confirmatory factor analysis (CFA). The analysis begins with an exploratory factor analysis (EFA), using the principal component extraction method with varimax orthogonal rotation, to identify the structure of the variables. Then, confirmatory factor analysis (CFA) is conducted to refine the scales derived from the EFA and to assess whether the number of factors and the loadings of the measured variables align with the theoretical developments [53]. The necessary research stages, that need to be executed before the application of structural modeling are illustrated in Figure 2.

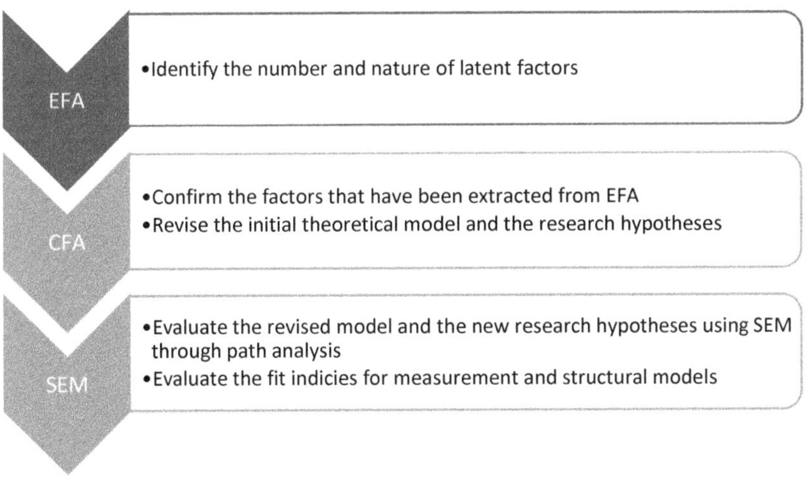

Figure 2. Research stages.

Based on the EFA, two latent constructs were introduced in the SSCM critical factors category ("firm-level critical sustainability factors" or FLCSF and "supply chain critical sustainability factors" or SCCSF). As a result, H1c from the theoretical model is not supported. Another two factors represented SSCM practices ("supply chain collaboration" or SCC and "supply chain strategic orientation" or SCSO) and three factors reflected the SSCM performance ("environmental" or ENV, "social" or SOC, and "economic" or ECO).

In order to investigate the relationships among dependent and independent variables, an SEM technique was applied [53]. Based on the results of the CFA, the study's theoretical framework has been adjusted to incorporate these changes, now including seven key factors i.e., FLCSF, SCCSF, SCC, SC, ECO, SOC, and ENV. The revised theoretical model is illustrated in Figure 3.

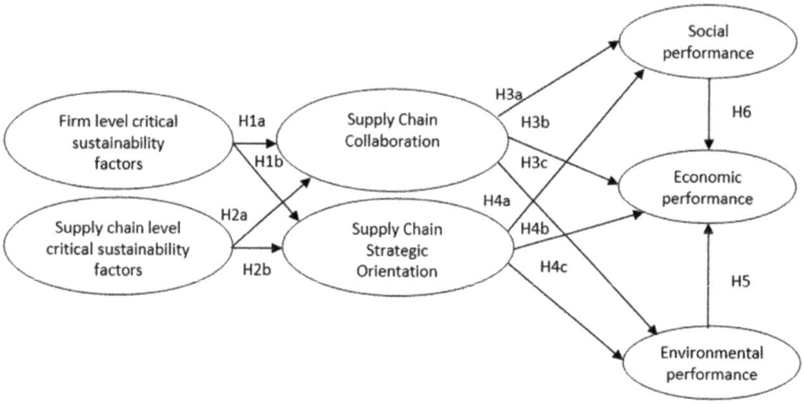

Figure 3. Revised theoretical model for SSCM in the food industry.

Given the changes to the theoretical model, the original hypotheses regarding the relationships between SSCM critical factors, SSCM practices, and SSCM performance were revised. These modified research hypotheses incorporate the confirmed latent constructs from CFA. Accordingly, new hypotheses were introduced to capture the proposed relationships.

Hypothesis 1a (H1a). *Firm level critical sustainability factors have a significant positive impact on supply chain collaboration.*

Hypothesis 1b (H1b). *Firm level critical sustainability factors have a significant positive impact on supply chain strategic orientation.*

Hypothesis 2a (H2a). *Supply chain level critical sustainability factors have a significant positive impact on supply chain collaboration.*

Hypothesis 2b (H2b). *Supply chain level critical sustainability factors have a significant positive impact on supply chain strategic orientation.*

Hypothesis 3a (H3a). *Supply chain collaboration has a significant positive impact on social performance.*

Hypothesis 3b (H3b). *Supply chain collaboration has a significant positive impact on economic performance.*

Hypothesis 3c (H3c). *Supply chain collaboration has a significant positive impact on environmental performance.*

Hypothesis 4a (H4a). *Supply chain strategic orientation has a significant positive impact on social performance.*

Hypothesis 4b (H4b). *Supply chain strategic orientation has a significant positive impact on economic performance.* **Hypothesis 4c (H4c).** *Supply chain strategic orientation has a significant positive impact on environmental performance.*

Hypothesis 5 (H5). *Environmental performance has a significant positive impact on economic performance.*

Hypothesis 6 (H6). *Social performance has a significant positive impact on economic performance.*

4.2. Structural Equation Model

The revised model and the new research hypotheses were evaluated using SEM through path analysis, which is a multivariate analytical approach providing insights into the causal sequence of variables within a network of relationships [56]. According to [57], the measurement and structural models are analyzed sequentially. The process started with a CFA before proceeding to test the hypothesized model [50]. The fit indices for both the measurement and structural models indicated that the data fits well, supporting the theoretical model with empirical evidence [53] (see Table 2). Once the fit of these models was confirmed, the SEM procedures, using the maximum likelihood method, were applied to estimate the causal relationships among latent factors, determining whether to accept or reject the developed hypotheses (H1–H6).

Table 2. Acceptable fit indices for the measurement and structural model.

Fit Indices	Measurement Model	Structural Model	Acceptable Fit Indices
Absolute fit indices			
Chi-square (CMIN or χ^2)	666.624	513.701	$0 \leq \chi^2 \leq 2$df
Degrees of freedom (df)	301.000	216.000	
Probability level	0.000 *	0.000 *	$p > 0.05$
Root mean square residual (RMR)	0.075	0.071	<0.08
Root mean square of approximation (RMSEA)	0.056	0.060	<0.08
Incremental fit indices			
Incremental fit index (IFI)	0.937	0.937	>0.90
Tucker–Lewis coefficient (TLI)	0.925	0.926	>0.90
Comparative fit index (CFI)	0.936	0.937	>0.90
Parsimonious fit indices			
Chi-square/degrees of freedom (χ^2/df)	2.215	2.378	Between 1 and 3
Normed fit index (NFI)	0.890	0.897	>0.50
Goodness of fit index (GFI)	0.889	0.901	>0.50
Adjusted goodness of fit index (AGFI)	0.861	0.874	>0.50

Note: * Acceptable when $n > 250$, the number of the measured variables range between 12 and 30, RMR < 0.08, RMSEA < 0.07, and CFI > 0.92 [48].

As detailed in Table 3, three hypotheses are fully accepted (H1, H2, and H6), two are partially accepted (H3 and H4) for two out of three unobserved factors, and one (H5) is rejected.

The following Figure 4 presents the SSCM structural equation model along with the estimated standardized parameters for the causal paths.

Table 3. Relationships between constructs.

Relationships	Standardized Regression Weights	Standard Error	p-Value	Hypothesis Test Result
H1a: FLCSF -> SCC	0.331	0.127	0.000	Accept hypothesis
H1b: FLCSF -> SCSO	0.548	0.083	0.000	Accept hypothesis
H2a: SCCSF -> SCC	0.356	0.177	0.000	Accept hypothesis
H2b: SCCSF -> SCSO	0.534	0.118	0.000	Accept hypothesis
H3a: SCC -> SOC	−0.027	0.043	0.685	Reject hypothesis
H3b: SCC -> ECO	0.230	0.089	0.003	Accept hypothesis
H3c: SCC -> ENV	0.166	0.077	0.029	Accept hypothesis
H4a: SCSO -> SOC	0.775	0.074	0.000	Accept hypothesis
H4b: SCSO -> ECO	−0.220	0.189	0.080	Reject hypothesis
H4c: SCSO -> ENV	0.481	0.108	0.000	Accept hypothesis
H5: ENV -> ECO	0.146	0.086	0.056	Reject hypothesis
H6: SOC -> ECO	0.248	0.178	0.015	Accept hypothesis

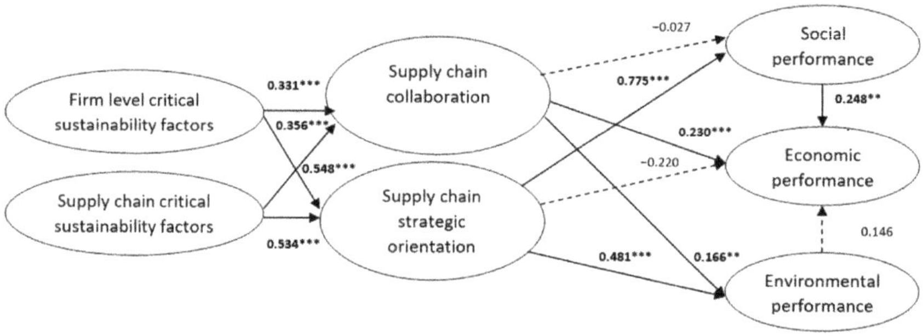

Figure 4. Structural equation model for SSCM in the food industry. Note: ** $0.01 \leq p\text{-value} \leq 0.05$; *** $p\text{-value} \leq 0.01$.

4.3. Discussion

The results of this study clearly indicate that critical sustainability factors at both the firm and supply chain levels significantly contribute to supply chain collaboration and strategic orientation. At the firm level, these results align with the work of [58], who emphasize that internal characteristics within the food sector firms can facilitate various supply chain relationships and collaboration opportunities. At the supply chain level, the results are consistent with the studies of [20,29], who highlight that fostering trust among supply chain members and enhancing information sharing can lead to better collaboration. Moreover, critical sustainability factors at both levels positively influence supply chain strategic orientation. For instance, addressing customer demands (as noted by [19,59] and sharing knowledge and information within the supply chain [12] are essential for developing a strategic orientation within the supply chain. Interestingly, external critical factors, such as legislative pressures significantly influence SMEs to integrate sustainable practices into their operations [60], in this study they have been excluded from the revised model. This is probably due to the fact that supply chain critical sustainability factors include items that are related to supply chain partnerships and this might have been perceived by the participants as external driving forces.

In line with previous research [61], this study provides empirical evidence of a positive link between SSCM practices and SSCM performance within the Greek food supply chain sector. The results indicate that SSCM practices, such as supply chain collaboration [62,63] and strategic orientation, significantly contribute to sustainability performance.

This study is based on the SSCM theory as proposed by [64]. The theory suggests that the triggers for SSCM involve stakeholder management and pressures, drivers, and barriers of SSCM. The supplier management for risk and performance element incorporates the multi-tier supplier management, the supplier selection and evaluation and the risk and performance outcomes. The third and last part of SSCM theory is related to the supply chain management for sustainable products, which involves supplier development, communication, and collaboration that target the development of sustainable products.

Consistent with SSCM theory as explored by [2,13,26,64], the results show that fostering collaboration through strengthened supply chain relationships can enhance both economic [61] and environmental performance in the food industry. However, the study found a non-significant negative relationship between supply chain collaboration and social performance, which contrasts with the findings of [22], who observed a significant link between SSCM practices and social performance. This discrepancy may stem from the fact that, while collaboration may bring economic and environmental benefits in the short term, social sustainability improvements typically require a longer timeframe to materialize.

Consistent with the findings of [22], who explored the connections between SSCM practices and SSCM performance, this study found a positive relationship between supply chain strategic orientation and both environmental and social performance. However, the results differ from [22] regarding the economic performance, where a non-significant negative relationship is observed. This suggests that the economic impacts of the strategic orientation may not be immediately apparent. Additionally, it indicates that employees and managers in the food industry recognize the environmental and social benefits but may be less aware of, or even skeptical about, potential positive economic outcomes. This result implies that when food supply chain firms adopt more sustainability-focused practices, there may be higher initial costs, which could negatively affect economic performance in the short term. This is also confirmed by [62], who found that the cost of sustainability may hinder the adoption of SSCM practices in multi-tier food supply chains.

The structural equation model also examined the interconnections of the sustainability performance dimensions. The results indicated that social performance has a significant positive influence on economic performance. This aligns with [40], who found a positive correlation between corporate social performance and corporate financial performance. Additionally, this result supports the case study of exemplars presented by [35], which demonstrated that social performance is positively linked to economic performance. The social performance indicators suggest a focus on safety (product, health, and safety), leading to the conclusion that social performance is connected to safety management. This implies that improving safety aspects (i.e., social performance) can enhance economic performance by lowering costs related to safety issues. On the other hand, environmental performance was found to have a positive but statistically insignificant effect on economic performance. This is consistent with [19], who observed that environmental performance does not directly impact costs in the food industry. Although the case study conducted by [35] suggests that improvements in environmental performance can yield economic benefits through short-term cost reductions, the respondents expressed a different opinion. It is worth noting that 87% of the sample consisted of SMEs, which often lack the resources to prioritize costly environmental improvements. Therefore, it is reasonable to conclude that Greek firms in the food supply chain still have considerable progress to make in raising environmental awareness and recognizing the long-term benefits and opportunities that could indirectly enhance economic performance. A potential regional factor that might explain this result is the prolonged Greek economic crises (encompassing the debt crisis, capital controls, and COVID-19), which have compelled companies fighting for survival to reduce or even eliminate costs deemed "unnecessary" or considered as "luxuries".

In order to facilitate the understanding of the results and provide a visualization for professionals in the SSCM field, a summary of the key results is presented in Table 4.

Table 4. Key results.

Key Result Area	Description
SSCM critical factors	Internal characteristics within the food sector firms facilitate supply chain relationships and collaboration opportunities Trust and enhanced information sharing among supply chain members lead to better collaboration
SSCM practices and performance	Addressing customer demands and sharing knowledge and information within the supply chain are essential for developing a strategic orientation Fostering collaboration through strengthened supply chain relationships enhances economic and environmental performance Social sustainability improvements due to improved collaboration typically require a longer timeframe to materialize
SSCM performance	Supply chain strategic orientation positively influences environmental and social performance Social performance has a significant positive influence on economic performance Greek firms in the food supply chain still have considerable progress to make in raising environmental awareness and recognizing the long-term benefits and opportunities that could indirectly enhance economic performance.

5. Conclusions

This study contributes to the existing sustainable supply chain management research by introducing a measurement tool that encompasses three key SSCM concepts: critical factors, practices, and performance. The first component examines the crucial elements firms need to consider for successful SSCM implementation. The second focuses on factors related to actual SSCM practices, while the third addresses factors necessary for improving SSCM performance. These constructs offer valuable insights specifically within the context of the food industry in Greece. A structural equation model was developed to test the relationships between these three constructs, offering insights that future studies can either validate or challenge.

A key conclusion from this research is that food firms aiming to effectively implement sustainable supply chain management must focus on three crucial areas, which form the foundation of the factors requiring attention. The study demonstrated a positive influence of both firm-level and supply chain-level critical sustainability factors on SSCM practices, such as supply chain strategic orientation and collaboration. It also highlighted the positive impact of effectively implementing SSCM practices on SSCM performance. Additionally, the research revealed a positive link between social and economic performance in food supply chains. The study presents a theoretically developed model, which was empirically tested and proved to be reliable and valid for exploring the relationships among critical factors, practices, and performance in SSCM. This model can serve as a self-assessment or benchmarking tool for firms and supply chain and sustainability professionals in the food industry.

In line with other research studies [65], this paper also contributes to two UN Sustainable Development Goals (SDGs). In terms of Zero Hunger (SDG 2), this study provides a model that helps decision-makers design sustainable practices and food systems from a supply chain perspective. Regarding Climate Action (SDG 13), the developed SSCM model is expected to assist in the reduction of vulnerability and strengthen climate resilience. The link between the results of this study and the UN SDGs, highlights the SSCM's ability to alleviate environmental, social, and economic pressures, while advancing sustainable development [66].

This research specifically focused on food enterprises in Greece, and, likely different SSCM critical factors, SSCM practices, and SSCM performance metrics may emerge in other industries or regions. Future studies could broaden their scope by including diverse industries and countries, and using larger sample sizes to improve the generalizability of the results. Finally, since this study was conducted during the COVID-19 pandemic, a valuable research opportunity lies in replicating the survey in the post-pandemic era. This could provide insights into how the pandemic may have influenced SSCM practices and performance within the context studied.

Author Contributions: Conceptualization, T.D.M. and K.G.; methodology, T.D.M., K.G., P.I. and G.T.; validation, T.D.M., K.G., P.I. and G.T.; formal analysis, T.D.M.; investigation, T.D.M.; resources, T.D.M.; data curation, T.D.M. and K.G.; writing—original draft preparation, T.D.M.; writing—review and editing, T.D.M., K.G., P.I. and G.T.; visualization, T.D.M.; supervision, K.G., P.I. and G.T; project administration, T.D.M. All authors have read and agreed to the published version of the manuscript.

Funding: This research received no external funding.

Institutional Review Board Statement: The study was conducted based on the principles of the Committee for Research Ethics of the University of Macedonia, that was established according to Chapter E′ (Articles 21–27) of Law 4521 (Government Gazette vol. A '38/2-3-2018).

Informed Consent Statement: Informed consent was obtained from all subjects involved in the study.

Data Availability Statement: The original contributions presented in the study are included in the article, further inquiries can be directed to the corresponding author.

Conflicts of Interest: The authors declare no conflicts of interest.

References

1. Ansari, Z.N.; Kant, R. A state-of-art literature review reflecting 15 years of focus on sustainable supply chain management. *J. Clean. Prod.* **2017**, *142*, 2524–2543. [CrossRef]
2. Seuring, S.; Müller, M. From a literature review to a conceptual framework for sustainable supply chain man-agement. *J. Clean. Prod.* **2008**, *16*, 1699–1710. [CrossRef]
3. Hassini, E.; Surti, C.; Searcy, C. A literature review and a case study of sustainable supply chains with a focus on metrics. *Int. J. Prod. Econ.* **2012**, *140*, 69–82. [CrossRef]
4. Beske-Janssen, P.; Johnson, M.P.; Schaltegger, S. 20 years of performance measurement in sustainable supply chain management—What has been achieved? *Supply Chain Manag. Int. J.* **2015**, *20*, 664–680. [CrossRef]
5. Yakovleva, N. Measuring the Sustainability of the Food Supply Chain: A Case Study of the UK. *J. Environ. Policy Plan.* **2007**, *9*, 75–100. [CrossRef]
6. Aramyan, L.H.; Lansink AG, O.; Van Der Vorst, J.G.; Van Kooten, O. Performance measurement in agri-food supply chains: A case study. *Supply Chain Manag. Int. J.* **2007**, *12*, 304–315. [CrossRef]
7. Zailani, S.; Jeyaraman, K.; Vengadasan, G.; Premkumar, R. Sustainable supply chain management (SSCM) in Ma-laysia: A survey. *Int. J. Prod. Econ.* **2012**, *140*, 330–340. [CrossRef]
8. Gualandris, J.; Golini, R.; Kalchschmidt, M. Do supply management and global sourcing matter for firm sustaina-bility performance? An international study. *Supply Chain Manag. Int. J.* **2014**, *19*, 258–274. [CrossRef]
9. Zhu, Q.; Sarkis, J. Relationships between operational practices and performance among early adopters of green supply chain management practices in Chinese manufacturing enterprises. *J. Oper. Manag.* **2004**, *22*, 265–289. [CrossRef]
10. Matos, S.; Hall, J. Integrating sustainable development in the supply chain: The case of life cycle assessment in oil and gas and agricultural biotechnology. *J. Oper. Manag.* **2007**, *25*, 1083–1102. [CrossRef]
11. Vance, L.; Cabezas, H.; Heckl, I.; Bertok, B.; Friedler, F. Synthesis of Sustainable Energy Supply Chain by the P-graph Framework. *Ind. Eng. Chem. Res.* **2012**, *52*, 266–274. [CrossRef]
12. Beske, P.; Land, A.; Seuring, S. Sustainable supply chain management practices and dynamic capabilities in the food industry: A critical analysis of the literature. *Int. J. Prod. Econ.* **2014**, *152*, 131–143. [CrossRef]
13. Carter, C.R.; Liane Easton, P. Sustainable supply chain management: Evolution and future directions. *Int. J. Phys. Distrib. Logist. Manag.* **2011**, *41*, 46–62. [CrossRef]
14. Pohlmann, C.R.; Scavarda, A.J.; Alves, M.B.; Korzenowski, A.L. The role of the focal company in sustainable development goals: A Brazilian food poultry supply chain case study. *J. Clean. Prod.* **2020**, *245*, 118798. [CrossRef]
15. Aramyan, L.; Ondersteijn, C.J.; Van Kooten, O.; Lansink, A.O. Performance indicators in agri-food production chains. *Frontis* **2006**, 47–64.
16. Ghadge, A.; Ghadge, A.; Kaklamanou, M.; Kaklamanou, M.; Choudhary, S.; Choudhary, S.; Bourlakis, M. Im-plementing environmental practices within the Greek dairy supply chain: Drivers and barriers for SMEs. *Ind. Manag. Data Syst.* **2017**, *117*, 1995–2014. [CrossRef]
17. Waddock, S.A.; Graves, S.B. The corporate social performance-financial performance link. *Strat. Manag. J.* **1997**, *18*, 303–319. [CrossRef]
18. Thach, L. Social Sustainability in the Wine Community: Managing for Employee Productivity and Satisfaction. Wine Business Monthly. 2002. Available online: https://www.winebusiness.com (accessed on 21 October 2020).

19. Pullman, M.E.; Maloni, M.J.; Carter, C.R. Food for thought: Social versus environmental sustainability practices and performance outcomes. *J. Supply Chain Manag.* **2009**, *45*, 38–54. [CrossRef]
20. Saeed, M.A.; Kersten, W. Drivers of Sustainable Supply Chain Management: Identification and Classification. *Sustainability* **2019**, *11*, 1137. [CrossRef]
21. Sajjad, A.; Eweje, G.; Tappin, D. Sustainable Supply Chain Management: Motivators and Barriers. *Bus. Strat. Environ.* **2015**, *24*, 643–655. [CrossRef]
22. Hong, J.; Zhang, Y.; Ding, M. Sustainable supply chain management practices, supply chain dynamic capabilities, and enterprise performance. *J. Clean. Prod.* **2018**, *172*, 3508–3519. [CrossRef]
23. Mathivathanan, D.; Kannan, D.; Haq, A.N. Sustainable supply chain management practices in Indian automotive industry: A multi-stakeholder view. *Resour. Conserv. Recycl.* **2018**, *128*, 284–305. [CrossRef]
24. Brassard, M.; Ritter, D. *The Memory Jogger II: A Pocket Guide of Tools for Continuous Improvement and Effective Planning*; Goal QPC Incorporated: Methuen, MA, USA, 1994.
25. Walker, H.; Jones, N. Sustainable supply chain management across the UK private sector. *Supply Chain Manag. Int. J.* **2012**, *17*, 15–28. [CrossRef]
26. Pagell, M.; Wu, Z. Building a more complete theory of sustainable supply chain management using case studies of 10 exemplars. *J. Supply Chain Manag.* **2009**, *45*, 37–56. [CrossRef]
27. Chkanikova, O.; Mont, O. Corporate Supply Chain Responsibility: Drivers and Barriers for Sustainable Food Retailing. *Corp. Soc. Responsib. Environ. Manag.* **2015**, *22*, 65–82. [CrossRef]
28. Giunipero, L.C.; Hooker, R.E.; Denslow, D. Purchasing and supply management sustainability: Drivers and barriers. *J. Purch. Supply Manag.* **2012**, *18*, 258–269. [CrossRef]
29. Golini, R.; Moretto, A.; Caniato, F.; Caridi, M.; Kalchschmidt, M. Developing sustainability in the Italian meat supply chain: An empirical investigation. *Int. J. Prod. Res.* **2016**, *55*, 1183–1209. [CrossRef]
30. Pagell, M. Understanding the factors that enable and inhibit the integration of operations, purchasing and logistics. *J. Oper. Manag.* **2004**, *22*, 459–487. [CrossRef]
31. Faisal, N.M. Sustainable supply chains: A study of interaction among the enablers. *Bus. Process Manag. J.* **2010**, *16*, 508–529. [CrossRef]
32. Grimm, J.H.; Hofstetter, J.S.; Sarkis, J. Critical factors for sub-supplier management: A sustainable food supply chains perspective. *Int. J. Prod. Econ.* **2014**, *152*, 159–173. [CrossRef]
33. Wittstruck, D.; Teuteberg, F. Understanding the Success Factors of Sustainable Supply Chain Management: Empirical Evidence from the Electrics and Electronics Industry. *Corp. Soc. Responsib. Environ. Manag.* **2012**, *19*, 141–158. [CrossRef]
34. Wolf, J. Sustainable supply chain management integration: A qualitative analysis of the German manufacturing in-dustry. *J. Bus. Ethics* **2011**, *102*, 221–235. [CrossRef]
35. Mastos, T.; Gotzamani, K. Sustainable Supply Chain Management in the Food Industry: A Conceptual Model from a Literature Review and a Case Study. *Foods* **2022**, *11*, 2295. [CrossRef] [PubMed]
36. Wolf, J. The relationship between sustainable supply chain management, stakeholder pressure and corporate sustain-ability performance. *J. Bus. Ethics* **2014**, *119*, 317–328. [CrossRef]
37. Vachon, S.; Klassen, R.D. Environmental management and manufacturing performance: The role of collaboration in the supply chain. *Int. J. Prod. Econ.* **2008**, *111*, 299–315. [CrossRef]
38. Ahi, P.; Searcy, C. An analysis of metrics used to measure performance in green and sustainable supply chains. *J. Clean. Prod.* **2015**, *86*, 360–377. [CrossRef]
39. Rao, P.; Holt, D. Do green supply chains lead to competitiveness and economic performance? *Int. J. Oper. Prod. Manag.* **2005**, *25*, 898–916. [CrossRef]
40. Orlitzky, M.; Schmidt, F.L.; Rynes, S.L. Corporate Social and Financial Performance: A Meta-Analysis. *Organ. Stud.* **2003**, *24*, 403–441. [CrossRef]
41. Ahi, P.; Searcy, C. Measuring social issues in sustainable supply chains. *Meas. Bus. Excel.* **2014**, *19*, 33–45. [CrossRef]
42. Diabat, A.; Kannan, D.; Mathiyazhagan, K. Analysis of enablers for implementation of sustainable supply chain management—A textile case. *J. Clean. Prod.* **2014**, *83*, 391–403. [CrossRef]
43. Tajbakhsh, A.; Hassini, E. Performance measurement of sustainable supply chains: A review and research questions. *Int. J. Prod. Perform. Manag.* **2015**, *64*, 744–783. [CrossRef]
44. Sellitto, M.A.; Hermann, F.F. Influence of Green Practices on Organizational Competitiveness: A Study of the Electrical and Electronics Industry. *Eng. Manag. J.* **2019**, *31*, 98–112. [CrossRef]
45. Malhotra, M.K.; Grover, V. An assessment of survey research in POM: From constructs to theory. *J. Oper. Manag.* **1998**, *16*, 407–425. [CrossRef]
46. Oksenberg, L.; Kalton, G. New strategies for pretesting survey questions. *J. Off. Stat.* **1991**, *7*, 349.
47. Yan, T.; Kreuter, F.; Tourangeau, R. Evaluating survey questions: A comparison of methods. *J. Off. Sta-Tistics* **2012**, *28*, 503.
48. Armstrong, J.S.; Overton, T.S. Estimating nonresponse bias in mail surveys. *J. Mark. Res.* **1977**, *14*, 396–402. [CrossRef]
49. Podsakoff, P.M.; MacKenzie, S.B.; Lee, J.-Y.; Podsakoff, N.P. Common method biases in behavioral research: A critical review of the literature and recommended remedies. *J. Appl. Psychol.* **2003**, *88*, 879. [CrossRef]

50. Singh, P.J. Empirical assessment of ISO 9000 related management practices and performance relationships. *Int. J. Prod. Econ.* **2008**, *113*, 40–59. [CrossRef]
51. Kim, D.; Kumar, V.; Kumar, U. Relationship between quality management practices and innovation. *J. Oper. Manag.* **2012**, *30*, 295–315. [CrossRef]
52. Singh, P.J.; Power, D.; Chuong, S.C. A resource dependence theory perspective of ISO 9000 in managing organi-zational environment. *J. Oper. Manag.* **2011**, *29*, 49–64. [CrossRef]
53. Hair, J.F., Jr.; Black, W.C.; Babin, B.J.; Anderson, R.E.; Tatham, R.L. *Multivariate Data Analysis*, 5th ed.; Pearson Prentice-Hall: New York, NY, USA, 2006.
54. Fornell, C.; Larcker, D.F. Evaluating structural equation models with unobservable variables and measurement error. *J. Mark. Res.* **1981**, *18*, 39–50. [CrossRef]
55. Lam, L.W. Impact of competitiveness on salespeople's commitment and performance. *J. Bus. Res.* **2012**, *65*, 1328–1334. [CrossRef]
56. Fynes, B.; Voss, C. A path analytic model of quality practices, quality performance, and business performance. *Prod. Oper. Manag.* **2001**, *10*, 494–513. [CrossRef]
57. Fynes, B.; Voss, C.; de Búrca, S. The impact of supply chain relationship quality on quality performance. *Int. J. Prod. Econ.* **2005**, *96*, 339–354. [CrossRef]
58. Matopoulos, A.; Vlachopoulou, M.; Manthou, V.; Manos, B. A conceptual framework for supply chain collaboration: Empirical evidence from the agri-food industry. *Supply Chain Manag. Int. J.* **2007**, *12*, 177–186. [CrossRef]
59. Paloviita, A. Consumers' Sustainability Perceptions of the Supply Chain of Locally Produced Food. *Sustainability* **2010**, *2*, 1492–1509. [CrossRef]
60. Moursellas, A.; De, D.; Wurzer, T.; Skouloudis, A.; Reiner, G.; Chaudhuri, A.; Manousidis, T.; Malesios, C.; Evangelinos, K.; Dey, P.K. Sustainability Practices and Performance in European Small-and-Medium Enterprises: Insights from Multiple Case Studies. *Circ. Econ. Sustain.* **2023**, *3*, 835–860. [CrossRef]
61. Kuwornu, J.K.; Khaipetch, J.; Gunawan, E.; Bannor, R.K.; Ho, T.D. The adoption of sustainable supply chain management practices on performance and quality assurance of food companies. *Sustain. Futur.* **2023**, *5*, 100103. [CrossRef]
62. Oyedijo, A.; Kusi-Sarpong, S.; Mubarik, M.S.; Khan, S.A.; Utulu, K. Multi-tier sustainable supply chain management: A case study of a global food retailer. *Supply Chain Manag. Int. J.* **2024**, *29*, 68–97. [CrossRef]
63. Barbosa, M.W. Uncovering research streams on agri-food supply chain management: A bibliometric study. *Glob. Food Secur.* **2021**, *28*, 100517. [CrossRef]
64. Seuring, S.; Aman, S.; Hettiarachchi, B.D.; de Lima, F.A.; Schilling, L.; Sudusinghe, J.I. Reflecting on theory de-velopment in sustainable supply chain management. *Clean. Logist. Supply Chain* **2022**, *3*, 100016. [CrossRef]
65. Jouzdani, J.; Govindan, K. On the sustainable perishable food supply chain network design: A dairy products case to achieve sustainable development goals. *J. Clean. Prod.* **2021**, *278*, 123060. [CrossRef]
66. Leal Filho, W.; Viera Trevisan, L.; Paulino Pires Eustachio, J.H.; Dibbern, T.; Castillo Apraiz, J.; Rampasso, I.; Anholon, R.; Gornati, B.; Morello, M.; Lambrechts, W. Sustainable supply chain management and the UN sustainable devel-opment goals: Exploring synergies towards sustainable development. *TQM J.* **2023**. *ahead-of-print*. [CrossRef]

foods

MDPI

Article

MycoTWIN Working Group Discussion: A Multi-Actor Perspective on Future Research Directions for Mycotoxins and Toxigenic Fungi Along the Food and Feed Chain

Martina Loi [1], Antonio Moretti [1], Vincenzo Lippolis [1], Hayrettin Özer [2], Ceyda Pembeci Kodolbas [2], Elif Yener [2], İlknur Demirtaş [2], Pilar Vila-Donat [3], Lara Manyes [3] and Veronica M. T. Lattanzio [1,*]

[1] Institute of Sciences of Food Production, National Research Council of Italy, Via Amendola 122/O, 70126 Bari, Italy; martina.loi@ispa.cnr.it (M.L.); antonio.moretti@ispa.cnr.it (A.M.); vincenzo.lippolis@ispa.cnr.it (V.L.)
[2] TUBITAK MAM Food Institute Gıda Enstitüsü, 41470 Gebze, Turkey; hayrettin.ozer@tubitak.gov.tr (H.Ö.); ceyda.pembeci@tubitak.gov.tr (C.P.K.); elif.yener@tubitak.gov.tr (E.Y.); ilknur.demirtas@tubitak.gov.tr (İ.D.)
[3] Laboratory of Food Chemistry and Toxicology, Faculty of Pharmacy, University of Valencia, 46010 València, Spain; pilar.vila@uv.es (P.V.-D.); lara.manyes@uv.es (L.M.)
* Correspondence: veronicamariateresa.lattanzio@cnr.it

Abstract: Mycotoxin research is facing unprecedented challenges, starting from the urgent need to cope with the consequences of climate change, the global shortage of grain due to unstable political scenarios, and the major transformation of the supply chains after the COVID-19 pandemic. In this scenario, the mycotoxin contamination of human and animal foods is still unavoidable, thus representing a major challenge to global food security. Next to this, the shift to sustainable and circular food production might be accompanied by an increase in food safety issues involving mycotoxins, e.g., when new technologies are applied to reuse side streams from the food industry, it is not known if and how mycotoxins accumulate in these by-products. MycoTWIN is an EU-funded Horizon 2020 project which fosters knowledge transfer and scientific cooperation within the Mediterranean area, involving worldwide experts, decision makers, and stakeholders in the field of mycotoxigenic fungi and mycotoxins. The MycoTWIN project hosted working group meetings, whose aim was to propose operational plans and/or scientific strategic plans to shape the future research directions to better cope with these challenges. In the working group cycle "Future proof approaches for the management of toxigenic fungi and associated mycotoxins along the food chain", a multi-actor group was guided in co-creation exercises to elaborate on future research directions and propose relevant actions to be implemented for the present to long-term time periods. The discussion focused on three main topics relevant to the assessment and management of risks associated with mycotoxins and toxigenic fungi: (i) needs for the harmonization of molecular and chemical methods and data analysis, (ii) from lab research to marketable solutions: how to fill the gap, and (iii) gaps in data quality for risk assessment.

Keywords: mycotoxin; food security; food safety; mycotoxigenic fungi; future research directions; mycotoxin management

1. Introduction

In recent years, the food system has been facing concurrent, emergent, and unexpected challenges, such as climate change, geopolitical tensions, the global shortage of grains, societal changes, and a major transformation of the supply chains after the COVID-19 pandemic. The capability of producing enough, safe, and nutritious food and feed for the growing global population is seriously affected, especially in countries already facing food safety and security issues.

A recent survey covering the first half of 2024 detected high mycotoxin prevalence and associated risks across North and Latin America, Europe, Sub-Saharan Africa, and Asia, while moderate risks were noted in Northern Europe and Oceania [1]. Besides regulated mycotoxins, emerging ones (such as fusaric acid, enniatins, culmorin, apicidin, butenolide, fusaproliferin, *Alternaria* toxins, auofusarin, emodin, beauvericin (BEA), diacetoxyscirpenol (DAS), moniliformin (MON), and sterigmatocystin) are increasingly detected worldwide in staple cereal commodities. The toxicity of these compounds, as well as their potential to interact with each other and with other xenobiotics, poses an even greater risk to human and animal health [2].

Indeed, despite the efforts of the scientific community and stakeholders, mycotoxins still represent the third most notified hazard category in Europe by the Rapid Alarm Safety System for Food and Feed (RASFF), with f.i. a 10.5% increase in 2022 compared to 2021 [3].

Constantly active research is in progress to develop advanced multi-mycotoxin detection methods for complex samples, as well as prevention strategies (including predictive modeling) and mitigation strategies through physical and biological approaches; however, due to the evolving scenario, several challenges persist. Building capacity for mycotoxin management, especially in resource-limited regions, and harmonizing efforts for effective risk assessment and mitigation is crucial. Achieving transitions towards a healthier, safer, sustainable, and environmentally friendly food system as set out in the Farm2Fork EU strategy [4], and counteracting the global threat of mycotoxin contamination, requires a bold and holistic response by the research community. One solid approach to creating future European Food Safety strategies is to identify the priorities of common interest and share the results, information, and knowledge for institutions and policy makers through cooperation initiatives.

Horizon 2020 was designed to promote scientific excellence, foster innovation, and address major societal challenges. The program prioritized fields such as climate action, sustainability, health, and digitalization, with a strong emphasis on involving business operators, fostering partnerships, and encouraging international cooperation to solve complex global issues. As part of the H2020 scope, during the EU H2020-funded project MycoTWIN [5], working group meetings were organized with the aim of proposing operational plan and/or scientific strategic plans to shape the future research directions in the field of mycotoxigenic fungi and mycotoxins. During the working group cycle "Future proof approaches for the management of toxigenic fungi and associated mycotoxins along the food chain" organized by CNR-ISPA, three trending topics were identified and discussed by the MycoTWIN working with a multidisciplinary, holistic, and collaborative approach provided by the hosting EU Food Safety Platform [6] (see Section 2 for full details).

When selecting the topics of the working groups, the choice was driven by an extensive review of the existing literature, engaging with scientific networks/research consortia (H2020-funded FoodSafety4EU and EFSA-funded MYCOBOOST projects), monitoring white papers and reports of government agencies (FAO and EFSA), and by the strategic insights gained from the round table discussions held during the first MycoKey International conference, titled "Global Mycotoxin Reduction in the Food and Feed Chain" [6–8]. The topics' selection was conducted to ensure access to up-to-date information on the challenges, research priorities, pressing industry and regulatory issues, and emerging trends shaping the field of mycotoxin research.

Future research needs in the food safety domain are elaborated in recently issued research agendas setting the updated research and innovation framework (f.i. the Food 2030 agenda, and its pathway action "Food Safety Systems of the Future") [9], or the FoodSafety4EU Strategic Research and Innovation Agenda [10]. In this framework, the generation of and access to Findable, Accessible, Interoperable, and Reusable (FAIR) data are acknowledged as the fundamentals for improving food safety management and risk assessment. Thinking about the rapid advances in toxicology and in the assessment of chemical safety in humans (e.g., mycotoxin mixtures) as well as the spreading of big data (large amount, complex, structured and unstructured data) in this field, the availability of improved and harmonized approaches to generate, analyze, interpret, reuse, and integrate data about mycotoxin contamination at national/EU level is of utmost importance.

The same consideration applies to the design of predictive and modeling tools needed, for instance, to cope with climate change's impact on mycotoxin/toxigenic fungi distribution patterns.

Next to risk assessment, extensive and systematic mycotoxin/fungi monitoring either in the field or along the chain can contribute to the development and implementation of mitigation measures supporting risk management decisions. In this respect, space and funding for the translation of laboratory research into practical solutions, enabling food business operators to control and share data about contamination, can be found in the Horizon Europe program [11]. The established and emerging technologies for real-time mycotoxin/fungi monitoring, complemented by validated approaches to generate reliable data, are expected to support preventive measures and the reduction in food waste due to product non-compliance.

Based on all the above considerations, the following topics were selected to be elaborated by the working group: (i) Needs for harmonization of molecular and chemical methods and data analysis, (ii) From lab research to marketable solutions: how to fill the gap, and (iii) Gaps in data quality for risk assessment.

2. Methodology for Collaborative Exercise

2.1. Participants

Working group participants were selected among the MycoTWIN consortium, international experts, and the private sector according to diverse research expertise and capabilities, geographic origin, affiliation, and interests in the food safety domain. They represented: research institutions (National Research Council of Italy—Institute of Sciences of Food Production—CNR-ISPA—Bari, Italy; TUBITAK MRC Life Sciences—Gebze/Kocaeli, Turkey; and International Institute of Tropical Agriculture, The Institut National De Recherche Pour L'agriculture, L'alimentation Et L'environnement—INRAE—Bordeaux, France), universities (Valencia University—Valencia, Spain; Piacenza University—Piacenza, Italy; Parma University—Parma, Italy; Liège University—Liège, Belgium; Ghent University—Ghent, Belgium; and Cranfield University—Cranfield, UK), global food authorities (the European Commission, Joint Research Centre—JRC), the EU Food Safety Platform, and national and international companies (Bonifiche Ferraresi, Eurofins Tecna, and Syngenta).

The collaborative exercise was organized as a series of 2 working groups as described in Sections 2.2 and 2.3; 37 and 11 participants were involved in phase 1 and phase 2, respectively.

2.2. Working Group 1 (1st Phase)

In the preparatory work executed before the working group, based on an extensive review of the available strategic documents (see Section 1), three relevant topics in the mycotoxin field were selected by the authors, and a draft list of issues for the working group participants was prepared.

The first working group was held as a live event on 22 May 2022. For each proposed topic and its related draft list of issues, the discussion was structured as follows:

1. *Silent reflection*: Each participant considered the proposed draft list of issues and annotated amendments or additions of new issues to be included in the subsequent discussion.
2. *Plenary discussion*: The discussion was led by a facilitator, who moderated the discussion, and a reporter, who took notes of the amendments and/or new issues generated by the participants.
3. *Voting and ranking*: Once the participants agreed on the list of issues generated from the plenary discussion, they were asked to vote on them according to the priority of action, with a scale of 1 (low priority) to 10 (high priority). To calculate the weighted mean, the total scores for each issue were divided by the number of participants who

ranked that issue. The first six issues, prioritized according to the weighted mean, were subjected to another round of voting according to uncertainty using the same scale (1 to 10).

2.3. Working Group 2 (2nd Phase)

The second working group discussion was held as a hybrid event among MycoTWIN experts and external guests on 15 April 2024. It was structured as follows:

1. *Topic discussion*: The issues prioritized in the previous working group were briefly discussed, also reviewing the priority and uncertainty scores.
2. *Target action proposition*: For each prioritized issue, the participants were asked to propose a target action or strategic plan by specifying the related hazard. The participants' contributions were first collected on a digital board and then reviewed and rephrased until consensus was achieved.
3. *Interlinks*: After carrying out step 2 for each of the three topics, the participants were asked to discuss and indicate interlinks between the proposed actions and subtopics.

Common actions or actions with multiple impacts selected with the interlink exercise were discussed considering urgency and uncertainty and classified into the following:

(i) Present (one-year needs): Urgent issues;
(ii) Near future (three-year needs): Emerging issues;
(iii) Normal term (five-year needs): Mid-term issues;
(iv) Long term (eight-year needs): Far future issues.

This was performed with the aim to provide input for future research in the field of mycotoxins and toxigenic fungi.

Clustering issues within specified timeframes: Scores for the prioritized issues (Section 2.2, item 3), based on votes for both the priority of action and level of uncertainty, were normalized on a 0–5 scale, with the highest voted issue across all three topics serving as the benchmark (score 5). The issues were clustered within the four timeframes (present—one-year needs, near future—three-year needs, normal term—five year-needs, and Long term—eight-year needs) based on a qualitative evaluation that, however, took into account not only the normalized scores, but also the experts' evaluations of issues, i.e., the urgency (priority) of the identified needs, the uncertainty of the available solutions/methodologies to cope with them, and the possible interlinks among the issues.

2.4. Statistical Analyses

The priority of action and uncertainty rank scores (item 3, Section 2.2) were subjected to normality testing distribution using the Shapiro–Wilk test. Following the assessment of non-normal data distribution, the non-parametric Friedman test was applied, along with the post hoc Wilcoxon signed-rank test, to assess statistically significant differences among the voted issues. The Bonferroni-adjusted p-value was applied for pairwise comparisons. All the statistical analyses were performed using the IBM SPSS software (IBM SPSS Statistics for Windows, Version 29.0 (Armonk, NY, USA: IBM Corp, 2022).

2.5. Glossary

For the purpose of the working group meetings, the following glossary was adopted:

(i) High level of priority of action: Something given or meriting attention before competing alternatives since it represents a pre-requisite, or a key requisite, to undertake the other issues in the list;
(ii) High level of uncertainty: not fully investigated and researched; not based on well-established knowledge; values are unknown; possible outcomes are incomplete and ambiguous; unstable; hard to predict; difficult to estimate; unobvious impact.

3. Results

In working group 1 (first phase), for each selected topic, the current constraints, enablers, or requirements (from now on defined as "issues") were identified by the experts in co-creation exercises and ranked according to the priority of action and uncertainty, according to the methodology described in Section 2.2. Working group 1 resulted in a list of six prioritized issues for each of the three discussed topics. In working group 2 (second phase), the prioritized issues were further explored, identifying the main related hazards and proposing actions to cope with them. Finally, the issues and relevant proposed actions were clustered in four timeframes according to the priority of action, uncertainty, and interlinks identified between the issues.

3.1. *Topic 1: Needs for Harmonization of Molecular and Chemical Methods and Data Analysis*

Harmonized, quality data are essential to ensure the consistency, comparability, and reliability of results. They represent a mandatory requirement to perform large-scale monitoring, gain insight into mycotoxin and fungal contamination, support the establishment or revision of mycotoxin regulatory limits in food and feed, and manage recommendations and guidelines at various levels. Nonetheless, given the current increase in data generation driven by high-throughput and AI-based approaches, harmonizing molecular and chemical methods and data analysis has become a critical issue that requires immediate attention from the research community.

3.1.1. Topic 1 Issue Definition and Prioritization

During the first working group meeting, eleven issues related to constraints and enablers for method harmonization were identified (Table 1). The experts identified as major issues the need for standardization protocols/procedures (issues n. 1, 2, 7, 8, 9, and 10), need for reference material (issue n. 3), data generation and sharing (issues n. 4, 6, and 7), and method validation (issues n. 8 and 11). The ranked issues are shown in Table 1 (priority of action) and Table 2 (top six, uncertainty). Looking at the uncertainty ranking, the group's view revealed that there are still gaps in traditional issues that need to evolve in the new scenario. These include reference material—with emphasis on multi-mycotoxin or emerging/modified mycotoxin reference materials—sampling and the validation of new analytical approaches.

Table 1. Prioritized issues (priority of action) for Topic 1. Priority scores range from 1 (low priority) to 10 (high priority).

Needs for Harmonization of Molecular and Chemical Methods and Data Analysis *For Each Issue, What Is the Perceived Level of Priority of Action?*		
Issue n.	**Issue**	**Priority Score [1]**
1	Standardized sampling protocols	6.00 a
2	Standardized data collection and curation	5.48 a
3	Availability of certified reference materials (CRMs), standards (STDs), quality control materials (QCMs), and blanks/control samples	5.41 ab
4	Data management plan and training	5.41 ab
5	Fit-for-purpose validation programs and guidelines for official control and rapid testing	5.03 ab
6	Platform to design experiments, share, and collect data	5.03 ab
7	Simplified standard data format	4.79 ab
8	Measurement uncertainty protocols	4.64 ab
9	Criteria for calculating and evaluating Limit Of Quantification (LOQ)	4.25 b
10	Guidelines and acceptance criteria for matrix effects	4.14 b
11	Availability of interlaboratory ring trials (IRTs) and/or proficiency testing programs (PTPs)	4.11 b

[1] letters indicate statistically significant differences among issues (Bonferroni-adjusted p-value < 0.000758).

Table 2. Prioritized issues (priority of action and uncertainty) for Topic 1. Uncertainty scores from 1 (low uncertainty) to 10 (high uncertainty).

Needs for Harmonization of Molecular and Chemical Methods and Data Analysis. *For Each Proposition, What Is the Perceived Level of Uncertainty?*	
Issue	**Uncertainty Score [1]**
Availability of certified reference materials (CRMs), standards (STDs), quality control materials (QCMs), and blanks/control samples	3.91
Standardized data collection and curation	3.69
Standardized sampling protocols	3.39
Fit-for-purpose validation programs and guidelines for official control and rapid testing (auto-control)	3.36
Data Management plan and Training	3.35
Platform to design experiments, share, and collect data	2.99

[1] no statistically significant difference emerged among issues (Friedman's p-value = 0.477).

3.1.2. Definition of Hazards and Relevant Actions for Topic 1

The development of collaborative, data-driven platforms, infrastructures, and online repositories were identified as crucial actions to undertake. These platforms enable data sharing and use across different research fields and link molecular/predictive data to phenotype and bioanalysis, as also highlighted during the plenary discussion. In addition, simplified and harmonized data formats and management, also supported by artificial intelligence, were classified as emerging issues (Table 3).

Interestingly, while it is clear that new hazards have been identified for emerging mycotoxins and mycotoxins in biofluids and new commodities, such as fermented foods, novel foods, and side streams, the situation for toxigenic fungi is less clear as they are not regulated. A notable normal-term issue that emerged from the discussion regarded fungal taxonomy, for which there is the need for reference strains to define fungal chemotypes based on the mycotoxigenic potential rather than core features. Hence, fungal taxonomy should be revised according to the toxigenic potential, especially in light of climate change and the emergence of new commodities of interest.

3.2. Topic 2: From Lab Research to Marketable Solutions: How to Fill the Gap

Filling the gap between laboratory research and marketable solutions is a crucial task to ensure that scientific advancements are translated into tangible benefits for stakeholders and society. This process involves a multi-actor, multidisciplinary intervention to address the multifaceted challenges, different backgrounds, and perspectives. Indeed, often the information or efforts required to bring innovation into the market rely also on other stakeholders, such as large-scale and/or smallholder farmers, and observations made by extension agents, policy makers, and private industry, rather than on solely scientists [12].

Table 3. Hazards and relevant actions proposed by the MycoTWIN working group participants for each prioritized issue of Topic 1—needs for the harmonization of molecular and chemical methods and data analysis.

Time	Issue	Hazard	Action
Present (1-year needs)	Platform to design experiments, share, and collect data for feed models	All mycotoxins Re-occurring mycotoxins related to climate change (f.i. aflatoxins)	Starting from the already existing ones, to develop or integrate collaborative infrastructures/data-driven platforms/online repositories Include data for bioanalysis for evidence based on public health information
Near Future (3-year needs)	Standardized data collection and curation	Emerging mycotoxins Modified mycotoxins Regulated mycotoxins in new commodities of interest	To develop simplified standard data formats, taking into account big data and AI developments as well as the FAIR [1] principles To develop training modules on data collection and curation for different target users To develop committee-supervised, unanimous consensus on methods/approaches used to obtain data for taxonomic rearrangements
		Fungal chemotypes vs. fungal species	To enforce the definition of "chemotypes" rather than species (mycotoxin production vs. core genome) To promote investments in risk monitoring
	Standardized sampling protocols	Regulated mycotoxins in regulated and unregulated commodities of interest (f.i. fermented foods, novel foods, and side stream) Fungi	To develop and/or standardize sampling methods for new commodities To develop and/or standardize alternative sampling methods (f.i. based on the latest developments in dust sampling) To develop and/or feed infrastructures and culture collections, including methodologies to define and maintain reference strains
	Fit-for-purpose validation programs and guidelines for official control and rapid testing (auto-control)	Regulated mycotoxins in regulated commodities	To develop validation programs, including in-house validation protocols for rapid testing—co-development with food business operators To develop dedicated proficiency tests To propose an update of the Regulation 519/2014/EC (mycotoxin screening methods)—link with EFSA [2] to take into account risk assessment needs
		Fungi	For the classification of new species: create more links between biology/genetics and taxonomy (structured pathways)

Table 3. *Cont.*

Time	Issue	Hazard	Action
	Data Management plan and Training	Regulated and (re)emerging mycotoxins	To propose harmonized data formats To implement AI [3]/machine learning approaches to merge redundant data/integrate the existing data To develop joint training modules for AI [3] experts and analytical chemists
		Fungi	To develop AI/machine learning for modeling—investigate the link between genomic elements To develop AI-based approaches to predict the biology of a microorganism
Normal term (5-year needs)	Availability of CRMs [4], STDs [5], QCMs [6], and blanks/control samples	Emerging mycotoxins Modified mycotoxins Regulated mycotoxins in new commodities of interest (f.i. fermented foods, novel foods, and side streams)	To develop new protocols for CRM [4], STDs [5], and QCMs [6]—go for the fitness-for-purpose approach To develop standard methods/procedures for reference material characterization Define reference values (low vs. high exposure) for mycotoxin bioanalysis
		Fungal chemotypes (defining mycotoxigenic potential)	Define reference strains for fungi Define robust taxonomy classification through interlaboratory comparison—long-term perspective

[1] FAIR: Findable, Accessible, Interoperable, and Reusable; [2] EFSA: European Food Safety Authority; [3] AI: artificial intelligence; [4] certified reference materials (CRMs); [5] standards (STDs); [6] quality control materials (QCMs).

3.2.1. Topic 2 Issue Definition and Prioritization

Twelve issues related to constraints and enablers for bridging the gap between laboratory research and marketable solutions were identified (Table 4). The experts highlighted not only the need for improvements in technologies (issues n. 1, 3, 4, 6, 8, 9, 10, and 11), but also the need to bring awareness and a structured, improved communication (issue n. 2), and to engage with the regulatory bodies (issues n. 5 and 7) and companies (issues n. 2, 9, and 12). These latter were also listed among the most uncertain issues to tackle (Table 5), also due to the lack of well-established and trusted channels of communication between these actors of the food safety system, especially when thinking about local authorities and small enterprises/farmers. However, issues like the verification of the analytical process and full traceability were put on the basis of acceptance by regulatory bodies and trustful communication.

Table 4. Prioritized issues (priority of action) for Topic 2.

	From Lab Research to Marketable Solutions: How to Fill the Gap ***For Each Proposition, What Is the Perceived Level of Priority of Action?***	
	Subtopic/Issue/Proposition	**Priority Score [1]**
1	Cost-effective	5.75 a
2	Create awareness/information chain/communication	5.67 a
3	Easy to use: limited expertise and minimal operator manipulations/easy calibration/verification of functionality	5.64 a
4	High degree of reliability/validity/comparable to official methods already in use/brand-independent transferability	5.42 ab
5	Acceptance by regulatory bodies	5.3 ab
6	Enables full traceability of measurement process and quality control; data accessible at any time	5.18 ab
7	Compliant with regulations/standards/contracts below regulation	5.18 ab
8	High degree of reliability/validity/comparable to official methods already in use	5.1 ab
9	Secure supply of materials	4.9 ab
10	"Green" technology, sustainable, safe, and easy handling	4.82 ab
11	Enables process automation and speed, online capability	4.78 ab
12	Productability studies	4.26 b

[1] letters indicate statistically significant differences among issues (Bonferroni-adjusted p-value < 0.000758).

Table 5. Prioritized issues (priority of action and uncertainty) for Topic 2.

From Lab Research to Marketable Solutions: How to Fill the Gap ***For Each Proposition, What Is the Perceived Level of Uncertainty?***	
Subtopic/Issue/Proposition	**Uncertainty Score [1]**
Easy to use: limited expertise and minimal operator manipulations/easy calibration/verification of functionality	3.91
Enables full traceability of measurement process and quality control; data accessible at any time	3.66
Acceptance by regulatory bodies	3.55
Create awareness/information chain/communication	3.23
High degree of reliability/validity/comparable to official methods already in use	2.97
Cost-effectiveness	2.58

[1] no statistically significant difference emerged among issues (Friedman's p-value = 0.082).

3.2.2. Definition of Hazards and Actions for Topic 2

Among the actions to be undertaken in the present–near future, the experts highlighted the need for fitness-for-purpose approaches in method validation, performance, and verification protocols. They also emphasized the importance of promoting automated procedures and providing on-site training to improve reliability, validity, and quality control. Investments and cooperation were considered urgent actions as well.

As for normal-term improvements (over the next 5 years), the experts only identified technical improvements, also based on AI, to improve the ease of use and method reliability.

Concerning hazard identification, regulated mycotoxins and mycotoxin adducts/metabolites in the blood and urine emerged as important hazards, for which the gap from lab research to marketable solutions is still high.

3.3. *Topic 3: Gaps in Data Quality for Risk Assessment*

Continuous risk assessment is crucial to comprehensively understand and mitigate the potential adverse effects of mycotoxin contamination, thereby ensuring safe food and feed. High-quality data represent the foundation for each step of risk assessment, i.e., hazard identification and characterization, exposure assessment, and risk characterization.

Nonetheless, researchers consistently point out the limited availability, inconsistency, and lack of harmonization and integration in data concerning fungal and mycotoxin contamination and the need for robust methodologies and cutting-edge technologies for data collection, analysis, and interpretation [13].

3.3.1. Topic 3 Issue Definition and Prioritization

Nine issues were identified by the experts (Table 6) regarding protocols and guidelines for data and metadata collection and sharing (issues n. 1, 2, 3, 6, and 7), platforms for data storage and management (issues n. 4, 5, and 9), and privacy/reputation (issue 8). The most urgent issues identified regarded harmonizing data and promoting data sharing (Table 7).

Table 6. Prioritized issues (priority of action) for Topic 3.

	Gaps in Data Quality for Risk Assessment ***For Each Proposition, What Is The Perceived Level of Priority of Action?***	
	Issue	**Priority Score** [1]
1	Harmonized protocols and guidelines for sample metadata collections	5.86
2	Policies, education, and funding to support metadata and data sharing	5.73
3	Data collection and curation	5.63
4	Availability of open-source platforms for data storage and sharing	5.53
5	Availability of affordable and user-friendly data management software	5.39
6	More transparency in the protocols	5.31
7	Simplified standard data format	5.04
8	Privacy/reputation	4.96
9	Increased knowledge of data management plans	4.86

[1] no statistically significant difference emerged among issues (Friedman's p-value = 0.091).

Table 7. Prioritized issues (priority of action and uncertainty) for Topic 3.

Gaps in Data Quality for Risk Assessment ***For Each Proposition, What Is the Perceived Level of Uncertainty?***	
Prioritized Issue	**Uncertainty Score** [1]
Harmonized protocols and guidelines for sample metadata collections	5.00
Policies, education, and funding to support metadata and data sharing	4.79
More transparency in the protocols	4.03

Table 7. *Cont.*

Gaps in Data Quality for Risk Assessment *For Each Proposition, What Is the Perceived Level of Uncertainty?*	
Prioritized Issue	**Uncertainty Score [1]**
Availability of open-source platforms for data storage and sharing	3.74
Availability of affordable and user-friendly data management software	3.70
Data collection and curation	3.46

[1] no statistically significant difference emerged among issues (Friedman's p-value = 0.051).

3.3.2. Definition of Hazards and Actions for Topic 3

The actions identified by the experts in the near- and normal-term timeframe aimed at supporting data collection and curation, and platforms in which they should be transparently shared. These actions targeted users (involvement and training) and digital infrastructures (user-friendly and accessible software, workflows, and open access tools).

For the first time, long-term actions (8-year needs) were identified and focused on metadata. In the era of omics techniques, metadata are becoming as important as the data themselves, as they include the full description of how the data were obtained, processed, and annotated [14]. Besides actions aimed to establish harmonized protocols and formats, the experts highlighted the need to implement knowledge and awareness on metadata, their acquisition through New Approach Methodologies (in silico, in vitro, omics, predictive, and biomarker-based technologies), and their sharing among researchers, risk assessors, and risk managers.

4. Discussion

In relation to the three discussed topics, a number of strategic actions was proposed to address relevant hazards that included regulated and emerging mycotoxins as well as mycotoxigenic fungi (Tables 3, 8 and 9).

The issues emerging from Topic 1 were among the ones with a higher impact on the issues of the other topics. Indeed, method harmonization is often a prerequisite to implementing the transferability of research to the market (Topic 2) and to defining data used in risk assessment (topic 3). A consistent methodology, which includes standardized data collection and curation, the availability of reference material, and fit-for-purpose validation programs, enables one to easily use and implement novel methods into the market and verify them against the official ones (Figure 1). Sharing data was also regarded as a tool to create awareness and promote communication among research, private companies, and consumers, which in turn fosters transparency, collaboration, and trust-building (Figure 1). Regulatory bodies support the strategic goals of the European Union, encouraging innovation and scientific advancement. Their role in promoting policies, education, and funding to support metadata and data sharing was acknowledged (Figure 1). The statistical analyses revealed significant differences only among the priority scores of Topics 1 and 2, while no differences were found in their uncertainty scores or among all the Topic 3 scores. Overall, the closely grouped scores—including those in which statistically significant differences were observed—indicated that the working group participants assigned a consistent level of recognition to the issues for all the topics. Their role in promoting policies, education, and funding to support metadata and data sharing was acknowledged (Figure 1). It is important to note that the scores were assigned individually, reflecting the individual participants' experience and expertise. To reach a consensus, the issues were further elaborated within the working group taking into account the scoring results. By elaborating on (i) the urgency (priority) of the identified needs, (ii) the uncertainty of the available solutions/methodologies to cope with them, and (iii) the interlinks showing which actions were pre-requisite for others, the experts clustered the proposed actions in four timeframes (present, near, medium, and long term) to envisage a possible strategic action plan. The actions to be undertaken in the short–medium terms were identified mainly in Topics 1 and 2. The discussion highlighted that advanced research on regulated mycotoxins (and relevant fungi) and emerging ones is already in progress. The needed actions deal mainly with better exploitation/transfer and sharing of already available research results. Therefore, approaches and guidelines for the harmonization and standardization of data and protocols, which in turn will enable data sharing and joint exploitation, resulted to be of high priority.

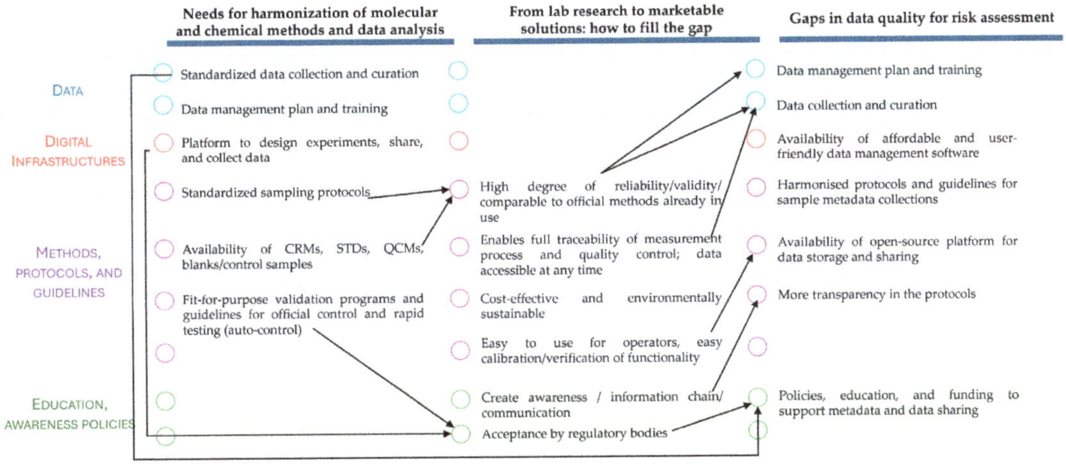

Figure 1. Simplified scheme of the topic interlinks proposed by the MycoTWIN working group. For each topic (column), the issues were categorized by color: Blue—issues related to data; Orange—issues related to digital infrastructures; Purple—issues related to methods, protocols, and guidelines; Green—issues related to education, awareness, and policies. The arrows indicate the interlinks among the issues of different topics.

A major impact will result from the increasing application of AI-based approaches in mycotoxin risk prediction, characterization, assessment, and management. In this respect, full compliance with the FAIR principles in data generation, collection, and storage will require actions in the medium–long term to cope with big data and issues related to Intellectual Property (IP) and data curation.

Common issues in Topics 1 and 3 highlighted the need for improving data collection, curation, and platforms/software for their management. Initiatives led by the World Health Organization (WHO) jointly with the Food and Agriculture Organization (FAO), and EFSA have made several progresses towards the definition of descriptors for mycotoxin contamination data [15]. Nonetheless, the occurrence of mycotoxigenic fungi in food and feed remains largely unregulated. Although there is a significant body of literature addressing the occurrence of these fungi in specific commodities and in specific geographical areas, the data reported in these studies are not systematically reported nor analyzed, often resulting in fragmented or inconsistent data. In addition, the current taxonomic definition for fungal species is not adequate to describe the toxigenic potential of a fungal strain in a specified matrix in a geographical area under a defined climatic condition. Harmonized approaches to identify mycotoxigenic fungi would facilitate more effective monitoring, risk assessment, and management of fungal contamination along the supply chain.

Table 8. Hazards and relevant actions proposed by the MycoTWIN working group participants for each prioritized issue of Topic 2—from lab research to marketable solutions: how to fill the gap.

Time	Issue	Hazard	Action
Present (1-year needs)	High degree of reliability/validity/comparable to official methods already in use	Regulated mycotoxins or under consideration for future regulation (recommended)	To develop fitness-for-purpose method verification strategy assessment
	Cost-effectiveness and environmental sustainability	Regulated mycotoxins or under consideration for future regulation (recommended)	To promote investments in research and innovation/collaborative research projects with academia and industries To develop environmentally safe kits (no plastic—eco-friendly)
Near Future (3-year needs)	Enables full traceability of measurement process and quality control; data accessible at any time	Regulated mycotoxins	To develop fully automated procedures with remote access to the data and integrate AI-based technologies
	Acceptance by regulatory bodies	Regulated mycotoxins	To develop fit-for-purpose validation/performance verification protocols for operators (industries) To provide on-site training tailored to specific industry/operator needs To develop guidelines for dossier/documents to be submitted to the regulatory bodies
		Fungi	*Note:* fungi are not regulated, so no need for acceptance by regulatory bodies. However, the early detection of fungi should be in the regulation/recommendation coupled with accurate prediction approaches and mitigation steps
	Create awareness/information chain/communication		To develop and pilot innovative/simple communication models to reach operators/consumers in rural areas with low levels of education To create awareness of the importance of the early detection of fungi to promote mitigatory practices
Normal term (5-year needs)	Easy to use: limited expertise and minimal operator manipulations Easy calibration/verification of functionality	Regulated mycotoxins in regulated commodities and unconventional matrices (blood/urine) Mycotoxin adducts/metabolites in unconventional matrices (blood/urine), e.g., Ochratoxin A in blood, aflatoxins–lysin adducts	To develop green and simplified sample preparation protocols, including protocols without extraction (e.g., Infrared Spectroscopy) To improve test readers/detectors with user-friendly interface and online data management and to integrate user-friendly AI—simplified AI interfaces and more intuitive interfaces hiding the technical complexities of machine learning Implement method validation/verification protocols to assure accuracy, completeness, consistency, and reliability of obtained data

Table 9. Hazards and relevant actions proposed by the MycoTWIN working group participants for each prioritized issue of Topic 3—gaps in data availability and quality for risk assessment.

Time	Subtopic	Hazard	Action
Near Future (3-year needs)	Data collection and curation	Emerging mycotoxins Modified mycotoxins Regulated mycotoxins in new commodities of interest	To develop training modules on data collection and curation according to the FAIR [1] principles for different target users To develop and/or make accessible infrastructures for data collection, taking into account big data and AI [2] developments
Normal term (5-year needs)	More transparency in the protocols	All mycotoxins and fungi	Propose methodologies and strategies to reinforce trust and IP [3]/ownership, like rights recognition and data anonymization
	Availability of open-source platforms for data storage and sharing	All mycotoxins and fungi	To map and propose pathways for the integration of the existing platforms to make data available and interoperable To develop approaches for data curation and data security and/or building awareness of the existing ones Propose clear definitions and roles in data responsibility and management
	Availability of affordable and user-friendly data management software	All mycotoxins and fungi	To develop user-friendly and cost-effective data management software To integrate in a user-friendly workflow open access tools for data management/processing
Long term (8-year needs)	Harmonized protocols and guidelines for sample metadata collections	Emerging mycotoxins Modified mycotoxins Regulated mycotoxins in new commodities of interest	To map the existing formats/protocols and guidelines and propose harmonized formats (f.i. commodity-dedicated) To develop harmonized protocols and guideline protocols for NAM [4] technologies (computational modeling, high-throughput screening, omics technologies, and mechanistic toxicology)
	Policies, education, and funding to support metadata and data sharing	All mycotoxins and fungi	To develop and implement training programs for risk assessors and risk managers about the latest scientific developments in relation to emerging risk identification, monitoring, and assessment To build awareness about in silico models for risk assessment (NAM [4]) To enable the scientific community to share data/acknowledge data To establish revenues arising from data access and sharing To set up links with the existing microbial collections and to establish an international data/microbial collection-sharing platform through liasons/bilateral agreements between national authorities To promote investments in data-driven innovation

[1] FAIR: Findable, Accessible, Interoperable, and Reusable; [2] AI: artificial intelligence; [3] IP: Intellectual Property; [4] NAM: New Approach Methodologies.

A cross-cutting need over the three topics resulted to be the availability of platforms and infrastructures for data storage, knowledge sharing, and access. This addresses not only mycotoxins (food safety and risk assessment) but also fungi, since further developments were advocated to establish and maintain international microbial collections as well as new and harmonized approaches for species classifications or re-definition based on their mycotoxigenic potential. Capitalizing and making the already existing platforms interoperable is an action that can be planned in the short term, while the achievement of fully open-source solutions is envisaged in the long term.

Long-term issues were identified within Topic 3 and deal with (i) the harmonization of data, metadata, and protocols for emerging risk assessment and toxicological assessment methodologies (NAMs) and (ii) long-term strategies to shape future policies and orienteering future funding schemes and promoting investments, also through education and training. The latter is expected to contribute to facilitating the relationship between risk assessors and risk managers with a view to the evolving food safety scenarios.

5. Conclusions

Starting from the available knowledge, a co-creation process was implemented, along with a workshop series, to generate a shared vision of the challenges related to the management of food safety risks, arising from mycotoxin and toxigenic fungi contamination along the food/feed chain. The multi-actor composition of the working groups allowed them to capture stakeholders' needs, fostering a multi-perspective approach in prioritizing and undertaking these challenges.

In line with the main objective of the MycoTWIN project, which was capacity building, this collaborative exercise might facilitate the incorporation of new researchers into the group and foster collaborative networks around topics of common interest, while the outcomes of the discussion are expected to serve the scientific community by guiding authors in developing new research projects.

By identifying common actions or actions with multiple impacts on the prioritized issues, the MycoTWIN working group selected and proposed three emerging research topics in the field of mycotoxigenic fungi and mycotoxins for the European framework.

(1) Further development of in silico models for emerging risk assessment (NAMs), including harmonized models and protocols for computational modeling, high-throughput screening, omics technologies, and mechanistic toxicology;
(2) To develop new and more robust approaches for the taxonomic classification of fungal species, also based on interlaboratory comparisons;
(3) To map the available resources and propose pathways for the integration and interoperability of the existing platforms/infrastructures connecting data and metadata from different sources (i.e., genomic analyses, transcriptomic studies, biological assays, expression profiles, and more).

If complemented by adequate measures for awareness building and training, future research in these domains is expected to provide input for future food safety R&I programs.

Author Contributions: Conceptualization, V.M.T.L. and M.L.; methodology, V.M.T.L. and M.L.; investigation, M.L.; resources V.M.T.L.; writing—original draft preparation, M.L. and V.M.T.L.; writing—review and editing, M.L., A.M., V.L., L.M., P.V.-D., H.Ö., C.P.K., E.Y., İ.D. and V.M.T.L.; visualization, M.L. and V.M.T.L.; supervision, A.M. and H.Ö.; project administration, A.M. and H.Ö.; funding acquisition, A.M. and H.Ö. All authors have read and agreed to the published version of the manuscript.

Funding: This work was financially supported by H2020-WIDESPREAD-2020-5 MycoTWIN "Enhancing Research and Innovation Capacity of TÜBITAK MAM Food Institute on Management of Mycotoxigenic Fungi And Mycotoxins" Grant Agreement-952337.

Institutional Review Board Statement: Not applicable.

Informed Consent Statement: Not applicable.

Data Availability Statement: The original contributions presented in the study are included in the article, further inquiries can be directed to the corresponding author.

Acknowledgments: The authors would like to thank Sarah De Saeger and Riccardo Baroncelli for their valuable suggestions during the second working group.

Conflicts of Interest: The authors declare no conflicts of interest.

References

1. DSM Mycotoxin Survey 2024. Available online: https://www.dsm.com/ (accessed on 29 October 2024).
2. Kolawole, O.; Siri-Anusornsak, W.; Petchkongkaew, A.; Elliott, C. A systematic review of global occurrence of emerging mycotoxins in crops and animal feeds, and their toxicity in livestock. *Emerg. Contam.* **2024**, *10*, 100305. [CrossRef]
3. RASFF. 2022 Annual Report Alert and Cooperation Network. Available online: https://food.ec.europa.eu/system/files/2023-10/acn_annual-report_2022.pdf (accessed on 14 June 2024).
4. European Environmental Agency. *Transforming Europe's Food System—Assessing the EU Policy Mix*; European Environmental Agency: Copenhagen, Denmark, 2022; Available online: https://www.eea.europa.eu/publications/transforming-europes-food-system (accessed on 17 June 2024).
5. MycoTWIN Project. Available online: https://www.mycotwin.eu/ (accessed on 20 June 2024).
6. FoodSafety4EU Project. Available online: https://foodsafety4.eu/ (accessed on 21 June 2024).
7. Dall'Asta, C.; De Boevre, M.; Dellafiora, L.; De Saeger, S.; Moretti, A.; Pinson-Gadais, L.; Susca, A. Boosting knowledge and harmonisation in the mycotoxin field through sustainable scientific alliances—MYCOBOOST. *EFSA Supp. Publ.* **2023**, *20*, 8420E. [CrossRef]
8. Leslie, J.F.; Lattanzio, V.; Audenaert, K.; Battilani, P.; Cary, J.; Chulze, S.N.; De Saeger, S.; Gerardino, A.; Karlovsky, P.; Liao, Y.-C.; et al. MycoKey Round Table Discussions of Future Directions in Research on Chemical Detection Methods, Genetics and Biodiversity of Mycotoxins. *Toxins* **2018**, *10*, 109. [CrossRef] [PubMed]
9. Food 2030: Green and Resilient Food System. Available online: https://research-innovation-community.ec.europa.eu/events/5XBL7D4zP2PsfvrXmAJPv1/overview (accessed on 19 July 2024).
10. Nastasia, B.; Duta, D.; Cito, N.; Rychlik, M.; Lattanzio, V.M.T. Strategic Research and Innovation Agenda for Food Safety in Europe. EU Food Safety Forum, Brussels. *Zenodo* **2023**. [CrossRef]
11. Horizon Europe Programme. Available online: https://research-and-innovation.ec.europa.eu/funding/funding-opportunities/funding-programmes-and-open-calls/horizon-europe_en (accessed on 22 June 2024).
12. Ortega-Beltran, A.; Bandyopadhyay, R. Aflatoxin Biocontrol in Practice Requires a Multidisciplinary, Long-Term Approach. Front. *Sustain. Food Syst.* **2023**, *7*, 1110964. [CrossRef]
13. Chhaya, R.S.; O'Brien, J.; Cummins, E. Feed to Fork Risk Assessment of Mycotoxins under Climate Change Influences-Recent Developments. *Trends Food Sci. Technol.* **2022**, *126*, 126–141. [CrossRef]
14. Cernava, T.; Rybakova, D.; Buscot, F.; Clavel, T.; McHardy, A.C.; Meyer, F.; Berg, G. Metadata Harmonization–Standards Are the Key for a Better Usage of Omics Data for Integrative Microbiome Analysis. *Environ. Microbiome* **2022**, *17*, 33. [CrossRef] [PubMed]
15. Mesfin, A.; Lachat, C.; Vidal, A.; Croubels, S.; Haesaert, G.; Ndemera, M.; Matumba, L. Essential Descriptors for Mycotoxin Contamination Data in Food and Feed. *Food Res. Int.* **2022**, *152*, 110883. [CrossRef] [PubMed]

foods

MDPI

Article

Physicochemical Characterization, Rheological Properties, and Antimicrobial Activity of Sodium Alginate-Pink Pepper Essential Oil (PPEO) Nanoemulsions

Mariah Almeida Lima [1,*], Juliana Carusi [2], Liliana de Oliveira Rocha [2], Renata Valeriano Tonon [3], Rosiane Lopes Cunha [4] and Amauri Rosenthal [3]

[1] Food Technology Department, Institute of Technology, University Federal Rural of Rio de Janeiro, Seropedica 23890-000, RJ, Brazil
[2] Department of Food Science and Nutrition, School of Food Engineering, University of Campinas (UNICAMP), Campinas 13083-862, SP, Brazil; jucarusi@unicamp.br (J.C.); lrocha@unicamp.br (L.d.O.R.)
[3] Embrapa Food Technology, Rio de Janeiro 23020-470, RJ, Brazil; renata.tonon@embrapa.br (R.V.T.); amauri.rosenthal@embrapa.br (A.R.)
[4] Department of Food Engineering and Technology, School of Food Engineering, University of Campinas (UNICAMP), Campinas 13083-862, SP, Brazil; rosiane@unicamp.br
* Correspondence: mariahalima@ufrrj.br

Abstract: Essential oils (EOs) have antimicrobial properties, but their low solubility in water and strong flavor pose challenges for direct incorporation into food, as they can negatively impact organoleptic properties. To overcome these issues, strategies such as oil-in-water (O/W) nanoemulsions have been developed to improve EO dispersion and protection while enhancing antimicrobial efficacy. The objective of this study was to create sodium alginate-pink pepper essential oil (PPEO) nanoemulsions using microfluidization. Various formulations were assessed for physicochemical, physical, and antimicrobial properties to evaluate their potential in food applications. The microfluidized emulsions and nanoemulsions had droplet sizes ranging from 160 to 443 nm, polydispersity index (PdI) ranging from 0.273 to 0.638, and zeta potential (ζ) ranging from −45.2 to 66.3 mV. The nanoemulsions exhibited Newtonian behavior and remarkable stability after 20 days of storage. Antimicrobial testing revealed effectiveness against *Staphylococcus aureus* and *Listeria monocytogenes*, with minimum inhibitory concentrations (MIC) of 200 μg/mL for both microorganisms and minimum bactericidal concentrations (MBC) of 800 μg/mL and 400 μg/mL, respectively, proving that encapsulation of PPEO in nanoemulsions significantly increased its antibacterial activity. These results present the possibility of using PPEO nanoemulsions as a more effective natural alternative to synthetic preservatives in food systems.

Keywords: pink pepper essential oil; nanoemulsion; microfluidization; antimicrobial activity; sodium alginate

1. Introduction

The demand for safe and high-quality foods is increasing among consumers. In this regard, the food sector has endeavored to offer safer and more natural alternatives as substitutes for synthetic and chemical antimicrobial agents, given the evidence indicating their adverse effects on human health [1–3]. Thus, natural preservatives derived from diverse plant components, including leaves, roots, fruits, and seeds, have been utilized to fulfill consumer demand for more natural foods [4]. Essential oils (EOs) are metabolites secondary to aromatic vegetal species that act naturally to protect against microbial and insect attacks [5]. They are acknowledged with the designation Generally Recognized as Safe (GRAS) for applications in foodstuffs. Their role as an ingredient is mainly as a substitute for synthetic preservatives due to their antioxidant and antimicrobial activities [6]. Pink pepper (*Schinus terebinthifolius* Raddi) is a plant indigenous to Brazil that is found in South and Central America, Europe, Asia, and Africa. Pink pepper essential oil (PPEO) has been suggested to have antitumor, antioxidant, and antimicrobial activities [7]. Prior research has established that the chemical constituents are primarily limonene, myrcene, α-pinene, and β-pinene, with a minor concentration of sesquiterpenes, including D-germacrene [8]. The Antimicrobial effects of PPEO are thought to be due to the presence of either monoterpene or sesquiterpene hydrocarbons, as well as their oxygenated derivatives, including limonene and α-pinene [9]. The effects of PPEO as an antioxidant and antimicrobial agent have been studied in foods such as cheese and meats [10–12]. However, its nanoencapsulated form has been little studied in terms of its antimicrobial activity against pathogenic microorganisms. Due to their volatility, flavor, and low solubility, the direct addition of these and other EOs becomes a challenge as they may negatively affect the organoleptic characteristics of foods [13].

To mitigate these issues, encapsulating EOs in nanoemulsions may provide a promising alternative, addressing the challenges associated with their direct application in foods and enhancing their effectiveness when used in active coatings [6]. Oil-in-water (O/W) nanoemulsions are defined as colloidal dispersions composed of nanometric oil droplets (d < 200 nm) dispersed in an aqueous phase [14,15]. Nanoemulsions are highly stable with regard to gravitational processes, including coalescence, creaming, flocculation, and sedimentation. Nevertheless, they are susceptible to Ostwald ripening [16,17]. The Ostwald ripening describes the phenomenon whereby larger droplets in a dispersed system, such as an emulsion, grow at the expense of smaller droplets. This occurs due to the difference in solubility between small and large droplets, resulting in a phase separation and a subsequent decrease in system stability [18]. The low water solubility of essential oils facilitates Ostwald ripening, which can occur immediately following emulsion formation. Consequently, the combination of water-insoluble oils, such as sunflower oil, and emulsifiers, like lecithin, with essential oils can reduce surface tension, which impedes droplet growth due to the entropy of mixing effect that opposes Ostwald ripening [19]. The nanometric size of the oil droplets in nanoemulsions results in an increase in surface area, thereby enhancing the efficacy of the added compounds. This, in turn, permits a reduction in concentration (10–50%) while minimizing the potential for sensory alterations [13,20].

The combination of emulsifiers, surfactants, and stabilizers can facilitate the formation of the emulsion, in addition to improving its properties and increasing its stability [21]. Medium or long-chain triglycerides, such as corn and sunflower oil, are defined as "ripening inhibitors", as they form a kinetic barrier, making EOs less soluble in water, reducing the rate at which smaller particles dissolve and reprecipitate into larger particles [22,23]. Soy lecithin is a natural emulsifier derived from the cell membrane of soybeans containing hydrophobic groups and alkyl side chains that make it amphiphilic, reducing the interfacial tension between the phases of the emulsion and improving its stability [24]. Polysorbates, also designated as Tweens, are non-ionic surfactants derived from sorbitan and are commonly utilized in the encapsulation of essential oils. Tweens 20 and 80, though derived from the same compound, possess distinct structures [25]. Tween 20 contains a side chain comprising lauric acid, whereas Tween 80 has oleic acid in its side chain, rendering it less hydrophilic than Tween 20 [26,27]. The structural differences between polysorbates allow them to stabilize nanoemulsions in different ways, which directly impacts the size and stability of the droplets.

Polysaccharide-based nanoemulsions, such as sodium alginate, incorporating EO as an antimicrobial agent, can be utilized to form edible active coatings. Using these coatings can increase shelf life and help prevent spoilage of food products. A number of studies report on various properties associated with polysaccharide-based nanoemulsions developed using microfluidization, as well as their applications in active films and coatings [28–31]. Khanzadi et al. [31] observed inhibition of microbial growth in trout fillets coated with sodium alginate nanoemulsion containing *Zataria multiflora* Boiss essential oil. Kazemeini, Azizian, and Adib [32] demonstrated that nanoemulsion comprising sodium alginate-

Trachyspermum ammi essential oil, when used as a coating, was capable of inhibiting the growth of *Listeria monocytogenes* in turkey fillets even after 12 days of storage at 4 °C.

Despite a few studies focused on the nanoencapsulation of pink pepper essential oil (PPEO) in nanoemulsions, no studies have specifically evaluated the antimicrobial potential of these nanoemulsions against foodborne pathogenic microorganisms, which can be of great interest to the food industry. Additionally, the combination of sunflower oil and lecithin with surfactants may prove an effective strategy for obtaining stable nanoemulsions, preventing destabilization phenomena such as Ostwald ripening and creaming. The aim of this study was to develop sodium alginate-pink pepper essential oil (PPEO) nanoemulsions) using sunflower oil and lecithin as stabilizers and to characterize their physicochemical and rheological properties after microfluidization. Furthermore, the antimicrobial potential of these nanoemulsions was assessed against foodborne pathogenic microorganisms, specifically *Salmonella* sp., *Escherichia coli*, *Listeria monocytogenes*, and *Staphylococcus aureus*, aiming their application as active coatings in animal-based products, such as cheeses.

2. Materials and Methods

2.1. Material

PPEO was acquired from Ferquima (São Paulo, Brazil). The company provided the composition of the essential oil: γ-terpinene (32%), α-phellandrene (20%), δ-limonene (15%), α-pinene (10%), and ρ-cimene (4%). Surfactants Tween 20 and Tween 80 were purchased from Dinamica (São Paulo, Brazil). Sodium alginate was obtained from GastronomyLab® (Brasília, Brazil). Mueller-Hinton Agar and Mueller-Hinton broth were obtained from OXOID® (Basingstoke, UK). Soy lecithin and sunflower oil were kindly donated by Bremil (Arroio do Meio, RS, Brazil) and Cargill Global (Mairinque, SP, Brazil), respectively. Nanoemulsions were produced using ultrapure water from a Milli-Q Plus system. The bacteria employed in the investigation of antimicrobial activity study included *Listeria monocytogenes* ATCC 7644, *Staphylococcus aureus* ATCC 6538, *Escherichia coli* ATCC 11229, and *Salmonella* sp. (LMA 32—Unicamp internal code).

2.2. Nanoemulsion Preparation

The oil-in-water nanoemulsions were developed by keeping the concentration for all components fixed in relation to the amount of water. Based on the study by Salvia-Trujillo et al. [33], the amount of PPEO, sodium alginate, and non-ionic surfactant (Tween 20 or Tween 80) remained constant at 1.0% (*w*/*w*). In some formulations, sunflower oil (0.5 or 0.75% *w*/*w*) or soy lecithin (0.2% *w*/*w*) was also added. These concentrations were chosen from our preliminary experiments to be published elsewhere, based on the work of Ghaderi et al. [34] and Asadinezhad et al. [16] with modifications, where the concentrations of sunflower oil varied from 0.25 to 1% (*w*/*w*) and lecithin from 0.1 to 0.2% (*w*/*w*), and the concentrations selected were based on the smallest droplet size and the best physical stability of the nanoemulsions. The identification and composition of the formulations are presented in Table 1. For identification, the samples were prefixed with CE—(coarse emulsion), and the nanoemulsions were prefixed with M—(microfluidized) before the sample name.

The aqueous phase was produced by solubilizing sodium alginate and an emulsifier in ultrapure water heated to 70 °C under magnetic stirring until complete dissolution, followed by cooling at refrigeration temperature. Non-ionic surfactant was added to the aqueous phase. The emulsion was obtained by combining the aqueous and oil phases (PPEO, PPEO + sunflower oil or lecithin) in an Ultra Turrax® T18 digital (S 18 N-19 G dispersing tool element with a diameter of 19 mm, IKA, Staufen, Germany) for a period of 5 min at 13,000 rpm at 20 °C. Nanoemulsion was produced by subjecting the coarse emulsion to a microfluidizer (LM20-20, Z-Type Interaction Chamber model H210Z, Microfluidics, Westwood, MA, USA) at 15,000 psi for a total of five cycles. Subsequently, the product was cooled in an ice-water bath at the outlet of the microfluidizer interaction chamber, ensuring that the sample temperature remained below 30 °C. Besides, the pH of all the samples was measured after microfluidization using a pH-meter (Metrohm 827 pH Lab, Metrohm, São Paulo, Brazil) with values of approximately 6.8 ± 0.2. Experiments were conducted in triplicate.

Table 1. Composition and identification of the different nanoemulsion formulations.

Sample	PPEO (%*w*/*w*)	Sodium Alginate (%*w*/*w*)	T20 (%*w*/*w*)	T80 (%*w*/*w*)	Sunflower Oil (%*w*/*w*)	SL (%*w*/*w*)
T20	1.0	1.0	1.0	-	-	-
0.5T20	1.0	1.0	1.0	-	0.5	-
0.75T20	1.0	1.0	1.0	-	0.75	-
SLT20	1.0	1.0	1.0	-	-	0.2
T80	1.0	1.0	-	1.0	-	-
0.5T80	1.0	1.0	-	1.0	0.5	-
0.75T80	1.0	1.0	-	1.0	0.75	-
SLT80	1.0	1.0	-	1.0	-	0.2

PPEO: pink pepper essential oil; T20: Tween 20; T80: Tween 80; and SL: soy lecithin.

2.3. *Physicochemical Properties of the Nanoemulsions*

2.3.1. Particle Size, Size Distribution, and Polydispersity Index

Particle size distribution of the emulsions and nanoemulsions was determined using a Zetasizer Nano-ZS (Malvern Instruments Ltd., Worcestershire, UK) at a wavelength of 633 nm and temperature 25 °C, with a backscatter detector of 173°. Prior to analysis, each sample was diluted with ultrapure water at a ratio of 1:99 in order to minimize multiple scattering effects. Size distribution was measured by setting the Mie theory, and a refractive index value of 1.46 was used for the disperse phase (pink pepper essential oil) and 1.33 for the continuous phase (water) [35]. Mean size and polydispersity index were evaluated based on the intensity particle size distribution. D10, D50, and D90 were derived from the volume-weighted distribution, indicating the size below which 10%, 50%, or 90% of all particles are located, respectively.

2.3.2. ζ-Potential

The zeta (ζ) potentials (mV) were measured by electrophoretic light scattering using a Zetasizer Nano-ZS instrument (Malvern Instruments Ltd., Worcestershire, UK). Prior to analysis, the emulsions were diluted in ultrapure water using a dilution factor 1:9 (sample to solvent). Measurements were made on the pH of the samples (6.8 ± 0.2).

2.3.3. Whiteness Index

The chromatic properties of emulsions and nanoemulsions were determined with a Hunter Lab colorimeter model UltraScanVIS (HunterLab, Reston, VA, USA) employing the CIE Lab scale (L*, a*, and b*) at room temperature. The device was calibrated with a standard white plate using the illuminant D65 and 10° observer angle settings, and the whiteness index (WI) value using the following Equation (1) [36]:

$$WI = 100 - \left((100 - L^*)^2 + (a^{*2} + b^{*2})\right)^{0.5} \quad (1)$$

where L* determines the brightness of the color, a* determines the position between green and red, and b* determines the position between blue and yellow.

2.3.4. Rheology Characterization

Emulsions and nanoemulsions flow curves were obtained using an AR1500ex rheometer (TA Instruments, Crawley, UK) equipped with double concentric cylinder geometry comprising an internal cylinder (external radius of 17.53 mm, internal radius of 16.02 mm) and an external cup (external radius of 18.45 mm, inner radius of 15.10 mm). A gap of

500 μm and the shear was adjusted within a range of 0 and 300 s^{-1}. In order to eliminate any potential effects of shear time, the flow curves were generated using up-down-up cycles. Emulsions and nanoemulsions were evaluated just after their preparation. Triplicate measurements were taken at 25 °C. The data obtained from the third flow curve were modeled using the Newtonian and power-law (Equation (2)) fluid models:

$$\sigma = k \cdot (\dot{\gamma})^n \quad (2)$$

In this context, σ represents shear stress (Pa), k denotes the consistency index (Pa·s^n), $\dot{\gamma}$ signifies the shear rate (s^{-1}) and n is the flow behavior (dimensionless) [37].

2.3.5. Emulsion Stability

The stability of the emulsions and nanoemulsions was evaluated in accordance with the methodology proposed by Ushikubo and Cunha [38], with modifications. Twenty-five milliliters of the emulsions were transferred to cylindrical glass tubes, which were then sealed and stored for a period of 30 days at a temperature of 25 °C. The percentage separation of the oil phase was expressed as a separation index (SI), calculated with Equation (3).

$$SI(\%) = \left(\frac{H}{H_0}\right) \times 100 \quad (3)$$

where H is the height of the separated oil phase, and H_0 is the total height of the sample.

The stability of the emulsion was also measured by the average droplet diameter. Measurements were performed every 10 days for 4 weeks at a temperature of 25 °C.

2.3.6. Optical Microscopy

The microstructure of emulsions and nanoemulsions was observed in an optical microscope (Axio Scope.A1, Carl Zeiss, Jena, Germany) with a 100× oil immersion objective lens. The images were captured using the AxioVision Rel. 4.8 software (Carl Zeiss, Germany). The microstructure of emulsion and nanoemulsions was observed after preparation.

2.3.7. Oil Retention

The total oil retained in emulsions and nanoemulsions was quantified via hydrodistillation in a Clevenger apparatus in triplicate, in accordance with the methodology delineated by Garcia, Tonon, and Hubinger [39], with modifications. Firstly, 100 mL of the sample was transferred to a 500 mL round-bottomed flask. The distillation process was conducted for a period of 60 min, and the distilled oil volume was directly read directly into the Clevenger apparatus. The percentage of oil retention was calculated according to Equation (4):

$$\text{Oil retention}(\%) = \left(\frac{V \times \rho}{m}\right) \times 100 \quad (4)$$

The volume of oil, denoted V, is that collected following distillation process. The term ρ represents the density of pink pepper essential oil or the mixture (PPEO and sunflower oil), while m denotes the mass of essential oil added to the sample.

2.4. Antimicrobial Activity

2.4.1. Agar Disk Diffusion

Firstly, the antimicrobial activity of non-encapsulated PPEO (control) and nanoemulsions was determined via the disk diffusion technique, as previously described by CLSI [40]. Additionally, solutions of Tween 20, Tween 80, and sodium alginate at a concentration of 1% (*w/w*) were also tested. The bacterial cultures were suspended in peptone water (0.1% *w/v*) to achieve a concentration of 10^8 CFU/mL, which is equivalent to a 0.5 Mc Farland standard. A volume of 100 μL of the inoculum was distributed with a sterilized swab onto the surface of Petri dishes with Mueller-Hinton Agar (MH-Oxoid®) as the growth medium. After that, the sterilized paper disks positioned on the plate received 10 μL of sample solution (non-encapsulated PPEO or microfluidized emulsion). The Petri dishes underwent incubation at 37 °C (*E. coli*, *Salmonella* sp., and *S. aureus*) and 30 °C (*L. monocytogenes*) for 24 h. After the incubation period, the presence of the inhibition zone was confirmed and quantified using a digital caliper.

2.4.2. Minimum Inhibitory and Bactericidal Concentrations

The broth technique was utilized to establish the minimal inhibitory concentration (MIC) and minimal bactericidal concentration (MBC), according to CLSI methodology [41]. The non-encapsulated PPEO (control) and nanoemulsions were diluted in Mueller-Hinton broth (MH-Oxoid®) supplemented with a Tween 80 solution (0.5% v/v) up to a concentration of 6400 μg/mL. Then, 5 mL of this solution was transferred to a glass tube containing 5 mL of pure Mueller-Hinton broth until 3200 to 25 μg/mL concentrations were obtained. A total of 250 μL of inoculum (10^8 CFU/mL) was added to each tube, which was then incubated at 37 °C (*E. coli*, *Salmonella* sp., and *S. aureus*) and 30 °C (*L. monocytogenes*) for 24 h. Subsequently, the tubes were evaluated, and the MIC was defined as the highest dilution without cell growth (compared to the negative control) through visual analysis.

MBC was obtained through inoculation of 10 μL from each tube that exhibited no growth in the MIC test onto Tryptic Soy Agar (TSA) plates. The lowest concentration that did not demonstrate growth was identified as the minimum bactericidal concentration (MBC) following a 24-h incubation period at 37 °C and 30 °C, respectively.

2.5. Statistical Analysis

The data were subjected to statistical analyses using the R i386 software (version 3.6.1; R Foundation for Statistical Computing, Vienna, Austria) to verify the behavior of the samples through analysis of variance (ANOVA). When a statistical difference was observed ($p < 0.05$), a Tukey test was employed for further investigation into the disparity between the corresponding means.

3. Results and Discussion

3.1. Particle Size Distribution, PdI and ζ-Potential

Microfluidization broke the primary emulsion droplets (micrometer size) into smaller sizes in microfluidized emulsions (mean size lower than 400 nm). This can be clearly observed in the z-average and D50 values (Table 2). There is still a lack of consensus on the size that defines what is classified as a nanoemulsion and an emulsion. A droplet diameter of less than 200 nm is indicative of nanoemulsion [15,18,42]. Considering this classification, only samples M-T20, M-T80, and M-0.5T20 cannot be classified as nanoemulsions, considering the z-average and D50 values. Mechanical stress caused by microfluidization promotes the formation of smaller droplets that are promptly enveloped by surfactant molecules, thereby reducing interfacial tension [43]. A lower mean size indicates a higher stability of nanoemulsions in comparison to coarse emulsions. Additionally, the size significantly affects the physicochemical, rheological, antimicrobial, and sensory properties and stability of nanoemulsion formulations [44].

The determination of particle size by DLS is predicated on the variation in the intensity of the laser beams that are scattered when they encounter the moving particles in a suspension [45]. This information is employed to ascertain the diffusion coefficient, which is used to calculate the hydrodynamic size in accordance with the Stoke-Einstein equation. DLS values are generated in relation to intensity; however, based on the intensity-weighted distribution, it is possible to generate volume-weighted and number-weighted distributions. According to the Rayleigh approximation, the intensity of the scattered light is proportional to the sixth power of the diameter (d^6), being dominated by large particles that contribute disproportionately to the measured intensity, even when in smaller quantities. In contrast, the volume distribution is proportional to the third power of the diameter (d^3), providing a more accurate representation of the material distribution and facilitating visualization of the smaller particles [46,47]. The particle size distribution showed a multimodal distribution in all nanoemulsions (Figure 1), which indicates a polydisperse distribution of particle sizes. The particle size distribution of the coarse emulsions can be found in the Supplementary Material (Figure S1).

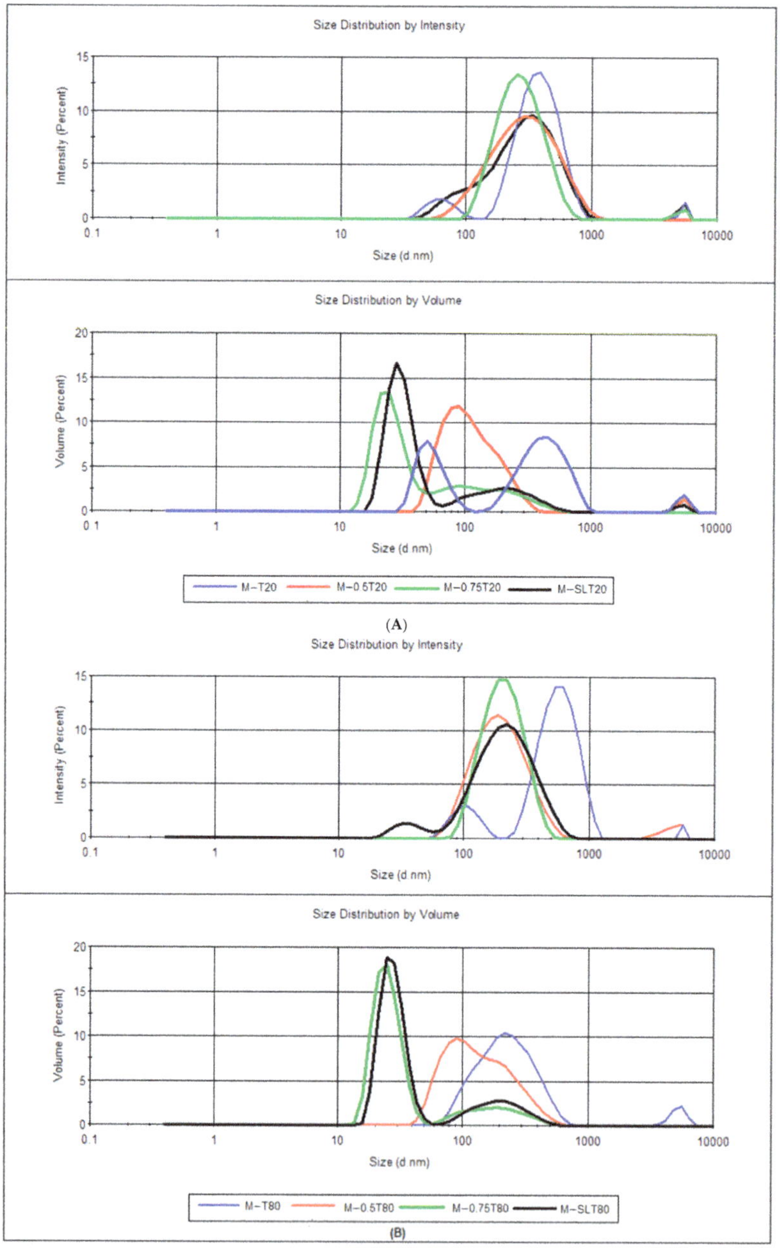

Figure 1. Particle size distribution (nm) of the sodium alginate-PPEO nanoemulsion expressed in intensity and volume stabilized by (**A**) Tween 20 and (**B**) Tween 80. M: microfluidized; T80: Tween 80; T20: Tween 20; 0.5: 0.5: 0.5% (*w*/*w*) sunflower oil; 0.75: 0.75% (*w*/*w*) sunflower oil; SL: soy lecithin.

Table 2. Mean particle size (z-average, D10, D50, and D90) and polydispersity index (PdI) of coarse emulsions (CE) and nanoemulsions (M).

Samples	Z-Average (nm)	Particle Size (nm)			PdI
		D_{90}	D_{50}	D_{10}	
CE-T20	1323 ± 156 [a]	1950 ± 1 [a]	804 ± 6 [b]	59.3 ± 5.5 [b]	0.891 ± 0.168 [a]
CE-T80	998 ± 29 [b]	1570 ± 101 [b]	1010 ± 10 [a]	87.7 ± 3.9 [a]	0.830 ± 0.140 [a]
M-T20	443 ± 10 [c]	468 ± 23 [c]	214 ± 5 [c]	103 ± 2 [c]	0.568 ± 0.091 [b]
M-T80	434 ± 27 [c]	498 ± 16 [c]	229 ± 8 [c]	113 ± 4 [c]	0.638 ± 0.073 [b]
M-0.5T20	221 ± 2 [d]	229 ± 15 [de]	106 ± 8 [e]	61.4 ± 3.6 [e]	0.303 ± 0.005 [c]
M-0.5T80	162 ± 2 [d]	298 ± 6 [d]	126 ± 7 [d]	65.4 ± 5.1 [d]	0.367 ± 0.044 [c]
M-0.75T20	187 ± 3 [d]	198 ± 12 [de]	30.1 ± 4.0 [f]	18.3 ± 2.6 [f]	0.294 ± 0.010 [c]
M-0.75T80	161 ± 4 [d]	175 ± 10 [de]	26.6 ± 3.5 [f]	18.4 ± 4.5 [f]	0.273 ± 0.007 [c]
M-SLT20	183 ± 3 [d]	257 ± 16 [de]	34.3 ± 4.1 [f]	22.7 ± 4.7 [f]	0.411 ± 0.060 [c]
M-SLT80	160 ± 5 [d]	213 ± 9 [de]	29.8 ± 2.6 [f]	21.0 ± 4.3 [f]	0.391 ± 0.039 [c]

Significant differences within each column are indicated by different letters ($p < 0.05$) by the Tukey test. CE: coarse emulsion; M: microfluidized; T80: Tween 80; T20: Tween 20; 0.5: 0.5% (*w*/*w*) sunflower oil; 0.75: 0.75% (*w*/*w*) sunflower oil; SL: soy lecithin. The data are expressed as means values ± standard deviation.

The small peak observed around 5000 nm could be aggregates of biopolymer molecules and/or surfactants, as demonstrated by Artiga-Artigas et al. [48]. When formulating a nanoemulsion with a low concentration of essential oil, the surfactant adsorbed at the droplet interface can lead to interaction between the biopolymer chains. Alternatively, these chains may exhibit a preference for binding to unbound surfactant molecules in preference to those situated on the oil droplet surface [49]. Sodium alginate interacts with the surfactant chains surrounding the oil droplets, resulting in the induction of steric and/or electrostatic repulsion between the droplet interfaces. This interaction effectively prevents destabilization phenomena, including coalescence and gravitational separation [48]. Silva et al. [50] investigated the interfacial tension between the components of an emulsion comprising sacha inchi oil, sodium alginate, and polysorbates (Tween 20 and 80). The times required to reach equilibrium tension were approximately 600 s for both Tween 20 and 80 (1.0 wt%), indicating that the equilibrium tension was achieved due to the steric stabilization capacity of sodium alginate in conjunction with the rapid diffusion of polysorbate molecules. Moreover, in a system comprising sodium alginate and Tween, the surfactant exerts a dominant influence on the surface of the drop, given its superior active surface area in comparison to sodium alginate.

According to Salvia-Trujillo et al. [33], z-average in non-monodispersed distributions can lead to misinterpretation, and for this, D10, D50, and D90 values are also displayed. These values allowed us to observe the effect of the addition of another surfactant (lecithin) and sunflower oil. Sunflower oil acted as an Ostwald ripening inhibitor and soy lecithin as a co-surfactant, resulting in the generation of nanoemulsions with diminished droplet dimensions, which is in agreement with other studies [51,52]. Due to the increased surface area resulting from reduced droplet size and the greater water solubility of EO, nanoemulsions become more susceptible to Ostwald ripening, which occurs when smaller droplets merge from large ones, driven by the molecular propagation of the dispersed phase through the continuous phase [16,53]. Medium or long-chain triglycerides act by reducing the water solubility of EO, stabilizing the nanoemulsion by minimizing the propensity for larger droplets to form from smaller droplets, which is crucial to maintaining the desired properties and effectiveness of the formulation [23,54]. Low molecular mass surfactants, including polysorbate (Tweens) and lecithin, are promptly adsorbed at the oil-water interface during emulsification, leading to a particle size reduction [33]. Lecithin played a crucial role as a co-surfactant in nanoemulsion formulation, contributing to the reduction in droplet size. Its complex composition, primarily consisting of phospholipids such as phosphatidylinositol (PI), phosphatidylethanolamine (PE), and phosphatidylcholine (PC), along with other lipids, contributes to its effectiveness. The lamellar structures formed by these phospholipids, including mono- and bilayers, create a layer of protection around the dispersed droplets in nanoemulsion, preventing coalescence or fusion of the droplets, resulting in a decrease in droplet size and enhanced nanoemulsion stability [52,55].

Polydispersity index values are presented in Table 2. PdI is a parameter that denotes the range of the particle size distribution. Values below 0.3 indicate a narrow size distribution, suggesting enhanced droplet uniformity since less variation in emulsion droplet size may indicate better stability against the phenomena of coalescence or Ostwald ripening [56]. Regarding PdI, the only samples exhibiting indices below 0.3 were those containing 0.75% *w/w* sunflower oil (M-0.75T20 and M-0.75T80), which demonstrated superior kinetic stability outcomes over a 30-day storage period. The observed reduction in polydispersity index values may be attributed to the incorporation of sunflower oil, as the increase in the oil phase may facilitate a more uniform distribution of the surfactant (Tween) within the emulsion. Furthermore, sunflower oil comprises natural emulsifiers, including lecithin, which, due to its amphiphilic nature, provides stability to nanoemulsions by forming a stable layer at the oil-water interface. This contributes to a reduction in interfacial tension and the promotion of the development of smaller, more uniform droplets [57]. Moreover, the presence of phosphate groups facilitates the establishment of repulsive electrostatic interactions, which are essential for preventing coalescence [58]. The remaining samples exhibited higher PdI values, exceeding 0.3, indicative of a significant variability in the size of the emulsion droplets. This often renders them less stable, as the extensive range of droplet size renders them susceptible to destabilization phenomena, resulting in visual or creaming phase separation. As observed by Artiga-Artigas et al. [48], high PdI values may result from the re-coalescence phenomena, which can occur after the microfluidization process.

Figure 2 presents ζ-potential values for sodium alginate-PPEO emulsions. All nanoemulsions displayed negative ζ-potentials, with a range of −35.0 mV to −66.3 mV. When the ζ-potential is less than −30 mV, the considerable electrical charge on the particles indicates that repulsive forces are the dominant force, contributing to the system stability. The negative ζ-potential values observed in every tested sample are attributed to the preferential adsorption of non-ionic surfactants, such as Tweens, onto the surface of the oil in conjunction with polymeric stabilizers, as is the case with sodium alginate [59]. The incorporation of sodium alginate into nanoemulsion formulations is instrumental in the electrostatic stabilization of these systems. This is due to the carboxylate and hydroxyl groups, which can readily dissociate in the hydrophilic phase and be adsorbed onto the oil droplet surface, thereby conferring sufficient negative charge to prevent destabilization phenomena [60]. Additionally, deprotonated alcohols (R-O-) and carboxylic acids ($R'CO^{2}$-) are responsible for an increase in electron density that consequently results in higher negative ζ-potential values [28]. Tweens are vulnerable to oxidative breakdown brought on by exposure to light, oxygen, and heat, resulting in the formation of hydroperoxides, aldehydes, and carboxylic acids. This process introduces new functional groups, such as carboxyl groups, which ionize in the water phase, altering the surface of nanoemulsion droplets and increasing negative charge. Furthermore, the presence of impurities and ionic contaminants can also contribute to negative zeta potential values [61–63].

The nanoemulsion formulated with lecithin exhibited the lowest ζ-potential values, with M-SLT20 and M-SLT80 displaying −65.9 and −66.3 mV, respectively. These values are the result of lecithin phosphate groups and their negative charges that contribute to reducing the charge on the surface of the droplets, thereby increasing the negative ζ-potential [58].

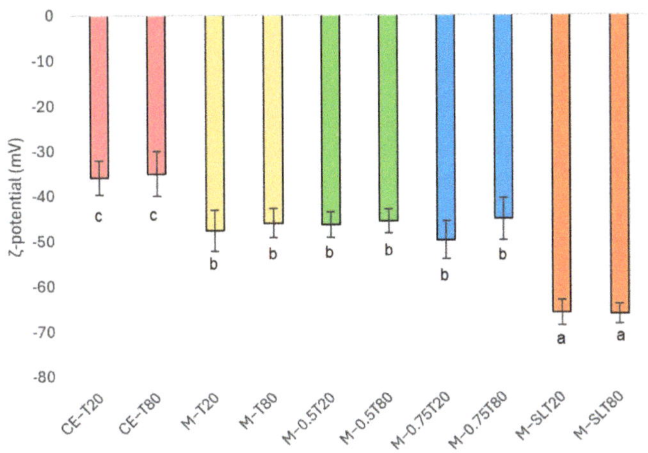

Figure 2. ζ-potential (mV) of coarse emulsions and nanoemulsions Significant differences within each column are indicated by different letters ($p < 0.05$) by the Tukey test. CE: coarse emulsion; M: microfluidized; T80: Tween 80; T20: Tween 20; 0.5: 0.5% (w/w) sunflower oil; 0.75: 0.75% (w/w) sunflower oil; SL: soy lecithin. The data are expressed as means values ± standard deviation.

3.2. Whiteness Index

The Whiteness index (WI) is an important parameter employed to evaluate the optical properties of emulsions, reflecting the degree of light scattering by droplets within these systems, directly impacting their appearance and stability. In general, nanoemulsions are described as slightly cloudy systems, attributed to the weak scattering of light by their small droplets. As the size of the droplets increases, light scattering becomes more intense, making the emulsions appear more opaque [64]. Figure 3 presents the WI values of coarse and microfluidized emulsions. Despite the reduction in droplet size, all microfluidized emulsions and nanoemulsions presented an opaque visual appearance after microfluidization, and the whiteness index of some nanoemulsions was even higher than that of primary emulsions (Figure 4). Regardless of the surfactant, emulsions showed a similar behavior. Nanoemulsions with a lower oil content, M-T20, M-T80, M-SLT20, and M-SLT80, showed the highest WI values, which can be attributed to some factors associated with the composition. Firstly, the high polydispersity in the particle distribution shows a population size on the micrometer scale that may be aggregated alginate particles. Larger particles exhibit more intense light scattering compared to smaller ones, leading to amplified luminosity, opacity, and an enhanced whiteness index measurement for the nanoemulsions [14,33]. Additionally, the inclusion of soy lecithin, which exhibits a pronounced color, alters the overall hue of the nanoemulsion [65] and contributes to an increment in the WI value.

Notably, the addition of sunflower oil, M-0.5T20, M-0.5T80, M-0.75T20, and M-0.75T80, reduced WI values. Smaller average droplet size, lower PdI, and narrower size distribution are some of several factors that may influence the reduction of WI values. In nanoemulsions, smaller droplets have a reduced capacity to scatter light effectively compared to larger droplets, resulting in decreased whiteness [64]. Therefore, the combination of these factors leads to a weakened light scattering effect, reducing the perception of whiteness in samples [66]. Light scattering is most effective when the droplet size is closer to 380–780 nm (the wavelength of visible light). However, nanoemulsions typically comprise droplet sizes well below 200 nm, which results in reduced light scattering and a corresponding decrease in the whiteness index [67]. A lower polydispersity index indicates a more uniform droplet size distribution within the nanoemulsion, which, in turn, reduces the variability in light scattering. Consequently, there is a decrease in diffuse reflection and an associated reduction in the whiteness index. These characteristics contribute to a smoother, more consistent appearance, which is why nanoemulsions are used in products such as creams, gels, or liquid foods. In fact, the specific type and composition of the oil utilized and nanoemulsions formulation have a notable impact on the whiteness index of the emulsions. In their study, Salvia-Trujillo et al. [33] produced nanoemulsions using ten distinct essential oils, and only three nanoemulsions achieved transparency after microfluidization (150 MPa/3 cycles). This finding highlights the significant impact of formulation composition on the optical and visual characteristics exhibited by nanoemulsions.

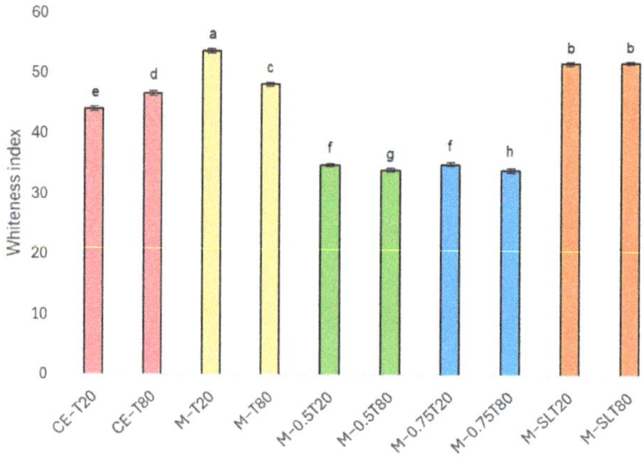

Figure 3. Whiteness index of coarse emulsions and nanoemulsions. Significant differences within each column are indicated by different letters ($p < 0.05$) by the Tukey test. CE: coarse emulsion; M: microfluidized; T80: Tween 80; T20: Tween 20; 0.5: 0.5% (*w*/*w*) sunflower oil; 0.75: 0.75% (*w*/*w*) sunflower oil; SL: soy lecithin. The data are expressed as means values ± standard deviation.

Figure 4. Images of sodium alginate-pink pepper essential oil (PPEO) emulsions and nanoemulsions. CE: coarse emulsion; M: microfluidized; T80: Tween 80; T20: Tween 20; 0.5: 0.5% (*w*/*w*) sunflower oil; 0.75: 0.75% (*w*/*w*) sunflower oil; SL: soy lecithin.

3.3. Rheological Properties

The viscosity and concentration of the dispersed phase play a crucial role in the formation of nanometric emulsion droplets [68]. The sodium alginate-PPEO emulsions flow curves, shown in Figure 5, were well-fitted to the power-law model (Table 3). The coarse emulsions exhibited a flow index (n) of approximately 0.72 and evidenced pseudoplastic fluids behavior. The value of n is less than 1, which indicates pseudoplasticity or shear-thinning behavior, which is defined by a decrease in viscosity as the shear rate increases [69]. After the microfluidization process, all samples, regardless of the formulation, showed low viscosity and Newtonian behavior. The microfluidization process may result in the deterioration of sodium alginate, which could lead to a reduction in the molecular weight of the alginate chains and the viscosity of PPEO nanoemulsions. The degradation may be caused by shear stress, cavitation, and turbulence, which can lead to chemical and physical changes in the material [68,70]. Furthermore, the decrease in droplet size promotes lower interaction between them, also contributing to Newtonian behavior.

The apparent viscosities of the samples after microfluidization were significantly lower than those observed for coarse emulsions (Table 3), which may be attributed to the reduced interaction between the droplets. The apparent viscosity values were obtained at a shear rate of 10.8 s^{-1}, as this allows observation of the influence of forces of attraction and repulsion between dispersed droplets on viscosity, thus facilitating understanding of the rheological behavior of the samples in states of slow flow or rest, which are quite common in the food industry.

The rheological properties of sodium alginate dispersions depend on their possibility of forming intermolecular bonds, such as hydrogen bonds, with the viscosity directly linked to polymer-solvent interactions, in addition to factors such as alginate size, structure, and flexibility [68,71]. For instance, larger alginate molecules or those with more complex structures may exhibit higher viscosities due to increased resistance to flow. Similarly, the flexibility of polymer chains can affect their ability to align and flow in response to shear forces, thereby influencing the viscosity [72]. Considering that the viscosity of coarse emulsions is much higher than that of nanoemulsions that have the same composition, the difference in rheological behavior may be associated with the greater interaction between alginate molecules under conditions of exposure to lower mechanical forces (unbroken alginate clusters) or greater ease of interaction between polysaccharide molecules on larger droplets (greater proximity between droplets).

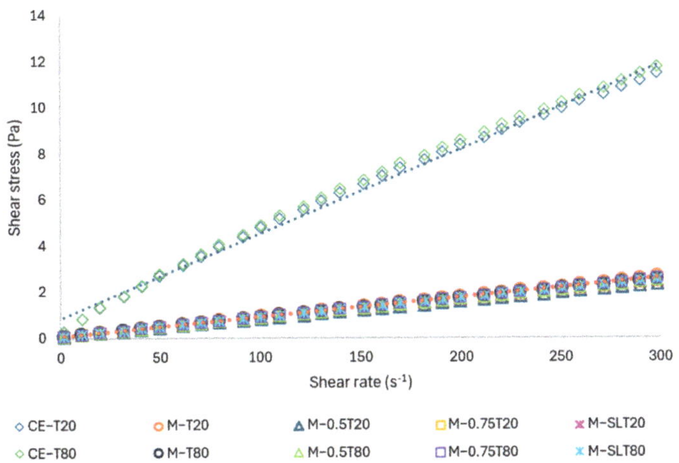

Figure 5. Flow curves of coarse emulsions and sodium alginate-PPEO nanoemulsions as a function of the shear rate. CE: coarse emulsion; M: microfluidized; Tween 80; T20: Tween 20; 0.5: 0.5% (w/w) sunflower oil; 0.75: 0.75% (w/w) sunflower oil; SL: soy lecithin.

Table 3. Viscosity (m Pa·s) and rheological properties of coarse emulsions (CE) and nanoemulsions (M) at 10.8 s^{-1} from the power-law model.

Samples	Viscosity (mPa·s)	n	k (Pa·s^n)	R^2
CE-T20	172.11 ± 3.18 [a]	0.727 ± 0.006 [b]	0.4239 ± 0.0101 [b]	0.998
CE-T80	172.08 ± 5.68 [a]	0.727 ± 0.005 [b]	0.4218 ± 0.0209 [b]	0.998
M-T20	35.33 ± 0.53 [b]	0.986 ± 0.002 [a]	0.0304 ± 0.0011 [a]	1.000
M-T80	33.45 ± 0.41 [b]	0.988 ± 0.001 [a]	0.0286 ± 0.0015 [a]	1.000
M-0.5T20	31.23 ± 0.43 [b]	0.990 ± 0.001 [a]	0.0261 ± 0.0001 [a]	0.999
M-0.5T80	32.94 ± 0.14 [b]	0.988 ± 0.001 [a]	0.0285 ± 0.0001 [a]	1.000
M-0.75T20	32.36 ± 0.35 [b]	0.988 ± 0.001 [a]	0.0274 ± 0.0004 [a]	1.000
M-0.75T80	33.92 ± 0.51 [b]	0.988 ± 0.005 [a]	0.0293 ± 0.0008 [a]	1.000
M-SLT20	34.54 ± 0.20 [b]	0.987 ± 0.001 [a]	0.0293 ± 0.0008 [a]	0.999
M-SLT80	33.84 ± 0.76 [b]	0.986 ± 0.002 [a]	0.0290 ± 0.0009 [a]	0.999

k: consistency coefficient, n: flow behavior index. Significant differences within each column are indicated by different letters ($p < 0.05$) by the Tukey test. CE: coarse emulsion; M: microfluidized; T80: Tween 80; T20: Tween 20; 0.5: 0.5% (w/w) sunflower oil; 0.75: 0.75% (w/w) sunflower oil; SL: soy lecithin. The data are expressed as means values ± standard deviation.

The low viscosity and Newtonian behavior of nanoemulsions represent characteristics that render these systems interesting for the food and pharmaceutical industries, allowing the addition of nanoemulsions to improve the texture and mouthfeel without affecting the rheological properties of foods. In the pharmaceutical field, these characteristics make nanoemulsions optimal delivery systems for controlled-release drugs, in addition to improving the absorption and spreadability of compounds added to cosmetics [73]. An understanding of the rheological properties of emulsions and nanoemulsions allows for the enhancement of their performance and functionality. Furthermore, it facilitates the development of novel, stable, and high-quality foods that align with the demands of the food market and consumers.

3.4. Emulsion Stability

The stability of emulsions and nanoemulsions is an important parameter for the effective encapsulation of bioactive compounds such as essential oils. The formation of an oil layer, gravitational separation, indicates instability. The curves showing the phase separation indices of the samples are shown in Figure 6.

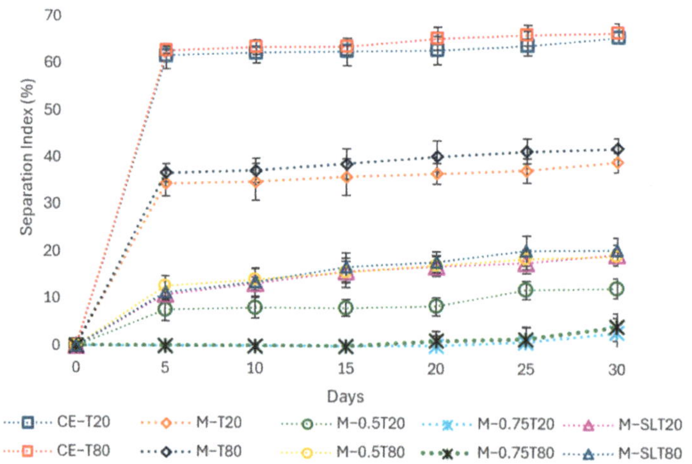

Figure 6. Separation index (SI) of coarse emulsions (CE) and nanoemulsions (M). CE: coarse emulsion; M: microfluidized; T80: Tween 80; T20: Tween 20; 0.5: 0.5% (w/w) sunflower oil; 0.75: 0.75% (w/w) sunflower oil; SL: soy lecithin. The data are expressed as means values ± standard deviation.

At the end of 30 days of storage, despite their formulations, the samples showed phase separation ranging from 3.02 to 62.65% (Figure 7). During the first 20 days of storage, only samples M-0.75T20 and M-0.75T80 had no phase separation. The greater stability of these samples may be related to the addition of sunflower oil combined with their reduced droplet size and PdI values.

Figure 7. Phase separation of sodium alginate-pink pepper essential oil (PPEO) microfluidized emulsions and nanoemulsions after 30 days of storage. M: microfluidized; T80: Tween 80; T20: Tween 20; 0.5: 0.5% (*w*/*w*) sunflower oil; 0.75: 0.75% (*w*/*w*) sunflower oil; SL: soy lecithin.

The coarse emulsions, CE-T20 and CE-T80, and the microfluidized emulsions, M-T20 and M-T80, showed higher phase separation rates, ranging from 34 to 62%. The high SI values indicate the low stability of the samples, which may be correlated to their larger droplet sizes and PdI. These samples showed instability right after microfluidization, which is the result of the droplets migrating to the top of the emulsion [18]. According to McClements [53], phase separation in emulsions is the result of an increase in droplet size caused by the phenomena of flocculation and coalescence. Guerra-Rosas et al. [74] observed different separation rates according to the essential oil used in the nanoemulsion. While the nanoemulsions with lemongrass and mandarin essential oil showed no phase separation over 56 days of storage, the samples with thyme and oregano essential oil showed separation rates of 6.5 and 17.08%, respectively, indicating that the composition plays a fundamental role in the stability of emulsions and nanoemulsions.

The physical stability of nanoemulsions is essential for their commercial use, ensuring that there are no changes in droplet size or phase separation [75]. Figure 8 presents the evolution over time of the mean droplet diameter in emulsions and nanoemulsions stored at 25 °C for 30 days. After microfluidization, all samples except M-0.75T20 and M-0.75T80 showed a significant increase in droplet size. In particular, samples M-T20 and M-T80 showed a significant increase in droplet size and phase separation in the first 24 h, with a stability index (SI) of about 40% after 30 days. Samples containing soy lecithin and 0.5% sunflower oil showed similar behavior, but the SI was reduced to about 20% at the end of the storage period. The destabilization of nanoemulsions is attributed to the Ostwald ripening phenomenon, where the droplet size increases due to the diffusion of smaller oil droplets into larger ones through the aqueous phase. The addition of Ostwald ripening inhibitors, such as sunflower oil and soybean lecithin, can be an effective strategy to improve the stability of nanoemulsions [76]. In this study, the addition of 0.75% sunflower oil conferred greater stability to samples M-0.75T20 and M-0.75T80, which showed only a slight increase in droplet size over 30 days and remained without phase separation for up to 20 days. A similar result was described by Wan et al. [77], who observed no significant changes in the droplet size of clove oil nanoemulsions with corn oil (75% *w*/*w*) over 30 days of storage at 4 and 25 °C. The stability of nanoemulsions is closely related to their small droplet size and PdI values, indicating that these nanoemulsions are resistant to droplet growth caused by coalescence, flocculation, and Ostwald ripening [78]. Ensuring this stability is critical to maintaining the integrity of the active ingredients and ensuring the efficacy of the nanoemulsions, resulting in superior-quality products.

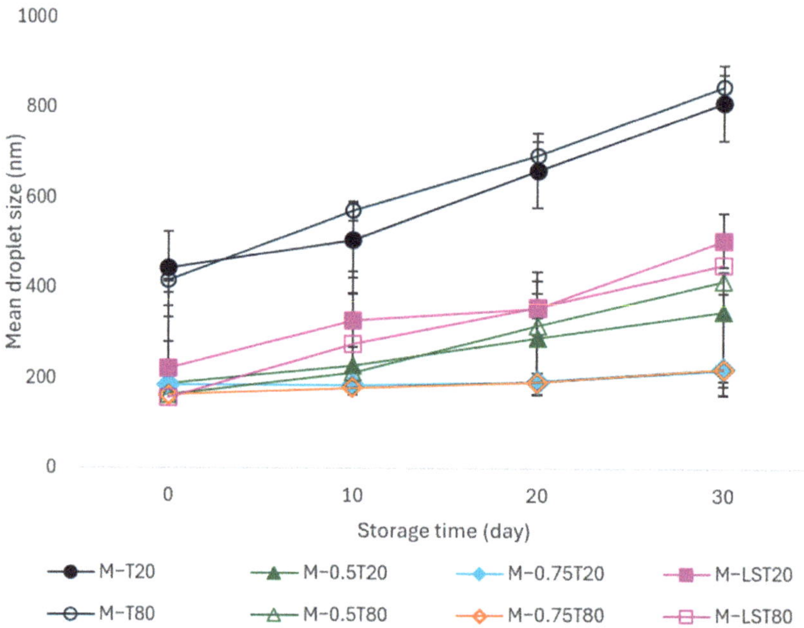

Figure 8. Storage stability of sodium alginate-pink pepper essential oil (PPEO) microfluidized emulsions and nanoemulsions at 25 °C. M: microfluidized; T80: Tween 80; T20: Tween 20; 0.5: 0.5% (w/w) sunflower oil; 0.75: 0.75% (w/w) sunflower oil; SL: soy lecithin. The data are expressed as means values ± standard deviation.

3.5. Microscopy

The optical microscopy images of the emulsions and nanoemulsions are shown in Figure 9. The coarse emulsions are shown in images A and B, and it is possible to see the larger droplet size and the aggregation of these droplets. Larger droplets and aggregation lead to the instability of the emulsions, which can induce creaming and phase separation. Sodium alginate-PPEO nanoemulsions can be observed in images B to J, and it is possible to observe a reduction in nanoemulsion droplet size after microfluidization. The samples M-T20 and M-T80 exhibited a larger droplet size distribution and lower pink pepper essential oil retention, and their micrographs (C and D) evidence a higher occurrence of aggregates. The limited accuracy of the images obtained is due to the tiny droplets since accurate observation using optical microscopy techniques diminishes notably when their size falls below 500 nanometers [79].

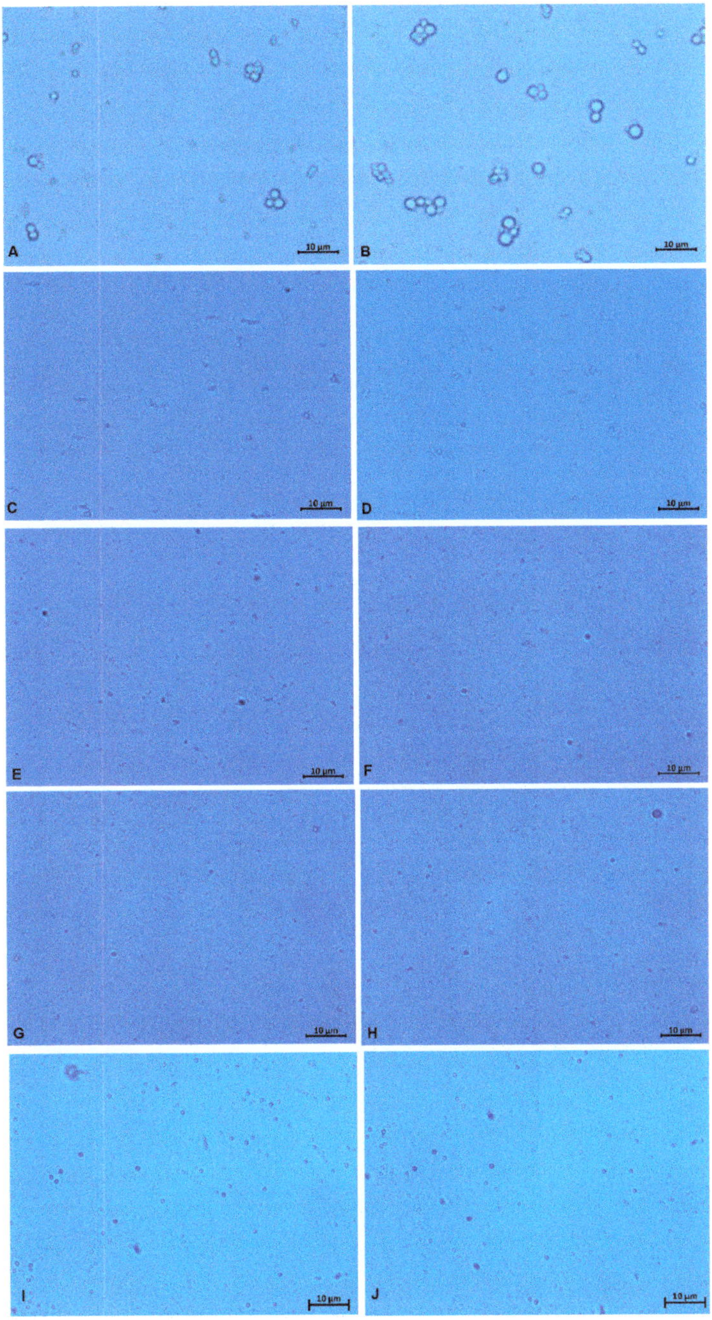

Figure 9. Optical microscopic images of coarse emulsion and sodium alginate-PPEO nanoemulsions. (**A**): CE-T20; (**B**): CE-T80; (**C**): M-T20; (**D**): M-T80; (**E**): M-0.5T20; (**F**): M-0.5T80; (**G**): M-0.75T20; (**H**): M-0.75T80; (**I**): M-SLT20, and (**J**): M-SLT80. CE: coarse emulsion; M: microfluidized; T80: Tween 80; T20: Tween 20; 0.5: 0.5% sunflower oil; 0.75: 0.75% sunflower oil; SL: soy lecithin.

3.6. Oil Retention

Figure 10 shows that the oil content decreased after the microfluidization process. Pink pepper oil-loaded nanoemulsions exhibited oil retention ranging between 68.48% and 82.16%, differing significantly ($p < 0.05$) from the coarse emulsion, which showed oil retention (OR) values greater than 90%. Nanoemulsions M-T80 and M-T20 exhibited the lowest OR values (around 70%). Oil retention is closely associated with droplet size, as higher OR values are associated with smaller droplet sizes [20]. Furthermore, when an essential oil has high water solubility, it displays a tendency to partition predominantly into the aqueous phase. This phenomenon results in a reduction in encapsulation efficiency because a portion of the essential oil may be lost to the aqueous phase and may not be effectively nano-encapsulated within the oil droplets of the nanoemulsion [24]. However, considering the above, coarse emulsions should have lower OR compared to M-T80 and M-T20 nanoemulsions. Although microfluidization produces stable and uniform nanoemulsions, it can also impact the oil retention within these systems. The high shear rates and elevated temperatures during the process may compromise droplet structure and increase diffusion rates, leading to the loss of encapsulated oil [80]. By understanding these challenges and optimizing process conditions, such as controlling shear rates and temperature, along with proper formulation, essential oil retention can be improved, thereby enhancing the effectiveness of nanoemulsions in various applications.

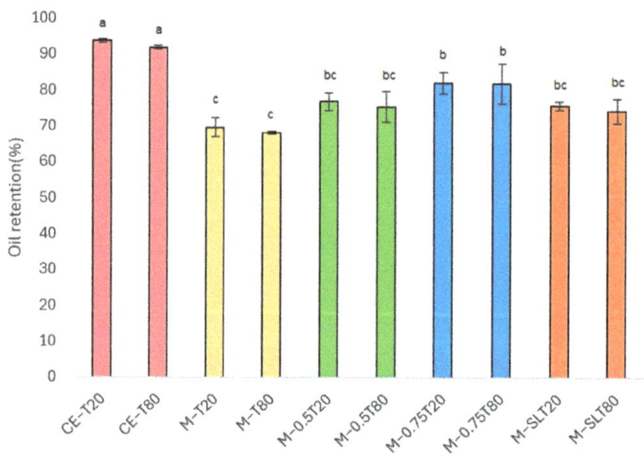

Figure 10. Oil retention of coarse emulsions (CE) and nanoemulsions (M). Significant differences within each column are indicated by different letters ($p < 0.05$) by the Tukey test. CE: coarse emulsion; M: microfluidized; T80: Tween 80; T20: Tween 20; 0.5: 0.5% (w/w) sunflower oil; 0.75: 0.75% (w/w) sunflower oil; SL: soy lecithin. The data are expressed as means values ± standard deviation.

The addition of soy lecithin to the composition of the emulsions (M-SLT20 and M-SLT80) increased the OR to values above 75%. This protective effect is due to the ability of lecithin to form hydrogen bonds between its phosphate groups and the hydroxyl groups of the essential oils [55]. Lecithin aliphatic phospholipid chains envelop the essential oil, effectively trapping it within the nanoemulsion structure [58,81]. Nanoemulsions containing 0.75% (w/w) of sunflower oil showed even higher encapsulation levels, ranging from 81.91% to 82.16%. Dávila-Rodrigues et al. [20] observed 82% retention of oregano essential oil after producing a nanoemulsion by high-frequency ultrasound. However, the amount of *Coriandrum sativum* essential oil in the Das et al. [82] nanoemulsion produced by ultrasonication ranged from 26 to 76%. The difference in the oil content of the nanoemulsions can be directly related to losses in the process since essential oils are volatile, and high-energy processes generate heat that can induce the volatilization of these oils.

According to Niu et al. [83], nanoemulsions with encapsulation efficiency above 80% are considered highly stable. The incorporation of essential oils into vegetable oil, such as sunflower oil, offers protection against microfluidization heating, minimizing its degradation and volatilization [84]. The lipophilic nature of sunflower oil acts as a protective matrix, increasing solubility and reducing exposure to aqueous phases where temperature fluctuations occur [85]. The protection mechanisms provided by both sunflower oil and lecithin contribute to the stability and integrity of the encapsulated essential oils in the nanoemulsion system.

The hydro distillation process, commonly used to extract essential oils from plant materials, can also be applied to extract oil from emulsified systems. This process involves three main physicochemical processes: hydrodiffusion, hydrolysis, and thermal decomposition, which can significantly affect the retention of oil in the system [86]. The oil retention rate is closely linked to the stability of the emulsions and the conditions of the process. As heating is applied, the stability of the emulsions may decrease, leading to coalescence and phase separation, which releases oil from the structure, ultimately reducing oil retention. Additionally, volatilization and thermal degradation of oil components can occur, leading to evaporation losses and alterations in the chemical composition of the oil [22,87,88]. Therefore, the greater the stability of the emulsion, the higher the amount of oil retained in the system. The retention of oil in the nanoemulsion is directly influenced by the interaction between the components of the formulation, such as surfactant, oil, polymer, and fabrication method, which will directly affect the stability and release properties of the nanoemulsion [22]. Therefore, the right choice of manufacturing process and formulation can guarantee the development of a nanoemulsion with the right properties for its purpose.

3.7. Antimicrobial Activity

The antimicrobial activity of pure PPEO (non-encapsulated) and its nanoemulsions was investigated against gram-positive bacteria, *Staphylococcus aureus*, *Listeria monocytogenes*, and gram-negative bacteria, and *Salmonella* sp. and *Escherichia coli*. Disk diffusion test results are shown in Table 4. Sodium alginate, Tween 20, and Tween 80 showed no antimicrobial activity against the microorganisms evaluated. PPEO and two nanoemulsions, M-0.75T80 and M-0.75T20, showed antimicrobial action against the gram-positive bacteria *L. monocytogenes* and *S. aureus*, while *E. coli* and *Salmonella* sp. were not sensitive (Figure 11). Pure PPEO was previously tested by Dannenberg et al. [9] against *L. monocytogenes*, *S. aureus*, *E. coli*, and *S.* Typhimurium and exhibited antimicrobial activity exclusively against the gram-positive bacteria *L. monocytogenes* and *S. aureus* with inhibition zones of 24.30 and 23.09 mm, respectively (similar to our results). The heightened susceptibility of gram-positive bacteria to PPEO is attributed to the facile penetration of nanometer-scale hydrophobic molecules through the peptidoglycan layer that constitutes the structure of gram-positive bacteria. However, the double layer of lipopolysaccharide found in gram-negative bacteria acts as a defense mechanism to the passage of antimicrobial molecules [89,90]. However, Salvia-Trujillo et al. [33] documented the inhibition of *E. coli* with sodium alginate nanoemulsion loaded with lemongrass and clove oil produced in a microfluidization process. The results demonstrated that the antimicrobial effect is predominantly influenced by the active compounds of essential oils rather than their concentration [91].

MIC and MBC values of samples that exhibited antimicrobial activity in the disk diffusion test are presented in Table 5. The differences in the inhibition halos and MIC and MBC values were greater than those exhibited by PPEO (non-encapsulated). The antimicrobial activity of nanoemulsions is associated with several factors, including oil retention (approximately 82% in emulsions with 0.75% sunflower oil), the small average droplet size of nanoemulsions (160.7–186.6 nm), and their stability. The reduction in droplet size of nanoemulsions enhances the functionality of EOs due to the increase in the surface area of the droplets [74]. Some authors have reported that encapsulation of essential oil (EO) can increase their antimicrobial activity [34,92,93]; however, in other cases, this activity may be reduced [75,94], indicating that the antimicrobial efficacy of nanoemulsions-based EO can be significantly affected due to the influence of EO components, formulation, and the nanoemulsion droplet size [90].

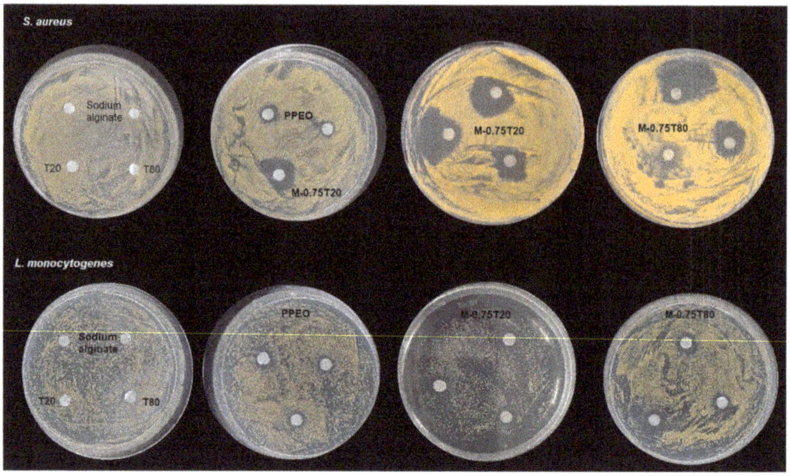

Figure 11. Antimicrobial activity of non-encapsulated PPEO and its encapsulated forms, as well as Tween 20, Tween 80, and sodium alginate solution against *S. aureus* and *L. monocytogenes*. PPEO: pink pepper essential oil (non-encapsulated); M: microfluidized; T80: Tween 80; T20: Tween 20; 0.75: 0.75% (*w*/*w*) sunflower oil.

Table 4. Diameters of inhibition zone (mm) demonstrated by non-encapsulated pink pepper essential oil and sodium alginate-PPEO nanoemulsions against both Gram-positive and Gram-negative bacterial strains.

	Inhibition Zone (mm)			
	S. aureus	*L. monocytogenes*	*E. coli*	*Salmonella* sp.
PPEO	14.13 ± 0.19 [a]	12.36 ± 1.90 [a]	ND	ND
M-T20	ND	ND	ND	ND
M-T80	ND	ND	ND	ND
M-0.5T20	ND	ND	ND	ND
M-0.5T80	ND	ND	ND	ND
M-0.75T20	26.18 ± 3.87 [b]	17.13 ± 0.73 [ab]	ND	ND
M-0.75T80	26.57 ± 6.99 [b]	23.28 ± 4.50 [b]	ND	ND
M-SLT20	ND	ND	ND	ND
M-SLT80	ND	ND	ND	ND

Significant differences within each column are indicated by different letters ($p < 0.05$); PPEO: pink pepper essential oil (non-encapsulated); M: microfluidized; T80: Tween 80; T20: Tween 20; 0.5: 0.5% (*w*/*w*) sunflower oil; 0.75: 0.75% (*w*/*w*) sunflower oil; SL: soy lecithin. The data are expressed as means values ± standard deviation; ND = Not Detected.

Table 5. Values for the minimal inhibitory concentration (MIC) and minimal bactericidal concentration (MBC) values (μg/mL) pink pepper essential oil and sodium alginate-PPEO nanoemulsions.

	L. monocytogenes		*S. aureus*	
	MIC (μg/mL)	MBC (μg/mL)	MIC (μg/mL)	MBC (μg/mL)
PPEO	200	800	400	1600
M-0.75T20	200	400	200	800
M-0.75T80	200	400	200	800

PPEO: pink pepper essential oil (non-encapsulated); M: microfluidized; Tween 80; T20: Tween 20; 0.75: 0.75% (*w*/*w*) sunflower oil.

4. Conclusions

Different sodium alginate-PPEO nanoemulsions were produced via microfluidization, and the findings demonstrated a notable influence of this process on their physicochemical, rheological, and antimicrobial properties. Sodium alginate-PPEO nanoemulsions showed small droplet size, multimodal particle size distribution, and high (absolute) and negative zeta potential values. Among the eight formulations outlined in the current study, M-0.75T80 and M-0.75T20 showed remarkable performance concerning the evaluated properties. The incorporation of sunflower oil into the formulation improved the stability of nanoemulsions during long storage and led to an increase in the PPEO retention within the nanoemulsion. The sodium alginate-PPEO nanoemulsion demonstrated remarkable antimicrobial activity against *L. monocytogenes* and *S. aureus*, which was associated with the formulation, oil retention, and droplet size.

Nevertheless, the production of nanoemulsions via microfluidization represents a significant challenge, primarily due to the elevated costs associated with the requisite equipment and the necessity for precise processing conditions to guarantee the stability and uniformity of emulsions on a large scale. These challenges need to be addressed to facilitate the commercial application of these nanoemulsions. Even so, the results obtained are noteworthy due to the promising advantages of using nanoemulsions as delivery systems, with potential for applications in food packaging. Future investigation is required to ascertain the efficacy of these nanoemulsions as coatings in enhancing food quality and prolonging the shelf life of foods. This will be crucial to elucidate their real potential for incorporation into food products.

Supplementary Materials: The following resources are available for online consultation at https://www.mdpi.com/article/10.3390/foods13193090/s1, Figure S1: Intensity and volume-particle size distribution (nm) of the alginate-PPEO coarse emulsion (CE) stabilized by Tween 20 and Tween 80.

Author Contributions: Conceptualization, M.A.L., L.d.O.R., R.V.T., R.L.C. and A.R.; methodology, M.A.L., J.C., L.d.O.R. and R.L.C.; validation, M.A.L. and J.C.; formal analysis, M.A.L. and J.C.; investigation, M.A.L., J.C. and R.L.C.; resources, L.d.O.R., R.V.T., R.L.C. and A.R.; data curation, M.A.L.; writing—original draft preparation, M.A.L.; writing—review and editing, L.d.O.R., R.V.T., R.L.C. and A.R.; visualization, M.A.L.; supervision, R.V.T., R.L.C. and A.R.; project administration, R.V.T., R.L.C. and A.R.; funding acquisition, R.V.T., R.L.C. and A.R. All authors have read and agreed to the published version of the manuscript.

Funding: This work was supported by Fundação Carlos Chagas Filho de Amparo à Pesquisa do Estado do Rio de Janeiro—FAPERJ (grant number E-26/202.710-2019), Fundação de Amparo à Pesquisa do Estado de São Paulo—FAPESP (grant number 2019/27354-3), and Conselho Nacional de Desenvolvimento Científico e Tecnológico—CNPq (grant number 311529/2021-6, 141604/2020-4 and 465335/2014-4).

Institutional Review Board Statement: Not applicable.

Informed Consent Statement: Not applicable.

Data Availability Statement: The data provided in this study are available on request from the corresponding author. The data are not publicly accessible due to privacy considerations.

Conflicts of Interest: The authors declare no conflict of interest.

References

1. Al-Zoreky, N.S.; Al-Taher, A.Y. Antibacterial Activity of Spathe from *Phoenix dactylifera* L. against Some Food-Borne Pathogens. *Ind. Crops Prod.* **2015**, *65*, 241–246. [CrossRef]
2. Hsouna, A.B.; Trigui, M.; Mansour, R.B.; Jarraya, R.M.; Damak, M.; Jaoua, S. Chemical Composition, Cytotoxicity Effect and Antimicrobial Activity of *Ceratonia siliqua* Essential Oil with Preservative Effects against *Listeria* Inoculated in Minced Beef Meat. *Int. J. Food Microbiol.* **2011**, *148*, 66–72. [CrossRef] [PubMed]
3. Kumar, N.; Singh, A.; Sharma, D.K.; Kishore, K. Toxicity of Food Additives. In *Food Safety and Human Health*; Academic Press: Cambridge, MA, USA, 2019; pp. 67–98. [CrossRef]

4. Hyldgaard, M.; Mygind, T.; Meyer, R.L. Essential Oils in Food Preservation: Mode of Action, Synergies, and Interactions with Food Matrix Components. *Front. Microbiol.* **2012**, *3*, 12. [CrossRef] [PubMed]
5. El Asbahani, A.; Miladi, K.; Badri, W.; Sala, M.; Addi, E.H.A.; Casabianca, H.; El Mousadik, A.; Hartmann, D.; Jilale, A.; Renaud, F.N.R.; et al. Essential Oils: From Extraction to Encapsulation. *Int. J. Pharm.* **2015**, *483*, 220–243. [CrossRef] [PubMed]
6. Almasi, H.; Azizi, S.; Amjadi, S. Development and Characterization of Pectin Films Activated by Nanoemulsion and Pickering Emulsion Stabilized Marjoram (*Origanum majorana* L.) Essential Oil. *Food Hydrocoll.* **2020**, *99*, 105338. [CrossRef]
7. Andrade, K.S.; Poncelet, D.; Ferreira, S.R.S. Sustainable Extraction and Encapsulation of Pink Pepper Oil. *J. Food Eng.* **2017**, *204*, 38–45. [CrossRef]
8. dos Santos Cavalcanti, A.; de Souza Alves, M.; da Silva, L.C.P.; dos Santos Patrocínio, D.; Sanches, M.N.; de Almeida Chaves, D.S.; de Souza, M.A.A. Volatiles Composition and Extraction Kinetics from Schinus Terebinthifolius and Schinus Molle Leaves and Fruit. *Rev. Bras. Farmacogn.* **2015**, *25*, 356–362. [CrossRef]
9. da Silva Dannenberg, G.; Funck, G.D.; da Silva, W.P.; Fiorentini, Â.M. Essential Oil from Pink Pepper (*Schinus terebinthifolius* Raddi): Chemical Composition, Antibacterial Activity and Mechanism of Action. *Food Control* **2019**, *95*, 115–120. [CrossRef]
10. de Souza Cordeiro, C.; Forato, L.A.; Bernardes Filho, R.; Nassu, R.T. Utilization of Zein-Based Coatings Containing Vegetable Oils on the Physicochemical Characteristics of Vacuum-Packaged Lamb Meat. *Ciênc. Rural* **2019**, *49*, e20190310. [CrossRef]
11. da Silva Dannenberg, G.; Funck, G.D.; dos Santos Cruxen, C.E.; de Lima Marques, J.; da Silva, W.P.; Fiorentini, Â.M. Essential Oil from Pink Pepper as an Antimicrobial Component in Cellulose Acetate Film: Potential for Application as Active Packaging for Sliced Cheese. *LWT-Food Sci. Technol.* **2017**, *81*, 314–318. [CrossRef]
12. Menegali, B.S.; Selani, M.M.; Saldaña, E.; Patinho, I.; Diniz, J.P.; Melo, P.S.; de Jesus Pimentel Filho, N.; Contreras-Castillo, C.J. Pink Pepper Extract as a Natural Antioxidant in Chicken Burger: Effects on Oxidative Stability and Dynamic Sensory Profile Using Temporal Dominance of Sensations. *LWT-Food Sci. Technol.* **2020**, *121*, 108986. [CrossRef]
13. McClements, D.J. Edible Nanoemulsions: Fabrication, Properties, and Functional Performance. *Soft Matter* **2011**, *7*, 2297–2316. [CrossRef]
14. McClements, D.J.; Rao, J. Food-Grade Nanoemulsions: Formulation, Fabrication, Properties, Performance, Biological Fate, and Potential Toxicity. *Crit. Rev. Food Sci. Nutr.* **2011**, *51*, 285–330. [CrossRef] [PubMed]
15. McClements, D.J. Nanoemulsions versus Microemulsions: Terminology, Differences, and Similarities. *Soft Matter* **2012**, *8*, 1719–1729. [CrossRef]
16. Asadinezhad, S.; Khodaiyan, F.; Salami, M.; Hosseini, H.; Ghanbarzadeh, B. Effect of Different Parameters on Orange Oil Nanoemulsion Particle Size: Combination of Low Energy and High Energy Methods. *J. Food Meas. Charact.* **2019**, *13*, 2501–2509. [CrossRef]
17. Guerra-Rosas, M.I.; Morales-Castro, J.; Cubero-Márquez, M.A.; Salvia-Trujillo, L.; Martín-Belloso, O. Antimicrobial Activity of Nanoemulsions Containing Essential Oils and High Methoxyl Pectin during Long-Term Storage. *Food Control* **2017**, *77*, 131–138. [CrossRef]
18. Tadros, T.; Izquierdo, P.; Esquena, J.; Solans, C. Formation and Stability of Nano-Emulsions. *Adv. Colloid Interface Sci.* **2004**, *108–109*, 303–318. [CrossRef]
19. Chang, Y.; McClements, D.J. Optimization of Orange Oil Nanoemulsion Formation by Isothermal Low-Energy Methods: Influence of the Oil Phase, Surfactant, and Temperature. *J. Agric. Food Chem.* **2014**, *62*, 2306–2312. [CrossRef]
20. Dávila-Rodríguez, M.; López-Malo, A.; Palou, E.; Ramírez-Corona, N.; Jiménez-Munguía, M.T. Antimicrobial Activity of Nanoemulsions of Cinnamon, Rosemary, and Oregano Essential Oils on Fresh Celery. *LWT-Food Sci. Technol.* **2019**, *112*, 108247. [CrossRef]
21. McClements, D.J.; Jafari, S.M. Improving Emulsion Formation, Stability and Performance Using Mixed Emulsifiers: A Review. *Adv. Colloid Interface Sci.* **2018**, *251*, 55–79. [CrossRef]
22. Pavoni, L.; Perinelli, D.R.; Bonacucina, G.; Cespi, M.; Palmieri, G.F. An Overview of Micro- and Nanoemulsions as Vehicles for Essential Oils: Formulation, Preparation and Stability. *Nanomaterials* **2020**, *10*, 135. [CrossRef] [PubMed]
23. Donsì, F.; Cuomo, A.; Marchese, E.; Ferrari, G. Infusion of Essential Oils for Food Stabilization: Unraveling the Role of Nanoemulsion-Based Delivery Systems on Mass Transfer and Antimicrobial Activity. *Innov. Food Sci. Emerg. Technol.* **2014**, *22*, 212–220. [CrossRef]
24. Zhao, S.; Wang, Z.; Wang, X.; Kong, B.; Liu, Q.; Xia, X.; Liu, H. Characterization of Nanoemulsions Stabilized with Different Emulsifiers and Their Encapsulation Efficiency for Oregano Essential Oil: Tween 80, Soybean Protein Isolate, Tea Saponin, and Soy Lecithin. *Foods* **2023**, *12*, 3183. [CrossRef] [PubMed]
25. Guo, L.; Fang, Y.; Liang, X.; Xu, Y.; Chen, J.; Li, Y.; Fang, S.; Meng, Y. Influence of Polysorbates (Tweens) on Structural and Antimicrobial Properties for Microemulsions. *Int. J. Pharm.* **2020**, *590*, 119939. [CrossRef]
26. Pavoni, L.; Perinelli, D.R.; Ciacciarelli, A.; Quassinti, L.; Bramucci, M.; Miano, A.; Casettari, L.; Cespi, M.; Bonacucina, G.; Palmieri, G.F. Properties and Stability of Nanoemulsions: How Relevant Is the Type of Surfactant? *J. Drug Deliv. Sci. Technol.* **2020**, *58*, 101772. [CrossRef]
27. Brandner, J.D. The Composition of NF-Defined Emulsifiers: Sorbitan Monolaurate, Monopalmitate, Monostearate, Monooleate, Polysorbate 20, Polysorbate 40, Polysorbate 60, and Polysorbate 80. *Drug Dev. Ind. Pharm.* **1998**, *24*, 1049–1054. [CrossRef]
28. Artiga-Artigas, M.; Acevedo-Fani, A.; Martín-Belloso, O. Improving the Shelf Life of Low-Fat Cut Cheese Using Nanoemulsion-Based Edible Coatings Containing Oregano Essential Oil and Mandarin Fiber. *Food Control* **2017**, *76*, 1–12. [CrossRef]

29. Chen, H.; Zhong, Q. Physical and Antimicrobial Properties of Self-Emulsified Nanoemulsions Containing Three Synergistic Essential Oils. *Int. J. Food Microbiol.* **2022**, *365*, 109557. [CrossRef]
30. Huang, M.; Wang, H.; Xu, X.; Lu, X.; Song, X.; Zhou, G. Effects of Nanoemulsion-Based Edible Coatings with Composite Mixture of Rosemary Extract and ε-Poly-l-Lysine on the Shelf Life of Ready-to-Eat Carbonado Chicken. *Food Hydrocoll.* **2020**, *102*, 105576. [CrossRef]
31. Khanzadi, S.; Keykhosravy, K.; Hashemi, M.; Azizzadeh, M. Alginate Coarse/Nanoemulsions Containing *Zataria Multiflora* Boiss Essential Oil as Edible Coatings and the Impact on Microbial Quality of Trout Fillet. *Aquac. Res.* **2020**, *51*, 873–881. [CrossRef]
32. Kazemeini, H.; Azizian, A.; Adib, H. Inhibition of Listeria Monocytogenes Growth in Turkey Fillets by Alginate Edible Coating with Trachyspermum Ammi Essential Oil Nano-Emulsion. *Int. J. Food Microbiol.* **2021**, *344*, 109104. [CrossRef]
33. Salvia-Trujillo, L.; Rojas-Graü, A.; Soliva-Fortuny, R.; Martín-Belloso, O. Physicochemical Characterization and Antimicrobial Activity of Food-Grade Emulsions and Nanoemulsions Incorporating Essential Oils. *Food Hydrocoll.* **2015**, *43*, 547–556. [CrossRef]
34. Ghaderi, L.; Moghimi, R.; Aliahmadi, A.; McClements, D.J.; Rafati, H. Development of Antimicrobial Nanoemulsion-Based Delivery Systems against Selected Pathogenic Bacteria Using a Thymol-Rich *Thymus Daenensis* Essential Oil. *J. Appl. Microbiol.* **2017**, *123*, 832–840. [CrossRef]
35. Okuro, P.K.; Viau, M.; Kermarrec, A.; Cunha, R.L.; Meynier, A.; Berton-Carabin, C. Lipid Recovery from Concentrated Emulsions by Freezing–Thawing as an Alternative to Solvent-Based Extraction: A Case Study on Mayonnaise. *Eur. J. Lipid Sci. Technol.* **2022**, *124*, 2200101. [CrossRef]
36. Vargas, M.; Cháfer, M.; Albors, A.; Chiralt, A.; González-Martínez, C. Physicochemical and Sensory Characteristics of Yoghurt Produced from Mixtures of Cows' and Goats' Milk. *Int. Dairy J.* **2008**, *18*, 1146–1152. [CrossRef]
37. Trujillo-Cayado, L.A.; Natera, A.; García, M.C.; Muñoz, J.; Alfaro, M.C. Rheological Properties and Physical Stability of Ecological Emulsions Stabilized by a Surfactant Derived from Cocoa Oil and High Pressure Homogenization. *Grasas Aceites* **2015**, *66*, e087. [CrossRef]
38. Ushikubo, F.Y.; Cunha, R.L. Stability Mechanisms of Liquid Water-in-Oil Emulsions. *Food Hydrocoll.* **2014**, *34*, 145–153. [CrossRef]
39. Garcia, L.C.; Tonon, R.V.; Hubinger, M.D. Effect of Homogenization Pressure and Oil Load on the Emulsion Properties and the Oil Retention of Microencapsulated Basil Essential Oil (*Ocimum basilicum* L.). *Dry. Technol.* **2012**, *30*, 1413–1421. [CrossRef]
40. CLSI. *Performance Standards for Antimicrobial Disk Susceptibility Tests M02–A12; Approved Standard*, 12th ed.; Clinical and Laboratorial Standards Institute: Wayne, PA, USA, 2015.
41. CLSI. *Methods for Dilution Antimicrobial Susceptibility Tests for Bacteria That Grow Aerobically M07–A10; Approved Standard*, 10th ed.; Clinical and Laboratorial Standards Institute: Wayne, PA, USA, 2015.
42. McClements, D.J.; Jafari, S.M. General Aspects of Nanoemulsions and Their Formulation. In *Nanoemulsions: Formulation, Applications, and Characterization*; Elsevier Inc.: Amsterdam, The Netherlands, 2018; pp. 3–20. ISBN 9780128118399.
43. Maa, Y.-F.; Hsu, C.C. Performance of Sonication and Microfluidization for Liquid–Liquid Emulsification. *Pharm. Dev. Technol.* **1999**, *4*, 233–240. [CrossRef]
44. Haghju, S.; Beigzadeh, S.; Almasi, H.; Hamishehkar, H. Chitosan Films Incorporated with Nettle (*Urtica dioica* L.) Extract-Loaded Nanoliposomes: I. Physicochemical Characterisation and Antimicrobial Properties. *J. Microencapsul.* **2016**, *33*, 438–448. [CrossRef]
45. Malvern Instruments Size Theory. In *Zetasizer Nano User Manual MAN 0485*; Malvern Instruments Ltd.: Worcestershire, UK, 2013; pp. 1–8.
46. Yeap, S.P.; Lim, J.; Ngang, H.P.; Ooi, B.S.; Ahmad, A.L. Role of Particle–Particle Interaction Towards Effective Interpretation of Z-Average and Particle Size Distributions from Dynamic Light Scattering (DLS) Analysis. *J. Nanosci. Nanotechnol.* **2018**, *18*, 6957–6964. [CrossRef] [PubMed]
47. Lim, J.; Yeap, S.P.; Che, H.X.; Low, S.C. Characterization of Magnetic Nanoparticle by Dynamic Light Scattering. *Nanoscale Res. Lett.* **2013**, *8*, 381. [CrossRef] [PubMed]
48. Artiga-Artigas, M.; Acevedo-Fani, A.; Martín-Belloso, O. Effect of Sodium Alginate Incorporation Procedure on the Physicochemical Properties of Nanoemulsions. *Food Hydrocoll.* **2017**, *70*, 191–200. [CrossRef]
49. Goddard, E.D. Polymer/Surfactant Interaction: Interfacial Aspects. *J. Colloid Interface Sci.* **2002**, *256*, 228–235. [CrossRef]
50. e Silva, K.F.C.; da Silva Carvalho, A.G.; Rabelo, R.S.; Hubinger, M.D. Sacha Inchi Oil Encapsulation: Emulsion and Alginate Beads Characterization. *Food Bioprod. Process.* **2019**, *116*, 118–129. [CrossRef]
51. Donsì, F.; Marchese, E.; Maresca, P.; Pataro, G.; Vu, K.D.; Salmieri, S.; Lacroix, M.; Ferrari, G. Green Beans Preservation by Combination of a Modified Chitosan Based-Coating Containing Nanoemulsion of Mandarin Essential Oil with High Pressure or Pulsed Light Processing. *Postharvest Biol. Technol.* **2015**, *106*, 21–32. [CrossRef]
52. Taghavi, E.; Mirhosseini, H.; Tan, C.P.; Tan, T.B.; Ngadin, A.A.; Lani, M.N.; Biabanikhankahdani, R.; Anarjan, N. Formulation and Functionalization of Linalool Nanoemulsion to Boost Its Antibacterial Properties against Major Foodborne Pathogens. *Food Biosci.* **2021**, *44*, 101430. [CrossRef]
53. Mcclements, D.J. Critical Review of Techniques and Methodologies for Characterization of Emulsion Stability. *Crit. Rev. Food Sci. Nutr.* **2007**, *47*, 611–649. [CrossRef]
54. Chang, Y.; McLandsborough, L.; McClements, D.J. Physicochemical Properties and Antimicrobial Efficacy of Carvacrol Nanoemulsions Formed by Spontaneous Emulsification. *J. Agric. Food Chem.* **2013**, *61*, 8906–8913. [CrossRef]
55. Artiga-Artigas, M.; Lanjari-Pérez, Y.; Martín-Belloso, O. Curcumin-Loaded Nanoemulsions Stability as Affected by the Nature and Concentration of Surfactant. *Food Chem.* **2018**, *266*, 466–474. [CrossRef]

56. Ghadetaj, A.; Almasi, H.; Mehryar, L. Development and Characterization of Whey Protein Isolate Active Films Containing Nanoemulsions of Grammosciadium Ptrocarpum Bioss. Essential Oil. *Food Packag. Shelf Life* **2018**, *16*, 31–40. [CrossRef]
57. Ashok, N.A.; Sontakke, M. Plant-Based Emulsifiers: Sources, Extraction, Properties and Applications. *Pharma Innov.* **2023**, *12*, 8–16. [CrossRef]
58. Zhang, H.Y.; Arab Tehrany, E.; Kahn, C.J.F.; Ponot, M.; Linder, M.; Cleymand, F. Effects of Nanoliposomes Based on Soya, Rapeseed and Fish Lecithins on Chitosan Thin Films Designed for Tissue Engineering. *Carbohydr. Polym.* **2012**, *88*, 618–627. [CrossRef]
59. Nambam, J.S.; Philip, J. Competitive Adsorption of Polymer and Surfactant at a Liquid Droplet Interface and Its Effect on Flocculation of Emulsion. *J. Colloid Interface Sci.* **2012**, *366*, 88–95. [CrossRef] [PubMed]
60. Heurtault, B.; Saulnier, P.; Pech, B.; Proust, J.E.; Benoit, J.P. Physico-Chemical Stability of Colloidal Lipid Particles. *Biomaterials* **2003**, *24*, 4283–4300. [CrossRef]
61. McClements, D.J.; Xiao, H. Potential Biological Fate of Ingested Nanoemulsions: Influence of Particle Characteristics. *Food Funct.* **2012**, *3*, 202–220. [CrossRef]
62. Kozuch, B.; Weber, J.; Buske, J.; Mäder, K.; Garidel, P.; Diederichs, T. Comparative Stability Study of Polysorbate 20 and Polysorbate 80 Related to Oxidative Degradation. *Pharmaceutics* **2023**, *15*, 2332. [CrossRef]
63. Okuro, P.K.; Gomes, A.; Cunha, R.L. Hybrid Oil-in-Water Emulsions Applying Wax(Lecithin)-Based Structured Oils: Tailoring Interface Properties. *Food Res. Int.* **2020**, *138*, 109798. [CrossRef]
64. Acevedo-Fani, A.; Salvia-Trujillo, L.; Rojas-Graü, M.A.; Martín-Belloso, O. Edible Films from Essential-Oil-Loaded Nanoemulsions: Physicochemical Characterization and Antimicrobial Properties. *Food Hydrocoll.* **2015**, *47*, 168–177. [CrossRef]
65. McClements, D.J. Colloidal Basis of Emulsion Color. *Curr. Opin. Colloid Interface Sci.* **2002**, *7*, 451–455. [CrossRef]
66. Chantrapornchai, W.; Clydesdale, F.; McClements, D.J. Influence of Droplet Characteristics on the Optical Properties of Colored Oil-in-Water Emulsions. *Colloids Surfaces A Physicochem. Eng. Asp.* **1999**, *155*, 373–382. [CrossRef]
67. Chantrapornchai, W.; Clydesdale, F.M.; McClements, D.J. Understanding Colors in Emulsions. In *Color Quality of Fresh and Processed Foods*; ACS Symposium Series; American Chemical Society: Washington, DC, USA, 2008; pp. 364–387.
68. Branco, I.G.; Sen, K.; Rinaldi, C. Effect of Sodium Alginate and Different Types of Oil on the Physical Properties of Ultrasound-Assisted Nanoemulsions. *Chem. Eng. Process.-Process Intensif.* **2020**, *153*, 107942. [CrossRef]
69. Quemada, D. Rheological Modelling of Complex Fluids. I. The Concept of Effective Volume Fraction Revisited. *Eur. Phys. J. Appl. Phys.* **1998**, *1*, 119–127. [CrossRef]
70. Salvia-Trujillo, L.; Rojas-Graü, A.; Soliva-Fortuny, R.; Martín-Belloso, O. Physicochemical Characterization of Lemongrass Essential Oil-Alginate Nanoemulsions: Effect of Ultrasound Processing Parameters. *Food Bioprocess Technol.* **2013**, *6*, 2439–2446. [CrossRef]
71. Peres, G.L.; Leite, D.C.; da Silveira, N.P. Ultrasound Effect on Molecular Weight Reduction of Amylopectin. *Starch-Stärke* **2015**, *67*, 407–414. [CrossRef]
72. Feng, L.; Cao, Y.; Xu, D.; Wang, S.; Zhang, J. Molecular Weight Distribution, Rheological Property and Structural Changes of Sodium Alginate Induced by Ultrasound. *Ultrason. Sonochem.* **2017**, *34*, 609–615. [CrossRef]
73. Ozogul, Y.; Karsli, G.T.; Durmuş, M.; Yazgan, H.; Oztop, H.M.; McClements, D.J.; Ozogul, F. Recent Developments in Industrial Applications of Nanoemulsions. *Adv. Colloid Interface Sci.* **2022**, *304*, 102685. [CrossRef]
74. Guerra-Rosas, M.I.; Morales-Castro, J.; Ochoa-Martínez, L.A.; Salvia-Trujillo, L.; Martín-Belloso, O. Long-Term Stability of Food-Grade Nanoemulsions from High Methoxyl Pectin Containing Essential Oils. *Food Hydrocoll.* **2016**, *52*, 438–446. [CrossRef]
75. Chang, Y.; McLandsborough, L.; McClements, D.J. Fabrication, Stability and Efficacy of Dual-Component Antimicrobial Nanoemulsions: Essential Oil (Thyme Oil) and Cationic Surfactant (Lauric Arginate). *Food Chem.* **2015**, *172*, 298–304. [CrossRef]
76. Nouri, M.; Baghaee-Ravari, S.; Emadzadeh, B. Nano-Emulsified Savory and Thyme Formulation Show Limited Efficacy to Suppress Pectobacterium Carotovorum Subsp. Carotovorum Compared with Pure Oil. *Ind. Crops Prod.* **2021**, *161*, 113216. [CrossRef]
77. Wan, J.; Zhong, S.; Schwarz, P.; Chen, B.; Rao, J. Influence of Oil Phase Composition on the Antifungal and Mycotoxin Inhibitory Activity of Clove Oil Nanoemulsions. *Food Funct.* **2018**, *9*, 2872–2882. [CrossRef] [PubMed]
78. Gharibzahedi, S.M.T.; Mohammadnabi, S. Characterizing the Novel Surfactant-Stabilized Nanoemulsions of Stinging Nettle Essential Oil: Thermal Behaviour, Storage Stability, Antimicrobial Activity and Bioaccessibility. *J. Mol. Liq.* **2016**, *224*, 1332–1340. [CrossRef]
79. Liang, R.; Xu, S.; Shoemaker, C.F.; Li, Y.; Zhong, F.; Huang, Q. Physical and Antimicrobial Properties of Peppermint Oil Nanoemulsions. *J. Agric. Food Chem.* **2012**, *60*, 7548–7555. [CrossRef]
80. Mahdi Jafari, S.; He, Y.; Bhandari, B. Nano-Emulsion Production by Sonication and Microfluidization—A Comparison. *Int. J. Food Prop.* **2006**, *9*, 475–485. [CrossRef]
81. Kumar, M.; Ahuja, M.; Sharma, S.K. Hepatoprotective Study of Curcumin-Soya Lecithin Complex. *Sci. Pharm.* **2008**, *76*, 761–774. [CrossRef]
82. Das, S.; Singh, V.K.; Dwivedy, A.K.; Chaudhari, A.K.; Upadhyay, N.; Singh, P.; Sharma, S.; Dubey, N.K. Encapsulation in Chitosan-Based Nanomatrix as an Efficient Green Technology to Boost the Antimicrobial, Antioxidant and in Situ Efficacy of Coriandrum Sativum Essential Oil. *Int. J. Biol. Macromol.* **2019**, *133*, 294–305. [CrossRef]

83. Niu, F.; Pan, W.; Su, Y.; Yang, Y. Physical and Antimicrobial Properties of Thyme Oil Emulsions Stabilized by Ovalbumin and Gum Arabic. *Food Chem.* **2016**, *212*, 138–145. [CrossRef]
84. Donsì, F.; Sessa, M.; Mediouni, H.; Mgaidi, A.; Ferrari, G. Encapsulation of Bioactive Compounds in Nanoemulsion- Based Delivery Systems. *Procedia Food Sci.* **2011**, *1*, 1666–1671. [CrossRef]
85. López, P.L.; Enemark, G.K.G.; Grosso, N.R.; Olmedo, R.H. Oxidative Protection of Sunflower Oil Used in Industrial Process at High Temperature by Volatile Components from Origanum Vulgare and Humulus Lupulus Essential Oils. *Food Bioprocess Technol.* **2023**, *16*, 2813–2824. [CrossRef]
86. Oreopoulou, A.; Tsimogiannis, D.; Oreopoulou, V. Extraction of Polyphenols From Aromatic and Medicinal Plants: An Overview of the Methods and the Effect of Extraction Parameters. In *Polyphenols in Plants*; Academic Press: Cambridge, MA, USA, 2019; pp. 243–259. [CrossRef]
87. Karthik, P.; Ezhilarasi, P.N.; Anandharamakrishnan, C. Challenges Associated in Stability of Food Grade Nanoemulsions. *Crit. Rev. Food Sci. Nutr.* **2017**, *57*, 1435–1450. [CrossRef]
88. Cossetin, L.F.; Garlet, Q.I.; Velho, M.C.; Gündel, S.; Ourique, A.F.; Heinzmann, B.M.; Monteiro, S.G. Development of Nanoemulsions Containing Lavandula Dentata or Myristica Fragrans Essential Oils: Influence of Temperature and Storage Period on Physical-Chemical Properties and Chemical Stability. *Ind. Crops Prod.* **2021**, *160*, 113115. [CrossRef]
89. Nazzaro, F.; Fratianni, F.; De Martino, L.; Coppola, R.; De Feo, V. Effect of Essential Oils on Pathogenic Bacteria. *Pharmaceuticals* **2013**, *6*, 1451–1474. [CrossRef] [PubMed]
90. Donsì, F.; Ferrari, G. Essential Oil Nanoemulsions as Antimicrobial Agents in Food. *J. Biotechnol.* **2016**, *233*, 106–120. [CrossRef] [PubMed]
91. Moore-Neibel, K.; Gerber, C.; Patel, J.; Friedman, M.; Jaroni, D.; Ravishankar, S. Antimicrobial Activity of Oregano Oil against Antibiotic-Resistant Salmonella Enterica on Organic Leafy Greens at Varying Exposure Times and Storage Temperatures. *Food Microbiol.* **2013**, *34*, 123–129. [CrossRef]
92. Salvia-Trujillo, L.; Rojas-Graü, M.A.; Soliva-Fortuny, R.; Martín-Belloso, O. Impact of Microfluidization or Ultrasound Processing on the Antimicrobial Activity against Escherichia Coli of Lemongrass Oil-Loaded Nanoemulsions. *Food Control* **2014**, *37*, 292–297. [CrossRef]
93. Moghimi, R.; Aliahmadi, A.; McClements, D.J.; Rafati, H. Investigations of the Effectiveness of Nanoemulsions from Sage Oil as Antibacterial Agents on Some Food Borne Pathogens. *LWT-Food Sci. Technol.* **2016**, *71*, 69–76. [CrossRef]
94. Xue, J.; Michael Davidson, P.; Zhong, Q. Antimicrobial Activity of Thyme Oil Co-Nanoemulsified with Sodium Caseinate and Lecithin. *Int. J. Food Microbiol.* **2015**, *210*, 1–8. [CrossRef]

foods

MDPI

Article

Emissions Reduction Strategies for the Orange and Cherry Industries in New South Wales

Aaron T. Simmons [1,2,*], Marja Simpson [3], Paul-Antoine Bontinck [4], John Golding [5], Tim Grant [4], Jess Fearnley [3] and Steven Falivene [6]

1 NSW Department of Primary Industries, Muldoon St., Taree, NSW 2430, Australia
2 School of Business, University of New England, Elm Ave, Armidale, NSW 2350, Australia
3 NSW Department of Primary Industries, Orange Agricultural Institute, Orange, NSW 2800, Australia
4 Lifecycles, 2/398 Smith Street, Collingwood, VIC 3066, Australia; paul-antoine@lifecycles.com.au (P.-A.B.)
5 NSW Department of Primary Industries, Locked Bag 26, Gosford, NSW 2250, Australia; john.golding@dpi.nsw.gov.au
6 NSW Department of Primary Industries, P.O. Box 62, Dareton, NSW 2717, Australia
* Correspondence: aaron.simmons@dpi.nsw.gov.au

Abstract: The orange and cherry industries in New South Wales, Australia, are major horticulture industries with a high export value. Climate change has resulted in the carbon footprint of products being used by consumers to guide purchases meaning that products with a relatively high carbon footprint risk losing market access. The carbon footprint of cherry and orange production is unknown and there is no assessment of the success of climate change mitigation strategies to reduce the carbon footprint of their production and move production towards being carbon neutral. This study assesses the climate change mitigation potential of five management changes to on-farm cherry and orange production (revegetation, the use of nitrification inhibitors, renewable energy, green N fertilisers, and pyrolysis of orchard residues) over a 25-year period. for example, orchards in relevant growing regions. The results show that the carbon footprint of production can be reduced by 73 and 83% for cherries and oranges, respectively, when strategies that avoid emissions are included in their production. When strategies that sequester C from the atmosphere are also included, cherry and orange production becomes C negative in the first few years of the scenario. The economics of implementing these strategies are unfavourable, at present; however, our results indicate that the NSW cherry and orange industries can be confident in achieving emissions reductions in on-farm production to assure market access for their products.

Keywords: pyrolysis; green nitrogen; revegetation; carbon neutral; photovoltaic

1. Introduction

Citrus and cherry production are important horticultural industries for Australia and the state of New South Wales (NSW), Australia. Globally, in 2021, Australia was the 25th largest producer of oranges with an annual production of 456,719 tonnes compared to the largest producer, Brazil, with an annual production of 16,214,982 tonnes [1]. NSW was the largest orange-producing state in Australia with the majority grown in the Riverina region. In 2021–2022, 260,557 tonnes of oranges were produced in NSW with a significant proportion being exported to Asia as the primary destination with a gross value of AUD 39.5 m [2]. There are two main varieties grown in NSW, Navel and Valencia, with a smaller area being planted to Valencia. Australian cherries are exported to more than 30 countries. In 2021, Australia was the 19th largest cherry producer globally, with an annual production of 21,309.94 tonnes. Türkiye was the largest producer with an annual production of 689,834 tonnes (FAO 2023). NSW was a major contributor to Australia's cherry industry, producing more than 4812 tonnes in 2021–2022 with a gross value of AUD 62.6 m with the main growing regions located in Orange and Young [2]. Cherries are considered a high value crop and NSW exported 950 tonnes in 2022 and 612 tonnes of cherries in 2021 with Asia being a major market [2]. For horticulture, as for any agricultural system, greenhouse gas (GHG) emissions are emitted during production and are a concern for the industry and consumers. Sources of direct and indirect GHG emissions in the perennial horticulture industry include the production and use of liquid fuels (e.g., diesel) and electricity, the production and use of nitrogenous fertilisers, and plant-protection products. Studies examining the GHG emissions associated with perennial horticultural crops are limited; however, one study assessed the 23 most commonly grown annual horticultural crops in Australia [3]. That study showed that 65% of total GHG emissions were emitted from energy generation for irrigation and processing, 17% from on-farm emissions associated with fertiliser use, 10% from agro-chemicals, 7% from fossil fuel use on the farm, and 1% from on-farm machinery. Although N_2O emissions associated with horticultural production are relatively low, these emissions are expected to increase due to an increase in fertiliser use to meet global food and fibre demands [4].

The concern by consumers has led to a consumer preference for low-GHG-emission produce [5], and the introduction of initiatives for products to declare their carbon footprint (e.g., the European Commission's Product Environmental Footprint) will facilitate consumers in choosing low-GHG-emissions foods. At present, there is no requirement for Australian horticulturists to undertake carbon accounting for their businesses; however, as the market demand for low-GHG-emissions foods increases, growers may need to provide a carbon footprint for their produce and demonstrate an ongoing reduction in their carbon footprints to maintain market access.

There are, at present, opportunities to reduce the GHG emissions intensity of horticultural systems. For example, nitrification inhibiters (NIs), urease inhibitors, and biochar can reduce N_2O emissions and improve the yield [6,7], and have been included as a mitigation strategy for agricultural N_2O emissions by the Intergovernmental Panel on Climate Change (IPCC) [8]. Amongst NIs, 3,4-dimethylpyrazole phosphate (DMPP) is widely used and has been shown to be effective in reducing NO_3 leaching and N_2O emissions when granular nitrogen (N) fertilisers are coated or DMPP is included in fertigation systems [9]. The research demonstrates that using DMPP can reduce N_2O emissions associated with inorganic fertiliser use by up to 100% [10]; although, other research reported lower reductions [11,12] or no reductions at all [13]. Using DMPP can increase the nitrogen use efficiency (NUE) in irrigated production systems, and the improvement of NUE is dependent on the N application rate [14]. Adding biochar, derived from pyrolysis (i.e., thermal degradation of biomass in an oxygen-limited environment) to soils can also reduce N_2O emissions associated with the use of fertilisers; however, this is unlikely to reduce it to the same extent as the use of NIs [6].

Biochar itself is also a recognized climate change mitigation strategy [8] because it takes C that was sequestered in vegetation from the atmosphere and stabilises it. Pyrolysis is recognised by the IPCC as a carbon dioxide removal (CDR) strategy [15]. Vegetation is a common source of biomass for pyrolysis, and orchards generate biomass from mature trees that are removed during replanting and from on-going maintenance, such as pruning.

Revegetation is also a climate change mitigation strategy because trees sequester carbon from the atmosphere and, although revegetation of parts of a farm may reduce the area that is planted with trees, it may also offer co-benefits, such as windbreaks to decrease fruit damage and provide a habitat for natural substances to improve biological pest control outcomes [16].

GHG emissions associated with the production and combustion of fossil fuels for orchard operations have the potential to be mitigated by electrifying farm machinery when electricity is generated from renewable sources [17]. Using renewable electricity to power pumps for irrigating orchards can also reduce the emissions associated with horticultural production [18].

Pathways for emissions reductions in the NSW horticulture sector to ensure the demand for low-emissions products in international markets are yet to be assessed. The purpose of this study is to evaluate five GHG mitigation strategies for two key horticultural crops in NSW, cherries and oranges, and to identify the potential pathways to carbon neutrality for the perennial horticulture sector more broadly.

2. Materials and Methods

2.1. Boundaries

A system boundary of cradle-to-farm-gate was used. This meant that the emissions associated with the production, transport, and use of all inputs, such as fertiliser, plant protection products, and orchard operations, were included in the assessment. A temporal boundary of 20 years was used to account for annual increases in carbon sequestered in vegetation, as described in Section Revegetation.

2.2. System Descriptions

2.2.1. Lapin Cherries

An irrigated cherry orchard cv. Lapin in the Orange region of NSW with a size of 20 ha was used for the study. Cherry orchards in the regions of study are redeveloped when trees reach an age of approximately 20 years old; therefore, it was assumed that 5% of the orchard was redeveloped on an annual basis. The orchard was assumed to have a density of 1000 trees ha^{-1}.

2.2.2. Navel Oranges

An irrigated orange orchard cv. Washington Navel in the Sunraysia region of NSW with a size of 20 ha was used. The orchard was assumed to have a density of 519 trees ha^{-1}. Orange orchards in the regions of study are redeveloped when trees reach an age of approximately 20 years old; therefore, it was assumed that 5% of the orchard was redeveloped on an annual basis.

2.3. Climate Change Mitigation Strategies

Climate change mitigation strategies can be categorised into two groups: avoided emissions or carbon sequestration. Avoided emissions are strategies that reduce the GHGs emitted relative to current practices and are not subject to reversals. Carbon sequestration strategies remove carbon from the atmosphere and retain carbon either temporarily or permanently, depending on the strategy, and some strategies can also be subject to reversals.

2.3.1. Avoided Emissions

Nitrification Inhibition

The use of the nitrification inhibitor 3,4-dimethylpyrazone phosphate (DMPP) when applying inorganic nitrogenous fertilisers via fertigation to orchards was assumed to reduce N_2O emissions associated with fertiliser use. The research demonstrates that the use of DMPP can reduce N_2O emissions in the citrus and cherry orchards by 55% based on a recent meta-analysis [19]. The use of nitrification inhibitors can also increase the efficiency of N fertiliser [14] that can reduce leaching and the atmospheric deposition of N from volatilised fertiliser; however, the impacts of these effects were not included in this study due to limited evidence.

Renewable Energy

Using renewably generated electricity to power tractors was also used as an emissions reduction strategy. Emissions associated with tractor production assumed the tractor had a 211 kW motor with a 62 kWh battery. Two batteries were used for the tractor so that one could be charged while the other was in use. A 20 kW solar array was used to ensure the batteries could be recharged in one day and allow for the consistent daily use of the tractor. Emissions associated with the production of the tractor and photovoltaic panels were amortised over the lifetime of the tractor (assumed to be 7000 h). The electric tractor was only used for maintenance activities (e.g., spraying and mowing) in the orchard and diesel-powered tractors that we required were retained for more intensive activities, such as tree stump pulling.

Replacing grid-sourced electricity with renewable electricity for pumping irrigation water was also used as an emissions reduction strategy. For cherry production, it was assumed that 154 W was required to move 1000 L of water [20]; therefore, it was assumed that a 20 kW solar array had the capacity to pump 0.13 ML of water per hour. For orange production, it took 309 W to move 1000 L of water from the pump; therefore, a 40 kW solar array power was assumed to provide the capacity to pump 0.13 ML of water per hour. The emissions associated with the production of solar panels to power the pump was amortised over the lifetime of the pump that was assumed to be 15 years.

Green N

The generation of N for use in fertilisers is GHG intensive [21]; therefore, an assumption that nitrates, in the form of nitric acid, were generated using a low-GHG-emissions process (e.g., the proprietary technology of Nitricity) was included to estimate reductions in scope 3 emissions from using green N.

2.3.2. Carbon Sequestration

Pyrolysis

Pyrolysis, the heating of organic materials in the absence of oxygen can stabilise organic carbon in the form of charcoal. The 5% of trees that was removed each year for redevelopment was assumed to be chipped and pyrolysed. It was assumed that 30% of pyrolysed dry matter was recovered as biochar with a carbon content of 80% (S. Joseph, pers. comm.). The remaining carbon in the biomass was assumed to be emitted as CO_2 after syngas created by the pyrolysis process was combusted converting CH_4 to CO_2. It was assumed that biochar was applied to soils where it would slowly degrade releasing CO_2 back into the atmosphere [22], and this was included as an emission. We assumed that high-temperature pyrolysis was used and, because we were uncertain where biochar would be applied to the soils, a soil temperature of 15 °C was assumed, resulting in 18% of carbon stored in the biochar being released back into the atmosphere over a 100-year period [22]. The loss function that describes this was linear; therefore, we assumed that 0.18% of carbon in the biochar was released back into the atmosphere annually.

To estimate the C in tree biomass that was available for stabilisation in the biochar each year, we used published values. For the cherry orchard, the mass of a tree on a dry matter basis was assumed to be 82 kg tree^{-1} [23] based on the values for 20-year-old *Prunus cerasus* trees planted at a rate of 292 trees ha^{-1}. Multiplying the mass of a tree by the planting density assumed in the present study, and assuming a 50% carbon content of biomass, produced a carbon mass of 41,000 kg ha^{-1}, or 2050 kg carbon available for stabilisation in pyrolysis when 5% of the orchard was removed per annum. For the orange orchard, a carbon mass of 102 kg tree^{-1} was assumed [24]. This was based on the values provided for 16-year-old trees of *Citrus sinensis* cv. Tarocco Sciré planted at a rate of 494 trees ha^{-1}. Multiplying the carbon mass of a tree by the planting density produced a carbon mass of 52,224 kg carbon ha^{-1} with 5% (2611 kg) of carbon available for stabilisation in pyrolysis per annum due to the rate of re-planting.

Revegetation

Revegetation was used to strategically revegetate areas of the farm to native trees to sequester atmospheric carbon, and where revegetation occurred, it was assumed to displace orchard production. The calculation of the area that was planted to shelterbelts was performed assuming that the 20 ha orchard had the dimensions of 400 × 500 m with a 15 m shelterbelt on the northern, western, and southern boundaries, and one shelterbelt through the centre of the orchard on an east–west axis. This resulted in 3.45 ha (or 17%) of the orchard being dedicated to shelterbelts. The planting of shelterbelts was assumed to occur when 1 ha was redeveloped; therefore, the area of the orchard that was dedicated to shelterbelts increased linearly over time from 1% when the first block was developed to the maximum of 17% when the final block was redeveloped in the 20-year period.

The carbon sequestered via revegetation was estimated using the Australian government Full Carbon Accounting Model (FullCAM) as set out in the methods for estimating sequestration under the Environmental Plantings method [25] for the emissions reduction fund parameterised with GPS co-ordinates of Balranald, NSW, and Orange, NSW, for the Riverina and Orange regions, respectively [25].

2.4. Data Sources

The data for on-farm operations, inputs, and yield of a Valencia orange orchard were obtained from NSW DPI gross margins of production [26], as were the operations and inputs for a cherry orchard [27]. These gross margins were developed by regional experts in the relevant production system and represent best-practice management. The data assumptions for yields, inputs, and orchard operations are presented in the Supplementary Materials.

2.5. GHG Emissions and Sequestration Calculations

GHG emissions associated with the use of nitrogenous fertilisers and the use of lime were calculated according the Australian National Greenhouse Gas Inventory [28], and GHG emissions associated with the combustion of liquid fuels on-farm were calculated using the appropriate emissions factor from the Australian National Greenhouse Accounts [29]. Pre-farm GHG emissions (i.e., emissions associated with the production and transport of inputs, such as fertilisers and plant protection products) were obtained from life cycle inventory databases, ecoinvent v3.9 [30], and AusLCI [20]. The ecoinvent database is the largest transparent unit-process life cycle inventory database globally available and he AusLCI is a scientifically robust, standardised, and transparent database specific to Australian agriculture production.

Though we assumed that 5% of the orchard was redeveloped each year, the carbon sequestered in the fruit trees was not included in our calculations. This was because when the carbon stocks were assessed across the entire orchard, there was no annual change as was demonstrated for forestry estates that were rotationally harvested [31].

The global warming potential values for a time horizon of values for 100 years (GWP_{100}) for CH_4 from fossil sources, CH_4 from natural sources, and N_2O were collected from the IPCC AR5 report and were 30, 28, and 265, respectively [32].

The emissions intensity of production was calculated by dividing the total yield of the orchard by the sum of GHG emissions and sequestration.

3. Results

3.1. Yield

A 1 ha block of the orchard was redeveloped each year and shelterbelts were added each time a 1 ha block was redeveloped. This resulted in a reduction in the area of land dedicated to fruit production and resulted in the total productivity of cherry and orange orchards declining over time. At the end of the 20-year period, the total productivity of the cherry declined from 245 t to 203 t annum^{-1} and the total productivity of the orange orchard declined from 501 to 416 t annum^{-1} (Figure 1).

3.2. Emissions Intensity

The emissions intensities of producing 1 kg of cherries and oranges using current production methods, as described above, were 137 and 128 g CO_2-e, respectively. The greatest contributor to the emissions intensity of cherries was the emissions associated with the production of fertiliser, and for oranges the greatest contributor to GHG emissions intensity was electricity associated with irrigation (Table 1).

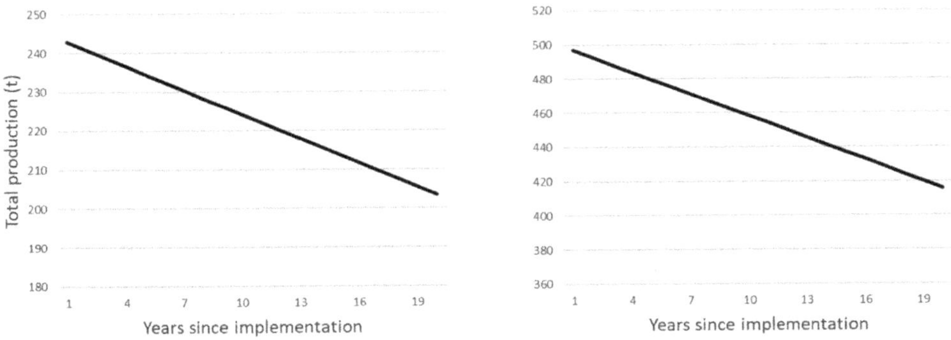

(**a**) (**b**)

Figure 1. Total production (t) over a 20-year period for (**a**) cherry cv. 'Lapin' and (**b**) orange cv. 'Washington' orchards of 20 ha when shelterbelts to sequester atmospheric CO_2 that reduce the productive area by 17% are planted on one ha annum^{-1}.

Table 1. Emissions intensity (g CO_2-e) of producing 1 kg of cherries cv. 'Lapin' and oranges cv. 'Washington Navel' under current production business as usual (BAU) and with emissions reductions strategies implemented (mitigation) and the % reduction in GHG emissions for each emissions category as calculated using life cycle assessment.

	Emissions Source Category	Cherry cv. 'Lapin' BAU (g CO_2-e/kg)	Cherry cv. 'Lapin' Mitigation (g CO_2-e/kg)	Reduction	Orange cv. 'Washington Navel' BAU (g CO_2-e/kg)	Orange cv. 'Washington Navel' Mitigation (g CO_2-e/kg)	Reduction
Pre-farm	Diesel production	1.1	0.5	59%	0.8	0.4	48%
	Fertiliser production	72.4	18.9	74%	14.5	5.2	64%
	Irrigation system	0.3	0.3	0%	0.3	0.3	0%
	Transport	0.7	0.7	0%	0.4	0.4	0%
	Plant protection products	3.2	3.2	0%	5.1	5.1	0%
	Irrigation electricity	5.2	0.3	94%	85.5	0.9	99%
	Total pre-farm	82.8	23.8	71%	106.5	12.2	89%
On-farm	Fertiliser emissions	26.1	4.4	83%	5.6	3.3	42%
	Tractor operations	28.0	6.5	77%	15.5	6.0	61%
	Total on-farm	54.1	10.9	80%	21.1	9.3	56%

3.3. Emissions Reductions Strategies

Applying emissions reductions strategies of nitrification inhibitors, renewable energy, and green N reduced the emissions intensities gf cherry and orange production by 75 and 83%, respectively. The greatest emissions reductions, expressed as a % of emissions under current production, occurred when pumps for irrigation were moved from grid-powered to solar-powered electricity (Table 1). For both crops, emissions reductions associated with replacing nitrates with 'green' nitrates and the use of a nitrification inhibitor to reduce emissions associated with fertiliser use and replacing diesel-fuelled tractors with electric tractors also made considerable contributions to emissions reductions outcomes (Table 1).

3.4. Carbon Sequestration

The two carbon sequestration methods included in this study further reduced the emissions intensity of producing 1 kg of cherries and oranges. When carbon sequestered in shelterbelts was included, the emissions intensities in year 1 of the 20-year scenario were 79 and 86% lower than the emissions intensity under current production for cherries and oranges, respectively. Furthermore, the production of cherries became carbon neutral in year 16 and the production of oranges became carbon neutral from year 10 onwards

(Figure 2). When the carbon sequestered in the vegetation and carbon sequestered in biochar were combined with the emissions reductions strategies, the emissions intensity of cherry production was reduced by 99% and became carbon neutral in the 2nd year of implementation, while orange production became carbon neutral in year 1 with a 113% reduction in GHG emissions relative to current production (Figure 2).

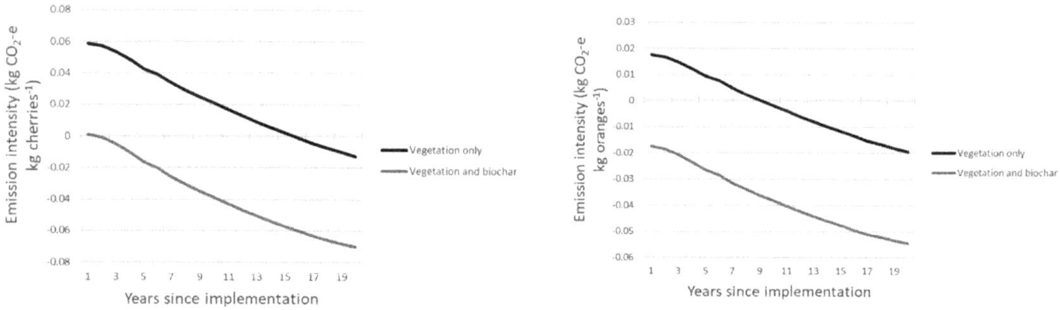

(**a**) (**b**)

Figure 2. Emissions intensities (kg CO_2-e) over a 20-year period for the production of 1 kg of (**a**) cherries cv. 'Lapin' and (**b**) oranges cv. 'Washington Navel' after the implementation of emissions reductions strategies with either C sequestration in vegetation included or C sequestered in vegetation and biochar included.

3.5. Total Emissions

The cumulative total GHG emissions for each orchard over the 20-year period was calculated (Figure 3). Where current management was assumed for the 20-year period, the total GHG emissions were 1372 and 1278 t CO_2-e for cherry and orange production, respectively. Where all emissions reduction strategies were implemented and carbon sequestration in trees was included, the cherry orchard was responsible for emitting 1273 less t CO_2-e than current management, and when carbon sequestration in vegetation and biochar were included, the orchard was responsible for 1536 less t CO_2-e. For orange production, where emissions reductions and carbon sequestered in trees were included, 1290 less t CO_2-e was emitted, and when carbon sequestered in biochar was added, 1617 less t CO_2-e was emitted.

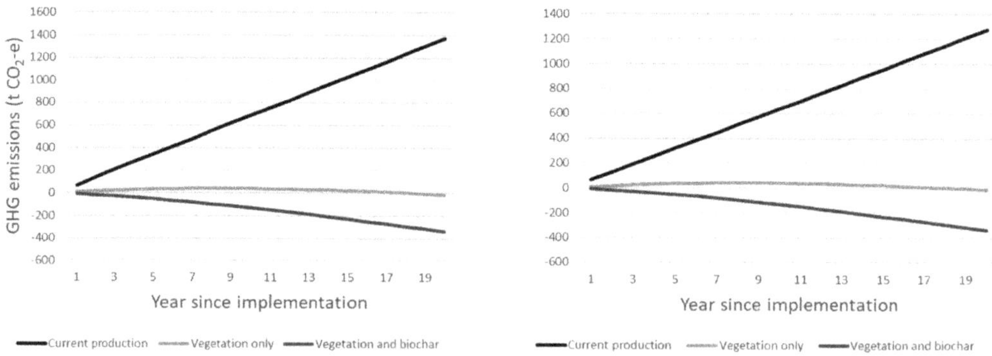

(**a**) (**b**)

Figure 3. Total GHG emissions over a 20-year period for a 20 ha orchard of (**a**) cherries cv. 'Lapin' and (**b**) oranges cv. 'Washington' under current production methods, with emissions reduction strategies and C sequestered in vegetation included in the calculations, and with emissions reduction strategies and C sequestered in vegetation and biochar included in the calculations.

4. Discussion

The results of this study suggest that it is possible for cherry and orange orchards in NSW to implement changes to their production systems so that they positively contribute to the climate change problem by being a carbon sink. A mix of different emissions reduction strategies can be utilised; however, our analysis suggests that the pyrolysis of trees, when the redevelopment of the orchard occurs, is essential for carbon neutrality to be achieved (Figure 3). This is because carbon sequestered from the atmosphere and stored in trees is stabilised as biochar when it is pyrolysed, thereby removing and storing carbon from the atmosphere in a long-term capacity. The pyrolysis of plant materials is a recognised carbon dioxide-removal strategy [15]; however, this research is the first to demonstrate the importance of this strategy in horticultural production achieving carbon neutrality. No research has identified the pathways to carbon neutrality for cherry production; however, one study assessed the pathways to carbon neutrality for orange production [33]. That study demonstrated that Chinese citrus production could become a carbon sink by optimising nitrogen fertiliser use, replacing 50% of chemical fertilisers with organic fertilisers and the use of cover crops as green manures to supply N. Those strategies are not relevant to the systems we assessed here because N fertiliser is not overapplied, as occurs in the Chinese systems; the supply of organic fertilisers to replace chemical fertilisers are constrained and the environment in which oranges are grown is not conducive to growing green manure crops. Nevertheless, both that study and the present study recognise that a more sustainable production of N is required for carbon neutrality to be achieved.

This study assessed the potential for five strategies to reduce the GHG emissions associated with cherry and orange production in NSW. However, the scope for greater emissions reductions exist but were not included in the present study due to a lack of data and/or a lack of technological readiness. For example, the emissions from diesel production and combustion were reduced by between 48 and 77%, depending on crop type; however, these emissions can be reduced further if high-horsepower machinery required for the redevelopment of orchards is electrified. There is also the scope to increase carbon sequestration by pyrolysing orchard prunings; however, the mass of prunings from an orchard on an annual basis cannot be quantified for this study. Potential also exists to cut back shelterbelts when orchards are redeveloped and the carbon stored in shelterbelts is pyrolysed. We assumed that nitrates were sourced from green production systems; however, systems to produce ammonia, another form of N used in fertilisers, are being developed and would further reduce the emissions associated with the production of inputs, such as urea (e.g., Jupiter Ionics). Emissions associated with fertiliser use can be reduced further where the use of NI improves NUE [11,12] allowing the mass of N fertiliser applied to crops to be reduced. Finally, maximising the use of integrated disease and pest management has the potential to reduce the emissions associated with the production and application of plant protection products.

The pathways to carbon neutrality assessed here are relatively mature with the technological readiness of four of the five climate change mitigation strategies assessed here considered high (i.e., technological readiness level (TRL) of 7–9). For example, nitrification inhibitors, such DMPP, are commercially available for producers to use, photovoltaic technologies are widely used globally to generate electricity, the planting of trees to sequester atmospheric carbon is a common method to generate carbon credits [25], and methods also exist to generate carbon credits by pyrolyzing organic materials [34]. The remaining strategy of using 'green' nitrates is of a lower TRL and is not commercially available at present; however, companies with proprietary technology (e.g., Nitricity) are scaling up production to meet the projected global demands for green N. Electric tractors are close to being produced [35] and companies that specialise in the retrofitting of electric motors to diesel tractors exist (e.g., Janus technologies).

The economics of moving citrus and cherry production to achieve carbon neutrality is uncertain and we recognise that it is unlikely to be cost effective at present; however, the imperative to reduce GHG emissions is likely to increase as consumers choose lower-emissions products [5] and global initiatives (e.g., the European commission product environmental footprint [36]) roll out to support consumers in making more environmentally friendly choices. The Australian horticultural sector does not have a target for carbon neutrality at present, unlike other agricultural sectors that are under greater pressure from markets due to their relatively high emissions intensity (e.g., the red meat sector that has set a carbon neutrality target of 2030 [37]). The results from this study indicate that the horticulture industry can set an ambitious target of carbon neutrality with pathways to support the process. Including the economic feasibility of the emissions reduction pathways was out of scope in this exploratory study. Therefore, further research should investigate the economics of the mitigation pathways demonstrated in this study and the potential trade-offs that may need to be considered. For example, the cost of utilising NI may be offset to some extent by the need to apply less nitrogen. Investments in solar for water pumping is a significant initial capital cost but would reduce the operating costs for a business. As more businesses begin implementing emissions reductions strategies and compete for customers, the economics of investing in technologies that support the move, carbon neutrality may become more favourable.

5. Conclusions

The carbon neutrality of cherry and orange production in NSW is technically possible; however, the economics of implementing strategies to achieve this is uncertain and should be investigated in further studies. As governments are implementing policies for emission reduction targets, incentive schemes can increase the uptake of emission reduction strategies. The strategies assessed here are also applicable to other perennial horticultural crops and warrant further studies to assess the technical potential for other crops to achieve carbon neutrality. Such studies can inform the Australian horticultural industry in setting on-farm GHG emissions reductions targets to support ongoing social licence for production to ensure market access is maintained.

Supplementary Materials: The following supporting information can be downloaded at: https://www.mdpi.com/article/10.3390/foods12183328/s1. Table S1: Average yield, orchard activities and inputs for a cherry orchard in Orange NSW. Table S2: Average yield, orchard activities and inputs for an orange orchard in the Sunraysia region of NSW.

Author Contributions: Conceptualization, A.T.S. and J.G.; methodology, A.T.S.; validation, J.F. and S.F.; formal analysis, A.T.S., P.-A.B., T.G. and M.S.; writing—original draft preparation, A.T.S. and M.S.; writing—review and editing, all authors; visualization; A.T.S. All authors have read and agreed to the published version of the manuscript.

Funding: This research received no external funding.

Data Availability Statement: The data presented in this study are available on request from the corresponding author.

Conflicts of Interest: The authors declare no conflict of interest.

References

1. FAOSTAT. *FAO Statistical Databases*; FAO: Rome, Italy, 2023.
2. Hort Innovation. *Australian Horticulture Statistics Handbook*; Hort Innovation: Sydney, Australia, 2023.
3. Maraseni, T.N.; Cockfield, G.; Maroulis, J.; Chen, G. An assessment of greenhouse gas emissions from the Australian vegetables industry. *J. Environ. Sci. Health* **2010**, *45*, 578–588. [CrossRef] [PubMed]
4. FAO. *The Future of Food and Agriculture: Alternative Pathways to 2050*; Food and Agriculture Organization of the United Nations: Rome, Italy, 2018.
5. Thøgersen, J.; Nielsen, K.S. A better carbon footprint label. *J. Clean. Prod.* **2016**, *125*, 86–94. [CrossRef]
6. Ma, H.; Jia, X.; Yang, J.; Liu, J.; Shangguan, Z.; Yan, W. Inhibitors mitigate N_2O emissions more effectively than biochar: A global perspective. *Sci. Total Environ.* **2023**, *859*, 160416. [CrossRef] [PubMed]
7. Gu, J.; Nie, H.; Guo, H.; Xu, H.; Gunnathorn, T. Nitrous oxide emissions from fruit orchards: A review. *Atmos. Environ.* **2019**, *201*, 166–172. [CrossRef]

8. de Coninck, H.; Revi, A.; Babiker, M.; Bertoldi, P.; Buckeridge, M.; Cartwright, A.; Dong, W.; Ford, J.; Fuss, S.; Hourcade, J.-C.; et al. Strengthening and Implementing the Global Response. In *Global Warming of 1.5 °C. An IPCC Special Report on the Impacts of Global Warming of 1.5 °C above Pre-Industrial Levels and Related Global Greenhouse Gas Emission Pathways, in the Context of Strengthening the Global Response to the Threat of Climate Change, Sustainable Development, and Efforts to Eradicate Poverty*; Masson-Delmotte, V., Zhai, P., Pörtner, H.-O., Roberts, D., Skea, J., Shukla, P.R., Pirani, A., Moufouma-Okia, W., Péan, C., Pidcock, R., et al., Eds.; Cambridge University Press: Cambridge, UK; New York, NY, USA, 2018; pp. 313–444.
9. Di, H.J.; Cameron, K.C. Inhibition of ammonium oxidation by a liquid formulation of 3, 4-Dimethylpyrazole phosphate (DMPP) compared with a dicyandiamide (DCD) solution in six new Zealand grazed grassland soils. *J. Soils Sediments* **2011**, *11*, 1032–1039. [CrossRef]
10. Maris, S.C.; Teira-Esmatges, M.R.; Arbonés, A.; Rufat, J. Effect of irrigation, nitrogen application, and a nitrification inhibitor on nitrous oxide, carbon dioxide and methane emissions from an olive (*Olea europaea* L.) orchard. *Sci. Total Environ.* **2015**, *538*, 966–978. [CrossRef] [PubMed]
11. Wang, F.; Ge, S.; Lyu, M.; Liu, J.; Li, M.; Jiang, Y.; Xu, X.; Xing, Y.; Cao, H.; Zhu, Z.; et al. DMPP reduces nitrogen fertilizer application rate, improves fruit quality, and reduces environmental cost of intensive apple production in China. *Sci. Total Environ.* **2022**, *802*, 149813. [CrossRef] [PubMed]
12. Vilarrasa-Nogué, M.; Teira-Esmatges, M.R.; Pascual, M.; Villar, J.M.; Rufat, J. Effect of N dose, fertilisation duration and application of a nitrification inhibitor on GHG emissions from a peach orchard. *Sci. Total Environ.* **2020**, *699*, 134042. [CrossRef] [PubMed]
13. Guo, Y.; Ji, Y.; Zhang, J.; Liu, Q.; Han, J.; Zhang, L. Effects of water and nitrogen management on N_2O emissions and NH3 volatilization from a vineyard in North China. *Agric. Water Manag.* **2022**, *266*, 107601. [CrossRef]
14. Abalos, D.; Jeffery, S.; Sanz-Cobena, A.; Guardia, G.; Vallejo, A. Meta-analysis of the effect of urease and nitrification inhibitors on crop productivity and nitrogen use efficiency. *Agric. Ecosyst. Environ.* **2014**, *189*, 136–144. [CrossRef]
15. Babiker, M.; Berndes, G.; Blok, K.; Cohen, B.; Cowie, A.; Geden, O.; Ginzburg, V.; Leip, A.; Smith, P.; Sugiyama, M. *Cross-Sectoral Perspectives*; Cambridge University Press: Cambridge, UK, 2022.
16. Owen-Turner, J.; Hardy, S.; Australia, C. Windbreaks for Citrus. Citrus Fact Sheet, Compiled by Cittgroups Australia. 2006. Available online: https://www.dpi.nsw.gov.au/agriculture/horticulture/citrus/content/orchard-development/windbreaks.pdf (accessed on 18 July 2023).
17. Gorjian, S.; Ebadi, H.; Trommsdorff, M.; Sharon, H.; Demant, M.; Schindele, S. The advent of modern solar-powered electric agricultural machinery: A solution for sustainable farm operations. *J. Clean. Prod.* **2021**, *292*, 126030. [CrossRef]
18. Guno, C.S.; Agaton, C.B. Socio-Economic and Environmental Analyses of Solar Irrigation Systems for Sustainable Agricultural Production. *Sustainability* **2022**, *14*, 6834. [CrossRef]
19. Grace, P.R. Revised Emission Factors for Estimating Direct Nitrous Oxide Emissions from Nitrogen Inputs in Australia's Agricultural Production Systems: A meta-analysis. 2023; *in press*.
20. Grant, T.; Eady, S.; Cruypenninck, H.; Simmons, A. *AusLCI Methodology for Developing Life Cycle Inventory for Australian Agriculture*; Lifecycle Strategies Pty Ltd.: Melbourne, Australia, 2017.
21. Smith, C.; Hill, A.K.; Torrente-Murciano, L. Current and future role of Haber–Bosch ammonia in a carbon-free energy landscape. *Energy Environ. Sci.* **2020**, *13*, 331–344. [CrossRef]
22. Woolf, D.; Lehmann, J.; Ogle, S.; Kishimoto-Mo, A.W.; McConkey, B.; Baldock, J. Greenhouse Gas Inventory Model for Biochar Additions to Soil. *Environ. Sci. Technol.* **2021**, *55*, 14795–14805. [CrossRef] [PubMed]
23. Tozzini, L.; Lakso, A.; Flore, J. Estimating the carbon footprint of Michigan apple and cherry trees-lifetime dry matter accumulation. In Proceedings of the IX International Symposium on Modelling in Fruit Research and Orchard Management, Saint-Jean-sur-Richelieu, QC, Canada, 19–23 June 2011; Volume 1068, pp. 85–90.
24. Liguori, G.; Gugliuzza, G.; Inglese, P. Evaluating carbon fluxes in orange orchards in relation to planting density. *J. Agric. Sci.* **2009**, *147*, 637–645. [CrossRef]
25. Australian Government. Carbon Credits (Carbon Farming Initiative-Reforestation by Environmental or Mallee Plantings-FullCAM) Methodology Determination 2014. *F2018C00118* **2014**. Available online: https://www.legislation.gov.au/Details/F2018C00118 (accessed on 28 August 2023).
26. Falivene, S.; Creek, A. *NSW Citrus Farm Budget Handbook*; NSW DPI: Orange, NSW, Australia, 2018.
27. Fearnley, J.; Falivene, S. Gross margins for cherry production in NSW. 2023.
28. Australian Government. *Australian National Greenhouse Accounts National Inventory Report 2020*; Department of Industry, Science, Energy and Resources: Canberra, Australia, 2022.
29. Australian Government. *Australian National Greenhouse Accounts Factors 2020*; Department of Industry, Science, Energy and Resources: Canberra, Australia, 2022.
30. Wernet, G.; Bauer, C.; Steubing, B.; Reinhard, J.; Moreno-Ruiz, E.; Weidema, B. The ecoinvent database version 3 (part I): Overview and methodology. *Int. J. Life Cycle Assess.* **2016**, *21*, 1218–1230. [CrossRef]
31. Cintas, O.; Berndes, G.; Cowie, A.L.; Egnell, G.; Holmström, H.; Marland, G.; Ågren, G.I. Carbon balances of bioenergy systems using biomass from forests managed with long rotations: Bridging the gap between stand and landscape assessments. *GCB Bioenergy* **2017**, *9*, 1238–1251. [CrossRef]

32. Myhre, G.; Shindell, D.; Bréon, F.-M.; Collins, W.; Fuglestvedt, J.; Huang, J.; Koch, D.; Lamarque, J.-F.; Lee, D.; Mendoza, B.; et al. Anthropogenic and Natural Radiative Forcing. In *Climate Change 2013: The Physical Science Basis. Contribution of Working Group I to the Fifth Assessment Report of the Intergovernmental Panel on Climate Change*; Stocker, T.F., Qin, D., Plattner, G.-K., Tignor, M., Allen, S.K., Boschung, J., Nauels, A., Xia, Y., Bex, V., Midgley, P.M., Eds.; Cambridge University Press: Cambridge, UK; New York, NY, USA,, 2013; pp. 659–740.
33. Zhao, H.; Dong, Z.; Liu, B.; Xiong, H.; Guo, C.; Lakshmanan, P.; Wang, X.; Chen, X.; Shi, X.; Zhang, F.; et al. Can citrus production in China become carbon-neutral? A historical retrospect and prospect. *Agric. Ecosyst. Environ.* **2023**, *348*, 108412. [CrossRef]
34. Etter, H.; Vera, A.; Aggarwal, C.; Delaney, M.; Manley, S. Methodology for Biochar Utilization in Soil and Non-Soil Applications. 2021. Available online: https://verra.org/wp-content/uploads/2021/08/210803_VCS-Biochar-Methodology-v1.0-.pdf (accessed on 28 August 2023).
35. CNH Industrial. *CNH Industrial Presents First Electric Tractor Prototype with Autonomous Features*; CNH Industrial: London, UK, 2022.
36. Manfredi, S.; Allacker, K.; Pelletier, N.; Chomkhamsri, K.; de Souza, D.M. Product Environmental Footprint (PEF) Guide. 2012. Available online: https://scholar.google.com/citations?view_op=view_citation&hl=en&user=97Myh60AAAAJ&citation_for_view=97Myh60AAAAJ:eQOLeE2rZwMC (accessed on 28 August 2023).
37. Meat and Livestock Australia. *The Australian Red Meat Industry's Carbon Neutral by 2030 Roadmap*; Meat and Livestock Australia: North Sydney, Australia, 2020.

foods

MDPI

Article

Case Studies of Small-Medium Food Enterprises around the World: Major Constraints and Benefits from the Implementation of Food Safety Management Systems

Jocelyn C. Lee [1], Marina Neonaki [2], Athanasios Alexopoulos [3] and Theodoros Varzakas [2,*]

1 Independent Researcher, Benicia, CA 94510, USA; jlee@gourmetrail.com
2 Department Food Science and Technology, University of the Peloponnese, GR24100 Kalamata, Greece; neonakimar@yahoo.gr
3 Laboratory of Microbiology, Biotechnology and Hygiene, Department of Agricultural Development, Democritus University of Thrace, GR68200 Orestiada, Greece; alexopo@agro.duth.gr
* Correspondence: t.varzakas@uop.gr; Tel.: +30-272-104-5279

Abstract: Global food safety and security are key principles to be followed in the context of the implementation of food safety management systems. The objective of this paper is to assess the contemporary developments of Food Safety Management System standards (FSMS) worldwide and to identify the primary constraints and advantages associated with their implementation by small and medium-sized enterprises across different regions. The effectiveness of these systems has also been evaluated. 116 case studies have been employed across developing and developed regions worldwide across 27 primary food sectors. After the implementation of FSMS, there was a significant increase in the percentage of companies that have implemented the international FSMS, both in developed (16.7% to 63.9%) and developing countries (26.6% to 48.1%). Certification has also increased from 34.2% to 59.6% in the total sample, namely from 33.3% to 61.1% in developed countries and from 34.6% to 59.0% in developing countries. There was a significant increase in medium vs. small company size (57.1% to 62.3%, $p = 0.046$), only in developing countries. Food safety culture and manager leadership implementation were increased to over 80% after FSMS implementation in both developed and developing countries ($p < 0.001$). Training, resources, and technology adequacy were also increased in all companies ($p < 0.001$).

Keywords: food safety management system (FSMS); Good Manufacturing Best Practices; prerequisite programs (PRPs); HACCP; systems thinking; food safety culture; food security; developing countries; developed countries

1. Introduction

Food safety is an integral component of daily existence, and it is crucial to ensure its safety. Many nations lack appropriate microbiological criteria for food products, and even where they do exist, their enforcement is insufficient [1]. Hygiene is an essential component of food safety, yet consumer education regarding food hygiene is inadequate in many nations [2]. To ensure effective implementation and enforcement of food safety standards, governments should allocate adequate resources, strengthen their regulatory frameworks, and promote international cooperation and the exchange of information and best practices. Developing nations face numerous obstacles to the implementation of food safety standards, including inadequate infrastructure, limited financial resources, and a lack of understanding of food safety fundamentals. Recommendations include allocating adequate funds for the implementation and enforcement of food safety standards, strengthening regulatory frameworks, training personnel, educating consumers about food safety, and promoting international cooperation [3–5].

For preventing infectious diseases, promoting food safety, and ensuring the integrity of food products, food safety standards have become indispensable [6,7]. In developing nations, implementing food safety standards can reduce the incidence of infectious illnesses, improve the quality of life for the population, and provide access to international export markets. Implementing food safety standards in industrialized nations can reduce the incidence of infectious diseases, increase consumer confidence in food products, and facilitate international trade [1,2,4,5,8,9].

Even though many nations have made significant progress in implementing food standards, the results revealed both effective practices and obstacles in the implementation of food standards, as well as gaps in the implementation of HACCP and microbiological criteria. This study examined the global implementation of food safety standards with a focus on six crucial aspects: food standards, HACCP, prerequisites, microbiological criteria, food hygiene, and process controls. The findings revealed that many countries have implemented food standards, but implementation and enforcement differ. HACCP is widely recognized as a crucial aspect of food safety, but the lack of trained personnel to implement it is a significant obstacle. GMPs and GHPs are essential for ensuring food safety, but their enforcement in some nations requires improvement [10].

Microbiological standards are also essential for guaranteeing food safety, but many nations lack the necessary standards. Hygiene is an essential aspect of food safety, and process controls are required to guarantee the safety of food products. The implementation of food safety standards reduces the likelihood of infectious illness, improves product quality, boosts consumer confidence, and provides access to global markets. Governments were advised to allocate adequate funds for the implementation and enforcement of food safety standards, to strengthen regulatory frameworks, and to train qualified personnel. The food safety standards are a set of rules and guidelines designed to ensure the safety of food products for human consumption [6].

However, developing nations confront numerous obstacles, such as inadequate infrastructure, lack of water, polluted water, sanitation facilities, and financial resources. Food safety standards can be implemented in developing nations to reduce the incidence of foodborne illnesses, improve the quality of life for the population, and facilitate export market access [2,11]. However, developed nations are hindered by complex regulatory environments and limited resources. The implementation of food safety standards in developed nations can increase consumer confidence in food products, facilitate international trade, and reduce the incidence of infectious diseases [5,12–16].

This article presents the results of a global data analysis pertaining to the implementation of food standards, HACCP, prerequisites, microbiological criteria, food hygiene, and process controls. It was conducted to evaluate global food safety systems and to identify implementation flaws in food standards, HACCP, prerequisites, microbiological criteria, food hygiene, and process controls.

This study seeks to evaluate the global implementation of food safety standards and to identify the challenges and benefits of their implementation. It took nearly sixteen months of intensive research and analysis by a team of global food safety experts to produce a reliable, comprehensive, and inclusive article. The team performed a variety of tasks, including consulting with food safety specialists, designing, completing, compiling case study survey data, analyzing, and summarizing content. This article provides solutions and frameworks that can help policymakers, regulators, and industry stakeholders enhance the efficacy of food safety management systems.

Consequently, the objective of this paper is to assess the contemporary developments of Food Safety Management System standards (FSMS) worldwide and to identify the primary challenges and advantages associated with their implementation by small to medium-sized enterprises across different regions. The study aims to suggest potential ways to enhance or mitigate the impact of these systems in propelling countries forward. The effectiveness of food safety programs, including HACCP and food safety management systems, will also be evaluated. Lastly, the research seeks to establish a consensus on whether limitations of food safety management systems standards among small to medium-sized enterprises and their affordability may affect global safety and hinder the prevention of foodborne illnesses. This article provides solutions and frameworks that can help policymakers, regulators, and industry stakeholders enhance the efficacy of food safety management systems.

2. Food Safety Management Systems around the World

According to the WHO, foodborne illnesses affect a significant number of people globally, with approximately 600 million falling ill each year and resulting in 420,000 deaths [5,17]. Developing countries bear a higher burden of these illnesses and deaths due to challenges in managing food safety caused by limited infrastructure and resources. Food recalls are initiated when a product is deemed unsafe for consumption due to contamination or other hazards, and they can be initiated by food manufacturers, government agencies, or consumer complaints [4]. Developed countries have strict regulations and monitoring systems in place, such as the FDA in the US and EFSA in the EU, to ensure food safety. In contrast, developing countries may have less robust systems, leading to more frequent foodborne illnesses and recalls [12,13]. Recent examples of food recalls include salmonella contamination in pet treats, onions, and peaches in the US, listeria concerns in cheese and pork products in Europe, wheat flour contamination with metal pieces in India, and potential salmonella contamination in milk powder in Nigeria [18–23]. These examples emphasize the importance of strong food safety measures and monitoring systems worldwide to prevent and address contamination issues effectively.

Although there are several pillars and resources available to food processors (i.e., food safety legislation, policies, standards, and guidelines related to food safety systems implementation, controls and audits, worker training programs, etc.), foodborne illnesses continue to arise globally at an alarming frequency. As a result, this leads to expensive public health emergencies, product recalls and a reduction in consumer confidence in both food processors and brands [2,6,7].

A food safety management system refers to a methodical approach to managing and controlling food safety risks within a food establishment to ensure that food is safe for consumption. All businesses are obligated to establish, execute and sustain a food safety management system that is founded on the principles of Hazard Analysis Critical Control Point (HACCP) [6].

A food safety management system is a set of procedures designed to regulate food safety and guarantee that all food produced is of acceptable quality and safe for consumption. This entails outlining processes for each stage of the production process, from receiving supplies to delivering finished products. Furthermore, it is crucial that the food safety management system adheres to the principles of HACCP and each organization is responsible for developing procedures based on these principles. By law, food safety management systems are necessary to comply with food safety regulations.

Food standards are implemented to guarantee safety and quality in the food industry. They offer guidance to farmers and processors on how to handle food hygienically. These standards also establish the maximum levels of additives, contaminants, pesticides, and veterinary drug residues that are safe for consumption. Additionally, they specify the proper ways to measure, package, and transport food to ensure its safety. Food Standards play a crucial role in informing consumers about the nutritional content of food and allergen labeling, allowing them to make informed choices about what they consume [2,6].

Food safety management systems (FSMS) have evolved to improve food safety practices and ensure the quality and safety of food products. The introduction of Hazard Analysis and Critical Control Points (HACCP) in the 1970s was a significant milestone, establishing preventive measures for identifying and controlling hazards. HACCP has become a global standard adopted by many countries. In the 2000s, the ISO 22000 standard [24] integrated HACCP principles with quality management systems, further enhancing food safety practices worldwide (Figure 1) [24–27].

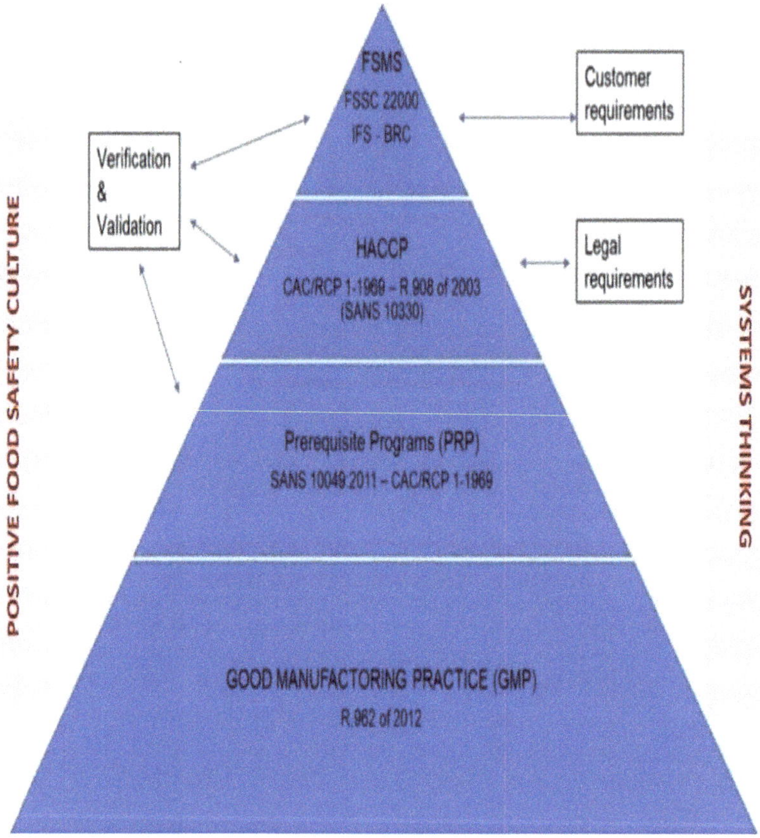

Figure 1. The FSMS Building Blocks and the Systems Thinking Approach (Adapted from "*What are Food Safety Management Systems?*", https://chimerasystems.co.za/services/fsms/), accessed on 31 July 2022, Lee, J.C. et al. 2021 [6,27].

Developed countries have well-established FSMS, regulatory frameworks, and training programs, enabling them to effectively control food safety risks and reduce foodborne illnesses. In contrast, developing countries face challenges due to limited resources, weak regulations, and low awareness of food safety. However, capacity-building initiatives aim to address these challenges and improve food safety practices in developing nations [23].

The impact of FSMS in reducing foodborne illnesses is evident in both developed and developing countries. However, more efforts are needed, especially in developing countries with greater food safety challenges, to ensure the effective implementation of good food safety practices and regulations.

2.1. Managerial Systematic Approach to Food Safety

Developing countries face challenges in implementing traceability programs due to limited resources, inadequate infrastructure, low literacy rates, and a lack of automation. This results in difficulties in tracking and tracing food products, which can increase the risk of foodborne illnesses. In contrast, developed countries have well-established traceability systems supported by advanced technologies and reliable infrastructure, enabling more effective implementation and comprehensive reporting [11,28].

Traceability programs in the food industry offer numerous benefits, including improved food safety, enhanced supply chain management, increased consumer confidence, compliance with government regulations, and reduced risks and wastage. These programs also create opportunities for value addition and improve market access for reliable food production intended for global export. However, the implementation of traceability programs entails certain drawbacks, such as higher operational costs, increased time requirements, and complexities associated with implementation and data management [29–32].

HACCP serves to establish, execute, and confirm the effectiveness of a Food Safety Management System. The fundamental basis for a practical and functional HACCP program in the prevention of foodborne illness outbreaks lies in the implementation of PRPs. A managerial approach that considers the entire system is essential when developing, implementing, and verifying HACCPs and PRPs. (Figure 2).

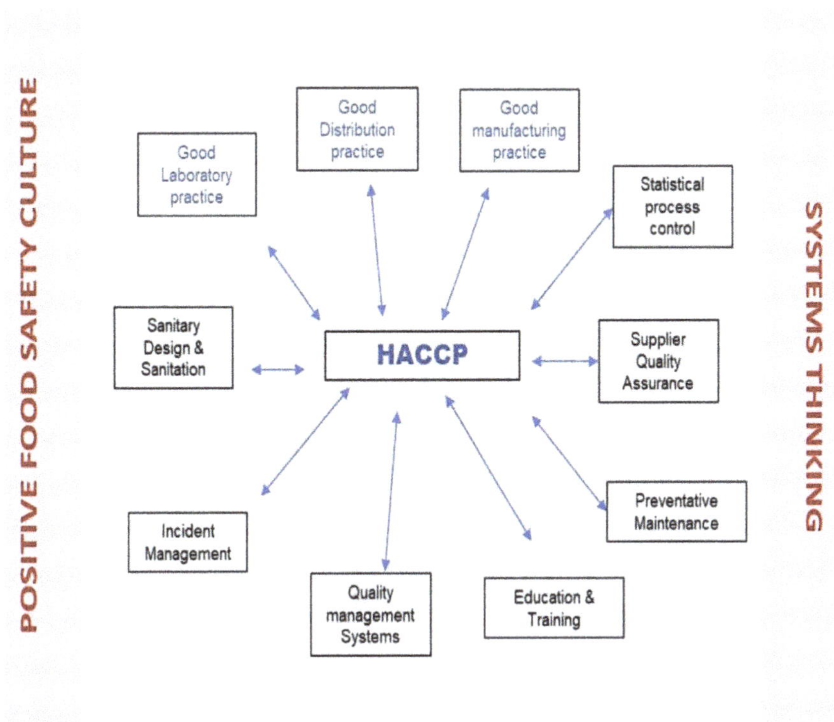

Figure 2. Managerial Systematic Approach to Food Safety Using HACCP and PRPs (Adapted from What is a Management System? Available online: https://images.app.goo.gl/Aqez5sSfpbvjQqmK7 (accessed 20 June 2021). Lee, J.C. et al. 2021 [6,33].

2.1.1. Prerequisites: FSMS Foundation in Developing and Developed Countries

Various studies highlight the significance of prerequisite programs as the foundation of HACCP-based food safety management systems in both developing and developed countries. These programs play a crucial role in preventing food contamination throughout the entire food production process, encompassing aspects such as food handling, storage, hygiene, sanitation, pest control, training, and maintenance (Figure 3) [6,7,34]. Moreover, organizational Food Safety Culture Works in Symbiosis with PRPs and HACCPs.

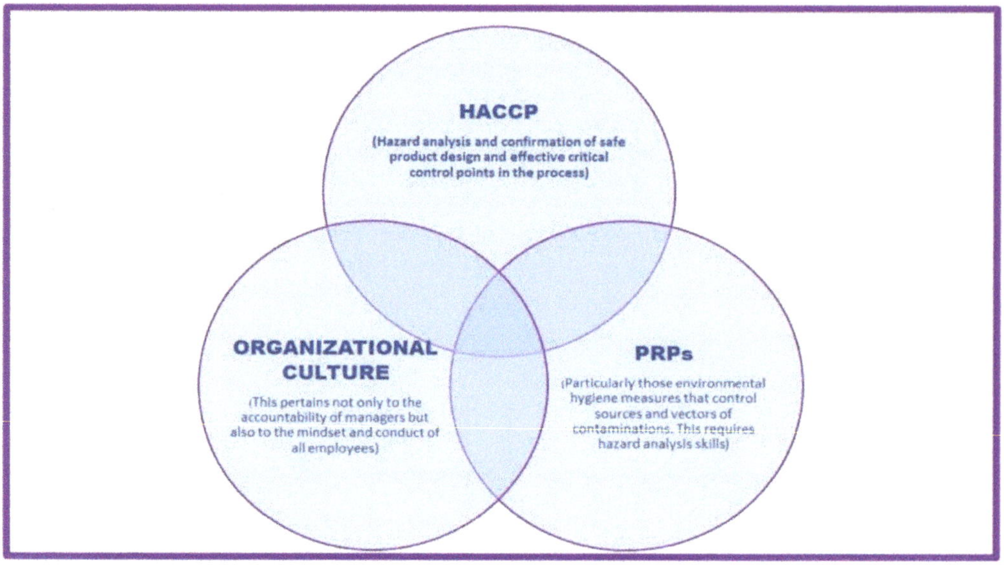

Figure 3. Organizational Food Safety Culture Works in Symbiosis with PRPs and HACCP (Adapted from: C.A. Wallace, and S.E. Mortimore, 2016) [7,35].

Developed countries, such as the United States and the European Union, have established guidelines and regulations for prerequisite programs through regulatory agencies like the FDA and EFSA. In contrast, implementing and complying with these programs may pose challenges in developing countries due to limited resources, cultural factors, and lack of knowledge and awareness.

Nevertheless, studies demonstrate that implementing prerequisite programs in developing countries can lead to significant improvements in food safety and a reduction in foodborne illnesses. For instance, a study conducted in Ghana revealed that the implementation of a food safety management system incorporating prerequisite programs resulted in a notable decrease in foodborne illnesses. Similarly, a study conducted in China demonstrated that the implementation of prerequisite programs in food processing plants contributed to a reduction in the prevalence of microbial contaminants in food products, thus mitigating the risk of foodborne illnesses [1,11,34].

Overall, prerequisite programs play a critical role in preventing food contamination and ensuring food safety. Their effectiveness in both developing and developed countries may vary depending on the extent of implementation and adherence to regulatory guidelines and standards.

2.1.2. FSMS and Systems Thinking

Managerial systems thinking involves taking a comprehensive approach to identifying the factors and interactions that contribute or could contribute to failures in a safety management system. In today's world, many challenges are intricate and cannot be adequately addressed from a single viewpoint. Therefore, it is necessary from a single viewpoint. Therefore, it is necessary to examine the system as a whole and identify areas for restructuring. By combining systems thinking with food safety expertise and abilities, complex problems can be addressed effectively. Figure 4 illustrates the interaction between FSMS, necessary good management practices, positive food safety culture, and systems thinking [6,7].

Figure 4. Organizational Food Safety Culture Works in Symbiosis with PRPs and HACCP (Adapted from: C.A. Wallace, W.H. Sperber, and S.E. Mortimore, 2018), Lee, J.C. et al. 2021 [6,36].

2.1.3. Positive Food Safety Culture

Food safety culture promotes food product and market safety in the food sector. Food safety culture impacts both developed and developing countries, but economic conditions, regulatory guidelines, and consumer awareness vary.

Food safety legislation, enforcement procedures, and authorities to safeguard the public are normally in place in developed countries. However, poor infrastructure, poverty, weak food safety rules, and limited resources make it more difficult for developing nations to establish a food safety culture. Despite such constraints, food safety awareness in developing nations has led to the establishment of various efforts and programs to improve food safety culture.

This study on the food safety culture element of FSMS and awareness in developing countries found that food safety culture is low due to low awareness and poor knowledge of food safety, inadequate infrastructure and regulatory standards, challenges implementing food safety guidelines, and insufficient financial resources to develop a robust food safety culture. However, developed countries have built robust infrastructure, resilient regulations, and enforcement mechanisms to control food safety concerns and promote a food safety culture.

Moreover, UNDP, WFP, WHO, and FAO initiatives have promoted food safety culture in developing countries and improved food safety management systems. These programs train, advise, and fund developing nations to improve food safety [7,37–41].

2.1.4. Food Safety, Food Security, and Food Sustainability

Food safety and food security are closely related concepts that impact the health and well-being of individuals worldwide. Food safety ensures that food is prepared and consumed without harm to consumers, while food security ensures access to sufficient, safe, and nutritious food that meets dietary needs and preferences. Factors influencing food safety and security include environmental conditions, agricultural practices, trade policies, food processing, storage, distribution, consumer behavior, and public health interventions. Food safety and food security are interdependent, with unsafe food causing diseases and decreased productivity. Developing and developed nations face challenges in food safety, with microbial contamination, pesticide residues, and aflatoxins being more prevalent in developing countries. To improve food safety and security, both developing and developed nations must adopt a comprehensive strategy that involves all food chain stakeholders [42–46].

Ensuring global food security and sustainability is a major challenge for both developing and developed countries. Food security focuses on providing access to food for all, while sustainability aims to meet current needs without compromising the ability to meet future needs. In developing countries, food insecurity is caused by factors like poverty, inadequate infrastructure, climate change, and political instability, leading to malnutrition and health issues. Developed countries face challenges such as food waste, unsustainable farming practices, and excessive resource use. Efforts to promote food sustainability include adopting sustainable agriculture practices, reducing food waste, and encouraging plant-based diets. To address food security, measures like providing food aid, supporting smallholder farming, and improving education and healthcare access are implemented. Achieving food security and sustainability requires a comprehensive approach tailored to the unique challenges of each country. It involves ensuring equal access to nutritious food, implementing sustainable agricultural practices, and reducing waste across the food system [47–51].

2.1.5. Food Fraud and Food Defense

Food fraud and food defense are major challenges globally, affecting both developed and developing countries. Food fraud involves intentional adulteration or misrepresentation of food products for financial gain, while food defense focuses on safeguarding food products against deliberate contamination.

Developing countries face greater vulnerabilities due to inadequate regulatory frameworks, weak enforcement, and poor infrastructure. Examples include the sale of misbranded and adulterated food products in India and the 2008 melamine scandal in China, where contaminated milk powder caused public outrage and exposed flaws in their food safety system [52,53].

Developed countries also experience food fraud incidents, such as the European horsemeat scandal and the U.S. honey laundering case, emphasizing the need for robust preventive measures. Food defense is a concern in developed countries, as the food supply-chain infrastructure may be targeted by terrorists, posing a risk to national security [34].

Both developed and developing countries must address these challenges by implementing effective regulations, strengthening enforcement mechanisms, improving infrastructure, and enhancing surveillance and monitoring systems. These measures are crucial to ensure the authenticity and safety of food products [54,55].

2.1.6. Rewording the Guidelines Related to Microbiological Standards

Academics and researchers generally concur that food quality can be viewed from two perspectives: objective and subjective. The objective aspect pertains to the physical characteristics of the product and involves quality control and food technology. On the other hand, the subjective aspect pertains to consumer's assessments and evaluations of the perceived quality characteristics of the product. Although food quality is a multifaceted and intricate concept, it is intrinsically linked to food safety. Food safety is the most crucial factor in ensuring that consumers purchase products that meet their safety expectations, and it is therefore subject to regulation. To ensure the quality and safety of food products, it is crucial to have monitoring systems in place at critical stages of the productions process. The use of accurate and quick analysis methods is essential to ensure that the product meets quality and safety standards and complies with labeling requirements. The prompt detection of spoilage agents, including bacteria, pathogens, and other microbial contaminants during food production and processing, is essential to minimize spoilage and ensure a safe food supply.

If the test results fail to meet the required criteria, food producers must follow the measures outlined by regulations. When repeated testing results are consistently satisfactory, food production companies must take immediate action to prevent the occurrence of microbial hazards. Companies must ensure that their products meet the specific criteria set for each type of food, without exception, depending on their characteristics [6].

As an example, *Cronobacter* spp., an opportunistic bacterium, can be found in various food types. While powdered infant formula (PIF) is the primary source of *Cronobacter*-related illnesses [22,23], the bacteria have also been detected in other foods such as fruits, vegetables, cereals, spices, starches, dry foods, and meat products. Several studies have investigated the prevalence and molecular characteristics of *Cronobacter* in different food sources, including studies conducted in Korea on poultry and vegetable products (published in 2019) and on powdered infant formula (published in 2012), a study in China on spices (published in 2018), and a study in England and Wales on clinical cases of *Cronobacter sakazakii* in infants (published in 2016). These studies contribute to our understanding of the occurrence and potential risks associated with *Cronobacter* in various food items [56–61].

Table 1 captures the main microorganisms of concern, toxins, and metabolites found in specific food categories [6].

Table 1. Examples of Micro-organisms, Parasite, Toxins, and Metabolites [6].

Category	Examples of Micro-Organisms, Parasites, Toxins and Metabolites
Raw meat: Carcasses of cattle, sheep, goats, pigs, and horses	*Salmonella* spp., *Escherichia coli*, *E.coli O157:H7*, *Yersinia enterocolitica*, *Campylobacter jejuni*, *Listeria monocytogenes*, *Clostridium botulinum* and *C. perfringens*, *Staphylococcus aureus* Parasites: *Toxoplasma*, *Trichinella*, *Taenia* and *Sarcocystis* Mycotoxins: Obtained from the animal through animal feed
Carcasses of broilers and turkeys	*Salmonella* spp., *Escherichia coli*, *Campylobacter* spp., *Listeria monocytogenes*, *Clostridium botulinum* and *C. perfringens*, *Staphylococcus aureus*, *Bacillus cereus*
Milk and dairy products	*Salmonella* spp., *Escherichia coli*, *E.coli O157:H7*, *Campylobacter* spp., *Listeria monocytogenes*, *Bacillus cereus*, *Staphylococcus aureus*, *Brucella* spp. Mycotoxins (aflatoxins M_1 and M_2), toxins from *S. aureus* and *C. botulinum* (mainly from yogurt containing fruits or nuts) shelfish
Egg products	*Enterobacteriaceae* (mainly *Salmonela* spp.), *L. monocytogenes*
Fishery products	*Staphylococcus* spp., *Clostridium botulinum*, *Vibrio* spp., *Vibrio parahaemolyticus*, histamine-producing bacteria (*Proteus morganii*), *Enterobacter* spp., *Citrobacter* spp., *Salmonella* spp., *Erysipelothrix rhusiopathiae* Parasites: *Diphyllobothrium latum*, *Clonorchis sinensis*, *Pseudoterranova* spp. Toxins: *Scombrotoxin, Ciguatera and Histamine*
shellfish	*Escherichia coli*, Toxins: *Amnesic shelfish toxin* (ASP), *Diarrhoetic shelfish toxin* (DSP), *Neurotoxic shellfish toxin* (NSP), *Paralytic shelfish toxin* (PSP), *Ciguatera*
Vegetables, fruits, and products thereof	*Escherichia coli*, *Salmonella* spp., *Shigella* spp., *Listeria* spp., *Staphylococcus* spp. (mainly in mushrooms) Toxins: *Mycotoxins* (Patulin mainly in apples and apple juice, pears and peaches), *aflatoxins* (mainly in figs), toxins from *Clostridium botulinum*
Cereals and nuts	*Salmonella* spp., *B. cereus*, *S. aureus* *Mycotoxins* (aflatoxins)
Milk powder & powdered infant formula, fruit, vegetables, cereals, starches, dry food & meat products	*Cronobacter* spp.

2.1.7. Constraints and Benefits: Third-Party Food Safety Schemes

Globally, third-party food safety schemes are growing in popularity because they enable food businesses to demonstrate their commitment to food safety to customers, regulators, and other stakeholders. In implementing and complying with third-party food safety schemes, however, both developing and developed nations face challenges and opportunities.

Constraints: Implementing third-party food safety programs can be expensive, particularly for small-scale and low-income food businesses. Investing in infrastructure, training, audits, and certification may be necessary. Some third-party food safety schemes may have excessively stringent requirements, making it difficult for small and medium-sized businesses to comply.

Complexity: Third-party food safety schemes may have complex requirements, and a lack of technical knowledge among businesses may impede their implementation and compliance.

Access: Access to third-party food safety schemes may be limited in some regions due to transportation or geographical barriers, particularly in developing nations.

Accrediting Bodies: Businesses and regulatory authorities may have little assurance of quality and consistency if certification bodies in different nations lack transparency, trust, and harmonization.

Benefits: Improved food safety third-party food safety programs are designed to ensure that food products are safe and free of contaminants, providing customers with confidence. Third-party food safety schemes provide assurance to consumers regarding the safety and quality of the food products they consume. Compliance with third-party food safety programs can provide access to new markets and competitive advantages over non-compliant competitors.

Continuous Improvement: Third-party food safety programs encourage food businesses to continually improve their processes and performance to maintain certification.

Harmonization: Third-party food safety schemes can promote harmonization in food safety standards and provide a common language across borders, resulting in a positive approach to global food security, public health, and commerce [62–68].

3. Materials and Methods

3.1. Sample Size

The sample of this study consisted of 116 small and medium-sized companies from all over the world. Thirty-six (36) companies from developed countries and 80 (80) companies from developing countries.

3.2. Content of Analyses-Case Study Survey Data Collection by Global Food Safety Professionals

The objective of this research is to examine the impact of food safety standards on the food industry in both developing and developed regions around the world. To achieve this, nearly 116 surveys were conducted across 27 primary food sectors. The surveys were designed to collect relevant data while ensuring complete confidentiality of the company information.

The surveys were divided into two categories—one that collected data before the implementation of food safety standards and the other that collected data after the implementation of food safety standards. This division allowed for a comparative analysis of the impact of food safety standards on the food industry. To ensure that the data collected was comprehensive and representative, the surveys were divided into developing regions and developed regions. Additionally, they were conducted across various food sectors to obtain a broad understanding of the impact of food safety standards in the food industry.

Overall, the surveys were carefully designed and conducted to gain valuable insights into the effects of food safety standards on the food industry. The analysis of this data will provide useful information for policymakers and stakeholders in developing and implementing future food safety standards (See Sample Case Studies Survey, Appendix A).

In addition to the surveys, our team of food safety professionals provided valuable insights on regional food safety standards. These expert opinions were instrumental in enhancing the study and providing a more comprehensive understanding of the impact of food safety standards on the global food industry (See Acknowledgements and Thanks Section).

It took nearly 16 months of intensive research and analysis by a team of active global food safety experts to produce a reliable comprehensive, and inclusive case study. The team performed a variety of tasks, including consulting with food safety specialists, designing, completing, compiling case study survey data, analyzing, and summarizing content. A dedicated share drive was opened for this research which only the key authors had overall access to. When food safety professionals forwarded their Case Study surveys, the lead author reviewed each survey and contacted the contributor to correct or clarify the survey data. All revised surveys with comments and replies are retained as authentic records. The lead author virtually collaborated with most of the team food safety professionals upon request. All the while, the confidentiality of company information was maintained.

The authors organized the data by food sector, developed or developing region, and divided the data into before and after implementation of the Food Safety Management System FSMS. The organized data were divided into data reports and then converted into illustrative bar graphs.

The topic contents of this FSMS study are presented in a similar order to the elements of Implementation of Food Safety Management Systems Case Study Surveys. See Appendix A, Sample Case Study Survey.

3.3. *Case Studies' Confidentiality Integrity*

Case Study Survey Disclosure:

Your case study contribution(s) data will be included within a collective statistical analysis to be part of our research paper "Global Implementation of Food Safety Management System Case Studies by Small-Medium Enterprises: Major Constraints and Benefits".

Confidentiality will be maintained. Therefore, this survey/questionnaire does not ask for company name, city, etc.

We ask non-specific general questions to gather pertinent information to be able to analyze and present the collective data in an informative format.

"The aim of the research paper is to evaluate the young history of Food Safety Management Systems Standards and its major constraints and benefits arising from its implementation by small to medium size enterprises around the world. How to better improve or alleviate the constraints and to catapult the success of world food safety best management practices programs, HACCP and food safety management systems. Finally, to arrive at a consensus as to whether the implementation of Food Safety Management Systems Standards in small to medium food enterprises has so far impacted global foodborne illness (for better or worse)." See Supplementary materials Sample Case Study Survey M1.

The following topic sections focus on worldwide FSMS elements and practices; before and after implementation of FSMS results for Developing and Developed Regions by Food Sector Category and the constraints and benefits of results of the Global Case Study Surveys visa vi the specific data and input contributed by reliable Food Safety Professionals around the world.

The Global Food Sector Chart illustrates the numerous food sectors represented by the Case Study Surveys, grouped by primary sectors for data analysis (Figure 5).

We followed the case study method "exploring a real-life, contemporary bounded system (a case) or multiple bounded systems (cases) over time, through detailed, in-depth data collection involving multiple sources of information as reported by Creswell [69] and Crowe et al. [70].

3.4. *Analytical Framework*

The processing of the statistical data was carried out with the statistical package IBM SPSS v.28. Scale reliability was examined with Cronbach's alpha index.

Descriptive statistics are presented with absolute and relative frequencies (N, %) for categorical variables and median, interquartile range values for Likert scale variables.

Dichotomous variables of FSMS Element Details before and after FSMS implementation were compared with the McNemar test, while Likert scale variables of attitudes towards constraints and incentives for FSMS implementation were compared with the Wilcoxon sign rank test [71].

Figure 5. The Global Food Sector Chart illustrates the numerous food sectors represented in our data.

4. Results and Discussion

In Table 2, the specific food sectors, and countries of the companies of the sample are presented. The most represented sectors in the sample were processed fruit and vegetables (13.8%), snacks and baked food (12.1%), dairy (9.5%), food service (9.5%), and ingredients (9.5%). Most companies were established in South Asia (22.4%), Africa (20.7%), East and West Asia (14.7%), Latin America and the Caribbean (13.8%), and Europe (12.1%).

Table 2. Descriptives for company characteristics in the sample of this study.

	Total Sample (N = 116)		Developed (N = 36)		Developing (N = 80)	
	N	%	N	%	N	%
Sectors						
Agri	6	5.2	2	5.6	4	5.0
Beverage.Water	5	4.3	2	5.6	3	3.8

Table 2. *Cont.*

	Total Sample (N = 116)		Developed (N = 36)		Developing (N = 80)	
	N	%	N	%	N	%
Canning/Bottling/Packaging	6	5.2	2	5.6	4	5.0
Coffee/Condiments/Sugar	8	6.9	0	0.0	8	10.0
Dairy	11	9.5	3	8.3	8	10.0
Services (Education, sanitary, transport, storage etc.)	7	6	2	5.6	5	6.3
Food service	11	9.5	6	16.7	5	6.3
Fruit.Veg-Processed	16	13.8	5	13.9	11	13.8
Ingredients	11	9.5	0	0.0	11	13.8
Meat	8	6.9	3	8.30	5	6.3
Snacks.BakedGoods	14	12.1	4	11.1	10	12.5
Other	13	11.2	7	19.4	6	7.5
Countries						
Africa (Algeria, Cameroon, Ethiopia, Ghana, Nigeria, Pakistan, South Africa)	24	20.7	0	0.0	24	30.0
UK and Australia	7	6.0	7	19.4	0	0.0
Latin America and the Caribbean (Brazil, Jamaica, Mexico, Trinidad)	16	13.8	0	0.0	16	20.0
North America (USA, Canada)	12	10.3	12	33.3	0	0.0
South Asia (India)	16	13.8	0	0.0	16	20.0
Other South Asia (Iran, Pakistan)	10	8.6	0	0.0	10	12.5
East and West Asia (Myanmar, Vietnam, China, Jordan, UAE)	17	14.7	7	19.4	10	12.5
Europe	14	12.1	10	27.8	4	5.0

4.1. Developed Countries vs. Developing Countries

Attitudes towards the constraints and benefits of implementing FSMS standards before and after implementation in developed and developing countries are presented in Figure 6a–c. After implementation, there was a significant increase in the percentage of companies that have implemented the international FSMS both in developed (16.7% to 63.9%) and developing countries (26.6% to 48.1%). Certification has also increased from 34.2% to 59.6% in the total sample, namely from 33.3% to 61.1% in developed countries and from 34.6% to 59.0% in developing countries. The increase in ISO 22000 implementation from 34.2% to 46.8% in the total sample presented a more significant increase for developing countries (35.5% to 48.7%, $p = 0.008$). Only developed countries presented a significant increase in SQF implementation (5.7% to 22.9%, $p = 0.014$) and only developing countries presented a significant increase in Food Standard Agency implementation (17.1% to 27.6%, $p = 0.005$). All companies have increased the rate of change or upgrade of FSMS standards (51.4% to 75.7%, $p < 0.001$). The implementation of quality standards increased from 51.4% to 78.9% ($p < 0.001$), the implementation of the earlier version of FSMS from 34.2% to 52.3% ($p < 0.001$) and the implementation of GFSI from 9.5% to 46.7% ($p < 0.001$). There was also an increase for all companies both in State (59.8% to 64.3%, $p = 0.025$) and Federal (33.9% to 38.4%, $p = 0.025$) Regulatory Agencies Inspections. There was a significant increase in medium vs. small company size (57.1% to 62.3%, $p = 0.046$), only in developing countries, while a smaller, non-significant shift was also detected in developed countries (41.2% to 44.1%, $p > 0.05$).

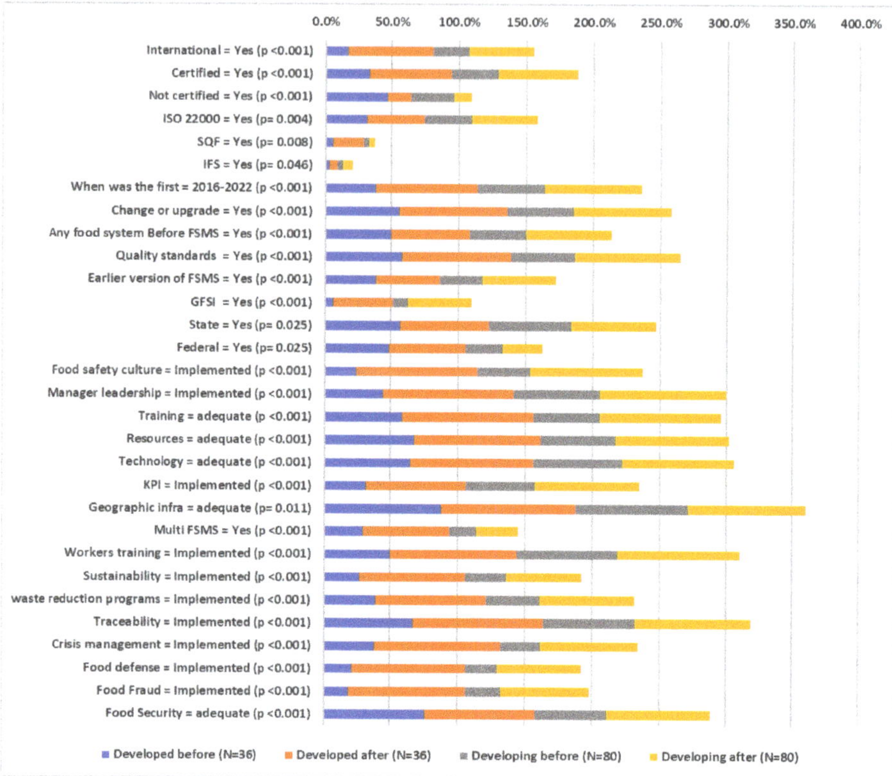

(**a**)

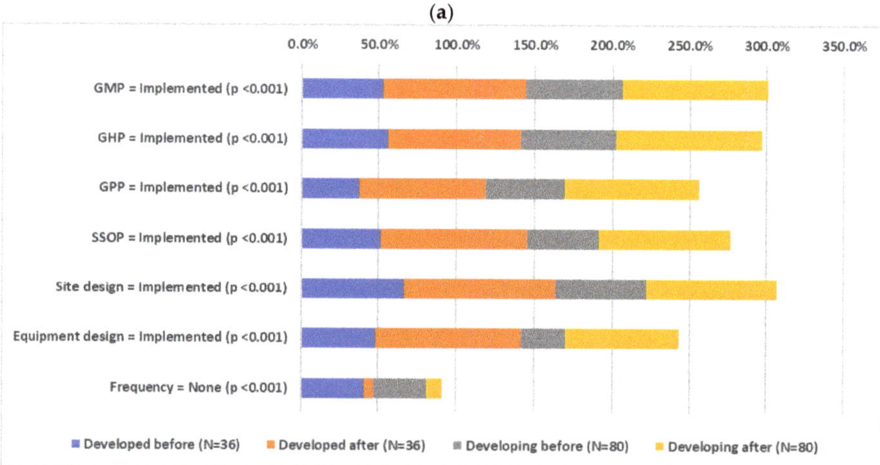

(**b**)

Figure 6. *Cont.*

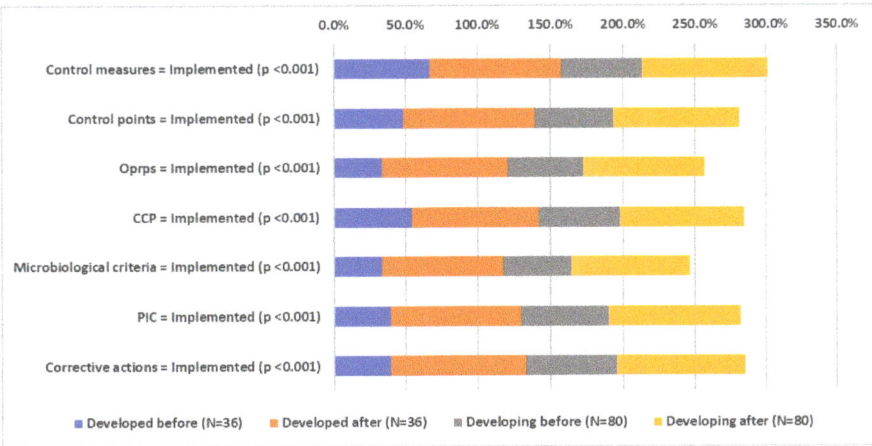

(**c**)

Figure 6. (**a**) Attitudes (FSMS element details) towards the constraints and benefits of implementing FSMS standards before and after implementation in developed and developing countries. Only attitudes with significant changes are presented. (**b**). Attitudes (Prerequisites) towards the constraints and benefits of implementing FSMS standards before and after implementation in developed and developing countries. Only attitudes with significant changes are presented. (**c**) Attitudes (HACCP) towards the constraints and benefits of implementing FSMS standards before and after implementation in developed and developing countries. Only attitudes with significant changes are presented.

Food safety culture and manager leadership implementation were increased to over 80% after FSMS implementation in both developed and developing countries ($p < 0.001$). Training, resources, and technology adequacy were also increased in all companies ($p < 0.001$). Production yield adequacy remained at the same level (78.5% to 79.4%) before and after FSMS implementation. Findings are in agreement with the study of Nguyen et al. [72], who indicated that mandatory and voluntary regulations and standards play a pivotal role in global food chains by ensuring comprehensive, forward-looking strategies that prioritize risk management and ongoing enhancement within the Food Safety Management System (FSMS).

Key Performance Indicator (KPI), Multi-FSMS, Workers Training, Sustainability Programs, Food & Materials Waste, Reduction Programs, Lot Identification Traceability, Crisis Management, Food Defense TACCP Plan and Food Fraud VACCP Plan implementations also increased after FSMS implementation for all companies participating in the sample ($p < 0.001$). These results agree well with the study of Fotopoulos et al. [73], who reported that essential elements like a company's characteristics (preliminary measures, machinery, and validation processes) and workforce attributes (staff availability, dedication, education, and willingness) are key factors when establishing an efficient HACCP system. As per the results, these underlying factors also greatly influence the system's objectives concerning the recognition, evaluation, and management of food-related safety risks.

Attitudes towards food security in the region of the company's activities were more positive after the implementation of FSMS in developing countries (53.1% to 78.1%, $p < 0.001$). Prerequisites implementations (GMP, GHP, GPP, SSOP, Site design, Equipment design) have increased by over 80% after FSMS implementation in all companies ($p < 0.001$), while internal audits implementations (annual, quarterly, monthly) have significantly increased to over 90% of the companies ($p < 0.001$). HACCP management system contents (control measures, control points, operational prerequisite, programs oPRPs, critical control points, processed based microbiological criteria and testing practices, monitor, verify, validate, record keeping, responsible person in charge PIC and deviations, corrective actions) have all significantly increased after FSMS implementation ($p < 0.001$). Our results agreed with those reported by Chen et al. [74], who indicated that the rate of defects in chaga products showed a noticeable decrease, and there was a significant reduction in the instances of process flow irregularities needing rectification ($p < 0.05$). These changes suggested an enhancement of safety and quality standards. With continued implementation over an extended duration, the advantages of the food safety management system would become even more pronounced, leading to substantial enhancements across a broader range of monitored aspects. Findings also agree with the study of Sun et al. [75], who indicated that once the prerequisite programs (PRPs) of Good Hygiene Practices (GHP) and Good Manufacturing Practices (GMP) are implemented, the Hazard Analysis and Critical Control Points (HACCP) system can be employed to control food safety risks. Moreover, the study by Cusato et al. [76] indicated the implementation of a food safety system in a dairy processing plant located in the State of São Paulo. After the implementation of the food safety system, a significant reduction in the yeast and mold count was observed ($p < 0.05$).

Relative to constraints, there was a decrease in companies reporting that FSMS implementation is an expensive and complicated task, i.e., there are economic, technological, and legislation constraints ($p < 0.001$), while there was an increase in companies reporting the non-familiarity of FSMS to customers and consumers ($p < 0.001$). In developing countries, the "lack of clarity" constraints decreased after FSMS implementation ($p = 0.007$).

There was a positive shift in attitudes towards the incentives of avoiding duplication between processes in developed countries ($p = 0.013$), improving quality of management in developing countries ($p = 0.007$) and providing evidence of legal compliance in both developed and developing countries ($p = 0.033$). Yet, after the implementation of FSMS companies in developing countries presented a more negative attitude towards the incentives of reducing product losses ($p = 0.033$), enhancing export competitiveness ($p < 0.001$), and improving company image ($p < 0.001$).

4.2. Differences per Geographical Region before and after FSMS Implementation

The differences before and after FSMS implementation for companies from different geographical regions of developed and developing countries are presented in Figure 7. UK and Australian companies did not present any significant shift in the application and attitudes toward FSMS standards; thus, they are not presented here. European companies reported significant increases in changes/upgrades of the FSMS standards, application of Quality standards, GFSI implementations, Food safety culture and manager leadership applications, multi-FSMS, Sustainability, waste reduction programs, crisis management, food defense, and food fraud systems integration. Also, significant increases in GMP, GHP and Equipment design prerequisites application, as well as all HACCP systems, were detected in European companies.

North American companies (US and Canada) indicated a significant increase in the implementation of International FSMS, Certification and SQF integration, changes which were not detected in Europe. Quality standards and GFSI integration were also increased for the North American sample, as well as Food safety culture, Manager leadership, Training, KPI, Workers' training, Sustainability, Traceability and Crisis Management. Moreover, prerequisites and HACCP implementations have increased to 80–100% after FSMS implementation in US and Canadian companies.

African companies reported significant increases in certification, changes/upgrades of the FSMS standards, application of Quality standards, GFSI implementations, Food safety culture, and manager leadership applications, adequacy of Training and Resources, KPI implementation, Waste reduction programs, Traceability, Crisis management, food defense, and food fraud systems integration, as well as food security adequacy. Also, significant increases in prerequisites and HACCP systems application were reported in African companies. Similar trends were reported in Latin American countries.

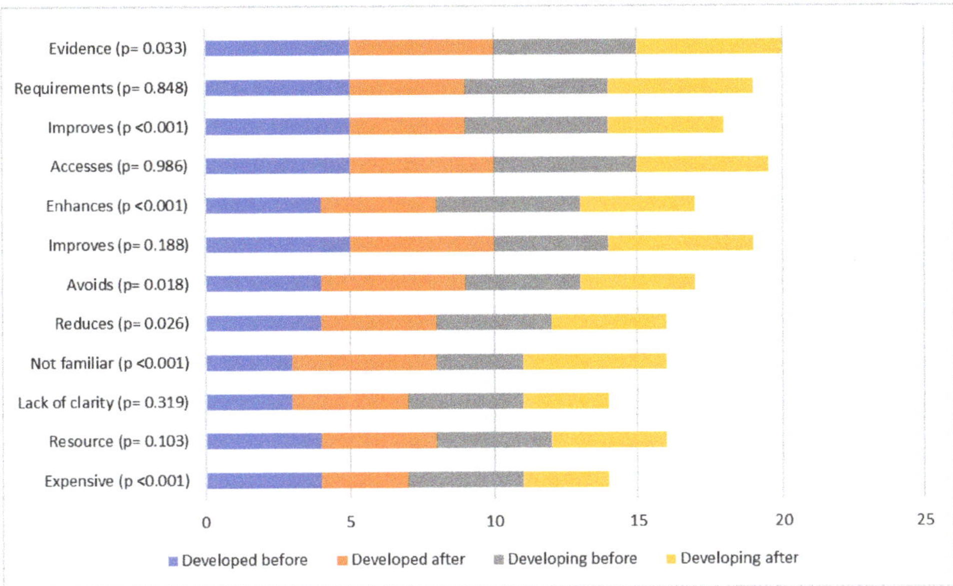

Figure 7. Median values of enterprise attitudes before and after implementation of FSM regarding specific constraints. Probability (p) values concern changes in attitude when both developed and developing countries are considered (Wilcoxon sign rank test, $p < 0.01$).

For Indian companies, on the other hand, international FSMS application and certification have presented a significant increase, larger than the ones reported for African or Latin American companies and similar to the levels of companies from developed countries. Yet, Indian companies presented high rates of most prerequisites and HACCP applications, even before FSMS integration, and significant increases were detected in SSOP, Site design, Equipment design prerequisites as well as HACCP Microbiological criteria and PIC that had lower integration rates before FSMS implementation. Other companies in South Asia (Iran, Pakistan) presented similar differences to Indian companies, apart from HACCP integration which increased from 30–50% before to approximately 90% of companies after FSMS implementation.

Findings are aligned with the study by Grover et al. [77], who explored the use of quality management tools for identifying and prioritizing challenges faced by small food establishments during the adoption of preventive controls mandated by the FSMA legislation.

European private food safety standards exert considerable influence on global food and agricultural supply chains. They act as the link between consumer expectations and producers to bring the safety standards of exporting countries to par with, and sometimes well above, EU legislation, as reported by Rao et al. [78]. Private standards also allow the retail sector to govern agri-food supply chains from a distance, without taking on additional legal or financial obligations. The most observable effects of private food safety standards are characterized by changes in governance patterns, supply chain structures, access to international markets, and increasing demands with respect to non-safety attributes of food products.

Finally, East, and West Asian companies established in Myanmar, Vietnam, China, Jordan and UAE reported similar changes to Indian companies yet greater significant increases in KPI implementation, Geographic infrastructure adequacy, Multi FSMS, Workers' training, Sustainability, Waste reduction programs, Traceability, Crisis management, Food Defense, and Food Fraud systems implementation. Such as in Iran and Pakistan, companies established in East and West Asia increased the integration of HACCP systems from 20–50% to 80% after FSMS implementation.

Attitudes towards constraints of FSMS implementation as an expensive and complicated task were reduced for North American, African, Latin American, and Asian companies (except for Iran and Pakistan). The constraint of non-familiarity of FSMS to customers and consumers increased for North American, European, African, Latin American, and Asian companies (except for Iran and Pakistan). For North American and African companies, the incentives of reduction in product loss and meeting customers' requirements were decreased, while the incentive of providing evidence for legal compliance increased. European companies were the only geographical sector that increased the belief in avoiding duplication between processes. Companies from Latin America and India presented a positive shift in the belief that FSMS standards implementation improves the quality of management.

4.3. Differences per Sector before and after FSMS Implementation

The processed fruit and vegetables, snacks and baked goods, and ingredients sectors presented a significant improvement in International FSMS implementation. The dairy, meat, and ingredients sectors expanded the rates of certifications. The snacks and baked goods and dairy sectors increased the rates of upgrades and changes in standards. Quality standards application has increased in the processed fruit and vegetables, snacks and baked goods, and dairy sectors, while GFSI implementation has increased also for ingredients and meat sectors. Moreover, all sectors presented significant increases in the implementation of Food safety culture, Manager leadership, and Training adequacy. Implementation of workers' training, Sustainability, waste reduction programs, Traceability, Crisis Management, Food Defense, and Food Fraud systems has increased for all sectors as well. Significant large improvements were also observed in prerequisites and HACCP systems applications for sectors that presented lower integration before FSMS implementation, and smaller non-significant improvements have also been detected for the ingredients and meat sectors, even though prerequisites and HACCP integration was sufficient even before the FSMS (Figure 8).

No significant change in the incentives to implement FSMS Standards was detected within different sectors before and after FSMS implementation. The sector of snacks and baked goods, as well as the ingredients sector, perceived the FSMS after implementation as less expensive and complicated, compared to their attitudes before implementation. Yet, the sectors of processed fruit and vegetables, snacks and baked goods, food service, ingredients, and meat have reported greater constraints of non-familiarity with FSMS standards to consumers and customers after FSMS implementation. These findings well agree with the study of Wilcock et al. [79] who indicated that the main motivating factors for HACCP implementation were the likelihood of future regulation, the value of HACCP for marketing, and avoiding food safety problems. They also mentioned that successful implementation depended heavily on management commitment. They also used semi-structured in-depth interviews from different food sectors, including meat, wine, baked goods, flavors, and minimally processed fruits and vegetables. Finally, the implementation of HACCP led to increased food safety and product quality.

All indicators are very good as shown in Table 3 (>0.7). A low Cronbach's alpha index appears for Constraints, although it is still marginally acceptable because it is >0.5.

Table 3. Cronbach alpha for all parameters tested.

	Cronbach's Alpha	
	Before	**After**
FSMS Element Details	0.740	0.718
Prerequisites	0.712	0.827

Table 3. *Cont.*

	Cronbach's Alpha	
	Before	**After**
HACCP	0.927	0.938
Constraints	0.559	0.503
Incentives	0.828	0.828

Figure 8. Differences per sector before and after FSMS implementation.

4.4. Limitations of the Study

- Financial

Developing nations face numerous obstacles to the implementation of food safety standards, including inadequate infrastructure, limited financial resources, and a lack of understanding of food safety fundamentals. Recommendations include allocating adequate funds for the implementation and enforcement of food safety standards, strengthening regulatory frameworks, training personnel, educating consumers about food safety, and promoting international cooperation [3–5].

- Infrastructure

Developing nations confront inadequate infrastructure, lack of water, polluted water, sanitation facilities, and financial resources. Food safety standards can be implemented in developing nations to reduce the incidence of foodborne illnesses, improve the quality of life for the population, and facilitate export market access [2,11]. However, developed nations are hindered by complex regulatory environments and limited resources. [5,12–16].

Developing countries face challenges due to limited resources, weak regulations, and low awareness of food safety. However, capacity-building initiatives aim to address these challenges and improve food safety practices in developing nations [23].

- Prerequisite Programs

Implementing and complying with these fundamental PRP programs may pose challenges in developing countries due to limited resources, cultural factors, and lack of knowledge and awareness. [1,11,34].

- Traceability Programs

Developing countries face challenges in implementing traceability programs due to limited resources, inadequate infrastructure, low literacy rates, and a lack of automation. This results in difficulties in tracking and tracing food products, which can increase the risk of foodborne illnesses. [11,28].

- Food Safety Culture (FSC)

This study on the food safety culture element of FSMS and awareness in developing countries found that food safety culture is low due to low awareness and poor knowledge of food safety, inadequate infrastructure, and regulatory standards, challenges implementing food safety guidelines, and insufficient financial resources to develop a robust food safety culture. The significance of the elements of food safety culture lies within the human factor and specific training including parameters such as leadership, commitment, communication, risk awareness, work environment, and management system, styles, and processes [80–82]. A more recent work developed an assessment tool that can be used to assess food workers' perceptions of their restaurant's food safety culture [83]. The highest-rated construct was Resource Availability, i.e., the availability of resources to maintain good hand hygiene. The second highest-rated construct was Employee Commitment, which assessed workers' perceptions of their coworkers' commitment to food safety.

The last two constructs were related to management. Leadership assessed the existence of food safety policies, training, and information sharing. The Management Commitment assessed whether food safety was a priority in practice. Finally, they assessed Workers' Beliefs about Food Safety Culture. We should not forget that food safety culture needs to be established and maintained according to The Commission Regulation (EU) No. 2021/382 (European Commission, 2021), amending the Regulation (EC) No. 852/2004 (European Commission, 2004). The Reg. (EU) No. 2021/382 did not detail methodologies for FSC implementation, investigation, or improvement, which have been recently evaluated by the European Commission Notice 2022/C 355/01 (European Commission, 2022). It contains an example of indicators for FSC assessment, to guide companies towards a successful FSC adoption and dissemination, and a checklist to evaluate FSC for official control activities [84].

4.5. Areas for Further Research

A much more systematic approach needs to be adopted to investigate how the FSMS interacts with other variables, such as food safety and quality culture. It is also important to see the human factor involvement in these systems and how effective training and communication could help towards a better understanding and maturation of the systems employed by food firms. A strong culture always helps firms to make decisions and act fast on recalls, or customer complaints. Models, guidelines, assessment tools and customization examples need to be developed for FSC evaluation, as well as strategies for its dissemination within companies and along the supply chain. Of course, this requires effective management commitment and effective communication across all company levels.

5. Conclusions

Attitudes towards the constraints and benefits of implementing FSMS standards before and after implementation in developed and developing countries showed that after implementation there was a significant increase in the percentage of companies that have implemented the international FSMS both in developed (16.7% to 63.9%) and developing countries (26.6% to 48.1%). Certification has also increased from 34.2% to 59.6% in the total sample, namely from 33.3% to 61.1% in developed countries and from 34.6% to 59.0% in developing countries. Food safety culture and manager leadership implementation were increased to over 80% after FSMS implementation in both developed and developing countries ($p < 0.001$).

Key Performance Indicator (KPI), Multi-FSMS, Workers Training, Sustainability Programs, Food & Materials Waste, Reduction Programs, Lot Identification Traceability, Crisis Management, Food Defense TACCP Plan and Food Fraud VACCP Plan implementations also increased after FSMS implementation for all companies participating in the sample ($p < 0.001$). Attitudes towards food security in the region of the company's activities were more positive after the implementation of FSMS in developing countries (53.1% to 78.1%, $p < 0.001$. There was a decrease in companies reporting that FSMS implementation is an expensive and complicated task, showing economic, technological, and legislation constraints ($p < 0.001$), while there was an increase in companies reporting the non-familiarity of FSMS to customers and consumers ($p < 0.001$).

Regarding differences in geographical regions, European companies reported significant increases in changes/upgrades of the FSMS standards, application of quality standards, GFSI implementations, food safety culture and manager leadership applications. North American companies (US and Canada) indicated a significant increase in the implementation of International FSMS, Certification and SQF integration, changes which were not detected in Europe. For Indian companies, international FSMS application and certification have presented a significant increase, larger than the ones reported for African or Latin American companies.

No significant change in the incentives to implement FSMS Standards was detected within different sectors before and after FSMS implementation. The sectors of processed fruit and vegetables, snacks and baked goods, food service, ingredients, and meat have reported greater constraints of non-familiarity of FSMS standards to consumers and customers after FSMS implementation.

According to the findings of the study, there are still gaps in the implementation of food safety standards, particularly in the areas of HACCP and microbiological standards. To ensure the effective implementation and enforcement of food safety standards, governments should allocate adequate resources and strengthen their regulatory frameworks. In this context, companies should improve food safety culture by empowering different dynamics with the aid of models, guidelines and assessment tools.

Supplementary Materials: The following supporting information can be downloaded at: https://www.mdpi.com/article/10.3390/foods12173218/s1, Sample Case Study Survey [M1].

Author Contributions: Conceptualization, J.C.L. and T.V.; methodology, J.C.L. and T.V.; software, M.N.; validation, J.C.L., A.A. and T.V.; formal analysis, J.C.L. and T.V.; investigation, J.C.L., M.N. and T.V.; resources, J.C.L. and T.V.; data curation, J.C.L. and T.V.; writing—original draft preparation, J.C.L., M.N., A.A. and T.V.; writing—review and editing, J.C.L., M.N., A.A. and T.V.; visualization, J.C.L. and T.V.; supervision, J.C.L. and T.V.; project administration, J.C.L. and T.V.; funding acquisition, J.C.L. and T.V. All authors have read and agreed to the published version of the manuscript.

Funding: This research received no external funding.

Data Availability Statement: Not applicable.

Acknowledgments: Dear Food Safety Colleagues, We really value your input on our review paper about international best practices and standards for food safety. Thanks to your hard work, we now have a research paper that is reliable, thorough, and insightful. Collaboration is necessary to ensure the safety of food around the world. It is an honor to be a part of a group that is so committed to ensuring the health and safety of consumers. Thank you very much for contributing to this research paper with your case studies data, review content and insightful comments. All your contributions ensured the success of this paper. From the initial concept to the final check of data, our teamwork resulted in a successful project. Your commitment to contributing to this review paper is admiral. Your hard work has resulted in a thorough report that will improve food safety and highlight our expertise. Let's keep collaborating and exchanging information to ensure a more secure and nutritious food supply chain around the world. To everyone who contributed so diligently, thank you.

Contributors, listed by Country in alphabetical order.
7 = Author's Content Contributor 8 = Case Studies Contributor 9 = Technical Advisor
Food Safety Consultant, Jocelyn C. Lee
Hrabi Nouredine, Algeria 7, 8
Ajay Shah, Australia 7, 8
Colin Christmas, Brazil 9
Carla Otsuki, Brazil 8
MPOPO Stéphane Hervé, Cameroon 7, 8
Sebastian Joseph, Canada 8
Desta Tamene Letik, Ethiopia 7, 8
Francois Bourdichon, France 8, 9
Joshua Owiredu, Ghana 8
Dimitris Drivas, Greece 7, 8
Anna Petrea, Greece 8
Vasilis Tsiminis, Greece 8
Sarika Agarwal, India 7, 8
Sonali B Gaikwad, India 7, 8
Nayan Maskai, India 7, 8
Wasi Asghar, India 8
Shuchi Arya, India 7, 8
Pranesh Badami, India 7, 8
Mahmood reza Akbarian, Iran 8
Valiollah Farhad Kasravi, Iran 8
Abdul Moiz, Italy 7, 8
Marva Hewitt, Jamaica WI 7, 8
Eva Inam Al Zein, Jordan 8
Jumana Faris, Jordan 8
Michael Manneh, Lebanon | USA 8
Mario Enchande-Bachtold, Mexico 8
Saint Yi Htet, Myanmar 8
Kitty Appels, Netherlands 7, 8
Solomon Oyeniran, Nigeria 7, 8
Abid Hussain, Pakistan 8
Nuno F. Soares, Portugal 7
Veselina Pelagic, Serbia 7, 8
Milosh Rodic, Serbia 7, 8
Dejana Kulesevic, Serbia 7, 8
Justice Bhekizulu Sifelani, South Africa 7, 8
Devi Yankataso, Trinidad-Tobago 8
Hrabi Nouredine, Tunisia 7, 8
Lavanya Nidunapu, United Arab Emirates 7, 8
Louise Roberts, United Kingdom 8
Jocelyn C. Lee, USA 7, 8
Steve Sayer, USA 8
John Duffill, Viet Nam 8

Conflicts of Interest: The authors declare no conflict of interest.

Appendix A

Acronyms Glossary

BRCGS	Brand Reputation Compliance Global Standard
CA	Corrective Action
CAR	Corrective Action Report
CL	Control Limit
CM	Control Measure
CCP	Critical Control Point
EMP	Environmental Monitoring Program
FDP	Food Defense Plan
FF	Food Fraud
FSMS	Safety Management System
FSC	Food Safety Culture
FSC	Food Sector Category
FSSC 22000	Food Safety System Certification
FSP	Food Safety Plan FSP
GAP	Good Agricultural Practices
GFSI	Global Food Safety Initiative
GMP	Good Manufacturing Practices
GHP	Good Hygiene Practices
GSP	Good Standardization Practices
HACCP	Hazard Analysis Critical Control Point
IFS	International Featured Standards
ISO 22000	International Standards Organization
Food Safety System	
oPRP	Operational Prerequisite Program
PDCA	Plan, Do, Check, Act
PRP	Prerequisite Program
RCA	Root Cause Analysis
SOP	Standard Operating Procedure
SQFI	Safe Quality Food Institute
SSOP	Sanitation Standard Operating Procedure
TACCP	Threat Analysis Critical Control Point
VACCP	Vulnerability Analysis Critical Control Point

References

1. Waddington, H.; Snilstveit, B.; Hombrados, J.; Vojtkova, M. Preventing Foodborne Diseases in Developing Countries: A Systematic Review. 2013. Available online: https://www.who.int/tdr/publications/documents/preventing-foodborne-diseases.pdf (accessed on 9 August 2022).
2. Global Food Safety Partnership, World Bank Group. *Food Safety in Low and Middle-Income Countries*; The World Bank Group: Washington, DC, USA, 2018.
3. Alli, I.; Okorie, N. Food safety and developing countries: The benefits of technology transfer. *J. Food Saf.* **2015**, *35*, 312–330.
4. Anang, F.M.; Rusul, M.; Atiliwi, M.C. Food safety management in developing countries: An overview. *Afr. J. Food Sci.* **2018**.
5. World Health Organization (WHO). The WHO's Website Includes Information on Food Safety, Global Foodborne Diseases, and International Food Safety Standards. Available online: https://www.who.int/foodsafety/en/ (accessed on 26 December 2022).
6. Lee, J.C.; Daraba, A.; Voidarou, C.; Rozos, G.; Enshasy, H.A.E.; Varzakas, T. Implementation of Food Safety Management Systems along with Other Management Tools (HAZOP, FMEA, Ishikawa, Pareto). The Case Study of Listeria monocytogenes and Correlation with Microbiological Criteria. *Foods* **2021**, *10*, 2169. [CrossRef] [PubMed]
7. Daraba, A.; Varzakas, T.; Lee, J.C. Implementation of FSMS and Correlation with Microbiological Criteria, Systems Thinking and Food Safety Culture, Food Safety Magazine (Dec 2022–Jan 2023 Issue), Process Control Column. Available online: https://digitaledition.food-safety.com/december-2022-january-2023/column-process-control/ (accessed on 8 August 2023).
8. Codex Alimentarius. Food Safety and Trade: How Food Safety Standards Can Facilitate International Trade. 2020. Available online: http://www.fao.org/fao-who-codexalimentarius/sh-media/news-detail/en/c/281286/ (accessed on 9 August 2022).
9. FAO. What Is Food Safety? Available online: http://www.fao.org/food-safety/about/what-is-food-safety/en/ (accessed on 30 May 2023).

10. Boscheri, E.; Parisi, S.; Ferrante, M. Food safety management systems: A need for harmonization and recognition. *Int. J. Food Sci. Technol.* **2018**.
11. Aung, M.M.; Chang, Y.S.; Chang, K.W. Food safety management systems and HACCP implementation in developing countries: A review of the literature. *Food Control* **2013**, *32*, 187–194.
12. Food and Drug Administration (FDA)—United States: The Official Website of the FDA, the Regulatory Agency Responsible for Ensuring the Safety, Efficacy, and Security of Food, Drugs, and Medical Devices in the United States. Available online: https://www.fda.gov/ (accessed on 30 May 2023).
13. European Food Safety Authority (EFSA)—Europe: The Website of EFSA, an Independent Scientific Agency That Provides Scientific Advice and Risk Assessment on Food Safety to the European Union. Available online: https://www.efsa.europa.eu/ (accessed on 30 May 2023).
14. Canadian Food Inspection Agency (CFIA)—Canada: The Official Website of CFIA, the Federal Agency Responsible for Safeguarding Food, Animals, and Plants in Canada. Available online: https://www.inspection.gc.ca/ (accessed on 30 May 2023).
15. Centers for Disease Control and Prevention (CDC)—United States: The CDC's Website Provides Information on Food Safety, Foodborne Illnesses, Outbreaks, and Preventive Measures in the United States. Available online: https://www.cdc.gov/foodsafety/index.html (accessed on 30 May 2023).
16. Food Standards Agency (FSA)—United Kingdom: The Official Website of FSA, the Government Department Responsible for Food Safety and Standards in the United Kingdom. Available online: https://www.food.gov.uk/ (accessed on 10 April 2023).
17. World Health Organization. Estimates of the Global Burden of Foodborne Diseases. 2015. Available online: https://www.who.int/foodsafety/areas_work/foodborne-diseases/ferg/en/ (accessed on 10 April 2023).
18. FoodSafety.gov. Recalls, Market Withdrawals, & Safety Alerts. 2021. Available online: https://www.foodsafety.gov/recalls (accessed on 9 August 2022).
19. European Commission. Rapid Alert System for Food and Feed (RASFF). 2021. Available online: https://ec.europa.eu/food/safety/rasff_en (accessed on 9 August 2022).
20. Business Standard. Pillsbury's Chakki Fresh Atta Recalled on Metal Contamination Fears. 2021. Available online: https://www.business-standard.com/article/companies/pillsbury-s-chakki-fresh-atta-recalled-on-metal-contamination-fears-121031300241_1.html (accessed on 8 August 2022).
21. Vanguard. NAFDAC Recalls 2 Batches of Infant Formula Milk Powder. 2021. Available online: https://www.vanguardngr.com/2021/02/nafdac-recalls-2-batches-of-infant-formula-milk-powder (accessed on 31 July 2022).
22. Available online: https://www.fda.gov/food/outbreaks-foodborne-illness/fda-investigation-cronobacter-infections-powdered-infant-formula-february-2022 (accessed on 31 July 2022).
23. FDA Outbreak Investigation (April 2023). Salmonella-Flour. Available online: https://www.fda.gov/food/outbreaks-foodborne-illness/outbreak-investigation-salmonella-flour-april-2023 (accessed on 30 May 2023).
24. *ISO 22000*; Food Safety Management System. International Organization for Standardization: Geneva, Switzerland, 2018.
25. Food and Agriculture Organization (FAO)—United Nations: The FAO's Website Provides Resources on Food Safety, Food Quality, and International Food Standards. Available online: http://www.fao.org/food/food-safety-quality/home/en/ (accessed on 10 April 2023).
26. What Are Food Safety Management Systems? Available online: https://chimerasystems.co.za/services/fsms/ (accessed on 10 April 2021).
27. Kumar, R. *Food Safety Management—Understanding and Implementing ISO 22000 and FSSC 22000*; CRC Press: Boca Raton, FL, USA, 2017.
28. Butts, J. Evolution of food safety systems: The past, present, and future. *J. Food Nutr. Res.* **2019**.
29. Ayodele, O.J.; Adetunji, M.O. Constraints to implementing food traceability in developing countries: A review. *Cogent Food Agric.* **2019**, *5*, 1597787.
30. Pourazadi, E.; Daneshmand, A.R.; Fathi SA, A.; Elshorbagy, A. Traceability as a Key Element for Ensuring of Food Safety in Supply Chain Management: A Review. *Qual. Eng.* **2019**, *31*, 567–588.
31. Schiefer, G.; Rickert, U. Traceability in food and agricultural supply chains–challenges and opportunities for developing countries. *J. Chain. Netw. Sci.* **2019**, *19*, 13–25.
32. Shailja, K.; Thomas, B. An examination of international food traceability programs. *J. Food Distrib. Res.* **2018**, *49*, 92–105.
33. What Is a Management System? Available online: https://images.app.goo.gl/Aqez5sSfpbvjQqmK7 (accessed on 20 June 2021).
34. Huang, L.; Wang, G.; Ma, J. The effectiveness of food safety management systems and prerequisite programs in food processing plants in China. *Food Control* **2016**, *69*, 54–61.
35. Wallace, C.A.; Mortimore, S.E. Chapter 3: HACCP. In *Handbook of Hygiene Control in the Food Industry*; Woodhead Publishing: Sawston, UK, 2016.
36. Wallace, C.A.; Sperber, W.H.; Mortimore, S.E. *Overview of a World Class Food Safety Programme in Food Safety for the 21st Century*, 2nd ed.; John Wiley & Sons Ltd.: Hoboken, NJ, USA, 2018; pp. 163–186.
37. Food and Agriculture Organization. Food Safety Culture and Awareness in Developing Countries. 2017. Available online: https://sdgs.un.org/un-system-sdg-implementation/food-and-agriculture-organization-fao-44217 (accessed on 10 April 2023).

38. Food and Agricultural Organization of the United Nations. Food Safety Systems and Food Security: A Review of the Critical Issues. 2017. Available online: https://www.fao.org/publications/home/fao-flagship-publications/the-state-of-food-security-and-nutrition-in-the-world (accessed on 10 April 2023).
39. Grace, D. Food Safety in Low and Middle-Income Countries. *Int. J. Environ. Res. Public Health* **2015**, *12*, 10490–10507. [CrossRef]
40. United Nations Development Programme. Strengthening Food Safety in Developing Countries. Available online: https://www.fao.org/publications/home/fao-flagship-publications/the-state-of-food-security-and-nutrition-in-the-world (accessed on 10 April 2023).
41. Käferstein, F. Foodborne diseases in developing countries: Aetiology, epidemiology and strategies for prevention. *Int. J. Environ. Health Res.* **2003**, *13*, 161–168.
42. Food Safety—World Health Organization (WHO). Available online: https://www.who.int/news-room/fact-sheets/detail/food-safety (accessed on 30 April 2023).
43. Food Security | USDA. Available online: https://www.usda.gov/topics/food-and-nutrition/food-security (accessed on 30 April 2023).
44. Food Safety in Food Security and Food Trade | IFPRI. Available online: https://www.ifpri.org/publication/food-safety-food-security-and-food-trade (accessed on 30 April 2023).
45. Food Safety in Developing Countries: An Overview—GOV.UK. Available online: https://www.gov.uk/research-for-development-outputs/food-safety-in-developing-countries-an-overview (accessed on 30 April 2023).
46. Food Safety Legislation in Some Developing Countries. Available online: https://www.intechopen.com/chapters/61873 (accessed on 30 April 2023).
47. Godfray, H.C.J.; Beddington, J.R.; Crute, I.R.; Haddad, L.; Lawrence, D.; Muir, J.F.; Pretty, J.; Robinson, S.; Thomas, S.M.; Toulmin, C. Food Security: The Challenge of Feeding 9 Billion People. *Science* **2010**, *327*, 812–818. [CrossRef] [PubMed]
48. Gustavsson, J.; Cederberg, C.; Sonesson, U.; van Otterdijk, R.; Meybeck, A. *Global Food Losses and Food Waste: Extent, Causes, and Prevention*; Food and Agriculture Organization of The United Nations: Rome, Italy, 2011.
49. Ingram JS, I.; Muir, J.F. Aligning Food Security and Sustainable Agriculture: Policy Imperatives and Research Challenges. *Front. Sustain. Food Syst.* **2019**, *3*, 1–6.
50. Lal, R. *Soil Health and Sustainability: Managing the Biotic Component of Soil Quality*; Elsevier: New York, NY, USA, 2018.
51. Tilman, D.; Balzer, C.; Hill, J.; Befort, B.L. Global Food Demand and the Sustainable Intensification of Agriculture. *Proc. Natl. Acad. Sci. USA* **2011**, *108*, 20260–20264. [CrossRef]
52. Chen, X.; Li, D. Food Fraud in China: Background, Challenges, and Prospects. *Compr. Rev. Food Sci. Food Saf.* **2018**, *17*, 806–821.
53. Henson, S.; Jaffee, S. Understanding Developing Country Strategic Responses to the Enhancement of Food Safety Standards. *Dev. Policy Rev.* **2013**, *31*, 647–665. [CrossRef]
54. Lelieveld, H.L.; Houben, R. *Food Safety Management: A Practical Guide for the Food Industry*; CRC Press: Boca Raton, FL, USA, 2018.
55. Aung, M.M.; Chang, Y.S.; Ching, T.Y. *Food Fraud and Vulnerability Assessment: Tools and Techniques*; John Wiley & Sons: New York, NY, USA, 2016.
56. Yan, Q.; Yang, Y. Cronobacter species in foods: Current state and perspectives. *J. Food Prot.* **2019**, *82*, 1732–1744.
57. Osaili, T.M.; Shaker, R.R.; Al-Omary, W.; Al-Nabulsi, A.A.; Holley, R.A. Cronobacter species and their significance in foodborne illnesses. *Food Control* **2018**, *93*, 237–247.
58. Kim, Y.K.; Lee, Y.; Kim, J.H.; Oh, S.W. Prevalence of Cronobacter spp. in Korean retail food: Poultry and vegetable products. *Int. J. Food Microbiol.* **2019**, *295*, 81–85.
59. Li, C.; Ling, N.; Zhang, X.; Zhang, S.; Deng, X.; Fang, W. Prevalence and molecular characteristics of *Cronobacter* spp. isolated from spices in China. *Food Control* **2018**, *84*, 116–123. [CrossRef]
60. Chon, J.W.; Choi, H.J.; Kim, J.H.; Kim, D.H.; Kwon, J.H.; Choi, I.S.; Seo, K.H. Cronobacter sakazakii in powdered infant formula marketed in Korea. *Food Control* **2012**, *28*, 225–229.
61. Almajed, F.S.; Forsythe, S.J. Isolates from neonates and Cronobacter sakazakii clinical infants in England and Wales: Their molecular, phenotypic characteristics, and genotypic diversity. *Front. Microbiol.* **2016**, *7*, 1564.
62. Food Safety Magazine. FSMA, Food Safety Audits and Certifications 2016. Available online: https://www.food-safety.com/articles/4995-fsma-food-safety-audits-and-certifications (accessed on 10 April 2023).
63. World Bank. Food Safety Handbook 2020. Available online: https://documents1.worldbank.org/curated/en/450921587054767474/pdf/Food-Safety-Handbook-A-Practical-Guide-for-Building-a-Robust-Food-Safety-Management-System.pdf (accessed on 10 April 2023).
64. Food Standards Agency. The Use of Third-Party Assurance Schemes in Relation to Food Safety. 2013. Available online: https://www.food.gov.uk/sites/default/files/media/document/835-1-1534_GSB_CR2435_3rd_Party_Assurance_Scheme_R2_V8_FCA.pdf (accessed on 10 April 2023).
65. Global Food Safety Initiative. Enhancing Food Safety Through Third-Party Certification. 2011. Available online: http://mygfsi.com/wp-content/uploads/2019/09/Third-Party-Certification-GFSI-White-Paper.pdf (accessed on 10 April 2023).
66. World Health Organization. Food Safety and Foodborne Illness. 2011. Available online: https://www.who.int/health-topics/foodborne-diseases#tab=tab_1 (accessed on 10 April 2023).
67. Escanciano, C.; Santos-Vijande, M.L. Reasons and constraints to implementing an ISO 22000 food safety management system: Evidence from Spain. *Food Control* **2014**, *40*, 50–57.

68. Abebe, G.K.; Bahn, R.A.; Chalak, A.; Yehya, A.A.K. Drivers for the implementation of market-based food safety management system: Evidence from Lebanon. *Food Sci. Nutr.* **2020**, *8*, 1082–1092. [CrossRef]
69. Creswell, J.W. *Qualitative Inquiry and Research Design: Choosing among Five Approaches*; Sage: Thousand Oaks, CA, USA, 2013.
70. Crowe, S.; Cresswell, K.; Robertson, A.; Huby, G.; Avery, A. The case study approach. *BMC Med. Res. Methodol.* **2011**, *11*, 100. [CrossRef] [PubMed]
71. Field, A. *Discovering Statistics Using IBM SPSS Statistics*, 5th ed.; University of Sussex, UK SAGE Publications Ltd.: Brighton, UK, 2017.
72. Nguyen, T.T.B.; Li, D. A systematic literature review of food safety management system implementation in global supply chains. *Br. Food J.* **2022**, *124*, 3014–3031. [CrossRef]
73. Fotopoulos, C.V.; Kafetzopoulos, D.P.; Psomas, E.L. Assessing the critical factors and their impact on the effective implementation of a food safety management system. *Int. J. Qual. Reliab. Manag.* **2009**, *26*, 894–910. [CrossRef]
74. Chen, H.; Liou, B.K.; Hsu, K.C.; Chen, C.S.; Chuang, P.T. Implementation of food safety management systems that meets ISO 22000:2018 and HACCP: A case study of capsule biotechnology products of chaga mushroom. *J. Food Sci.* **2021**, *86*, 40–54. [CrossRef]
75. Sun, Y.-M.; Ockerman, H.W. A review of the needs and current applications of hazard analysis and critical control point (HACCP) system in foodservice areas. *Food Control* **2005**, *16*, 325–332. [CrossRef]
76. Cusato, S.; Gameiro, A.H.; Corassin, C.H.; Sant, A.S.; Cruz, A.G.; Faria, J.d.A.F.; de Oliveira, C.A.F. Food Safety Systems in a Small Dairy Factory: Implementation, Major Challenges, and Assessment of Systems' Performances. *Foodborne Pathog. Dis.* **2013**, *10*, 6–12. [CrossRef]
77. Grover, A.K.; Chopra, S.; Mosher, G.A. Food safety modernization act: A quality management approach to identify and prioritize factors affecting adoption of preventive controls among small food facilities. *Food Control* **2016**, *66*, 241–249. [CrossRef]
78. Rao, M.; Bast, A.; de Boer, A. European private food safety standards in global agri-food supply chains: A systematic review. *Int. Food Agribus. Manag. Rev.* **2021**, *24*, 739–754. [CrossRef]
79. Wilcock, A.; Ball, B.; Fajumo, A. Effective implementation of food safety initiatives: Managers', food safety coordinators' and production workers' perspectives. *Food Control* **2011**, *22*, 27–33.
80. Zanin, L.M.; Stedefeldt, E.; Luning, P.A. The evolvement of food safety culture assessment: A mixed-methods systematic review. *Trends Food Sci. Technol.* **2021**, *118*, 125–142.
81. Manning, L.; Luning, P.A.; Wallace, C.A. The evolution and cultural framing of food safety management systems—Where from and where next? *Compr. Rev. Food Sci. Food Saf.* **2019**, *18*, 1770–1792. [CrossRef]
82. De Boeck, E.; Jacxsens, L.; Bollaerts, M.; Uyttendaele, M.; Vlerick, P. Interplay between food safety climate, food safety management system and microbiological hygiene in farm butcheries and affiliated butcher shops. *Food Control* **2016**, *65*, 78–91. [CrossRef]
83. Kramer, A.; Hoover, E.R.; Hedeen, N.; DiPrete, L.; Tuttle, J.; Irving, D.J.; Viveiros, B.; Nicholas, D.; Monroy, J.A.; Moritz, E.; et al. Development of an Empirically Derived Measure of Food Safety Culture in Restaurants. *J. Food Prot.* **2023**, *86*, 100043. [CrossRef]
84. Marconi, F.; Sartoni, M.; Nuvoloni, R.; Torracca, B.; Gagliardi, M.; Zappalà, G.; Guidi, A.; Pedonese, F. Food safety culture in food companies: Evaluation of the perception of food safety culture in three Tuscan food companies. *Ital. J. Food Saf.* **2023**, *12*, 11012. [PubMed]

foods

MDPI

Article

Novel Plant-Protein (Quinoa) Derived Bioactive Peptides with Potential Anti-Hypercholesterolemic Activities: Identification, Characterization and Molecular Docking of Bioactive Peptides

Feyisola Fisayo Ajayi [1,†], Priti Mudgil [1,†], Amie Jobe [2], Priya Antony [2], Ranjit Vijayan [2,3,4], Chee-Yuen Gan [5] and Sajid Maqsood [1,4,*]

1 Department of Food Science, College of Agriculture and Veterinary Medicine, United Arab Emirates University, Al-Ain 15551, United Arab Emirates; 202190079@uaeu.ac.ae (F.F.A.); priti.d@uaeu.ac.ae (P.M.)
2 Department of Biology, College of Science, United Arab Emirates University, Al-Ain 15551, United Arab Emirates; 201450107@uaeu.ac.ae (A.J.); 201990021@uaeu.ac.ae (P.A.); ranjit.v@uaeu.ac.ae (R.V.)
3 The Big Data Analytics Center, United Arab Emirates University, Al-Ain 15551, United Arab Emirates
4 Zayed Center for Health Sciences, United Arab Emirates University, Al-Ain 15551, United Arab Emirates
5 Analytical Biochemistry Research Centre (ABrC), Universiti Sains Malaysia, SAINS@USM Campus, Bayan Lepas 11900, Malaysia; cygan@usm.my
* Correspondence: sajid.m@uaeu.ac.ae
† These authors contributed equally to this work.

Abstract: Hypercholesterolemia remains a serious global public health concern. Previously, synthetic anti-hypercholesterolemic drugs were used for ameliorating this condition; however, long-term usage presented several side-effects. In this regard, natural products as an adjunct therapy has emerged in recent times. This study aimed to produce novel bioactive peptides with anti-hypercholesterolemic activity (cholesterol esterase (CEase) and pancreatic lipase (PL)) from quinoa protein hydrolysates (QPHs) using three enzymatic hydrolysis methods (chymotrypsin, protease and bromelain) at 2-h hydrolysis intervals (2, 4, and 6 h). Chymotrypsin-generated hydrolysates showed higher CEase (IC_{50}: 0.51 mg/mL at 2 h) and PL (IC_{50}: 0.78 mg/mL at 6 h) inhibitory potential in comparison to other derived hydrolysates and intact quinoa proteins. Peptide profiling by LC-MS QTOF and in silico interaction with target enzymes showed that only four derived bioactive peptides from QPHs could bind in the active site of CEase, whereas twelve peptides could bind in the active site of PL. Peptides QHPHGLGALCAAPPST, HVQGHPALPGVPAHW, and ASNLDNPSPEGTVM were identified to be potential CEase inhibitors, and FSAGGLP, QHPHGLGALCAAPPST, KIVLDSDDPLFGGF, MFVPVPH, and HVQGHPALPGVPAHW were identified as potential PL inhibitors on the basis of the maximum number of reactive residues in these bioactive peptides. In conclusion, QPHs can be considered as an alternative therapy for the treatment of hypercholesterolemia.

Keywords: quinoa protein; protein hydrolysate; bioactive peptides; cholesterol esterase; pancreatic lipase

1. Introduction

The rise in diet-related metabolic disorders such as hyperlipidemic or hypercholesterolemia, cardiovascular diseases, hypertension, and other related metabolic diseases has been identified as the major cause of mortality globally, and thus has become a serious public health issue. The consumption of processed food and animal products has been associated with the occurrences of these health issues. Cholesterol esters are normally found in the human diet, which the human intestine cannot sufficiently absorb, thereby producing free fatty acids and cholesterol that elevate the level of blood cholesterol. It is a known fact that rising levels of cholesterol and triglycerides in the blood serum results in hyperlipidemia or hypercholesterolemia [1]. According to [2], the molecular mechanism for understanding the physiological development of hyperlipidemia is quite difficult. However, two enzymes (cholesterol esterase (CEase) and pancreatic lipase (PL) have been associated with hyperlipidemia and hypercholesterolemia complications, and their inhibitory potentials have been targeted by scientists in order to develop anti-hypercholesteremic drugs [1,3]. For many years, synthetic drugs have been used to lower blood cholesterol production and absorption. However, synthetic drugs (mainly the CEase or PL inhibitors) are often accompanied with harmful consequences on human health and metabolism due to long-term usage. Therefore, there is a high demand for natural products that can prevent these malignant diseases.

Nowadays, generating bioactive peptides from plant-based proteins have gained more attention owing to their biological and functional activities, which is highly significant in alleviating chronic diseases, as well as in promoting human health. Moreover, the effectiveness of the derived peptides depends on their sequence, structure, and inherent amino acid composition. Quinoa proteins are a very good source of essential amino acids [4]. Apart from providing significant nutrients, they also possess several peptide sequences with specific physiological benefits [5]. Protein hydrolysates from quinoa have been reported to be a potential ingredient in designing functional foods and nutraceuticals because of their multifunctional bioactivity, including antioxidant, antihypertensive, antidiabetic, dipeptidyl peptidase IV (DPP-IV) inhibitory, and angiotensin-converting enzyme (ACE) inhibitory properties [6,7].

Similarly, several bioactive peptides from protein hydrolysates have demonstrated the capacity to inhibit the digestion of dietary lipids in in-vitro and cultured cell assays [8]. Peptides that possess anti-hyperlipidemic activity operate in the gastrointestinal tract, hepatocytes, and adipocytes by binding cholesterol micelles and bile acids, thereby hindering their absorption. According to Boachie, Yao and Udenigwe [8], their operation affects the activity of some regulatory enzymes, particularly the 3-hydroxy-3-methylglutaryl CoA reductase (HMGCoAR), which is well-known to be primarily responsible for cholesterol biosynthesis, and this enzyme inhibition is targeted to decrease the rate of cholesterol biosynthesis.

Several scientific investigations have shown that natural peptides found in plant proteins can inhibit cholesterol absorption or HMGCoAR activity [9]. However, protein hydrolysates and bioactive peptides derived from quinoa have not been studied for their anti-hypercholesterolemic effect via the inhibition of CEase and PL enzymes to date. Therefore, this study investigated the role and binding mechanism of quinoa derived bioactive peptides for inhibiting enzymatic markers i.e., CEase and PL that are responsible for hypercholesterolemia. For effective inhibition of the cholesterol metabolizing enzymes, the potential of generated bioactive peptides to bind to their specific hotspots on the active site of the enzyme depends on the cleavage sites where the proteolytic enzymes will act on the protein [10]. As such, different enzymes can result in generating varying amino acids sequences in peptides, which is vital in demonstrating their bioactivity. Moreover, numerous studies have demonstrated the potential of anti-hypercholesterolemic peptides derived using specific enzymes (alcalase, pepsin, trypsin, flavourzyme, chymotrypsin, etc.) in inhibiting CEase and PL active sites [11–13].

Accordingly, quinoa protein isolates were hydrolyzed using three different enzymes (chymotrypsin, protease, and bromelain) in order to get a diverse sequence of peptides which might be potential inhibitors of CEase and PL. The generated quinoa hydrolysates were characterized biochemically via the degree of hydrolysis, and their CEase and PL inhibitory potentials were determined. Furthermore, peptides generated from quinoa protein isolates were characterized and identified using LC-MS-QTOF. A molecular docking approach was carried out to elucidate the molecular binding mechanism between potent anti-hypercholesterolemia peptides with the target enzymes (CE and PL) through the in silico structural activity relationship.

2. Materials and Methods

2.1. Materials (Quinoa Seeds) and Chemicals

Quinoa seeds used in this study were purchased from the local market (Al-Ain, United Arab Emirates). Enzymes such as protease (Type XIV; ≥3.5 units/mg solid derived from *Streptomyces griseus*), α-chymotrypsin (EC 3.4.21.1 from bovine pancreas; C4129 40 units/mg), and bromelain (B4882: 3–7 units/mg solid; EC 3.4.22.33 from pineapple stem), HPLC grade reagents (acetonitrile and methanol), and other chemicals such as formic acid *p*-nitrophenyl butyrate, *o*-phthaldialdehyde, trizma base, β-mercaptoethanol, SDS, and sodium tetra-borate were procured from Sigma Aldrich (St. Louis, MO, USA). Other analytical grade reagents used were purchased from the UAE (BDH Middle East, Dubai, United Arab Emirates).

2.2. Preparation of Quinoa Protein Isolate

Firstly, quinoa seeds were oven dried (40 °C for 3 h), dry-blended (IKA A11, Guangzhou, China), and passed through a screen (3 μm pore size) to obtain a fine powder. The methodology described by [6] as adopted from [14] was used to obtain quinoa protein isolate (QPI). In sum, defatted quinoa flour was dispersed in 0.015 mol L^{-1} NaOH (1:10 ratio) maintained at a pH of 10–11 and was stirred for 2 h on an electric shaker (24 ± 2 °C), and the resulting slurry was stored at 4 °C overnight. The slurry was homogenized for consistency and centrifuged twice at 4 °C at a speed of 10,000× *g* for 20 min after overnight storage. Whatman No. 1 paper was used to filter the obtained supernatant, and the pH of the filtrate reduced to 4.5 using the pH drop method with 1.0 mol L^{-1} HCl solution to precipitate quinoa protein. Afterwards, precipitated proteins were separated by centrifuging at 4 °C for 10 min at the speed of 10,000× *g*, and were subsequently washed twice with deionized water. The pH of the resultant pellets was adjusted to pH 7.0 (with 1.0 mol L^{-1} NaOH) and stored at −20 °C until further use. The Kjeldahl method was adopted to determine the protein content of quinoa protein isolate (QPI) slurries, which was found to be 44%.

2.3. Quinoa Protein Hydrolysate Production

The methodology as described by [6] was used in the production of quinoa protein hydrolysates (QPHs). The obtained QPI slurry was thoroughly homogenized (T 25 digital ULTRA-TURRAX® homogenizer IKA®-Werke GmbH, Staufen, Germany). The protein content of the slurry was set to 4% with deionized water, and the content was divided into four batches for treatment with bromelain, chymotrypsin, and protease, respectively, while the last part was kept as control. Then, 1 mol/L NaOH and 5N HCl were used to adjust the pH of the treated protein slurries for the bromelain enzyme to 7.0, the chymotrypsin enzyme to 7.8, and the protease enzyme to 8.0, respectively. Furthermore, enzyme concentration was pre-calculated, suspended in 1 mL of distilled water resulting in a 1% enzyme/substrate ratio, and subsequently added to the QPI slurry. The samples were then distributed into 50 mL falcon tubes in triplicate for each period of hydrolysis and incubated (water bath) for up to 6 h at the speed of 100 ramps/min at a controlled temperature of 50 °C. Samples were taken every 2 h from all enzymatic incubations, and the reaction was inactivated at 95 °C for 600 s. Generated QPHs were centrifuged at 4 °C at a speed of 10,000× *g* for 10 min, and supernatants were collected and stored at −20 °C for further analysis. All of the analyses were performed within two weeks of the production of hydrolysates.

2.4. Degree of Hydrolysis (DH%)

The previously published o-phthaldialdehyde (OPA) methodology of [15] was used for the determination of DH values.

$$Degree\ of\ hydrolysis(DH\%) = \frac{h}{h_{tot}} \times 100 \quad (1)$$

Herein, h_{tot} = total number of peptide bonds per protein equivalent; and h = number of hydrolyzed bonds, and calculated using h = (SerineNH2 − β)/α. Where h_{tot}, β, and α, β and h_{tot} values for quinoa proteins were used as 8.0, 0.40, and 1.00 mEq/g protein, respectively.

2.5. Anti-hypercholesterolemia Assays

2.5.1. Determination of Pancreatic Lipase (PL) Inhibitory Activity of QPHs

The PL-inhibitory activity of generated hydrolysates was measured using the methodology of [13]. Here, PL (20 µL) and *p*-nitrophenyl butyrate (25 µL) were mixed with each sample (50 µL) in a sodium chloride: sodium phosphate buffer (pH 7.2, 100 mM), and then incubated in a 96-well microplate reader. The quantity of the resultant reaction was modified with the above buffer to 150 µL and incubated (37 °C for 30 min). After the incubation period, the released *p*-nitrophenyl for each sample was measured on a plate reader (Epoch 2, BioTek, Winooski, VT, USA) at 405 nm. The PL-inhibitory activity was calculated using the equation below

$$\%\ Enzyme\ inhibition = \left(1 - \left(\frac{C-D}{A-B}\right)\right) \times 100 \qquad (2)$$

Here: A = the absorbance of control; B = the absorbance of control reaction blank; C = the absorbance of sample, and D = the absorbance of the sample blank. The IC_{50} values were determined by plotting the percentage inhibition against the concentration of the test compound and expressed in mg (eqv protein)/mL.

2.5.2. Determination of Cholesterol Esterase Inhibitory Activity of QPHs

The CEase-inhibitory activity was determined as per the procedure documented by [13]. Briefly, QPHs (25 µL), substrate (50 µL) containing 5 mM *p*-nitrophenyl butyrate in sodium phosphate (100 mM): NaCl buffer (pH 7.2, 100 mM) was placed in a 96-well microtiter plate. Subsequently, the mixtures were incubated with 50 µL of porcine pancreatic CEase (5 µg/mL) at 37 °C for 180 min. After that, the *p*-nitrophenol released from the enzymatic hydrolysis of *p*-nitrophenyl butyrate was read at 405 nm. The percentage CEase inhibition was determined using Equation (2) above. The IC_{50} values were determined by plotting the percentage inhibition against the concentration of the test compound and expressed as mg (eqv protein)/mL.

2.6. Peptide Identification

2.6.1. Sequencing of Peptides Implied in PL and CEase-Inhibitory Activities Using Liquid Chromatography-Mass Spectrometry Quadrupole Time-of-Flight (LC-MS Q T-O-F)

For the identification of peptides in selected hydrolysates, QC-6 (chymotrypsin-generated at 6 h hydrolysis), LC-MS Q T-O-F was conducted as described previously by [16]. The identification of peptides was performed as follows: Advance Bio Peptide Map, C18 column (2.1 × 100 mm, 2.7 µm particles; Agilent, Santa Clara, CA, USA) was used for the peptide separation whereas the LC-MS Q T-O-F (model 6520, Agilent, Santa Clara, CA, USA) was used for the analysis. The mobile phases used were: (A) deionized water containing 0.1% formic acid, and (B) acetonitrile containing 0.1% (*v*/*v*) formic acid with a flow rate of 15 µL/min. The mobile phase gradient of the HPLC system was as follows: (a) 0–5 min, 10% B; (b) 5–115 min, 10–95% B, (c) 115–120 min, 95% B, (d) 120–135 min, 95–10% B, and (e) 140–150 min, 10% B. The electrospray ionization-quadrupole-time-of flight system (ESI-QTOF) condition was: (a) mass range: 70–2000 m/z; (b) collision energy: 6V/100 Da (offset-2); (c) flow rate: 15 µL/min; (d) ion spray sources: 3.5 kV; (e) drying gas: Nitrogen at a temperature of 350 °C with a flow rate of 10 L/min; (f) nebulizer pressure: 3 psig; (g) fragmentor voltage: 110 V; and (h) fragmentation mode: collision induced dissociation (CID). Data generated from the mass spectrometry approach were analyzed using PEAKS studio version 6.0 [17]. The precursor was selected on the basis of a minimum charge of 2 and a maximum charge of 10. From the generated peptides, only peptides with average local confidence (ALC) > 70% were chosen for further analysis.

2.6.2. Potential Biologically Active Peptides Selection Using PeptideRanker

Potential bioactive peptides were screened using the PeptideRanker web server (http://bioware.ucd.ie/, accessed on 26 July 2022) [18]. Peptides showing a score of more than 0.5 were regarded as potentially biologically active and subjected to further in silico analysis.

2.6.3. Selection of Identified PL and CEase Inhibitory Peptides Using Peptide Ranker

The preliminary insight into the molecular mechanism of peptides for their PL and CEase inhibitory activities were explored through an in silico docking investigation using the online program Pepsite2, accessible at http://pepsite2.russellab.org (accessed on 16–19 August 2022) [19]. Herein, 3D structures of CEase (PDB code: 1AQL) and PL (PDB code: 1ETH) were introduced from the Protein Data Bank through http://www.rcsb.org/pdb/ (accessed on 16–19 August 2022). Peptide inputs were then entered jointly with a protein receptor in PDB format. The best interaction of peptides with enzymes showing p-values < 0.05 and the interaction with key hotspots on enzymes playing a significant role in enzyme inhibition were taken into consideration for further analysis.

2.7. Molecular Docking

2.7.1. Preparation of Protein Structure

The 3D structures of human and bovine CEase, PDB IDs 1F6W and 1AQL respectively, and PL (PDB IDs: 1LPB and 1ETH) were obtained from the Protein Data Bank (PDB) [20]. These proteins were processed and optimized prior to docking using the Schrödinger Suite's Protein Preparation Wizard (Schrödinger Suite 2021-1: Protein Preparation Wizard; Schrödinger LLC, New York, NY, USA). This workflow prepares a PDB structure for docking by adding and optimizing hydrogen bonds, assigning bond orders, simplifying multimeric complexes, creating disulfide bonds, deleting unwanted water molecules, adjusting ionization states, fixing disoriented groups, filling missing loops and sidechains and finally, energy minimizing and optimizing to produce a geometrically stable structure [21].

2.7.2. Active Site Identification and Grid Generation

The active sites of CEase (PDB IDs: 1F6 W and 1AQL) and PL (PDB IDs: 1LPB and 1ETH) that are directly involved in ligand-binding were identified from the literature The active site of lipase is located in the N-terminal domain and their catalytic activity is mediated by the catalytic triad Ser152, Asp176 and His263 Similarly, CEase also shares a comparable catalytic triad composed of Ser194, Asp320 and His435. For the docking analysis, a receptor grid suitable for peptide docking was generated for the minimized protein structures enveloping the active site residues. No constraints were applied when default parameters for van der Waals scaling factor (1.00) and charge cut-off (0.25) were used. The OPLS 2005 force field was used for structural representation [22].

2.7.3. Peptide Docking and Binding Free Energy Calculation

The Peptide Docking panel of Schrödinger Maestro [23] was employed to perform the docking of peptides to the protein structures. Using this tool, small peptides lower than 16 amino acids can be docked and scored with either Glide Score or molecular mechanics-generalized Born surface area (MM-GBSA) approach. Each peptide was docked to the protein receptor using Glide with increased sampling in several docking runs. Subsequently, OPLS molecular mechanics force field was used for pose optimization. After carrying out conformer clustering, ten representative peptide poses were chosen. Lastly, selected poses were re-scored and ranked based on GlideScore empirical scoring function [23,24].

The binding free energy of the best docked poses obtained after peptide docking were evaluated using the MM-GBSA approach. Binding free energy calculations were performed using Schrödinger Prime with OPLS 2005 force field combined with VSGB 2.0 implicit solvent model. The peptide was minimized, and the receptor was treated as rigid for the MM-GBSA calculations.

2.8. Statistical Analysis

All hydrolysates were produced in three batches representing three replicates. All analyses were carried out in triplicate. Data analysis was performed with SPSS 24.0 statistical software (SPSS INC., Chicago, IL, USA) using one-way analysis of variance. Means separation was conducted using Tukey's multiple range test, and a p-value of 5% was defined as statistically significant ($p < 0.05$).

3. Results & Discussion

3.1. Degree of Hydrolysis (DH%)

The DH and enzyme specificity are among the measurable factors to be controlled because they define the biological activities of the resulting peptides. Studying the hydrolysis of protein reveals the capability of an enzyme to degrade proteinaceous substrate, and this serves as a pointer for their proteolytic reaction [25]. Noteworthy, the protein profile obtained from intact quinoa (QPI), and quinoa protein hydrolysates (QPHs) from the three enzymes understudied after 2, 4, and 6 h of hydrolysis is depicted in Supplementary Figure S1.

In this work, the effects of different enzymes (chymotrypsin, protease, and bromelain) at 2 h-interval reaction periods on the DH of quinoa protein hydrolysates (QPHs) were reported (Table 1).

Table 1. DH, CEase and PL IC_{50} inhibitory activities of quinoa protein hydrolysates generated by bromelain, chymotrypsin, and protease after hydrolysis reaction.

Sample	DH (%)	CEase (mg/mL)	PL (mg/mL)
QPI	-	1.01 ± 0.03^{e}	10.4 ± 0.14^{d}
QB-2	34.99 ± 0.92^{a}	0.67 ± 0.03^{c}	1.97 ± 0.03^{b}
QB-4	41.91 ± 0.36^{b}	0.64 ± 0.02^{bc}	1.03 ± 0.01^{a}
QB-6	51.91 ± 2.03^{c}	0.63 ± 0.01^{bc}	0.90 ± 0.00^{a}
QC-2	36.01 ± 3.08^{a}	0.51 ± 0.01^{a}	2.90 ± 0.11^{c}
QC-4	44.09 ± 0.47^{b}	0.55 ± 0.00^{ab}	1.51 ± 0.06^{ab}
QC-6	66.05 ± 1.60^{d}	0.55 ± 0.01^{ab}	0.78 ± 0.04^{a}
QP-2	45.05 ± 1.00^{b}	0.80 ± 0.02^{d}	3.32 ± 0.52^{c}
QP-4	46.01 ± 0.80^{b}	0.88 ± 0.02^{d}	1.53 ± 0.03^{ab}
QP-6	63.3 ± 2.66^{d}	0.87 ± 0.01^{d}	1.08 ± 0.02^{ab}

Keynote: QPI: Intact quinoa protein isolate; QB-2, QC-2 and QP-2 denotes bromelain, chymotrypsin, and protease 2 h generated hydrolysate; QB-4, QC-4 and QP-4 denotes bromelain, chymotrypsin and protease 4 h generated hydrolysate; QB-6, QC-6 and QP-6 denotes bromelain, chymotrypsin and protease 6 h generated hydrolysate, respectively. DH: Degree of hydrolysis; CE: Cholesterol esterase; and PL: Pancreatic lipase. Different small letters in the column indicates significant different amongst unhydrolysed and hydrolysed QPHs.

The results obtained showed a characteristic similar hydrolysis pattern among all the enzymes, where the DH values significantly increased with the progression of hydrolysis time up to 6 h ($p < 0.05$). These results are comparable with those obtained by [26], where they reported increased DH values of quinoa protein hydrolysates as a function of increased hydrolysis time. Moreover, varying DH values were demonstrated among the QPHs generated using the three enzymes of study. The DH levels in the range of 36.01 to 66.05%, and 45.05 to 63.3%, and 34.99 to 51.91% were obtained for chymotrypsin, protease, and bromelain generated QPHs, respectively.

Based on the enhanced DH values with progressive hydrolysis time, chymotrypsin-produced hydrolysates after 6 h (QC-6) displayed the highest value of 66.05%. However, it was not noted that the DH value obtained in protease-produced hydrolysates after 6 h (QP-6) was not significantly different from the DH value of QC-6h ($p > 0.05$). Bromelain-produced hydrolysates after 6 h (QB-6) showed an intermediate DH value of 51.91%, and the lowest protein hydrolysis was demonstrated in both bromelain-produced hydrolysates (QB-2) and chymotrypsin-produced (QC-2) after 2 h of bio-catalysis ($p > 0.05$). In general, chymotrypsin was found to be the most effective enzyme in hydrolyzing quinoa proteins as opposed to the other two proteolytic enzymes (bromelain and protease). The hydrolytic pattern of chymotrypsin indicates greater proteolytic efficiency of the enzyme, which may be attributed to the size and action of the released peptides. As a result, the abundance of small-sized peptides with enhanced effectiveness against metabolic markers could be produced from chymotrypsin at prolonged hydrolysis periods. This is consistent with previous studies that suggested that higher DH due to disintegration of more peptide bonds could be attributed to the higher activity of the enzyme [27].

The enzymatic hydrolysis of quinoa proteins using bromelain and chymotrypsin exhibited a rapid and steady increase in the DH values over a period of 6 h, an indication that the larger peptides that were released at a lower hydrolysis time were further disintegrated into smaller peptides as the hydrolysis time progressed. We hypothesize that a higher degree of hydrolysis due to the increased time of hydrolysis is often related to further enzyme activity on the reaction substrates. In contrast, the progression of hydrolysis-reaction from 2 h to 4 h ($p > 0.05$) of protease-generated hydrolysates did not show a very rapid increase in the displayed DH. However, after progressing the bio-catalysis reaction to 6 h, a significantly higher DH ($p < 0.05$) was recorded, as presented in Table 1. A similar hydrolysis reaction where increasing hydrolysis up to 4 h was found to be slower than between 4 to 6 h of hydrolysis [6].

The level of DH values obtained in this study remains comparable to those of other studies. For instance, the hydrolysis of quinoa proteins after 2 h of hydrolysis by pepsin pancreatin and papain was noticed to be in the range of 15–35% [7]. Similarly, 4 h of hydrolysis by Alcalase reported a DH value of 48% [14]. The variations noticed between DH produced by different enzymes after a similar time of hydrolysis could be ascribed to the specificity and substrate affinity of each enzyme towards the quinoa protein substrate [16].

3.2. Cholesterol Esterase Inhibitory Activity

QPHs were analyzed for their ability to inhibit CEase. CEase is a known polymeric enzyme existing in the bile that initiates dietary cholesterol esters' hydrolytic reaction, thereby releasing cholesterol and free fatty acids [28]. Inhibiting CEase can indirectly prevent cholesterol absorption by the human body by diminishing cholesterol's release from dietary lipids. The CEase inhibitory potential of generated QPHs and QPI were measured in terms of IC_{50} (half maximal inhibitory concentrations), and the results are illustrated in Table 1. The low IC_{50} value of hydrolysates connotes high CEase inhibitory activity. Upon hydrolysis, the QPHs derived from bromelain showed an increased CEase inhibition, with progression of the bio-catalysis reaction from 2 to 6 h (IC_{50} values = 0.67 mg/mL to 0.64 mg/mL to 0.63 mg/mL, respectively). Therefore, bromelain-generated hydrolysates with high CEase inhibitory activity can be developed at longer hydrolysis times. On the other hand, the CEase inhibitory potential of chymotrypsin and protease derived hydrolysates decreased with the increased bio-catalysis reaction time from 2 to 6 h.

From this study, it was evident that chymotrypsin-generated QPHs showed the highest CEase inhibitory activity with IC_{50} values of 0.51 mg/mL, 0.55 mg/mL, and 0.55 mg/mL after 2, 4, and 6 h of hydrolysis, respectively. However, there was no significant difference ($p > 0.05$) observed between hydrolysates generated at 4 h and 6 h hydrolysis time. Similar results were obtained by [6], where chymotrypsin exhibited higher activity compared to other enzymes used, and this implies that it is more effective in hydrolyzing quinoa protein. This was attributed to the broad specificity of chymotrypsin enzymes to generate small-sized peptide and free amino acids. In addition, the variation exhibited in the IC_{50} values of hydrolysates with respect to different time periods could be explained with the ability of specific enzymes to either degrade or generate CEase inhibitory peptides during hydrolysis [29]. The protease-generated hydrolysates showed a lower CEase inhibitory activity in comparison to bromelain and chymotrypsin generated hydrolysates. A possible explanation could be that the cleavage sites that the enzyme binds to was not fully exposed, which consequently reduced the inhibition effect. However, the protease hydrolyzed under hydrostatic pressure processing showed an increased inhibition rate of up to 49.1% [30].

Unhydrolyzed quinoa protein, i.e., intact quinoa protein (QPI), showed the least CEase inhibitory activity (IC_{50} value of 1.01 mg/mL). A study of the CEase-inhibitory activity of amaranth protein recently reported in our lab also showed a significant low CEase-inhibitory effect of amaranth protein isolate in comparison to IC_{50} values recorded for hydrolysates [13]. The hydrolysate with the utmost CEase inhibitory IC_{50} value of 0.51 mg/mL (chymotrypsin-derived for 2 h) was approximately two times lower than the IC_{50} value (1.01 mg/mL) of QPI, indicating that derived hydrolysates can therefore inhibit CEase two-fold more than their intact protein counterpart. Furthermore, the results suggested an overall high CEase inhibitory activity of QPHs compared to intact quinoa protein, which is evident from the lower IC_{50} values observed. This is similar to the study of Jafar, et al. [31] that reported the higher CEase inhibitory activity of protein hydrolysates derived from camel whey compared to intact protein from Camel whey.

3.3. *Pancreatic Lipase (PL) Inhibitory Activity*

Pancreatic lipase is the primary enzyme that contributes largely to the digestion of dietary lipids, and the only effective way to adjust lipid absorption is by inhibiting the enzyme [32]. The pancreatic lipase inhibitory activity IC_{50} values for QPHs (bromelain, chymotrypsin, protease) and un-hydrolyzed quinoa protein are depicted in Table 1. Results from this study showed that the PL-inhibitory IC_{50} values varied from 0.90 to 10.4 mg/mL, implying a substantial ability to inhibit PL activity. The PL-inhibitory activity of all the hydrolysates (0.90–3.32 mg/mL) exhibited significantly higher inhibition than the QPI, indicating that hydrolysis efficiently increased the inhibitory activity of quinoa protein. Hydrolysate QC-6 demonstrated the highest PL-inhibitory activity (0.78 mg/mL), followed by QB-6, recording IC_{50} values of 0.90 mg/mL, respectively. All chymotrypsin and bromelain generated hydrolysates showed higher PL-inhibitory activities than the protease generated hydrolysates. The different inhibitory activity observed among hydrolysates generated from different enzymes (bromelain, chymotrypsin and proteases) could be explained with the specificity and varying degree of hydrolysis among other factors of the individual enzyme [5].

On the other hand, a significant effect ($p < 0.05$) was observed in the pancreatic lipase inhibitory activity of QPHs upon hydrolysis with different enzymes at different hydrolysis times. A significant increase in the PL inhibitory activity of QPHs was observed with the increase in hydrolysis time progression, which is shown with lower IC_{50} values. For instance, chymotrypsin derived hydrolysate with the most potent PL-inhibitory activity IC_{50} values decreased from 2.90 to 0.78 mg/mL as the hydrolysis time increased. In the past, various studies have demonstrated an increase in the enzyme inhibitory potential of protein hydrolysates with the increase in the duration of hydrolysis. This implies that the progressive time of hydrolysis would sufficiently release peptides that might bind the active site of pancreatic lipase, thus inhibiting lipase enzymes. In addition, hydrolysates generated from chymotrypsin showed the maximum inhibitory activity in comparison to bromelain and proteases. This result upholds the potent characteristic of chymotrypsin generated hydrolysates, as reported by [6,16].

Furthermore, overall, QPHs showed the highest pancreatic lipase inhibitory activity in comparison to unhydrolyzed samples, and this accounted for an approximately 110% reduction in the IC_{50} value. This result agrees with the findings of Mudgil, et al. [33], who reported an increased PL-inhibitory activity of milk proteins after a bio-catalysis reaction using bromelain, chymotrypsin, alcalase and papain. Similarly, [34] had previously shown the potential of quinoa as a coadjutant therapeutic agent of cardiovascular diseases. The authors also reported quinoa's ability to minimize lipid profile and glucose levels that are initiated by fructose which often leads to most of the well-known detrimental effects in humans. The use of derived hydrolysates from quinoa seeds comprising grains greater than 10% was reported to significantly lessen high levels of plasma and liver total cholesterol in nourished mice [35]. Consequently, QPHs can substitute synthetic drugs as anti-hyperlipidemia therapeutic agents, having shown excellent potential towards PL-inhibition.

3.4. Selection and Identification of Cholesterol Esterase Inhibitory Peptides from Selected QPHs

Based on the overall higher CEase and PL inhibitory activities of chymotrypsin generated hydrolysates obtained at the highest hydrolysis time of 6 h (QC-6), it was further chosen for peptide identification by LC-MS Q T-O-F. Additionally, the ABTS radical scavenging and anti-hemolytic activities shown by this specific hydrolysate reported in a previously published work by [6] made it an exciting hydrolysate to characterize its existing bioactive peptides. The identified peptides were classified as bioactive peptides by their scoring of >0.80 on a web server known as Peptide Ranker and further exploration into their in silico interaction with target enzymes using the Pepsite 2 web server [19].

In total, 136 peptides were identified (Supplementary Table S1), however, only 35 peptides were shortlisted to be biologically active based on a Peptide Ranker score greater than 0.8 (Table 2), and these peptides were subjected to an in-silico mode for further structure-activity analysis. The potential active binding sites of the identified peptides on the binding sites of CEase enzyme were accessed based on a statistical significance of 5% (Table 2). The significant variation ($p < 0.05$) implies that peptides derived from QC-6 significantly bound to the hotspot sites of CEase, resulting in non-competitive inhibition due to the modification in protein interaction networks or loss of enzyme activity that blocked its catalytic binding sites and substrate [19]. The results obtained from PepSite 2 showed that the peptides FFE, DFTF, DFLM, ML, CDCP, CYTF, QHPHGLGALCAAPPST, LR, RR, HVQGHPALPGVPAHW, AGLR, FTVM, LLPYH, ASNLDNPSPEGTVM, and HMLH had a significant ($p < 0.05$) binding effect on CEase (Table 2). However, QHPHGLGALCAAPPST, HVQGHPALPGVPAHW, and ASNLDNPSPEGTVM among the listed peptides were observed to be capable of binding the active sites of CEase. The most potent identified peptides HVQGHPALPGVPAHW and QHPHGLGALCAAPPST that could bind up to 13 and 14 bound residues were selected for mass spectrometry analysis. Two to sixteen amino acid residues were observed in the thirty-five peptides recorded (Table 2). This number is within the range of anti-hypercholesterolemic peptides reported by [36] for β-lactoglobulin peptides. Previous studies have suggested that peptides that can bind to more than eight residues have the ability to effectively inhibit the activity of antidiabetic and antihypertensive properties [37]. This connotes that the identified peptides (QHPHGLGALCAAPPST, HVQGHPALPGVPAHW, ASNLDNPSPEGTVM) from this study have the potential of inhibiting the activity of CEase.

Table 2. Pepsite 2 analysis of the selected peptides using cholesterol esterase as a protein receptor.

Peptide	*p*-Value	Peptide Residues	Bound Residues
FFE	0.04342	F1, F2, E3	Tyr7, Phe12
DFTF	0.03113	D1, F2, T3, F4	Phe235, Trp236, Tyr279, His283, Tyr284
DFLM	0.01757	D1, F2, L3, M4	Tyr7, Phe12
ML	0.04441	M1, L2	Phe235, Trp236, His283
CDCP	0.04734	C1, C3, P4	Tyr7, Phe12
CYTF	0.01539	C1, Y2, T3, F4	Tyr7, Phe12
FSAGGLP	0.3886	G5, L6, P7	Phe235, Trp236, His283, Tyr284
MLLL	0.05859	M1, L2, L3, L4	Tyr7, Phe12
MYLY	0.1116	M1, Y2, L3, Y4	Tyr7, Phe12

Table 2. *Cont.*

Peptide	*p*-Value	Peptide Residues	Bound Residues
QHPHGLGALCAAPPST	0.04033	Q1, H2, P3, G7, L9, C10, P14, S15	Gy107, Ala108 *, Ser194 *, Ala195 *, Ser220, Trp227 *, Phe235, Trp236, His283, Tyr284, Leu285, Phe324 *, Phe393, His435 *
KIVLDSDDPLFGGF	0.2461	I2, V3, D7, D8, P9, L10, F11	Ala108 *, Ser194 *, Trp227 *, Phe235, Trp236, His283, Tyr284, Leu285, Phe324 *, Phe393, His435 *
AHCGGLPY	0.2596	A1, H2, C3, G4	Tyr7, Phe12
LYNDWDLR	0.379	W5, D6, L7, R8	Tyr7, Phe12
PGGAR	0.2249	G3, A4, R5	Phe235, Trp236, His283, Tyr284
SGPAR	0.08677	P3, A4, R5	Tyr7, Phe12
LPLLR	0.04626	P2, L3, L4, R5	Tyr7, Phe12
KPGGTAGSALPRPAHW	0.3388	K1, P2, G3, L10	Phe235, Tyr236, Tyr279, His283, Tyr284, Phe324 *
FE	0.06821	F1, E2	Tyr7, Phe12
HDSF	0.09456	H1, D2, S3, F4	Tyr7, Phe12
LR	0.04441	L1, R2	Tyr7, Phe12
VYML	0.08002	V1, Y2, M3, L4	Tyr7, Phe12
RR	0.02794	R1, R2	Tyr7, Phe12
LRL	0.1121	L1, R2, L3	Tyr7, Phe12
MFVPVPH	0.3249	P4, V5, P6	Phe235, Trp236, His283, Tyr284
HVQGHPALPGVPAHW	0.02606	H1, V2, P6, A7, L8, P9, G10, P12, A13, H14	Ala108 *, Ser194 *, Ala195 *, Ser220, Trp227 *, Phe235, Trp236, His283, Tyr284, Leu285, Phe324 *, Phe393, His435 *
AGLR	0.02471	A1, G2, L3, R4	Tyr7, Phe12
HVASGAGPW	0.3417	V2, A3, P8, W9	Ala108 *, Ser194 *, Ala195 *, Trp227 *, Phe235, Trp236, His283, Tyr284, Phe324 *
FTVM	0.02389	F1, T2, V3, M4	Phe235, Trp236, His283, Tyr284
LLPYH	0.04167	L2, P3, Y4, H5	Tyr7, Phe12
MVLP	0.1188	V2, L3, P4	Phe235, Trp236, His283
GARR	0.1087	G1, A2, R3	Phe235, Trp236, His283, Tyr284

Table 2. Cont.

Peptide	p-Value	Peptide Residues	Bound Residues
ASNLDNPSPEGTVM	0.03203	D5, N6, P7, S8, E10, T12, V13, M14	Ala108 *, Ser194 *, Trp227 *, Phe235, Trp236, Arg239, Tyr279, His283, Tyr284, Leu285, Phe324 *, Phe393, His435 *
CVLSPL	0.2097	V2, S4, P5, L6	Phe235, Trp236, Tyr279, His283, Tyr284, Val353
HMLH	0.02838	H1, M2, L3, H4	Tyr7, Phe12
CALVGL	0.2557	C1, A2, V4	Tyr7, Phe12

* Hotspots that could inhibit the activity of cholesterol esterase if bounded by the peptide.

According to [38], the active binding sites of CEase, which is vital for its catalytic function, contain an oxyanion hole (glycine and alanine), the catalytic triad (histidine and serine), and esteratic site. Some generated bioactive peptides from QC-6 derived hydrolysates could bind to part of the active site containing the catalytic triad, specifically His435 and Ser194. For instance, QHPHGLGALCAAPPST, HVQGHPALPGVPAHW, ASNLDNPSPEGTVM, HVASGAGPW, and KPGGTAGSALPRPAHW could bind up to fourteen, thirteen, thirteen, nine, and six bound residues of CEase, respectively. This result is consistent with previous studies where peptides that have shown higher inhibition potential to His435 and Phe324 residues are categorized as CEase inhibitors because they play a major role in the binding of cholesterol [29]. Similarly, [32], stated that peptides derived from Camel milk whey hydrolysates that could bind to residues His435 and Phe325 are important CEase inhibitors. Likewise, peptide HVASGAGPW could only bind to residue Phe324, which also suggests a potential CEase inhibitor.

Furthermore, [29] had earlier predicted the binding site residues of nine inhibitors of CEase using molecular docking to human and bovine CEase. The observation from this peptide sequencing indicated that the identified peptide sequence was able to form a hydrogen bonding with Gly107 in human CEase and formed a hydrophobic interaction with Ala108 in both human and bovine CEase. In addition, the Trp227 and Phe 324 of bovine CEase was observed to form cation-π interactions and π-stacking with the Phe1 residue of the peptide. Similarly, here Ala108, Phe324, and Trp227 were the major residues with active hotspots, except for Gly107 residues with no active site. This suggests that the most potent peptides QHPHGLGALCAAPPST, HVQGHPALPGVPAHW and ASNLDNPSPEGTVM could bind Ala108, Ser194, Ala195, Trp227, Phe324 and His435 in the active site of CEase, which could suggest an excellent bioactive peptide with prospects of inhibiting CEase, thus limiting dietary cholesterol absorption, which is anticipated for a CEase inhibitor.

Moreover, the CEase inhibitory capacity can further be explained with existing hydrophobic amino acids, and residue near the C-terminal and N-terminal end in the sequenced peptides. A previous study reported the major role the hydrophobicity of peptides plays in the anti-hypercholesterolemic activity of peptides, which is mainly the binding of bile acids. Identified peptides with anti-hypercholesterolemic ability inhibit bile acids absorption in the ileum; as a result, they reduce the level of blood cholesterol concentration [30]. Hydrophobic amino acids, namely alanine, leucine, phenylalanine, proline, etc. have been directly associated with the hypercholesterolemic-lowering activity of peptides. Similarly, peptides with specific hydrophobic amino acids at their terminal ends have demonstrated antidiabetic inhibitory potential. In the present study, aliphatic (Leu and Ala) and aromatic (Phe, Tyr and Trp) hydrophobic amino acids were identified, which are desirable at the C-terminal of the peptide. A previous study identified peptide FDGEVK derived from β-lactoglobulin tryptic hydrolysate as a potential anti-hypercholesterolemic peptide due to the presence of two such C-terminal amino acids [36]. Therefore, bioactive peptides derived from quinoa proteins could potentially inhibit CEase enzymes.

3.5. Identification of PL Inhibitory Peptides

Peptides with PL inhibitory potentials are shown in Table 3. From this study, results show that the peptides ML, MLLL, QHPHGLGALCAAPPST, LPLLR, MFVPVPH, HVQGHPALPGVPAHW, FTVM, LLPYH, MVLP, and HMLH derived from QC-6 showed a significant ($p < 0.05$) binding effect to PL. In this study, the number of peptides derived from QC-6 was 35, and the number of amino acids ranged from 2 to 16. This value is comparable to the range reported by [33] for PL-inhibiting peptides derived from Camel milk protein hydrolysates and the range reported by [39] for pinto beans. However, this was lower than the amino acid value of twenty-three derived from Cumin seeds [40]. The results indicated two peptides (QHPHGLGALCAAPPST and HVQGHPALPGVPAHW) with the most significant ($p < 0.05$) potent PL-inhibitory potential, binding up to 14 and 13 important amino acids, suggesting superior linkages to the target enzyme (Table 3).

Table 3. Pepsite 2 analysis of the selected peptides using pancreatic lipase as a protein receptor.

Peptide	*p*-Value	Peptide Residues	Bound Residues
FFE	0.06672	F1, F2, E3	Phe78 *, Tyr115, Ser153 *, Pro181, Phe216 *
DFTF	0.1717	F2, T3, F4	Phe78 *, Tyr115, Ser153 *, Pro181, Phe216 *
DFLM	0.08332	F2, L3, M4	Phe78 *, Tyr115, Ser153 *, Pro181, Phe216 *, His264 *
ML	0.008899	M1, L2	Phe78 *, Tyr115, Ser153 *, Phe216 *
CDCP	0.2673	C1, D2, C3	Lys81, Glu84, Trp253
CYTF	0.1707	C1, Y2, T3, F4	Phe78 *, Tyr115, Pro181, Ile210, Phe216 *
FSAGGLP	0.3038	F1, S2, A3, G4	Gly77, Phe78 *, Tyr115, His152 *, Ser153 *, Leu154, Gly155, Ala179, Pro181, Phe216 *, His264 *
MLLL	0.009667	M1, L2, L3, L4	Gly77, Phe78 *, Tyr115, Ser153 *
MYLY	0.05041	M1, Y2, L3, Y4	Phe78 *, Tyr115, Ser153 *, Leu154, Ala179, Pro181, Ile210, Phe216 *, His264 *
QHPHGLGALCAAPPST	0.01253	Q1, H2, P3, H4, L9, C10, A11, A12, P13, P14, S15	Gly77, Phe78 *, Lys81, Tyr115, His152 *, Ser153 *, Leu154, Gly155, Ala179, Pro181, Phe216 *, His264 *, Trp253
KIVLDSDDPLFGGF	0.05873	K1, I2, V3, D5, S6, D7, P9, L10, F11, G13, F14	Gly77, Phe78 *, Ile79, Asp80, Lys81, Glu84, Tyr115, Ser153 *, Leu154, Pro181, Phe216 *, Trp253, Arg257, His264 *
AHCGGLPY	0.2518	G5, L6, P7, Y8	Gly77, Phe78 *, Tyr115, Ser153 *, Leu154, Pro181, Ile210, Phe216 *, His264 *
LYNDWDLR	0.5264	L1, Y2, N3, W5	Gly77, Phe78 *, Tyr115, His152 *, Ser153 *, Leu154, Gly155, Ala179, Pro181, Ile210, Phe216 *, His264 *

Table 3. *Cont.*

Peptide	*p*-Value	Peptide Residues	Bound Residues
PGGAR	0.2735	P1, G2, A4	Gly77, Phe78 *, Tyr115, Ser153 *, Leu154, Ala179, Glu180, Pro181, Phe216 *, His264 *
SGPAR	0.1096	G2, P3, A4	Phe78 *, Tyr115, Ser153 *, Pro181, Phe216 *, His264 *
LPLLR	0.01133	L1, P2, L3, L4	Phe78 *, Tyr115, Ser153 *, Ala179, Glu180, Pro181, Phe216 *, His264 *
KPGGTAGSALPRPAHW	0.1243	K1, P2, G3, A9, L10	Phe78 *, Lys81, Glu84, Tyr115, Ser153 *, Phe216 *, Trp253, His264 *
FE	0.1965	F1, E2	Phe78 *, Tyr115, Ser153 *, Phe216 *
HDSF	0.2314	H1, S3, F4	Phe78 *, Tyr115, Ser153 *, Leu154, Ala179, Phe216 *, His264 *
LR	0.2794	L1, R2	Lys81, Glu84, Trp253
VYML	0.1071	V1, Y2, M3, L4	Phe78 *, Tyr115, Ser153 *, Leu154, Ala179, Pro181, Ile210, Phe216 *, His264 *
RR	0.2074	R1, R2	Lys81, Glu84, Trp253
LRL	0.4811	L1, L3	Phe78 *, Tyr115, Ser153 *, Ala179, Glu180, Pro181, Phe216 *, His264 *
MFVPVPH	0.01305	F2, V3, P4, P6, H7	Gly77, Phe78 *, Tyr115, His152 *, Ser153 *, Leu154, Gly155, Ala179, Pro181, Phe216 *, His264 *
HVQGHPALPGVPAHW	0.02747	H5, P6, A7, L8, P9, H16	Gly77, Phe78 *, Lys81, Tyr115, His152 *, Ser153 *, Leu154, Gly155, Ala179, Pro181, Phe216 *, Trp253, His264 *
AGLR	0.2014		Gly77, Phe78 *, Tyr115, Ser153 *, Leu154, Pro181, Phe216 *, His264 *
HVASGAGPW	0.1145	H1, A6, G7, P8, W9	Gly77, Phe78 *, Lys81, Glu84, His152 *, Ser153 *, Leu154, Gly155, Ala179, Phe216 *, Trp253, His264 *
FTVM	0.02278	F1, T2, V3, M4	Phe78 *, Try115, Ser153 *, Leu154, Ala179, Pro181, Phe216 *, His264 *

Table 3. *Cont.*

Peptide	*p*-Value	Peptide Residues	Bound Residues
LLPYH	0.008942	L1, L2, P3, H5	Gly77, Phe78 *, Tyr115, Ser153 *, Leu154, Pro181, Phe216 *, His264 *
MVLP	0.006331	M1, V2, L3, P4	Phe78 *, Tyr115, Ser153 *, Ala179, Glu180, Pro181, Phe216 *, His264 *
GARR	0.1135	G1, A2, R3, R4	Lys81, Glu84, Trp253
ASNLDNPSPEGTVM	0.08445	A1, S2, N3, L4, N6, P7, S8, P9, E10, V13, M14	Phe78 *, Asp80, Lys81, Glu84, Tyr115, Ser153 *, Leu154, Ala179, Phe216 *, Trp253, Arg257, His264 *
CVLSPL	0.1236	L3, S4, P5, L6	Phe78 *, Tyr115, Ser153 *, Ala179, Glu180, Pro181, Phe216 *, His264 *
HMLH	0.008295	H1, M2, L3, H4	Gly77, Phe78 *, Tyr115, His152 *, Ser153 *, Leu154, Gly155, Ala179, Phe216 *
CALVGL	0.2596	A2, L3, V4, L6	Phe78 *, Tyr115, Ser153 *, Pro181, Ile210, Phe216 *, His264 *

* Hotspots that could inhibit the activity of lipase if bounded by the peptide.

Furthermore, these peptides have also demonstrated CEase inhibitory potential, as explained above, implying that these peptides have valuable anti-hypercholesterolemic properties. Other peptides (FSAGGLP, MFVPVPH, and ASNLDNPSPEGTVM) have shown a similar potent inhibitory potential binding up to 11, 11, and 12 important amino acids of PL, respectively. These peptides could be considered as potential PL inhibitors due to their maximum interaction with the binding sites of PL enzyme. Similarly, some peptides showed a high number of binding sites, thus possessing promising PL-inhibitory activity. Peptide HMCH could interact with nine bound residues, four peptides (LPLLR, FTVM, LLPYH, and MVLP) with eight bound residues, and peptide DFLM interacting with six bound residues. Among all the peptides that significantly ($p < 0.05$) bound to the active sites, only peptide ML interacted with a lower number of hotspots (four sites), and can be regarded as a poor PL inhibitor.

Furthermore, all six of the most potent peptides that interacted with maximum bound residues could bind Phe78, Ser153, Phe216, His264, and His152 from the hotspot sites of PL. The other peptides (HMCH, LPLLR, FTVM, LLPYH, and MVLP, DFLM) were capable of binding to similar active sites. Similar binding spots for PL inhibitory activity from cumin seeds have been reported by [40], and camel whey protein hydrolysates by [29]. Moreover, the peptides DFLM, KIVLDSDDPLFGGF, LPLLR, FTVM, LLPYH, MVLP, ASNLDNPSPEGTVM, and HMLH showed the ability to bind to similar hotspots (Phe78, Ser153, His152, Phe216, and His264). Furthermore, peptides showed the potential to indirectly inhibit PL by binding Try115, Leu154, Gly155, Ala179, and Pro181, which do not belong to the characteristic catalytic triad of pancreatic lipase.

It was observed that peptides that have the potential to bind a higher number of active sites of lipase were the hydrophobic amino acids. A majority of the peptides that exhibited the potential of binding a high number of binding sites of lipase consisted of hydrophobic amino acid residues such as proline and leucine. This is comparable to a previous study that showed proline and leucine to be the most active residue bound of lipase inhibiting peptides derived from pinto beans [39]. In addition, hydrogen bonding can also influence hydrophilic amino acid to bind generated bioactive peptides inhibitors to PL [40]. The PL inhibitory method of derived peptides from seed proteins has been reportedly measured against the mechanism of a synthetic lipase inhibitor (Orlistat), and a similar PL inhibitory mechanism observed between them [39,40]. Thus, novel QPHs derived from chymotrypsin at 6 h have PL-inhibitory peptides with the potential to bind active sites of lipase enzyme. This study provides novel PL inhibitory peptdies derived from quinoa that could be health-promoting in combating hypercholesterolemia.

3.6. Molecular Docking of Shortlisted Peptides in the Active Site of Human and Bovine CEase

To elucidate the binding pose and to identify the underlying intermolecular interactions of potential quinoa derived peptides in the active site of human and bovine CEase, molecular docking was performed. Table 4 details the GlideScore, MM-GBSA binding free energy, and the observed intermolecular interactions of the top ranked peptides. It was observed that the peptide HVASGAGPW docked to human CEase with a GlideScore of −9.07 kcal/mol and an MM-GBSA binding free energy of −64.32 kcal/mol, while the peptides AHCGGLPY, LYNDWDLR, MFVPVPH and FSAGGLP exhibited GlideScores of −8.95, −8.9, 8.3 and 7.4 kcal/mol, respectively.

Table 4. Docking score (GlideScore), MM-GBSA binding free energy, and interacting residues of human and bovine CEase with the identified peptides.

Peptide	GlideScore (kcal/mol)	MM-GBSA Based Binding Free Energy (kcal/mol)	Hydrogen Bonds	Hydrophobic Interactions	Salt-Bridge	π-π Interactions	Cation-π Interactions
CEase (Human)—PDB ID: 1F6W							
HVASGAGPW	−9.07	−64.32	Gly107, Asn121, Asn122, Asp434, His435, Gln440, Lys445	Ile69, Ala108, Leu110, Met111, Ala117, Phe119, Tyr123, Leu124, Tyr125, Ala195, Ile323, Phe324, Ala436, Ile439, Tyr453	Asp437		Arg423
AHCGGLPY	−8.95	−55.00	Ile69, Asn121, Leu274, Asp437	Ile69, Ala108, Leu110, Met111, Ala117, Leu124, Val272, Pro273, Leu274, Ala275, Met281, Leu282, Val285, Ile323, Phe324, Ile327, Ala436, Ile439, Tyr453	Lys445		
LYNDWDLR	−8.90	−63.67	Ile69, Asn121, Pro273, Asp434, Asp437	Ile69, Ala108, Met111, Ala117, Phe119, Leu124, Val272, Pro273, Leu274, Ala275, Met281, Ile323, Phe324, Ile327, Ala436	Arg423, Asp437, Lys445		
MFVPVPH	−8.37	−94.06	Asn121, His435	Tyr105, Ala108, Met111, Ala117, Phe119, Leu124, Tyr125, Ala195, Trp227, Met281, Leu282, Val285, Ile323, Phe324, Ile327, Leu392, Phe393, Ala436, Ile439, Tyr453	Lys445		Trp227
FSAGGLP	−7.44	−47.70	His435, Asp437	Ala108, Leu110, Met111, Ala117, Leu124, Tyr125, Ala195, Trp227, Val272, Met281, Leu282, Tyr284, Val285, Phe324, Leu392, Phe393, Ala436, Ile439			
HDSF	−7.89	−44.43	Gly107, Ala108, Glu193, Asp437, Gln440	Ala108, Tyr125, Ala195, Trp227, Met281, Leu282, Val285, Ile323, Phe324, Leu392, Phe393, Ala436, Ile439, Tyr453	Asp437		
LRL	−7.81	−49.25	Glu193, Ser194	Trp103, Tyr105, Ala108, Leu110, Met111, Tyr125, Ala195, Trp227, Val272, Met281, Leu282, Tyr284, Val285, Phe324, Leu392, Phe393, Ala436, Ile439	Glu193		

Table 4. *Cont.*

Peptide	GlideScore (kcal/mol)	MM-GBSA Based Binding Free Energy (kcal/mol)	Hydrogen Bonds	Hydrophobic Interactions	Salt-Bridge	π-π Interactions	Cation-π Interactions
LLPYH	−7.57	−47.61	Ile69, Glu193, His435, Asp437	Ile69, Ala108, Leu110, Met111, Leu124, Val272, Met281, Val285, Ile323, Phe324, Ile327, Ala436, Ile439	Asp437		
MYLY	−7.30	−46.79	Gly107, His435, Asp437	Ala108, Met111, Leu124, Ala195, Trp227, Met281, Leu282, Val285, Ile323, Phe324, Ile327, Leu392, Phe393, Ala436, Ile439	Asp437	Phe324	
CVLSPL	−7.64	−65.08	Gly107, Ser194, Ala436, Asp437	Ala108, Leu110, Ala117, Leu124, Ala195, Trp227, Val272, Met281, Leu282, Val285, Ile323, Phe324, Ile327Leu392, Phe393, Ala436, Ile439			
				CEase (Bovine)—PDB ID: 1AQL			
AHCGGLPY	−10.05	−72.14	Ser194, Asp434, Asp437	Ala108, Leu110, Met111, Ala117, Leu124, Tyr125, Trp227, Leu272, Leu274, Leu285, Leu323, Phe324, Met327, Leu392, Phe393, Tyr427, Ala436, Leu439, Tyr453	Asp434, Asp437		
MFVPVPH	−9.828	−87.97	Thr68	Ala67, Leu69, Tyr75, Tyr105, Ala108, Leu110, Met111, Ala113, Ala117, Leu124, Tyr125, Trp227	Arg63		
				Leu272, Leu274, Leu282, Leu285, Leu323, Phe324, Met327, Leu392, Phe393, Ala436, Leu439			
LPLLR	−9.218	−78.64	Glu193	Leu69, Trp103, Tyr105, Ala108, Leu110, Met111, Ala117, Leu124, Tyr125, Trp227, Leu272, Leu274, Leu282, Leu285, Leu323, Phe324, Met327, Leu392, Phe393, Ala436, Leu439	Glu193		
CYTF	−8.261	−62.93	Glu193, Ser194, His435	Ala108, Leu110, Met111, Leu124, Tyr125, Ala195, Trp227, Leu272, Leu274, Leu282, Leu285, Leu323, Phe324, Met327, Leu392, Phe393, Ala436, Leu439	Glu193		
FSAGGLP	−8.861	−73.84	Ala117	Tyr75, Tyr105, Ala108, Leu110, Met111, Ala113, Ala117, Leu124, Tyr125, Ala195, Leu272, Leu274, Leu282, Leu285, Leu323, Phe324, Met327, Leu392, Phe393, Leu439		Trp227	
LLPYH	−8.437	−76.32	Gly106, Glu193	Ala108, Leu110, Met111, Tyr125, Ala195, Trp227, Leu272, Leu274, Leu282, Leu285, Leu323, Phe324, Met327, Leu392, Phe393, Ala436, Leu439		His435	Phe324, His435

Table 4. *Cont.*

Peptide	GlideScore (kcal/mol)	MM-GBSA Based Binding Free Energy (kcal/mol)	Hydrogen Bonds	Hydrophobic Interactions	Salt-Bridge	π-π Interactions	Cation-π Interactions
CVLSPL	−8.980	−71.06	Ala113, Ala117, His435	Tyr75, Tyr105, Ala108, Leu110, Met111, Ala113, Ala117, Leu124, Tyr125, Ala195, Trp227, Leu272, Leu274, Leu282, Leu285, Leu323, Phe324, Met327, Leu392, Phe393, Ala436, Leu439			Trp227, Phe324
PGGAR	−8.680	−72.64	Glu193	Ala108, Leu110, Met111, Tyr125, Ala195, Trp227, Leu272, Leu274, Leu282, Leu285	Glu193		Trp227
				Leu323, Phe324, Met327, Leu392, Phe393, Ala436, Leu439			
HDSF	−8.348	−33.84	Ala108, Ser194, Ala436	Ala108, Leu124, Ala195, Trp227, Leu272, Leu282, Leu285, Leu323, Phe324, Leu392, Phe393, Tyr427, Ala436, Leu439	Asp434, His435		
HVASGAGPW	−8.884	−67.55	Arg63, Gly106, Asn118, Asn122, Asp434, His435, Ala436	Tyr75, Trp103, Met111, Ala113, Ala117, Phe119, Tyr123, Leu124, Tyr125, Leu323, Met424, Tyr427, Ala436, Leu439, Tyr453	Asp434, Asp437		

CEase is a member of the alpha/beta hydrolase fold family. In the human CEase protein, the active site region includes an oxyanion hole comprising Gly107-Ala108 and Ala195 residues, and a catalytic triad containing Ser194-His435-Asp320 residues; both regions are essential for the functional activity of the protein. The hydroxyl group of Ser194 functions as a nucleophile that facilitates the hydrolytic action [41]. Apart from this, two acidic amino acids, Asp434 and Glu437, are positioned close to the active site region which renders a negative charge to the whole catalytic domain of the protein.

Among the docked peptides, the top docked peptide HVASGAGPW fits perfectly in the active site and is stabilized by hydrogen and hydrophobic bonds. Interestingly, Gly107, Ala108, and Ala195 residues, of the oxyanion hole, and His435, from the catalytic triad, were involved in the strong binding of this peptide with human CEase. The peptide AHCGGLPY showed interactions with several residues surrounding the active site including Ala108. Similarly, other top-docked quinoa peptides such as LYNDWDLR, MFVPVPH and FSAGGLP have shown interactions with the residues involved in the active site (Figure 1). These interactions can further block the catalytic activity of human CEase, supporting their potential as human CEase inhibitors. Like human CEase, bovine CEase shares similar residues (Ser194-His435-Asp320) in the catalytic triad and oxyanion hole of the molecule [42]. Sequence analysis revealed that the residue identity of human CEase and bovine CEase is 80%, and the overall three-dimensional structure of the active site region closely resembles human CEase [43]. The docked peptide AHCGGLPY exhibited a GlideScore of −10.05 kcal/mol and an MM-GBSA-based binding free energy of −72.14 kcal/mol. The active site residues Ala108 and Ser194 formed hydrogen and hydrophobic interactions with this peptide. Following this, the peptides MFVPVPH, FSAGGLP, HVASGAGPW and LPLLR also exhibited good GlideScores and MM-GBSA values (Table 4; Figure 1). The top peptides identified were observed to interact with the critical residues involved in the active site of the enzyme, which directly contributes to the inhibitory activity of the peptides and ligand interaction diagrams, which are represented in Figure 2.

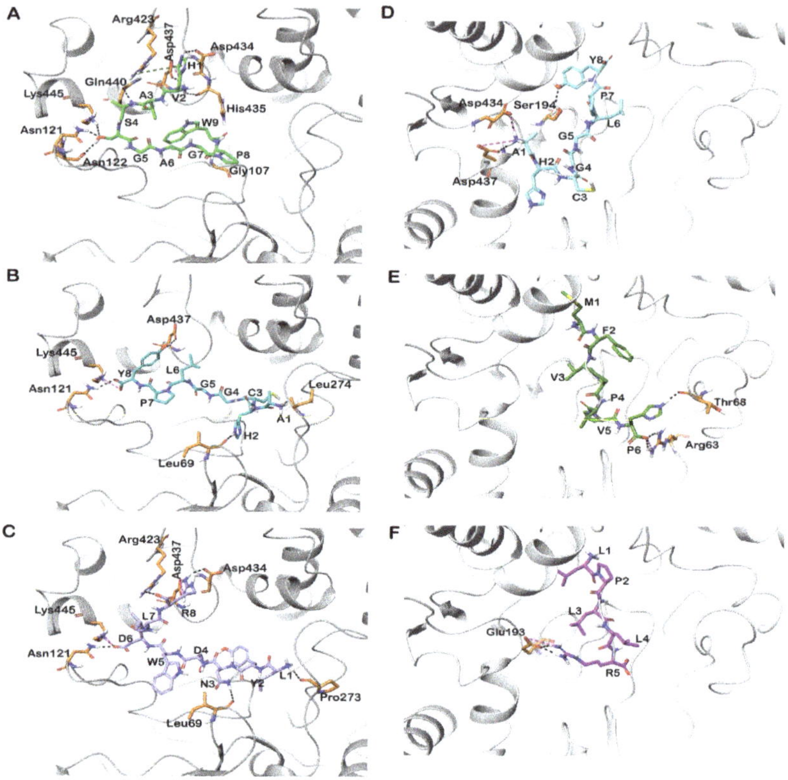

Figure 1. Docked poses of peptides in the active site of human and bovine CEase (PDB IDs: 1F6W and 1AQL). The protein is shown in cartoon representation and the docked peptides are shown in stick representation. The interactions of peptides (**A**) HVASGAGPW, (**B**) AHCGGLPY, (**C**) LYNDWDLR in the active site human CEase. Interactions of peptides (**D**) AHCGGLPY, (**E**) MFVPVPH and (**F**) LPLLR in the active site of bovine CEase. Hydrogen bonds, π-π stacking, π-cation interactions and salt bridges are represented in black blue, green and pink dashed lines. Keywords: Cholesterol esterase (CEase).

3.7. Molecular Docking of Identified Peptides in the Active Site of Human and Porcine PL

The binding pose and intermolecular interaction of quinoa peptides with PL was elucidated using molecular docking simulations. GlideScore and calculated binding energy, based on MM-GBSA, for the best-docked poses as well as intermolecular interactions observed in the docked protein-peptide complex are provided in Table 5. Similar to several esterases and lipases, the PL active site is positioned in the N-terminal domain, which has a catalytic triad (Ser152, Asp176, and His263), and access to this site is controlled by a surface loop. Movement of the lid induces conformational changes in the protein structure that eventually leads to oxyanion hole unmasking that is involved in the interfacial binding region [44].

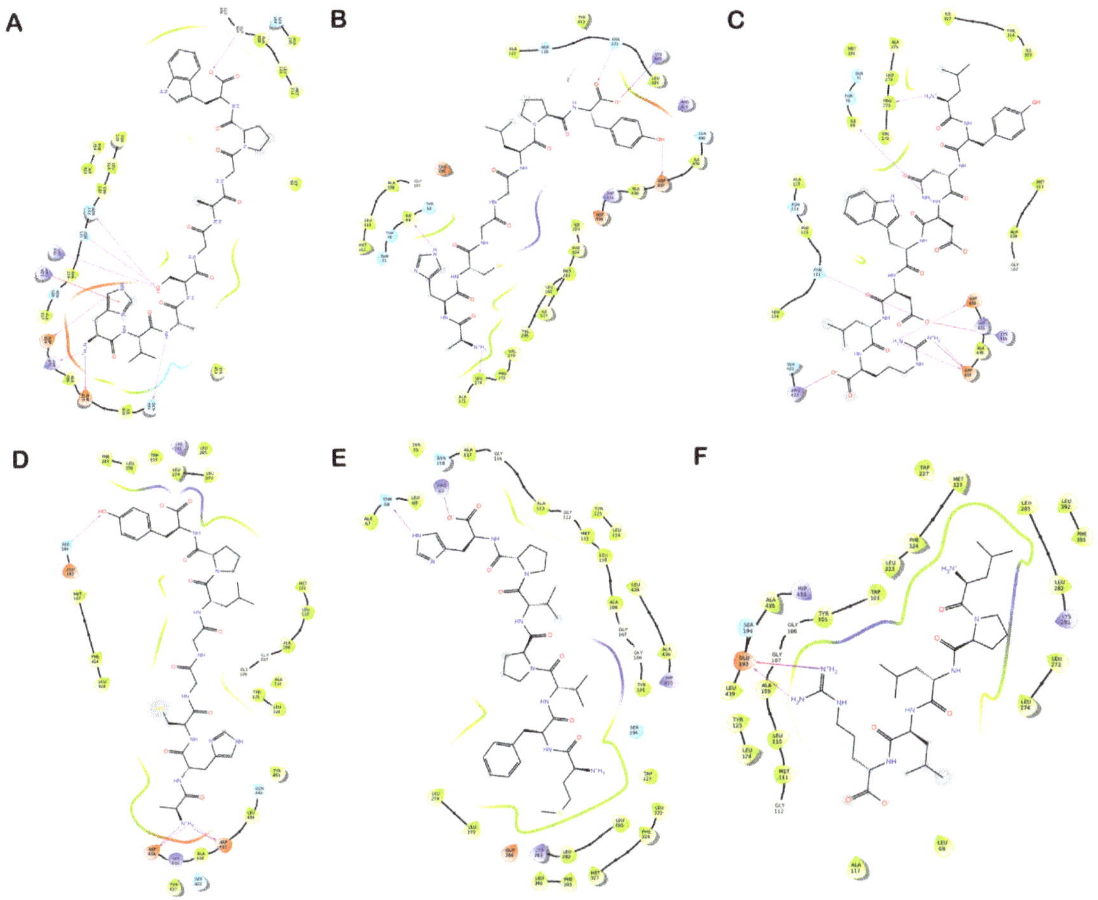

Figure 2. Ligand interaction diagram of inhibitory peptides with the residues of the enzyme CEase. The interactions of peptides (**A**) HVASGAGPW, (**B**) AHCGGLPY, and (**C**) LYNDWDLR in the active site human CEase. Interactions of peptides (**D**) AHCGGLPY, (**E**) MFVPVPH and (**F**) LPLLR in the active site of bovine CEase. Purple circles represent positively charged amino acids, red circles represent negatively charged amino acids, green circles represent hydrophobic amino acids, and light blue circles represent polar amino acids. Purple dashed arrows represent hydrogen bonds involving amino acid side chain and regular purple arrows represent hydrogen bonds involving the amino acid backbone. π-π interactions are represented with green lines.

The binding mode of the best-docked peptide, HVASGAGPW, within the active site of PL was analyzed. Results indicated that the peptide docked to the enzyme recorded a GlideScore of −11.74 kcal/mol and an MM-GBSA binding energy of −46.02 kcal/mol. The other top docked peptides AHCGGLPY, FSAGGLP, LLPYH and MFVPVPH also exhibited good GlideScores and binding energies (Table 5). It was observed that peptide AHCGGLPY recorded a GlideScore of −11.73 kcal/mol, and −61.96 kcal/mol for the calculated binding energy. FSAGGLP recorded −9.72 kcal/mol and −79.39 kcal/mol for the GlideScore and MM-GBSA binding energy, respectively. The GlideScores for peptides LLPYH and MFVPVPH were −9.72 kcal/mol and −8.82 kcal/mol, respectively, and the MM-GBSA binding energies were −77.18 kcal/mol and −70.12 kcal/mol, respectively.

Table 5. Docking score (GlideScore), MM-GBSA binding free energy, and interacting residues of human and porcine PL with the identified peptides.

Peptide	GlideScore (kcal/mol)	MM-GBSA Based Binding Free Energy (kcal/mol)	Hydrogen Bonds	Hydrophobic Interactions	Salt Bridge	π-π Interactions	Cation-π Interactions
LIP (Human)—PDB ID: 1LPB							
HVASGAGPW	−11.74	−46.02	Phe77, Asp79, Glu233, Lys238, Ala259, His263	Phe77, Ile78, Tyr114, Leu153, Ala178, Pro180, Ala207, Ile209, Leu213, Phe215 Val232, Cys237, Trp252, Phe258, Ala259, Ala260, Cys261, Leu264	Glu233		
AHCGGLPY	−11.73	−61.96	Phe77, Ser152, Asp205, Cys237	Phe77, Ile78, Tyr114, Leu153, Ala178, Pro180, Ala207, Ile209, Leu213, Phe215, Trp252, Phe258, Ala259, Ala260, Cys261, Leu264	Asp205		Phe215, Lys238
FSAGGLP	−9.726	−79.39	Phe77, Asp79, Tyr114, Thr115, Ser152	Phe77, Ile78, Tyr114, Leu153, Ala178, Pro180, Ile209, Phe215, Trp252, Ala259, Ala260, Leu264	Asp79		
LLPYH	−9.716	−77.18	Asp79, Tyr114	Phe77, Ile78, Tyr114, Leu153, Ala178, Pro180, Ile209, Leu213, Phe215, Trp252, Ala259, Ala260, Leu264, Tyr267	Asp79	Tyr114, Phe215	
MFVPVPH	−8.824	−70.12	Phe77, Lys238, Ala259, Asp205	Phe77, Ile78, Tyr114, Leu153, Ala178, Pro180, Ala207, Leu213, Phe215, Val232, Trp252, Phe258, Ala259, Ala260, Cys261, Leu264	Asp79	Phe215, His263	
HDSF	−8.305	−45.69	Asp79, Phe77	Phe77, Ile78, Tyr114, Leu153, Ala178, Pro180, Ile209, Leu213, Phe215, Trp252, Ala259, Ala260, Leu264	Asp79	Phe215	
CVLSPL	−8.170	−68.48	Phe77	Phe77, Ile78, Tyr114, Leu153, Ala178, Pro180, Ile209, Leu213, Phe215, Ala259, Ala260, Leu264	Asp79		
CYTF	−8.353	−74.89	Phe77, Asp79	Phe77, Ile78, Tyr114, Ala178, Pro180, Ile209, Val210, Leu213, Phe215, Ala259, Ala260, Leu264, Tyr267	Asp79		
DFTF	−7.716	−55.27	Phe77, Asp79, Ser152	Phe77, Ile78, Trp85, Tyr114, Leu153, Ala178, Pro180, Ile209, Phe215, Trp252, Ala259, Ala260, Leu264, Tyr267	Asp79		
MYLY	−8.085	−95.47	Phe77, Asp79	Phe77, Ile78, Tyr114, Leu153, Ala178, Pro180, Ile209, Val210, Leu213, Phe215, Trp252, Ala259, Ala260, Leu264	Asp79		
LIP (Porcine)—PDB ID: 1ETH							
AHCGGLPY	−10.95	−53.03	Gly77, Asp80, Asn213, Cys238	Phe78, Ile79, Trp86, Tyr115, Leu154, Leu214, Phe216, Cys238, Ile242, Phe259			
			Gln239, Lys240, Val260, Asn263	Val260, Ala261, Cys262, Leu265			

Table 5. *Cont.*

Peptide	GlideScore (kcal/mol)	MM-GBSA Based Binding Free Energy (kcal/mol)	Hydrogen Bonds	Hydrophobic Interactions	Salt Bridge	π-π Interactions	Cation-π Interactions
MFVPVPH	−8.583	−67.89	Val260, Cys262	Phe78, Ile79, Tyr115, Leu154, Ala179, Pro181, Ile210, Leu214, Phe216, Cys238, Ile242, Phe259, Val260, Ala261, Cys262, Leu265			
HVASGAGPW	−10.19	−80.44	Phe78, Asn213, Gln234, Gln239 Phe259, Cys262	Phe78, Ile79, Tyr115, Pro212, Leu214, Phe216, Trp253, Phe259, Val260, Ala261, Cys262, Leu265	Asp80	Trp253	
LPLLR	−8.602	−77.53	Phe78, Asp206	Phe78, Ile79, Tyr115, Ile210, Leu214, Phe216, Trp253, Phe259, Val260, Ala261, Cys262, Leu265	Asp80, Asp206		Phe216
CYTF	−8.164	−75.32	Phe78, Asp80	Phe78, Ile79, Tyr115, Leu154, Ala179, Pro181, Ile210, Phe216, Val260, Ala261, Leu265	Asp80	Phe78, Tyr115	
FSAGGLP	−8.144	−72.31	Asp206, Leu214, Phe259, Val260	Phe78, Tyr115, Leu154, Ala179, Pro181, Ile210, Leu214, Phe216, Cys238, Phe259, Val260, Ala261, Cys262	Asp206		
FFE	−8.202	−58.14	Phe78, Asp80	Phe78, Ile79, Tyr115, Leu154, Ala179, Pro181, Phe216, Trp253, Val260, Ala261, Leu265	Asp80	Phe216, Trp253	
CDCP	−8.166	−57.75	Phe78, Asp80	Phe78, Ile79, Tyr115, Ala179, Pro181, Ile210, Phe216, Trp253, Val260, Ala261, Leu265	Asp80		
HDSF	−7.751	−58.29	Phe78, Asp80	Phe78, Ile79, Tyr115, Leu154, Ala179, Pro181, Ile210, Phe216, Trp253, Val260, Ala261, Leu265	Asp80		
MYLY	−8.308	−76.41	Phe78, Asp80, Phe259	Phe78, Ile79, Tyr115, Leu154, Ala179, Pro181, Leu214, Phe216, Trp253, Phe259, Val260, Ala261, Leu265	Asp80	Phe216, His264	

All of the reported peptides exhibited interactions with critical amino acids in the catalytic binding pockets of the enzyme by forming several hydrogen and hydrophobic interactions (Figure 3). Similar to human PL, the active site of porcine PL comprises the catalytic domain, consisting of Ser153, Asp177, and His264. Similar to human PL, the peptides HVASGAGPW, AHCGGLPY, MFVPVPH and FSAGGLP exhibited good GlideScores and binding energies. Apart from these peptides, LPLLR and CYTF also docked well in the active site of the protein. It was observed that the peptide HVASGAGPW was well-docked in the catalytic region by exhibiting −10.19 kcal/mol and −80.44 kcal/mol for the GlideScore and binding energy, respectively. Followed by this, the peptides AHCGGLPY and MFVPVPH had GlideScores of −10.95 and −8.58 kcal/mol and a binding energy of −53.03 and −67.89 kcal/mol. The peptides were stabilized primarily by electrostatic and hydrophobic interactions, with residues in the active site of the enzyme (Figures 3 and 4).

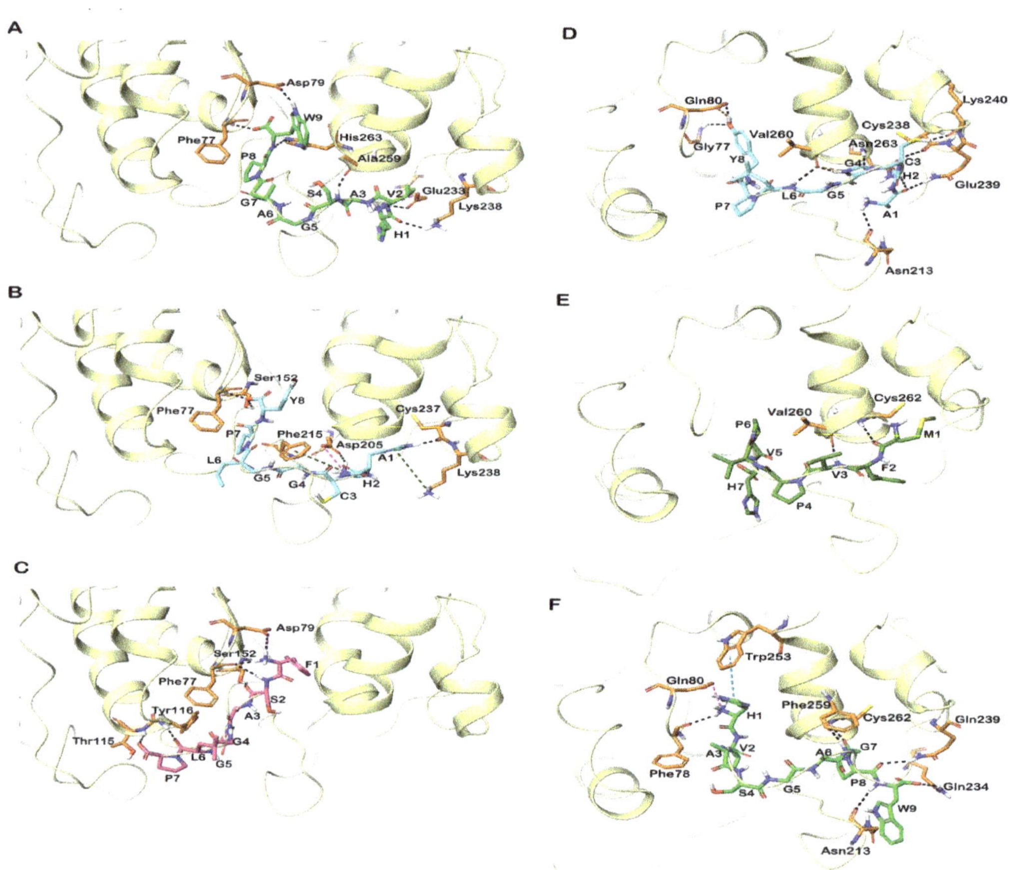

Figure 3. Docked poses of peptides in the active site of human and porcine PL (PDB ID: 1LPB, 1ETH). The protein is shown in cartoon representation and the docked peptides are shown in stick representation. The interactions of peptides (**A**) HVASGAGPW, (**B**) AHCGGLPY, and (**C**) FSAGGLP in the active site of human PL. Interactions of peptides (**D**) AHCGGLPY, (**E**) MFVPVPH and (**F**) HVASGAGPW in the active site of porcine PL. Hydrogen bonds, π-π stacking, π-cation interactions and salt bridges are represented in black-blue, green and pink dashed lines. Keyword: Pancreatic lipase (PL).

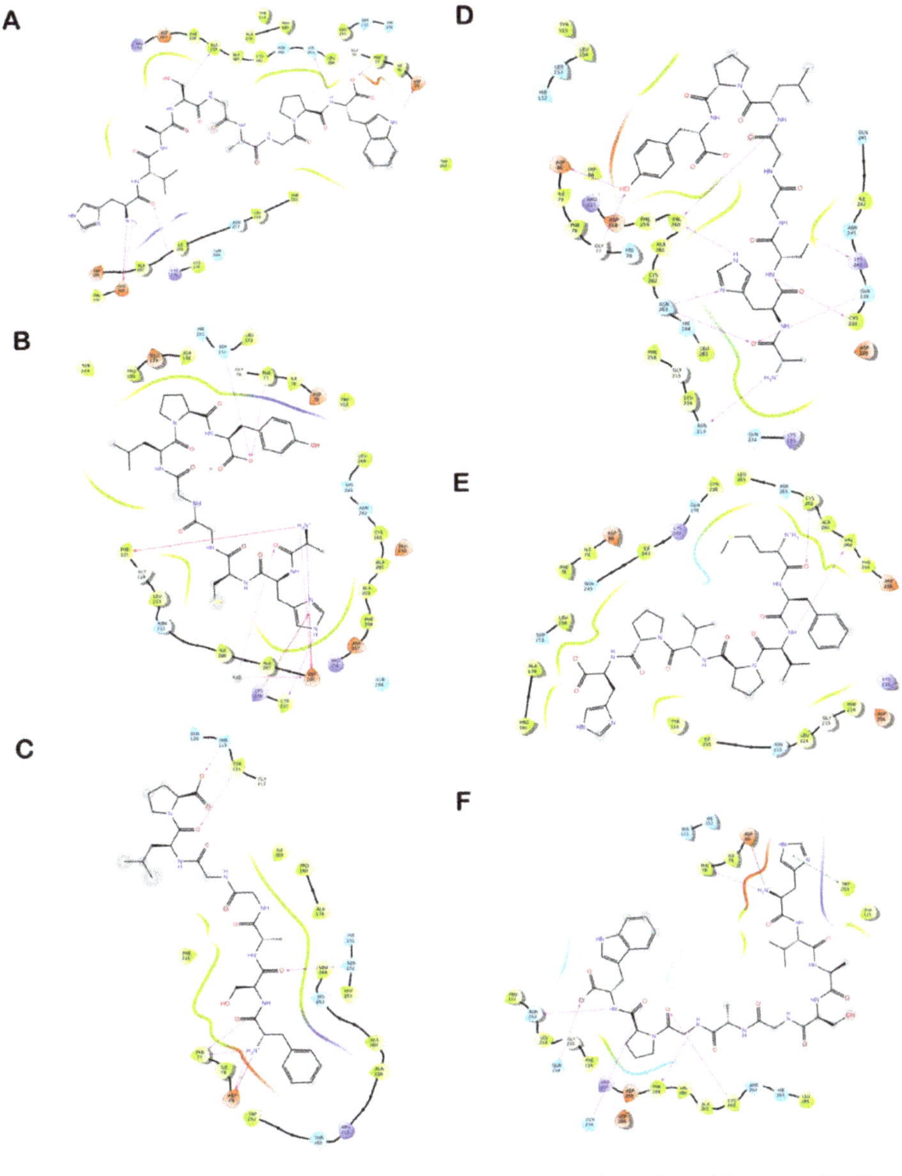

Figure 4. Ligand interaction diagram of inhibitory peptides with the residues of the enzyme PL. The interactions of peptides (**A**) HVASGAGPW, (**B**) AHCGGLPY, and (**C**) FSAGGLP in the active site of human PL. Interactions of peptides (**D**) AHCGGLPY, (**E**) MFVPVPH and (**F**) HVASGAGPW in the active site of porcine PL. Purple circles represent positively charged amino acids, red circles represent negatively charged amino acids, green circles represent hydrophobic amino acids, and light blue circles represent polar amino acids. Purple dashed arrows represent hydrogen bonds involving the amino acid side chain and regular purple arrows represent hydrogen bonds involving the amino acid backbone. π-π interactions are represented with green lines.

4. Conclusions

This study investigated the CEase- and PL-inhibitory potential of QPHs as a functional ingredient for producing nutraceuticals that can be used to manage hypercholesterolemia. This current study explored the generation of hydrolysates from quinoa proteins using three enzymatic approaches (chymotrypsin, proteases, and bromelain). The three enzymes explored produced hydrolysates with higher inhibitory activities towards CEase and PL compared to unhydrolyzed quinoa proteins. Nonetheless, chymotrypsin-generated hydrolysates at 6 h hydrolysis time exhibited higher CEase and PL inhibitory activity as confirmed by lower IC_{50} values. Findings showed that QPHs effectively inhibited the two enzymatic markers responsible for hypercholesterolemia by disrupting the enzyme-substrate interactions at the enzyme hotspot site. Novel peptide ASNLDNPSPEGTVM was found to have the best potential to be a CEase inhibitor, while FSAGGLP, KIVLDSDDPLFGGF, and MFVPVPH were found to be the potential PL inhibitors. The peptides QHPHGLGALCAAPPST and HVQGHPALPGVPAHW we observed to be potential inhibitors against both CEase and PL. In silico data showed that Gly107, Ala108, Ala195, and His435 were involved in the strong binding of these peptide with human CEase. Thus, peptides generated from quinoa seed protein may have a valuable application in ameliorating hypercholesterolemia based on predictions using software tools and molecular docking studies. However, further studies using in vitro, in vivo and cell line assays should be carried out to validate the potential anti-hypercholesterolemic activities observed in this work.

Supplementary Materials: The following supporting materials can be downloaded at: https://www.mdpi.com/article/10.3390/foods12061327/s1, Table S1: List of peptides identified in chymotrypsin hydrolysed quinoa protein hydrolysate QC6; Figure S1: Electrophoretic patters of quinoa proteins and quinoa protein hydrolysates.

Author Contributions: F.F.A.: Investigation, experimentation, writing-original draft; P.M.: Conceptualization, investigation, experimentation, validation, writing-review, and editing; A.J.: Peptide analysis, formal analysis, writing-review and editing; P.A.: Peptide analysis, writing-review and editing; R.V.: Supervision, peptide analysis, review and editing; C.-Y.G.: Conceptualization, data-curation, review and editing; S.M.: Conceptualization, resources, validation, supervision, review and editing. All authors have read and agreed to the published version of the manuscript.

Funding: Authors would like to thank UAE University for funding this study through Asian University Alliance (AUA) grant (12F002).

Data Availability Statement: The data presented in this study are available in the supporting materials.

Conflicts of Interest: The authors declare that they have no conflict of interest with regard to the preparation of this article and that they have no affiliations with or involvement in any entity or organization with any financial or non-financial interest in the subject matter discussed in this manuscript.

References

1. Adisakwattana, S.; Moonrat, J.; Srichairat, S.; Chanasit, C.; Tirapongporn, H.; Chanathong, B.; Ngamukote, S.; Mauml, K.; Sapwarobol, S. Lipid-Lowering mechanisms of grape seed extract (*Vitis vinifera* L.) and its antihyperlidemic activity. *J. Med. Plants Res.* **2010**, *4*, 2113–2120. [CrossRef]
2. Fox, K.M.; Wang, L.; Gandra, S.R.; Quek, R.G.; Li, L.; Baser, O. Clinical and economic burden associated with cardiovascular events among patients with hyperlipidemia: A retrospective cohort study. *BMC Cardiovasc. Disord.* **2016**, *16*, 13. [CrossRef] [PubMed]
3. Rajan, L.; Palaniswamy, D.; Mohankumar, S.K. Targeting obesity with plant-derived pancreatic lipase inhibitors: A comprehensive review. *Pharmacol. Res.* **2020**, *155*, 104681. [CrossRef] [PubMed]
4. Janssen, F.; Pauly, A.; Rombouts, I.; Jansens, K.J.A.; Deleu, L.J.; Delcour, J.A. Proteins of Amaranth (*Amaranthus* spp.), Buckwheat (*Fagopyrum* spp.), and Quinoa (*Chenopodium* spp.): A Food Science and Technology Perspective. *Compr. Rev. Food Sci. Food Saf.* **2017**, *16*, 39–58. [CrossRef] [PubMed]

5. Nongonierma, A.B.; Le Maux, S.; Dubrulle, C.; Barre, C.; FitzGerald, R.J. Quinoa (*Chenopodium quinoa* Willd.) protein hydrolysates with in vitro dipeptidyl peptidase IV (DPP-IV) inhibitory and antioxidant properties. *J. Cereal Sci.* **2015**, *65*, 112–118. [CrossRef]
6. Mudgil, P.; Omar, L.S.; Kamal, H.; Kilari, B.P.; Maqsood, S. Multi-functional bioactive properties of intact and enzymatically hydrolysed quinoa and amaranth proteins. *LWT* **2019**, *110*, 207–213. [CrossRef]
7. Shi, Z.; Hao, Y.; Teng, C.; Yao, Y.; Ren, G. Functional properties and adipogenesis inhibitory activity of protein hydrolysates from quinoa (*Chenopodium quinoa* Willd.). *Food Sci. Nutr.* **2019**, *7*, 2103–2112. [CrossRef]
8. Boachie, R.; Yao, S.; Udenigwe, C.C. Molecular mechanisms of cholesterol-lowering peptides derived from food proteins. *Curr. Opin. Food Sci.* **2018**, *20*, 58–63. [CrossRef]
9. Lammi, C.; Arnoldi, A.; Aiello, G. Soybean peptides exert multifunctional bioactivity modulating 3-hydroxy-3-methylglutaryl-coa reductase and dipeptidyl peptidase-IV targets in vitro. *J. Agric. Food Chem.* **2019**, *67*, 4824–4830. [CrossRef]
10. Gomez, H.L.R.; Peralta, J.P.; Tejano, L.A.; Chang, Y.-W. In silico and in vitro assessment of portuguese oyster (*Crassostrea angulata*) proteins as precursor of bioactive peptides. *Int. J. Mol. Sci.* **2019**, *20*, 5191. [CrossRef]
11. Awosika, T.O.; Aluko, R.E. Inhibition of the in vitro activities of α-amylase, α-glucosidase and pancreatic lipase by yellow field pea (*Pisum sativum* L.) protein hydrolysates. *Int. J. Food Sci. Technol.* **2019**, *54*, 2021–2034. [CrossRef]
12. Zhong, F.; Zhang, X.; Ma, J.; Shoemaker, C.F. Fractionation and identification of a novel hypocholesterolemic peptide derived from soy protein Alcalase hydrolysates. *Food Res. Int.* **2007**, *40*, 756–762. [CrossRef]
13. Ajayi, F.F.; Mudgil, P.; Gan, C.-Y.; Maqsood, S. Identification and characterization of cholesterol esterase and lipase inhibitory peptides from amaranth protein hydrolysates. *Food Chem. X* **2021**, *12*, 100165. [CrossRef] [PubMed]
14. Aluko, R.; Monu, E. Functional and bioactive properties of quinoa seed protein hydrolysates. *J. Food Sci.* **2003**, *68*, 1254–1258. [CrossRef]
15. Mudgil, P.; Jobe, B.; Kamal, H.; Alameri, M.; Al Ahbabi, N.; Maqsood, S. Dipeptidyl peptidase-IV, α-amylase, and angiotensin I converting enzyme inhibitory properties of novel camel skin gelatin hydrolysates. *LWT* **2019**, *101*, 251–258. [CrossRef]
16. Mudgil, P.; Kilari, B.P.; Kamal, H.; Olalere, O.A.; FitzGerald, R.J.; Gan, C.-Y.; Maqsood, S. Multifunctional bioactive peptides derived from quinoa protein hydrolysates: Inhibition of α-glucosidase, dipeptidyl peptidase-IV and angiotensin I converting enzymes. *J. Cereal Sci.* **2020**, *96*, 103130. [CrossRef]
17. Tran, N.H.; Qiao, R.; Xin, L.; Chen, X.; Liu, C.; Zhang, X.; Shan, B.; Ghodsi, A.; Li, M. Deep learning enables de novo peptide sequencing from data-independent-acquisition mass spectrometry. *Nat. Methods* **2019**, *16*, 63–66. [CrossRef]
18. Mooney, C.; Haslam, N.J.; Pollastri, G.; Shields, D.C. Towards the improved discovery and design of functional peptides: Common features of diverse classes permit generalized prediction of bioactivity. *PLoS ONE* **2012**, *7*, e45012. [CrossRef]
19. Trabuco, L.G.; Lise, S.; Petsalaki, E.; Russell, R.B. PepSite: Prediction of peptide-binding sites from protein surfaces. *Nucleic Acids Res.* **2012**, *40*, W423–W427. [CrossRef]
20. Sussman, J.L.; Lin, D.; Jiang, J.; Manning, N.O.; Prilusky, J.; Ritter, O.; Abola, E.E. Protein Data Bank (PDB): Database of three-dimensional structural information of biological macromolecules. *Acta Crystallogr. Sect. D Biol. Crystallogr.* **1998**, *54*, 1078–1084. [CrossRef]
21. Schrödinger-Suite. *Protein Preparation Wizard, Prime*; Schrödinger LLC: New York, NY, USA, 2021.
22. Banks, J.L.; Beard, H.S.; Cao, Y.; Cho, A.E.; Damm, W.; Farid, R.; Felts, A.K.; Halgren, T.A.; Mainz, D.T.; Maple, J.R. Integrated modeling program, applied chemical theory (IMPACT). *JCoCh* **2005**, *26*, 1752–1780. [CrossRef] [PubMed]
23. Friesner, R.A.; Banks, J.L.; Murphy, R.B.; Halgren, T.A.; Klicic, J.J.; Mainz, D.T.; Repasky, M.P.; Knoll, E.H.; Shelley, M.; Perry, J.K. Glide: A new approach for rapid, accurate docking and scoring. 1. Method and assessment of docking accuracy. *J. Med. Chem.* **2004**, *47*, 1739–1749. [CrossRef] [PubMed]
24. Halgren, T.A.; Murphy, R.B.; Friesner, R.A.; Beard, H.S.; Frye, L.L.; Pollard, W.T.; Banks, J.L. Glide: A new approach for rapid, accurate docking and scoring. 2. Enrichment factors in database screening. *J. Med. Chem.* **2004**, *47*, 1750–1759. [CrossRef] [PubMed]
25. Hanafi, M.A.; Hashim, S.N.; Chay, S.Y.; Ebrahimpour, A.; Zarei, M.; Muhammad, K.; Abdul-Hamid, A.; Saari, N. High angiotensin-I converting enzyme (ACE) inhibitory activity of Alcalase-digested green soybean (Glycine max) hydrolysates. *Food Res. Int.* **2018**, *106*, 589–597. [CrossRef]
26. Mahdavi-Yekta, M.; Nouri, L.; Azizi, M.H. The effects of hydrolysis condition on antioxidant activity of protein hydrolyzate from quinoa. *Food Sci. Nutr.* **2019**, *7*, 930–936. [CrossRef]
27. Ambigaipalan, P.; Shahidi, F. Date seed flour and hydrolysates affect physicochemical properties of muffin. *Food Biosci.* **2015**, *12*, 54–60. [CrossRef]
28. Mudgil, P.; Baba, W.N.; Kamal, H.; FitzGerald, R.J.; Hassan, H.M.; Ayoub, M.A.; Gan, C.-Y.; Maqsood, S. A comparative investigation into novel cholesterol esterase and pancreatic lipase inhibitory peptides from cow and camel casein hydrolysates generated upon enzymatic hydrolysis and in-vitro digestion. *Food Chem.* **2022**, *367*, 130661. [CrossRef]
29. Baba, W.N.; Mudgil, P.; Baby, B.; Vijayan, R.; Gan, C.-Y.; Maqsood, S. New insights into the cholesterol esterase-and lipase-inhibiting potential of bioactive peptides from camel whey hydrolysates: Identification, characterization, and molecular interaction. *J. Dairy Sci.* **2021**, *104*, 7393–7405. [CrossRef]
30. Chen, G.-W.; Lin, H.-T.V.; Huang, L.-W.; Lin, C.-H.; Lin, Y.-H. Purification and identification of cholesterol micelle formation inhibitory peptides of hydrolysate from high hydrostatic pressure-assisted protease hydrolysis of fermented seabass byproduct. *Int. J. Mol. Sci.* **2021**, *22*, 5295. [CrossRef]

31. Jafar, S.; Kamal, H.; Mudgil, P.; Hassan, H.M.; Maqsood, S. Camel whey protein hydrolysates displayed enhanced cholesteryl esterase and lipase inhibitory, anti-hypertensive and anti-haemolytic properties. *LWT* **2018**, *98*, 212–218. [CrossRef]
32. Mudgil, P.; Baby, B.; Ngoh, Y.-Y.; Vijayan, R.; Gan, C.-Y.; Maqsood, S. Identification and molecular docking study of novel cholesterol esterase inhibitory peptides from camel milk proteins. *J. Dairy Sci.* **2019**, *102*, 10748–10759. [CrossRef] [PubMed]
33. Mudgil, P.; Kamal, H.; Yuen, G.C.; Maqsood, S. Characterization and identification of novel antidiabetic and anti-obesity peptides from camel milk protein hydrolysates. *Food Chem.* **2018**, *259*, 46–54. [CrossRef] [PubMed]
34. Pasko, P.; Barton, H.; Zagrodzki, P.; Izewska, A.; Krosniak, M.; Gawlik, M.; Gawlik, M.; Gorinstein, S. Effect of diet supplemented with quinoa seeds on oxidative status in plasma and selected tissues of high fructose-fed rats. *Plant Foods Hum. Nutr.* **2010**, *65*, 146–151. [CrossRef]
35. Takao, T.; Watanabe, N.; Yuhara, K.; Itoh, S.; Suda, S.; Tsuruoka, Y.; Nakatsugawa, K.; Konishi, Y. Hypocholesterolemic effect of protein isolated from quinoa (*Chenopodium quinoa* Willd.) seeds. *Food Sci. Technol. Res.* **2005**, *11*, 161–167. [CrossRef]
36. Nagaoka, S.; Futamura, Y.; Miwa, K.; Awano, T.; Yamauchi, K.; Kanamaru, Y.; Tadashi, K.; Kuwata, T. Identification of novel hypocholesterolemic peptides derived from bovine milk beta-lactoglobulin. *Biochem. Biophys. Res. Commun.* **2001**, *281*, 11–17. [CrossRef] [PubMed]
37. Kamal, H.; Mudgil, P.; Bhaskar, B.; Fisayo, A.F.; Gan, C.-Y.; Maqsood, S. Amaranth proteins as potential source of bioactive peptides with enhanced inhibition of enzymatic markers linked with hypertension and diabetes. *J. Cereal Sci.* **2021**, *101*, 103308. [CrossRef]
38. Lin, G.; Shieh, C.-T.; Tsai, Y.-C.; Hwang, C.-I.; Lu, C.-P.; Chen, G.-H. Structure-reactivity probes for active site shapes of cholesterol esterase by carbamate inhibitors. *Biochim. Biophys. Acta (BBA)—Protein Struct. Mol. Enzymol.* **1999**, *1431*, 500–511. [CrossRef]
39. Ngoh, Y.-Y.; Choi, S.B.; Gan, C.-Y. The potential roles of Pinto bean (*Phaseolus vulgaris* cv. Pinto) bioactive peptides in regulating physiological functions: Protease activating, lipase inhibiting and bile acid binding activities. *J. Funct. Foods* **2017**, *33*, 67–75. [CrossRef]
40. Siow, H.-L.; Choi, S.-B.; Gan, C.-Y. Structure–activity studies of protease activating, lipase inhibiting, bile acid binding and cholesterol-lowering effects of pre-screened cumin seed bioactive peptides. *J. Funct. Foods* **2016**, *27*, 600–611. [CrossRef]
41. Singh, H.; Singh, J.V.; Gupta, M.K.; Singh, P.; Sharma, S.; Nepali, K.; Bedi, P.M.S. Benzoflavones as cholesterol esterase inhibitors: Synthesis, biological evaluation and docking studies. *Bioorg. Med. Chem. Lett.* **2017**, *27*, 850–854. [CrossRef]
42. Chen, J.C.-H.; Miercke, L.J.; Krucinski, J.; Starr, J.R.; Saenz, G.; Wang, X.; Spilburg, C.A.; Lange, L.G.; Ellsworth, J.L.; Stroud, R.M. Structure of bovine pancreatic cholesterol esterase at 1.6 Å: Novel structural features involved in lipase activation. *Biochemistry* **1998**, *37*, 5107–5117. [CrossRef] [PubMed]
43. Terzyan, S.; Wang, C.-S.; Downs, D.; Hunter, B.; Zhang, X.C. Crystal structure of the catalytic domain of human bile salt activated lipase. *Protein Sci.* **2000**, *9*, 1783–1790. [CrossRef] [PubMed]
44. Zhu, G.; Fang, Q.; Zhu, F.; Huang, D.; Yang, C. Structure and function of pancreatic lipase-related protein 2 and its relationship with pathological states. *Front. Genet.* **2021**, *12*, 1061. [CrossRef] [PubMed]

foods

Article

Waste Orange Peels as a Feed Additive for the Enhancement of the Nutritional Value of *Tenebrio molitor*

Konstantina Kotsou [1], Theodoros Chatzimitakos [1], Vassilis Athanasiadis [1], Eleni Bozinou [1], Christina Adamaki-Sotiraki [2], Christos I. Rumbos [2], Christos G. Athanassiou [2] and Stavros I. Lalas [1,*]

[1] Department of Food Science and Nutrition, University of Thessaly, Terma N. Temponera Str., 43100 Karditsa, Greece

[2] Laboratory of Entomology and Agricultural Zoology, Department of Agriculture, Crop Production and Rural Environment, School of Agricultural Sciences, University of Thessaly, Phytokou Str., 38446 Volos, Greece

* Correspondence: slalas@uth.gr

Abstract: Lately, additional attention is being placed on edible insects, since they constitute an excellent, cost-efficient source of proteins with a low ecological footprint. *Tenebrio molitor* was the first insect that was considered edible by EFSA in 2021. This species can replace conventional protein sources and thus, it has the potential to be used in many different food products. In the present study, a food by-product that is commonly produced (i.e., albedo orange peel waste) was used as a feed additive for *T. molitor* larvae, in an effort to further improve the circular economy and enhance the nutritional value of the insects. To this end, bran, which is commonly used as feed for *T. molitor* larvae, was fortified with the albedo orange peel waste (up to 25% *w*/*w*). Larval performance, in terms of larval survival and growth, as well as the larval nutritional value, i.e., the content of protein, fat, carbohydrates, ash, carotenoids, vitamins A and C, and polyphenols, was evaluated. Based on the results, the increase in the percentage of orange peel albedos in *T. molitor* feed resulted in a subsequent increase in the content of larvae in carotenoids and vitamin A up to 198%, in vitamin C up to 46%, and an increase in the protein and ash content up to 32% and 26.5%, respectively. Therefore, the use of albedo orange peel waste for feeding of *T. molitor* larvae is highly recommended, since it results in larvae with enhanced nutritional value and at the same time, the utilization of this feeding substrate further lowers the cost of insect farming.

Keywords: edible insects; *Tenebrio molitor*; orange albedo; proximate composition; carotenoids; vitamin A; vitamin C; proteins; fat; fatty acids

1. Introduction

As the world population increases (it is estimated to reach ~10 billion by 2050) the demand for food will increase. This will be accompanied by higher demand for protein-rich foods, in order to sustain a healthy life. Currently, the main source of protein is from animals. Around 350 million tons of meat are produced annually and the estimation is an increase of ~60 million tons by 2050 [1]. To sustain this livestock production, nearly 75% of the total arable land is used [2]. However, intensive livestock farming and its industrialization have a negative environmental impact, since the livestock sector is responsible for 14.5% of total greenhouse gas emissions [3]. Therefore, doubts are rising about the sustainability of livestock farming and the cover of the increasing needs. Hence, alternative sources of protein are being exploited, to address the issue and avoid malnutrition. To this end, insects have emerged as a promising option [4]. Insects have attracted the interest of researchers, primarily due to their high nutritional value. In addition, since their growth requires less space, water, and feed, as well as the fact that insects have a fast life cycle, they exhibit major economic and ecological advantages, rendering them not only a suitable alternative to animal-derived protein but also advantageous, when compared to chicken, pork, and beef meat [5].

To this end, many insects have been studied to examine whether they are edible. Among the ~2000 edible insect species, the yellow mealworm, *Tenebrio molitor* L. (Coleoptera: Tenebrionidae), is one of the most thoroughly investigated insect species and it is an excellent, cost-efficient source of proteins with a low ecological footprint [6]. This was the reason why it was considered edible by EFSA in 2021 and became the first, officially accepted, edible insect in Europe (European Commission 2021/882) [7], even though it was already used for animal feeds. *Tenebrio molitor* larvae can be eaten whole or processed to produce flour, extracts, and oils, which can be used to reinforce biscuits, snacks, pasta, etc. [6]. On average, *T. molitor* larvae contain about 20–50% of dry weight proteins, 20–45% fat, 5–20% carbohydrates, and 1–4% ash [8]. However, these values can vary a lot, depending on the developmental stage of the larvae, the sex, the growth environment, and mainly, their diet [9–11]. The optimum conditions for the survival and development of all life stages of *T. molitor* are 25 °C and 75% relative humidity, whereas its diet consists mainly of flour, bran, cereals, and bread [12–14].

In an effort to produce larvae with high nutritional value, various feeds have been exploited. For instance, Kröncke et al. [10] examined 25 substrates used as feed for *T. molitor* larvae and found that substrates such as rice protein flour, lupine flour, and potato flakes had a positive impact on the biomass growth of the larvae. Raspberry seed cake, dried sugar beet pulp, and coconut flour should be added in small amounts, while others (e.g., mustard flour and psyllium husks) must be avoided, since they can contain compounds that exhibit insecticide activity. Therefore, the selection of the proper substrate is of paramount importance. This is also strengthened by the fact that the mass production of the larvae is expensive, and as such, feeds of low cost are needed to make the production more profitable. To this end, a wide spectrum of organic side streams has been evaluated so far as an alternative option [15]. Indicatively, about 90 million tons of food waste are produced annually in the European Union [16]. Therefore, they could consist of an inexpensive, easily found source of nutrients for insects. Since these side-streams are unsuitable for human consumption, their use for insect rearing will not compete with the human diet, as is the case for other animals. Additionally, their disposal is a major environmental and economic issue. Therefore, valorizing organics side-streams, such as former foodstuff products [17], by-product meals from vegetable oil production [18], and by-products of the cereal and legume seed cleaning process [19], leads toward the development of a circular economy system that exhibits multiple benefits.

Taking into account that organic side-streams can be an excellent source of nutrients and bioactive compounds, they could be upcycled to promote larval growth and simultaneously increase their overall nutritional value to be used for human diets. In most studies, special emphasis is placed on the increase in protein content, while the increase in other nutrients, such as vitamins, is often overlooked. In the present study, a food by-product that is commonly produced (i.e., waste orange peel) was valorized as feed for *T. molitor* larvae, in an effort to enhance their nutritional value and propose an alternative way of valorizing the orange peels. The albedo of the peels was added to the bran feed and the larvae were fed to examine whether their nutritional value could be increased. Additionally, the contents of vitamins A and C were examined, since orange peel albedos are an excellent source of both vitamins [19]. Albedos were chosen, since they are less commonly employed for other uses, compared to flavedos, which are used for culinary, zesting, and pastry purposes.

2. Materials and Methods

2.1. Chemicals

All solvents used were of HPLC grade and were purchased from Carlo Erba (Val deReuil, France). Gallic acid, sodium anhydrous carbonate, 2,2-diphenyl-1-picryl-hydrazyl (DPPH), 2,4,6-tri-2-pyridinyl-1,3,5-triazine (TPTZ), and Folin–Ciocalteu reagent were received from Penta (Prague, Czech Republic). Hydrochloric acid, Bradford reagent, iron

(III) chloride, ascorbic acid, β-carotene, and trichloroacetic acid were purchased from Sigma-Aldrich (Steinheim, Germany).

2.2. Insects

Stock colonies of *T. molitor* were reared in plastic insect breeding trays (60 × 40 × 14.5 cm) (Beekenkamp Verpakkingen BV, Maasdijk, The Netherlands). The boxes were lacking the top cover to allow air circulation and box aeration. The stock colonies were maintained in a pilot-scale insect-rearing unit at the Laboratory of Entomology and Agricultural Zoology at the University of Thessaly, Greece under constant conditions (i.e., 27 ± 0.5 °C, 60 ± 5% relative humidity and continuous darkness).

To obtain newly emerged larvae, 250 g of adults were allowed to oviposit for 7 d in crates (60 × 40 × 14.5 cm) along with 2 Kg of wheat flour. Adults were placed on the top of a mesh to avoid cannibalism of eggs. Agar was provided to adults ad libitum as a moisture source. After the 7-day interval, adults were removed and the newly emerged larvae were left to feed on flour for 14 d before the initiation of the experiments. Larvae were separated from the flour by sieving. To determine the available amount of larvae, 6 subsamples (with at least 100 larvae) were taken and the initial individual larval weight as well as the number of larvae per weight, were estimated.

2.3. Feeding Trial

The insects that served as control were fed with wheat bran (OA0) (purchased from a local market, Volos, Greece) with a particle size smaller than 2 mm. For the different dietary treatments, orange albedo flour was added to the wheat bran at three rates (i.e., 10 (OA10), 17.5 (OA17.5), and 25% (OA25) *w*/*w*). In order to prepare the orange albedo flour, oranges (bought from a local market, Karditsa, Greece) were rinsed with water and dried with a paper towel. Then, the peel was removed, and the albedo was separated from the flavedo manually. The albedos were cut into smaller pieces (~2 × 2 cm) and placed in a freeze-dryer (Biobase BK-FD10P freeze-dryer (Jinan, China)) for 24 h. The dried albedos were crushed to a fine powder and kept frozen (−20 °C) until further use.

Plastic insect breeding trays (24 × 29.5 × 10 cm) were used as experimental units for the feeding trials. Each tray was filled with 500 g of feed (either wheat bran or bran fortified with albedo orange peel waste) and 2500 14 day-old larvae. For each dietary treatment, four tray replicates were prepared. During larval development, agar was provided to larvae three times per week as a moisture source, increasing the amount of agar provided over time, in order to ensure agar availability for larvae. Larvae were left to feed undisturbed. At the end of each week (from 3rd week onwards) the content of each tray was homogenized by gently pouring the content of the tray into an empty container and repeating this procedure five times. Two random subsamples were subsequently taken from each tray. Afterwards, for each subsample the number of larvae was determined, as well as the total larval weight using an analytical balance (Equinox EAB125i, Adam Equipment Inc., Fox Hollow Road, Oxford, England). Care was taken, so that each subsample contained at least 100 larvae. To determine the average individual larval weight, the total larval weight was divided by the number of the larvae in each subsample. At the end of the 6th week, the larvae were harvested in order to carry out the analyses described in Section 2.4. After harvesting, larvae were fasted for 24 h, separated into batches, weighed, and euthanized by freezing (−20 °C). The larvae harvested at the end of the trial were then placed in a freeze dryer for 24 h. The dried larvae were then crushed to a fine powder and stored in amber-glass vials at −20 °C until further analysis.

2.4. Larval Composition Analysis

2.4.1. Water Content Calculation

The water contained in the larvae was determined by subtracting the weight after the freeze-dryer process from the weight of the same batch, before freeze-drying.

2.4.2. Crude Protein Content

The protein content of the samples was determined using the Bradford assay [20]. To extract the proteins from the samples, 10 g of distilled water was added to 1 g of the sample. The pH of the distilled water was previously adjusted to 12, using NaOH solution (1 M). Extraction was carried out by stirring the mixture at 500 rpm for 1 h at room temperature. Next, the mixture was centrifuged for 5 min at 4500 rpm (at room temperature). The supernatant was collected and 10 g of distilled water (pH adjusted to 12) was added to the solid residue. The extraction procedure was repeated two more times (three times in total) and the supernatants were combined. For the determination of the protein content, 100 μL of the combined supernatants was transferred to an Eppendorf tube and 900 μL of Bradford reagent was added. After vortexing for 30 sec, the samples were left to incubate for 10 min at room temperature in the absence of light. Then, the absorbance of the samples was measured at 595 nm using a spectrophotometer (Shimadzu UV-1700 PharmaSpec Spectrophotometer, Kyoto, Japan). A standard calibration curve was prepared using bovine serum albumin.

2.4.3. Carbohydrates

The carbohydrates were determined using the phenol/sulfuric acid assay [21]. To extract carbohydrates, 10 g of distilled water was added to 1 g of the sample. Extraction was carried out by stirring the mixture at 500 rpm for 1 h at 50 °C. Next, the mixture was centrifuged for 5 min at 4500 rpm (at room temperature) and the supernatant was collected. The supernatant (0.22 mL) was transferred to a plastic tube and immediately concentrated in sulfuric acid (0.65 mL) and phenol solution (0.13 mL) (5% *w*/*v* in distilled water). The mixture was placed in a water bath at 90 °C for 5 min and then left to cool at room temperature for 5 min. Finally, the absorbance of the solution was measured at 495 nm using a spectrophotometer. A calibration curve was prepared using D(+)-glucose as a standard.

2.4.4. Ash

The crude ash content of *T. molitor* larvae was found using a gravimetric approach. In a porcelain crucible, an accurately weighted portion of ~5 g dried insect powder was added and placed in an oven. The temperature was increased at a rate of 5 °C/min up to 550 °C. Samples were left in the oven for about 5 h until no black residue could be observed. Then, the samples were placed in a desiccator and left to cool to room temperature. The weight of the crucibles was determined, and, after calculations, the ash content was determined.

2.4.5. Total Fat, Fatty Acids, and Calculated Oxidizability Value (COX)

To determine the fat content of the larvae, a defatting process was carried out. In a screw-capped, amber-glass vial, 1 g of sample was added along with 10 mL of hexane. The mixture was stirred at 600 rpm for 60 min at 40 °C. The mixture was centrifuged at 4500 rpm for 5 min. The supernatant was transferred to a pre-weighted flask and the defatting process was repeated two more times to the solid residue. Each time, the supernatants were transferred to the flask. The solvent was removed using a rotary evaporator and the weight of the extracted oil was calculated.

To determine the fatty acids contained in the extracted oil, fatty acid methyl esters (FAMEs) were prepared according to Commission Regulation (EC) No 796/2002 (Annex XB) [22]. The analysis of methyl esters with GC-FID was carried out according to a modified method described by Lalas et al. [23]. An Agilent Technologies (Santa Clara, CA, USA) Gas Chromatograph model 7890A, equipped with a capillary column Omegawax (30 m × 320 μm × 0.25 μm) (Supelco, Bellefonte, PA, USA) was used. Helium was the carrier gas at a flow rate of 1.4 mL/min. The column temperature program was as follows: initially isotherm for 5 min at 70 °C, an initial programmed rate of 20 °C/min up to 160 °C, then a second rate of 4 °C/min up to 200 °C, and a final rate of 5 °C/min up to 240 °C. The injector and flame ionization detector (FID) temperatures were maintained at 240 and 250 °C, respectively. The flow rate for hydrogen was 50 mL/min, for air 450 mL/min, and the makeup flow of helium was 50 mL/min. Samples of 1 μL were injected into the split mode (1:100). The individual peaks were identified by comparison of reference standards of FAME Mix C8–C24 (Sigma-Aldrich, St. Louis, MO, USA). The percentage composition of the samples was computed from the peak areas using the normalization method (without correction factors). The component percentages were calculated as mean values from triplicate GC-FID analysis [22,24,25].

The calculated oxidizability value (COX) was also measured using the method described by Fatemi and Hammond [25], as shown below:

$$COX = \frac{1\ (18:1,\ \%) + 10.3\ (18:2,\ \%) + 21.6\ (18:3,\ \%)}{100} \tag{1}$$

where 18:1 represents the % percentage of oleic acid, 18:2 of ω-6, linoleic acid, and C18:3 the percentage of ω-3, linolenic acid.

2.4.6. β-Carotene–Vitamin A

β-Carotene and vitamin A content were estimated using a previously reported method [26]. For the extraction step, 1 g of each sample was extracted using 10 mL of ethanol for 30 min at 300 rpm at room temperature, and then, they were placed in an ice bath for 5 min with occasional shaking. Then, the mixture was centrifuged for 5 min at 4500 rpm. The absorbance of the extract was read at 450 nm and the β-carotene content was calculated from a standard β-carotene curve using the interpolation method. Vitamin A was determined using a conversion factor proposed by the United States Department of Agriculture (USDA). Vitamin A: 1 International Unit (IU) = 0.60 μg β-carotene.

2.4.7. Vitamin C

The ascorbic acid content was determined using a modified chromatometric assay [27]. A quantity of 5 g of ground insect was weighed and placed in a beaker, followed by the successive addition of 27 mL distilled water: methanol mixture (60:40, *v*/*v*) and 3 mL of 10% *w*/*v* trichloroacetic acid solution. After vortexing for 1 min, 20 mL of hexane was added. Then, the mixture was further stirred for 30 min at room temperature, followed by centrifugation at 4500 rpm for 5 min. The lower aqueous phase was transferred to a centrifugal tube and centrifuged for 10 min at 10,000 rpm. In an Eppendorf tube, 1 mL of the aqueous layer was transferred, followed by the addition of 0.5 mL Folin–Ciocalteu reagent (10% *v*/*v*), and the mixture was incubated for 10 min at room temperature. Finally, the absorbance was measured at 760 nm. Quantification was carried out by preparing a calibration curve using ascorbic acid.

2.4.8. Total Polyphenol Content (TPC) Determination

For the determination of the total polyphenol content (TPC) of the extracts, a previously reported method was employed [28]. In brief, 100 μL of *T. molitor* extracts (prepared as described in Section 2.4.7) were mixed with an equal portion of the Folin–Ciocalteu reagent. After 2 min, 800 μL of Na_2CO_3 solution (5% *w*/*v*) was added and the solution was incubated at 40 °C for 20 min in the absence of light. Finally, the absorbance was measured at 740 nm. Quantification was carried out using a standard calibration curve with gallic acid. The TPC of the solution was expressed as mg of gallic acid equivalents (GAE) per L. Next, the extraction yield of total polyphenols (Y_{TP}) (expressed as mg GAE per g dry weight) was calculated using the following equation:

$$Y_{TP}\ (\text{mg GAE/g dw}) = \frac{C_{TP} \times V}{w} \tag{2}$$

where V is the volume of the extraction medium (in L) and w is the dry weight of the sample (in g).

2.4.9. Ferric-Reducing Antioxidant Power (FRAP) Assay

The FRAP activity was measured based on a previous study [29]. In brief, 0.05 mL of $FeCl_3$ solution (4 mM in 0.05 M HCl) was added to an equal portion of the sample (prepared as described in Section 2.4.7) and the solution was incubated for 30 min at 37 °C. Then, 0.90 mL of TPTZ solution (1 mM in 0.05 M HCl) was added, and after 5 min, the absorbance at 620 nm was recorded. A calibration curve was prepared using ascorbic acid as the standard compound. Results (P_R) were expressed as μmoL of ascorbic acid equivalents (AAE) per g dw, using the following equation:

$$P_R\ (\mu\text{moL AAE/g dw}) = \frac{C_{AA} \times V}{w} \quad (3)$$

where V is the volume of the extraction medium (in L) and w is the dry weight of the sample (in g).

2.5. Statistical Analysis

For the survival and growth study of the larvae, a total of 8 measurements (four replicate trays and 2 subsamples per tray) were taken. For the proximate composition analyses, all determinations were carried out in triplicates, using three samples in each, resulting in a total of nine measurements. Results were expressed as mean values of the nine measurements ± standard error of mean (SEM). Next, the Kolmogorov–Smirnov test was carried out to examine whether the results were normally distributed. Statistically significant differences ($p < 0.05$) between the samples were assessed with the Kruskal–Wallis test. The statistical analyses were performed using SPSS (version 29) (SPSS Inc., Chicago, IL, USA) software.

3. Results and Discussion

3.1. Survival and Growth of *T. molitor* Larvae

The first parameter to consider when examining a potential feed for insect larvae is whether it has an impact on their survival. To this end, the effect of the addition of different orange albedo powder proportions in wheat bran was examined, in terms of larval survival. Wheat bran was selected as the control feed since it is a commonly used substrate that promotes rapid growth and ensures high survival and high larval weight gain [30]. Larval survival was evaluated every week, starting from the end of the third week, until the end of the sixth week (before larvae pupate). The % survival rates of the larvae at each evaluation interval are given in Table 1. As can be seen, no significant differences ($p > 0.05$) were recorded among the control samples within the 4 weeks. This was also the case with all the examined additions of the orange albedo powder. Therefore, it can be concluded that the use of orange albedo powder is safe for the production of *T. molitor* larvae. This is an encouraging finding, since in a previous study, the use of the essential oils from orange peels was found to reduce the survival of *T. molitor* adults by 12.4% [31]. However, this was not proven to be a hindrance in our case, for two main reasons. First, the essential oil of the orange peel is located mainly in the outer part (i.e., flavedo), where the oil glands are located [32]. Since in our case, the albedo part was used (since the flavedo part is used as zest for culinary reasons), the presence of essential oils is minimal. Secondly, in order to produce the orange albedo powder, the albedos were subjected to freeze-drying, which is known to reduce the amount of essential oils in plant tissues [33]. Additionally, it is noteworthy that in all cases, the larvae exhibited a preference toward albedo powder, compared to the wheat bran, suggesting that the larvae, through self-selection, can better regulate their nutritional needs [10].

The next crucial parameter is to examine the effect of orange albedo powder on the growth of the *T. molitor* larvae. The individual larval weight for each dietary treatment is provided in Figure 1. It is noteworthy that the weight difference of the larvae upon birth was very low (<0.00010); therefore, all differences in the weight were attributed to the different treatments. The individual larval weight at the initiation of the trial was 1.07 mg.

Based on the results, not only did the orange albedo not hinder larval growth, but in most cases, it promoted their growth. More specifically, after 1 week of consuming the fortified feeds (third week), a significant ($p < 0.05$) weight increase was recorded for the larvae fed with wheat bran supplemented with 25% orange albedo. In the two other feeds (10% and 17.5% albedo), no significant increase was recorded compared to the control. By week four, the larvae that consumed the feed that contained 17.5% and 25% albedo were found to be significantly ($p < 0.05$) increased, when compared to the control, as an increase in the larval weight of up to 14.7% was recorded. The same trend was also observed at the next two evaluation intervals (weeks five and six). The larvae fed with the two abovementioned diets were found to be significantly ($p < 0.05$) larger, compared to the control larvae.

Table 1. Survival (%) of *Tenebrio molitor* larvae (±SEM) fed for six weeks with wheat bran (control) (OA0) and bran fortified with different rates of albedo orange peel waste (10 (OA10), 17.5 (OA17.5), and 25% (OA25)) (n = 8).

Diets	Survival (%)			
	Third Week	Fourth Week	Fifth Week	Sixth Week
OA0	86.64 ± 2.15	86.36 ± 2.96	82.24 ± 2.50	81.16 ± 2.79
OA10	84.48 ± 2.02	81.44 ± 2.56	80.80 ± 2.94	79.88 ± 2.14
OA17.5	85.16 ± 2.23	83.84 ± 2.21	82.44 ± 2.76	82.16 ± 2.21
OA25	94.32 ± 3.68	90.84 ± 4.21	88.60 ± 2.14	88.04 ± 2.98

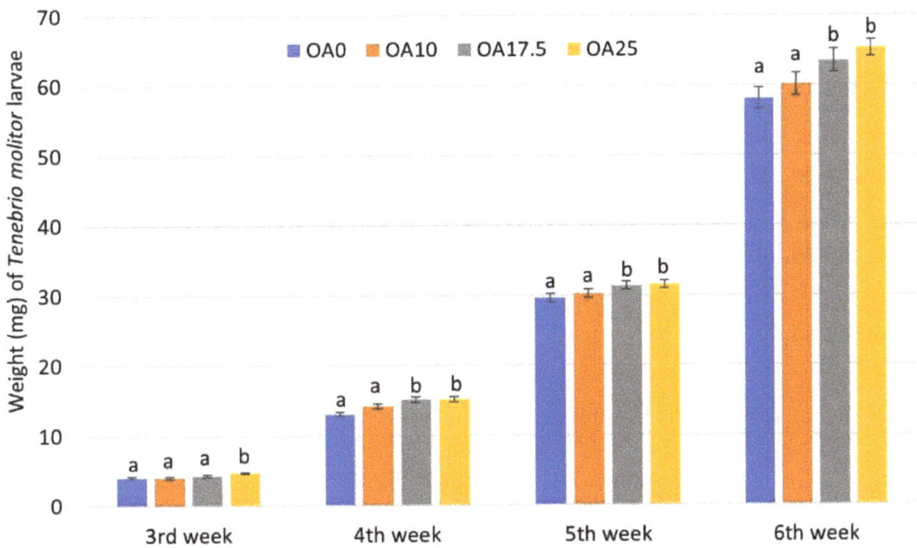

Figure 1. Individual weight (mg) of *Tenebrio molitor* larvae fed for six weeks with wheat bran (control) (OA0) and bran fortified with different rates of albedo orange peel waste (10 (OA10), 17.5 (OA17.5), and 25% (OA25)) (n = 8). Statistically significant differences ($p < 0.05$) are denoted with different superscript letters.

To date, quite a few by-products have been examined as feed for mealworm larvae [15]. For instance, feeding biscuits as a single substrate or supplementing them with used cereals or bread resulted in reduced larval growth and development [17]. On the contrary, chicken feed, rapeseed meal, wheat bran, and sunflower with willow leaves, were found to promote growth, with bran being the optimum [34]. As such, the increased growth of larvae upon any feed, supplemented with by-products, cannot be taken for granted. According to our results, the orange albedo can be used as an alternative option to supplement feed, since it does not reduce larval survival and it promotes their growth above a concentration of 17.5% *w*/*w*.

3.2. Evaluation of the Nutritional Value of the Larvae

Before examining the proximate composition of the larvae, the water content of the samples was evaluated. In all cases, it was found that the samples contained water ranging between 37 and 38%.

3.2.1. Proximate Composition

The main reason why insects have attracted the interest of researchers and have started to be exploited in Western countries as edible food is that they have a high protein content [35]. In principle, *T. molitor* larvae are known to have a high protein content. For instance, Costa et al. [36] estimated the protein content of *T. molitor* larvae to be 13.7 g per 100 g of their net weight, Nowak et al. [37] reported values between 13.68 and 22.32 g/100 g, and finally, Ghaly et al. [38] reported values between 24.3 and 27.6 g per 100 g of their net weight. Therefore, in the present study, the larvae fed with different diets were examined for their crude protein content. The proximate composition of the *T. molitor* larvae fed the different dietary treatments by the end of the sixth week are shown in Table 2. Our results illustrate that the increase in the percentage of the albedo in the feed caused a rather proportional increase in the protein content of the larvae. More specifically, the larvae fed with wheat bran containing 10% orange albedo contained 13.2% more proteins ($p < 0.05$) compared to the control. Likewise, a statistically significant increase ($p < 0.05$) of 24.6% and 32% was registered when the larvae were fed with 17.5% and 25% orange albedo added to the wheat bran. This is a considerable increase in the protein content, given that the orange albedo proved to be a very poor source of crude protein (~1.5%), compared to previous reports [39]. For instance, Mancini et al. [17] found that feeding *T. molitor* larvae with bread remains or soon-to-expire cookies resulted in a decrease in the protein content of the larvae, despite the ~10% protein content of the feed. In our case, orange albedo promoted larval growth and protein formation. Although this may seem contradictory, the higher protein content of the orange albedo-fed larvae may be due to the increased feed assimilability and feed-protein conversion rates and not due to the increased protein content of the feed [40]. Therefore, orange albedo proved to be an efficient means to increase the protein content of the larvae. Thus, the orange albedo-fed larvae can be considered more beneficial for human consumption, due to the increased protein content.

Table 2. Proximate composition of *Tenebrio molitor* larvae fed for six weeks with wheat bran (control) (OA0) and bran fortified with different rates of albedo orange peel waste (10 (OA10), 17.5 (OA17.5), and 25% (OA25)) (n = 9).

% Composition of Dry Weight	OA0	OA10	OA17.5	OA25
Crude protein	$32.25 \pm 0.94^{a,*}$	36.5 ± 1.28^{b}	40.18 ± 1.65^{c}	44.2 ± 2.08^{d}
Crude fat	42.16 ± 1.12^{a}	41.97 ± 1.98^{a}	41.12 ± 1.03^{a}	38.84 ± 0.81^{b}
Carbohydrates	22.8 ± 0.89^{a}	19.28 ± 0.75^{b}	15.91 ± 0.46^{c}	15.67 ± 0.56^{c}
Crude ash	1.06 ± 0.03^{a}	1.17 ± 0.03^{b}	1.29 ± 0.05^{c}	1.34 ± 0.06^{c}
Energy (kcal/100 g)	599.64 ± 23.07^{a}	600.85 ± 22.58^{a}	594.44 ± 24.45^{a}	590.24 ± 15.04^{a}

* Within each line, statistically significant differences ($p < 0.05$) are denoted with different superscript letters.

Next, the content of the mealworm larvae in carbohydrates was evaluated, despite their relatively low content that often results in overlooking this parameter [21]. As can be seen in Table 2, the increase in the content of orange albedo caused a decrease in carbohydrates. In this context, a decrease in the *T. molitor* larvae carbohydrate content of 15.4, 30.1, and 31.2% (significant for $p < 0.05$) was recorded for OA10, OA17.5, and OA25, respectively. This decrease in the carbohydrate content may be associated with the increased growth rate. Since the larvae are preparing gradually for metamorphosis, proteins and lipids are more important and therefore, more carbohydrates are needed to meet the energy demands for growth [10]. The relatively low carbohydrate content of the larvae may be beneficial for the human diet, since carbohydrates can easily be consumed by other sources and therefore, the low content assists in an easier diet balance. Moreover, out of the carbohydrates contained in the larvae, about 3% is due to soluble sugars, while a considerable amount comprises chitin, which serves as an excellent source of fiber [21].

Additionally, the crude ash content of the samples was determined (Table 2). The control larvae were found to contain 1.06% crude ash. As the content of the feed in orange albedo increased, an increase in the crude ash of the larvae was also recorded. More specifically, a statistically significant ($p < 0.05$) increase of 10, 21.6, and 26.5% was recorded in the crude ash content for OA10, OA17.5, and OA25, respectively. The above results corroborate that the larvae fed with the orange albedo are a better source of minerals, compared to the control larvae. The crude ash content of the samples was also found to be comparable to that of previous studies. For instance, Siemianowska et al. [41] reported an ash content of 1.55%, Costa et al. [36] reported a content of 1.23%, while in other studies, the content ranged between 0.9 and 3.81% [37], or between 1.8 and 2.2% [38]. The increased content of ash in the larvae can assist human nutrition [38].

Fat is another key nutrient that contributes to proper human nutrition [42]. In our results, the control larvae contained 42.16% crude fat (Table 2). When orange albedo was added to the feed at percentages of 10% and 17.5%, no significant ($p > 0.05$) difference was recorded when compared to the control sample. However, when orange albedo was added at 25%, the larvae were found to contain 3.32% less fat, as compared with the respective figures from the control larvae. This decrease in the fat content may be attributed to the low fat content of the orange albedo (~1.47% [43]). In previous studies, various percentages have been reported for the fat content of the larvae, ranging from 32.7% [44,45] up to 41% [46]. Therefore, the larvae in our case have comparable fat content, when compared to previous studies.

In all cases, the main fatty acids were found to be oleic acid (C18:1), ranging between 39.07% and 43.68%, followed by linoleic acid (C18:2), ranging between 27.34% and 29.68%, and finally, palmitic acid (C16:0), ranging between 21.39% and 23.22% (Table 3). Oleic acid has many beneficial properties for human health, such as the prevention–suppression of breast cancer and rheumatoid arthritis [47]. The *T. molitor* larvae were also rich in linoleic acid [36], which is one of the two essential fatty acids that the human body cannot synthesize from other food components [48]. Animal sources that are considered rich in omega-6 fatty acids are chicken and egg yolk. In particular, chicken fat may contain 18–23% and egg yolk 16% linoleic acid [49], whereas *T. molitor* larvae were found to contain nearly two times more linoleic acid [36]. It is noteworthy that the amount of saturated fatty acids (SFAs) was found to decrease in the presence of orange albedo in the feed. More specifically, in all cases, a statistically significant decrease ($p < 0.05$) of ~16% was recorded. As regards the monounsaturated fatty acids (MUFAs), although an increase of 2–3% was recorded in the larvae fed with orange albedo, this increase was not found to be significant ($p > 0.05$). However, a significant increase ($p < 0.05$) of up to 8.6% in the content of polyunsaturated fatty acids (PUFAs) was recorded in all cases of the larvae fed with orange albedo. Considering the above, the PUFA: SFA ratio was calculated for all samples to assess their impact on cardiovascular health [50]. The control larvae were found to have a ratio of 0.87, whereas for the orange albedo-fed larvae, in all cases, the ratio was above 1. Due to this, their consumption is likely to have a more positive effect on the cardiovascular system. In comparison, the pig, lamb, and cattle meat correspond to a ratio between 0.13 and 0.48, while fish, a well-renounced source of fatty acids, has a PUFA: SFA ratio of 0.8–1.6 [50], suggesting that the larvae are a comparable source of fatty acids with potential beneficial effects. The ccalculated oxidizability value (COX) was also measured for all samples. As can be seen from the results, a significant increase ($p < 0.05$) in the COX value was recorded in all cases of the larvae fed with orange albedo. Therefore, the oil extracted from the larvae exhibits better oxidative stability, and thus, has a longer shelf life.

Table 3. Fatty acid composition of *Tenebrio molitor* larvae fed for six weeks with wheat bran (control) (OA0) and bran fortified with different rates of albedo orange peel waste (10 (OA10), 17.5 (OA17.5), and 25% (OA25)) (n = 9).

Fatty Acid (%)	Diets			
	OA0	OA10	OA17.5	OA25
C12:0	0.16 ± 0.01 [b,c,*]	0.15 ± 0.01 [c]	0.19 ± 0.01 [a]	0.17 ± 0.01 [b]
C14:0	2.32 ± 0.11 [b]	2.71 ± 0.08 [a]	2.68 ± 0.06 [a]	2.70 ± 0.06 [a]
C16:0	21.39 ± 0.98 [a]	23.27 ± 1.16 [a]	23.03 ± 0.71 [a]	23.22 ± 1.46 [a]
C16:1	0.12 ± 0.01 [a,b]	0.10 ± 0.01 [c]	0.11 ± 0.01 [b,c]	0.12 ± 0.01 [a]
C18:0	0.13 ± 0.01 [d]	0.17 ± 0.01 [c]	0.19 ± 0.01 [b]	0.20 ± 0.01 [a]
C18:1	39.07 ± 2.89 [b]	44.1 ± 2.82 [a]	43.79 ± 0.18 [a]	43.68 ± 2.10 [a]
C18:2 (ω-6)	27.34 ± 0.57 [b]	29.28 ± 0.73 [a]	29.84 ± 1.1 [a]	29.68 ± 0.65 [a]
C20:0	1.36 ± 0.05	nd **	nd	nd
C18:3 (ω-3)	0.08 ± 0.01 [b]	0.11 ± 0.01 [a]	0.09 ± 0.01 [b]	0.11 ± 0.01 [a]
C22:0	6.04 ± 0.31 [a]	0.03 ± 0.01 [b]	0.02 ± 0.01 [b]	0.03 ± 0.01 [b]
C22:1	1.82 ± 0.13 [a]	0.09 ± 0.01 [b]	0.07 ± 0.01 [b]	0.08 ± 0.01 [b]
C24:0	0.16 ± 0.01	nd	nd	nd
∑ SFA [1]	31.57 ± 1.48 [a]	26.32 ± 1.26 [b]	26.11 ± 0.78 [b]	26.32 ± 1.54 [b]
∑ MUFA [2]	41.01 ± 3.02 [a]	44.29 ± 2.83 [a]	43.97 ± 0.18 [a]	43.88 ± 2.11 [a]
∑ PUFA [3]	27.42 ± 0.58 [b]	29.39 ± 0.74 [a]	29.93 ± 1.11 [a]	29.79 ± 0.66 [a]
PUFA: SFA ratio	0.87 ± 0.02 [b]	1.12 ± 0.03 [a]	1.15 ± 0.01 [a]	1.13 ± 0.04 [a]
MUFA: PUFA ratio	1.49 ± 0.08 [a]	1.51 ± 0.06 [a]	1.47 ± 0.05 [a]	1.47 ± 0.04 [a]
ω-6: ω-3 ratio	331.29 ± 6.96 [a]	261.62 ± 8.92 [b]	344.68 ± 9.69 [a]	277.82 ± 3.62 [b]
COX [4]	3.22 ± 0.09 [b]	3.48 ± 0.11 [a]	3.53 ± 0.12 [a]	3.52 ± 0.09 [a]

* Within each line, statistically significant differences ($p < 0.05$) are denoted with different superscript letters. ** nd: not detected. [1] SFAs, saturated fatty acids (%): SUM of C12:0, lauric acid; C14:0, myristic acid; C16:0, palmitic acid; C18:0, stearic acid; C20:0, arachidic acid; C22:0, behenic acid; C24:0, lignoceric acid. [2] MUFAs, monounsaturated fatty acids (%): SUM of C16:1, palmitoleic acid; C18:1, oleic acid; C22:1, erucic acid. [3] PUFAs, polyunsaturated fatty acids (%): SUM of C18:2, ω-6, linoleic acid; C18:3, ω-3, linolenic acid. [4] COX, calculated oxidizability value.

3.2.2. Content of *T. molitor* Larvae in Vitamins A and C

Vitamin A is a group of fat-soluble retinoids, including retinol and retinyl esters [51]. Vitamin A can be found either preformed or as a provitamin A, known as carotenoids. The latter are plant pigments that are convertible to vitamin A in the gut [52]. Vitamin A is well-known for its anti-inflammatory activity, whereas it also assists the functions of the immune system by supporting the growth and distribution of T cells, helps the preservation of epithelial and mucosal tissues, and promotes fetal development and the well-being of the reproductive system [53–56]. Given that fat-soluble vitamins, such as vitamin A, are detected in low amounts in edible insects, such as *T. molitor*, increasing their amounts would significantly enhance their nutritional value. In general, many insect species have been found to contain low amounts of vitamin A, so various attempts have been made to increase it [57]. For instance, Oonincx et al. [58] tried to increase the vitamin A content of the migratory locust, *Locusta migratoria* (L.) (Orthoptera: Acrididae), by enriching its diet with fresh carrots. The results showed that insects that consumed carrots had higher levels of vitamin A than those that consumed their standard diet; however, the increase was not as high as expected. A similar piece of research was also carried out on *T. molitor* larvae using commercial supplements; however, the larvae were found to contain marginally detectable vitamin A amounts [59].

Orange albedo is a good source of β-carotene and vitamin A. According to the results of our previous study, the orange albedo was found to contain ~36 μg/g dw of β-carotene [28]. Therefore, it was expected that the larvae would contain an increased amount of β-carotene and vitamin A after being fed with orange albedos. The respective results can be seen in Table 4. It is obvious that in both cases, as the amount of orange albedo in the feed increased, an increase in the amounts of β-carotene and vitamin A was also recorded. More specifically, when the larvae were fed a diet containing 10, 17.5, or 25% orange albedo, a significant ($p < 0.05$) increase of 85, 144, and 198%, respectively, was recorded in the β-carotene and vitamin A content. Therefore, orange albedos resulted in a significant increase in the β-carotene and vitamin A content of the larvae.

Table 4. Content of *Tenebrio molitor* larvae fed for six weeks with wheat bran (control) (OA0) and bran fortified with different rates of albedo orange peels waste (10 (OA10), 17.5 (OA17.5), and 25% (OA25)) in β-carotene, vitamin A, and vitamin C (n = 9).

Diets	β-Carotene (μg/g)	Vitamin A (μg RAE/100 g)	Vitamin C (μg/g)
OA0	$3.83 \pm 0.06^{a,*}$	11.41 ± 0.19^{a}	200.37 ± 6.44^{a}
OA10	7.10 ± 0.29^{b}	21.17 ± 0.86^{b}	251.44 ± 4.63^{b}
OA17.5	9.45 ± 0.01^{c}	28.19 ± 0.04^{c}	282.50 ± 2.88^{c}
OA25	11.43 ± 0.1^{d}	34.08 ± 0.29^{d}	292.90 ± 1.42^{d}

* Within each column, statistically significant differences ($p < 0.05$) are denoted with different superscript letters.

As regards the content of the larvae in vitamin C, an increase in the amount of vitamin C detected in the larvae was recorded, upon feeding with orange albedo (Table 4). When the larvae were fed a diet containing 10% orange albedo, a significant ($p < 0.05$) increase of 25% was recorded in the vitamin C content. In the case of 17.5% orange albedo, an increase of 40% (significant for $p < 0.05$) was recorded, whereas in the case of 25% albedo, an increase of 46% (significant for $p < 0.05$) was also recorded. Therefore, feeding with orange albedo also proved beneficial in this case.

3.2.3. Antioxidant Activity of the Larvae Extracts

Orange albedo contains many polyphenolic compounds that exhibit bioactivity [28]. As such, it was expected that when the larvae were fed with orange albedos, their content in polyphenols would increase. This hypothesis was validated by our results. It was found that when the larvae were fed with a diet containing 0 (control), 10, 17.5, and 25% orange albedo, the polyphenol content was found to be 59.67 ± 0.43, 69.01 ± 0.58, 73.64 ± 1.60, and 78.67 ± 1.07 mg/g polyphenols, corresponding to a significant ($p < 0.05$) increase of 15.6, 23.4, and 31.8% when compared to the control, respectively. Therefore, orange albedo significantly increased the polyphenol content of the larvae, bestowing them with potential benefits for human consumption. The reported values for the TPC were similar to that of previous reports [60,61]. Finally, the FRAP values of the larvae were measured. It was found that the control larvae exhibited a FRAP value of 119.39 ± 1.97 μmoL AAE/g. For the OA10 larvae, the FRAP value was found to be 140.01 ± 6.57 μmoL AAE/g (a statistically significant, $p < 0.05$, increase of 17.3%). In the case of OA17.5, the FRAP value was found to be 143.64 ± 2.22 μmoL AAE/g (a statistically significant, $p < 0.05$, increase of 20.3%), while in the case of OA25, the FRAP value was found to be 145.97 ± 2.68 μmoL AAE/g (a statistically significant, $p < 0.05$, increase of 22.2%). The enhanced content in polyphenols resulted in increased antioxidant activity, which is beneficial for human consumption, and it may prove beneficial for the protection of the nutrients contained in the larvae from oxidation. The FRAP value for the control larvae (OA0) was found to be nearly half when compared to that reported by Sánchez-Muros et al. [62]. However, this difference may be attributed to the different diets of the larvae (cereal brans were used in the previous study, compared to wheat bran only in our case).

4. Conclusions

Supplementation of the standard diet of *T. molitor* larvae (wheat bran) with orange albedo was found to be highly beneficial for larval growth and performance, as well as for their nutritional value. Without adversely affecting larval survival, the addition of 17.5% and 25% orange albedo in the feed resulted in a noticeable increase in larval growth. Moreover, the nutritional value of the larvae was significantly enhanced. Even with the smallest addition of orange albedo, a major increase in the crude protein content was recorded, accompanied by an increase in the ash content. On the contrary, the content of fat and carbohydrates decreased as the amount of orange albedo increased. In addition, a further enhancement of the nutritional value was recorded, as the amount of SFAs decreased and the amount of MUFAs and PUFAs increased. Moreover, increased contents of vitamins A and C, as well as polyphenols and antioxidant activity, were recorded. Overall, the use of 25% waste orange peel albedos is highly recommended for the rearing of *T. molitor* larvae, since it combines many advantages that simultaneously correspond to various quantitative and qualitative parameters, such as larval growth and larval composition. Apart from this, the addition of orange peel albedos to the diet of *T. molitor* will further reduce the cost of mass-rearing protocols, rendering the entire production procedure a more economically preferable option for insect rearing. Additionally, by feeding larvae with orange albedos, the circular economy is further promoted, since albedos constitute a by-product that is rarely reused, which constitutes an additional advantage in terms of improving the environmental compatibility indicators in insect farming.

Author Contributions: Conceptualization, S.I.L., C.G.A., C.I.R. and K.K.; methodology, S.I.L. and K.K.; software, T.C. and V.A.; validation, K.K., E.B. and V.A.; formal analysis, K.K., T.C., V.A. and E.B.; investigation, K.K., C.A.-S. and E.B.; resources, C.G.A. and S.I.L.; data curation, T.C., V.A. and S.I.L.; writing—original draft preparation, K.K. and T.C.; writing—review and editing, K.K., T.C., V.A., E.B., C.A.-S., C.I.R., C.G.A. and S.I.L.; visualization, K.K.; supervision, C.G.A. and S.I.L.; project administration, S.I.L. All authors have read and agreed to the published version of the manuscript.

Funding: This research received no external funding.

Institutional Review Board Statement: Not applicable.

Informed Consent Statement: Not applicable.

Data Availability Statement: All the data are contained within the article.

Conflicts of Interest: The authors declare no conflict of interest.

References

1. Boland, M.J.; Rae, A.N.; Vereijken, J.M.; Meuwissen, M.P.M.; Fischer, A.R.H.; van Boekel, M.A.J.S.; Rutherfurd, S.M.; Gruppen, H.; Moughan, P.J.; Hendriks, W.H. The future supply of animal-derived protein for human consumption. *Trends Food Sci. Technol.* **2013**, *29*, 62–73. [CrossRef]
2. Foley, J.A.; Ramankutty, N.; Brauman, K.A.; Cassidy, E.S.; Gerber, J.S.; Johnston, M.; Mueller, N.D.; O'Connell, C.; Ray, D.K.; West, P.C.; et al. Solutions for a cultivated planet. *Nature* **2011**, *478*, 337–342. [CrossRef]
3. Grasty, S.; FAO. *Reducing Enteric Methane and Livelihoods Win-Win Opportunities for Farmers*; FAO: Rome, Italy, 1999; Volume 14, ISBN 9789251079201.
4. Patel, S.; Suleria, H.A.R.; Rauf, A. Edible insects as innovative foods: Nutritional and functional assessments. *Trends Food Sci. Technol.* **2019**, *86*, 352–359. [CrossRef]
5. Premalatha, M.; Abbasi, T.; Abbasi, T.; Abbasi, S.A. Energy-efficient food production to reduce global warming and ecodegradation: The use of edible insects. *Renew. Sustain. Energy Rev.* **2011**, *15*, 4357–4360. [CrossRef]
6. Errico, S.; Spagnoletta, A.; Verardi, A.; Moliterni, S.; Dimatteo, S.; Sangiorgio, P. *Tenebrio molitor* as a source of interesting natural compounds, their recovery processes, biological effects, and safety aspects. *Compr. Rev. Food Sci. Food Saf.* **2022**, *21*, 148–197. [CrossRef] [PubMed]
7. European Union Commission. *Commission Implementing Regulation (EU) 2021/882 Authorising the Placing on the Market of Dried Tenebrio molitor Larva as a Novel Food under Regulation (EU) 2015/2283 of the European Parliament and of the Council, and Amending Commission Implementing Regulation (EU) 2017/2470 (Text with EEA Relevance)*; European Union Commission: Brussels, Belgium, 2021.
8. Melo, V.; Garcia, M.; Sandoval, H.; Jiménez, H.D.; Calvo, C. Quality proteins from edible indigenous insect food of latin America and Asia. *Emirates J. Food Agric.* **2011**, *23*, 283–289.
9. van Huis, A.; Oonincx, D.G.A.B. The environmental sustainability of insects as food and feed. A review. *Agron. Sustain. Dev.* **2017**, *37*, 1–14. [CrossRef]
10. Kröncke, N.; Benning, R. Self-Selection of Feeding Substrates by *Tenebrio molitor* Larvae of Different Ages to Determine Optimal Macronutrient Intake and the Influence on Larval Growth and Protein Content. *Insects* **2022**, *13*, 657. [CrossRef]
11. Oonincx, D.G.A.B.; Finke, M.D. Nutritional value of insects and ways to manipulate their composition. *J. Insects Food Feed.* **2021**, *7*, 639–659. [CrossRef]
12. Punzo, F.; Mutchmor, J.A. Effects of Temperature, Relative Humidity and Period of Exposure on the Survival Capacity of *Tenebrio molitor* (Coleoptera: Tenebrionidae). *Survival* **1980**, *53*, 260–270.

13. Cotton, R.T.; Ashby, W. Insect pests of stored grains and seed. In *Insects: The Yearbook of Agriculture*; US Gov. Printing Office: Washington, DC, USA, 1952; pp. 629–639.
14. Rumbos, C.I.; Karapanagiotidis, I.T.; Mente, E.; Psofakis, P.; Athanassiou, C.G. Evaluation of various commodities for the development of the yellow mealworm, *Tenebrio molitor*. *Sci. Rep.* **2020**, *10*, 11224. [CrossRef]
15. Van Peer, M.; Frooninckx, L.; Coudron, C.; Berrens, S.; Álvarez, C.; Deruytter, D.; Verheyen, G.; Van Miert, S. Valorisation potential of using organic side streams as feed for *Tenebrio molitor*, *Acheta domesticus* and *Locusta migratoria*. *Insects* **2021**, *12*, 796. [CrossRef] [PubMed]
16. Pfaltzgraff, L.A.; De Bruyn, M.; Cooper, E.C.; Budarin, V.; Clark, J.H. Food waste biomass: A resource for high-value chemicals. *Green Chem.* **2013**, *15*, 307–314. [CrossRef]
17. Mancini, S.; Fratini, F.; Turchi, B.; Mattioli, S.; Dal Bosco, A.; Tuccinardi, T.; Nozic, S.; Paci, G. Former foodstuff products in *Tenebrio molitor* rearing: Effects on growth, chemical composition, microbiological load, and antioxidant status. *Animals* **2019**, *9*, 484. [CrossRef] [PubMed]
18. Morales-Ramos, J.A.; Rojas, M.G.; Kelstrup, H.C.; Emery, V. Self-selection of agricultural by-products and food ingredients by *Tenebrio* molitor (Coleoptera: Tenebrionidae) and impact on food utilization and nutrient intake. *Insects* **2020**, *11*, 827. [CrossRef] [PubMed]
19. Rumbos, C.I.; Bliamplias, D.; Gourgouta, M.; Michail, V.; Athanassiou, C.G. Rearing *Tenebrio molitor* and *Alphitobius diaperinus* Larvae on Seed Cleaning Process Byproducts. *Insects* **2021**, *12*, 293. [CrossRef]
20. Bradford, M.M. A rapid and sensitive method for the quantitation of microgram quantities of protein utilizing the principle of protein-dye binding. *Anal. Biochem.* **1976**, *72*, 248–254. [CrossRef] [PubMed]
21. Son, Y.J.; Hwang, I.K.; Nho, C.W.; Kim, S.M.; Kim, S.H. Determination of carbohydrate composition in mealworm (*Tenebrio molitor* L.) larvae and characterization of mealworm chitin and chitosan. *Foods* **2021**, *10*, 640. [CrossRef] [PubMed]
22. European Union Commission. *Commission Regulation (EC) No 796/2002 of 6 May 2002 amending Regulation (EEC) No 2568/91 on the Characteristics of Olive Oil and Olive-Pomace Oil and on the Relevant Methods of Analysis and the Additional Notes in the Annex to Council Regulation (EEC) No 2658/87 on the Tariff and Statistical Nomenclature and on the Common Customs Tariff*; European Union Commission: Brussels, Belgium, 2002.
23. Lalas, S.; Gortzi, O.; Athanasiadis, V.; Dourtoglou, E.; Dourtoglou, V. Full Characterisation of Crambe abyssinica Hochst. Seed Oil. *J. Am. Oil Chem. Soc.* **2012**, *89*, 2253–2258. [CrossRef]
24. Ourailoglou, D.; Athanasiadis, V.; Bozinou, E.; Salakidou, C.; Evmorfopoulos, E.; Lalas, S. Manufacturing Process and Physicochemical Analysis of Kariki: A Traditional Cheese from The Island of Tinos, Greece. *Int. Food Res. J.* **2021**, *28*, 262–268. [CrossRef]
25. Fatemi, S.H.; Hammond, E.G. Analysis of oleate, linoleate and linolenate hydroperoxides in oxidized ester mixtures. *Lipids* **1980**, *15*, 379–385. [CrossRef]
26. Ocampo, E.T.M.; Libron, J.A.M.A.; Guevarra, M.L.D.; Mateo, J.M.C. Phytochemical screening, phenolic acid profiling and antioxidant activity analysis of peels from selected mango (*Mangifera* spp.) genotypes in the Philippines. *Food Res.* **2020**, *4*, 1116–1124. [CrossRef] [PubMed]
27. Jagota, S.K.; Dani, H.M. A new colorimetric technique for the estimation of vitamin C using Folin phenol reagent. *Anal. Biochem.* **1982**, *127*, 178–182. [CrossRef]
28. Athanasiadis, V.; Chatzimitakos, T.; Kotsou, K.; Palaiogiannis, D.; Bozinou, E.; Lalas, S.I. Optimization of the Extraction Parameters for the Isolation of Bioactive Compounds from Orange Peel Waste. *Sustainability* **2022**, *14*, 13926. [CrossRef]
29. Athanasiadis, V.; Palaiogiannis, D.; Poulianiti, K.; Bozinou, E.; Lalas, S.I.; Makris, D.P. Extraction of Polyphenolic Antioxidants from Red Grape Pomace and Olive Leaves: Process Optimization Using a Tailor-Made Tertiary Deep Eutectic Solvent. *Sustainability* **2022**, *14*, 6864. [CrossRef]
30. LeCato, G.; Flaherty, B. Description of eggs of selected species of stored-product insects (Coleoptera and Lepidoptera). *J. Kansas Entomol. Soc.* **1974**, *47*, 308–317.
31. Abdelatti, Z.A.S.; Hartbauer, M. Plant oil mixtures as a novel botanical pesticide to control gregarious locusts. *J. Pest Sci. (2004)* **2020**, *93*, 341–353. [CrossRef]
32. Yang, Y.; Wang, X.; Zhao, C.; Tian, G.; Zhang, H.; Xiao, H.; He, L.; Zheng, J. Chemical Mapping of Essential Oils, Flavonoids and Carotenoids in Citrus Peels by Raman Microscopy. *J. Food Sci.* **2017**, *82*, 2840–2846. [CrossRef]
33. Hazarika, U.; Gosztola, B. Lyophilization and its Effects on the Essential Oil Content and Composition of Herbs and Spices—A Review. *Acta Sci. Pol. Technol. Aliment.* **2020**, *19*, 467–473. [CrossRef]
34. Bordiean, A.; Krzyżaniak, M.; Stolarski, M.J. Bioconversion Potential of Agro-Industrial Byproducts by *Tenebrio molitor*—Long-Term Results. *Insects* **2022**, *13*, 810. [CrossRef]
35. Rumpold, B.A.; Schlüter, O.K. Nutritional composition and safety aspects of edible insects. *Mol. Nutr. Food Res.* **2013**, *57*, 802–823. [CrossRef]
36. Costa, S.; Pedro, S.; Lourenço, H.; Batista, I.; Teixeira, B.; Bandarra, N.M.; Murta, D.; Nunes, R.; Pires, C. Evaluation of Tenebrio molitor larvae as an alternative food source. *NFS J.* **2020**, *21*, 57–64. [CrossRef]
37. Nowak, V.; Persijn, D.; Rittenschober, D.; Charrondiere, U.R. Review of food composition data for edible insects. *Food Chem.* **2016**, *193*, 39–46. [CrossRef] [PubMed]
38. Ghaly, A.E.; Alkoaik, F.N. The yellow mealworm as a novel source of protein. *Am. J. Agric. Biol. Sci.* **2009**, *4*, 319–331. [CrossRef]

39. Ben-Shalom, N.; Pinto, R.; Berman, M. Polysaccharides and proteins in the orange albedo and in the aqueous extract of the tissue. *Can. Inst. Food Sci. Technol. J.* **1985**, *18*, 335–336. [CrossRef]
40. Yu, X.; He, Q.; Wang, D. Dynamic Analysis of Major Components in the Different Developmental Stages of *Tenebrio molitor*. *Front. Nutr.* **2021**, *8*, 689746. [CrossRef]
41. Siemianowska, E.; Kosewska, A.; Aljewicz, M.; Skibniewska, K.A.; Polak-Juszczak, L.; Jarocki, A.; Jędras, M. Larvae of mealworm (*Tenebrio molitor* L.) as European novel food. *Agric. Sci.* **2013**, *04*, 287–291. [CrossRef]
42. Kaur, N.; Chugh, V.; Gupta, A.K. Essential fatty acids as functional components of foods- a review. *J. Food Sci. Technol.* **2014**, *51*, 2289–2303. [CrossRef]
43. Baioumy, A.A.; Abedelmaksoud, T.G. Quality properties and storage stability of beef burger as influenced by addition of orange peels (albedo). *Theory Pract. Meat Process.* **2021**, *6*, 33–38. [CrossRef]
44. Ravzanaadii, N.; Kim, S.-H.; Choi, W.-H.; Hong, S.-J.; Kim, N.-J. Nutritional Value of Mealworm, *Tenebrio molitor* as Food Source. *Int. J. Ind. Entomol.* **2012**, *25*, 93–98. [CrossRef]
45. Ghosh, S.; Lee, S.M.; Jung, C.; Meyer-Rochow, V.B. Nutritional composition of five commercial edible insects in South Korea. *J. Asia. Pac. Entomol.* **2017**, *20*, 686–694. [CrossRef]
46. Caparros Megido, R.; Poelaert, C.; Ernens, M.; Liotta, M.; Blecker, C.; Danthine, S.; Tyteca, E.; Haubruge, É.; Alabi, T.; Bindelle, J.; et al. Effect of household cooking techniques on the microbiological load and the nutritional quality of mealworms (*Tenebrio molitor* L. 1758). *Food Res. Int.* **2018**, *106*, 503–508. [CrossRef]
47. Pastor, R.; Bouzas, C.; Tur, J.A. Beneficial effects of dietary supplementation with olive oil, oleic acid, or hydroxytyrosol in metabolic syndrome: Systematic review and meta-analysis. *Free Radic. Biol. Med.* **2021**, *172*, 372–385. [CrossRef]
48. Burr, M.; Miller, S. Fatty Acids in Nutrition. *J. Biol* **1932**, *97*, 1–9.
49. Nutter, M.K.; Lockhart, E.E.; Harris, R.S. The chemical composition of depot fats in chickens and turkeys. *Oil Soap* **1943**, *20*, 231–234. [CrossRef]
50. Chen, J.; Liu, H. Nutritional indices for assessing fatty acids: A mini-review. *Int. J. Mol. Sci.* **2020**, *21*, 5695. [CrossRef]
51. Huth, E.J. A Library for Internists VI Recommended by the American College of Physicians. *Ann. Intern. Med.* **1988**, *108*, 497–498. [CrossRef]
52. Solomons, N.W.; Orozco, M. Alleviation of vitamin A deficiency with palm fruit and its products. *Asia Pac. J. Clin. Nutr.* **2003**, *12*, 373–384. [PubMed]
53. Sommer, A. Vitamin A deficiency and clinical disease: An historical overview. *J. Nutr.* **2008**, *138*, 1835–1839. [CrossRef] [PubMed]
54. Liang, Y.; Yi, P.; Wang, X.; Zhang, B.; Jie, Z.; Soong, L.; Sun, J. Retinoic Acid Modulates Hyperactive T Cell Responses and Protects Vitamin A–Deficient Mice against Persistent Lymphocytic Choriomeningitis Virus Infection. *J. Immunol.* **2020**, *204*, 2984–2994. [CrossRef]
55. Ross, S.A.; McCaffery, P.J.; Drager, U.C.; De Luca, L.M. Retinoids in embryonal development. *Physiol. Rev.* **2000**, *80*, 1021–1054. [CrossRef]
56. Bastos Maia, S.; Rolland Souza, A.S.; Costa Caminha, M.d.F.; Lins da Silva, S.; Callou Cruz, R.d.S.B.L.; Carvalho dos Santos, C.; Batista Filho, M. Vitamin A and Pregnancy: A Narrative Review. *Nutrients* **2019**, *11*, 681. [CrossRef] [PubMed]
57. Finke, M.D. Complete nutrient composition of commercially raised invertebrates used as food for insectivores. *Zoo Biol.* **2002**, *21*, 269–285. [CrossRef]
58. Oonincx, D.G.A.B.; Van Der Poel, A.F.B. Effects of diet on the chemical composition of migratory locusts (*Locusta migratoria*). *Zoo Biol.* **2011**, *30*, 9–16. [CrossRef] [PubMed]
59. Finke, M.D. Complete nutrient content of four species of commercially available feeder insects fed enhanced diets during growth. *Zoo Biol.* **2015**, *34*, 554–564. [CrossRef] [PubMed]
60. Antonopoulou, E.; Panteli, N.; Feidantsis, K.; Mastoraki, M.; Koutsogeorgiou, E.; Grivaki, E.; Papagrigoriou, T.; Christias, S.; Chatzifotis, S.; Lazari, D.; et al. Carob (*Ceratonia siliqua*) as Functional Feed Is Beneficial in Yellow Mealworm (*Tenebrio molitor*) Rearing: Evidence from Growth, Antioxidant Status and Cellular Responses. *Antioxidants* **2022**, *11*, 1840. [CrossRef]
61. Keil, C.; Grebenteuch, S.; Kröncke, N.; Kulow, F.; Pfeif, S.; Kanzler, C.; Rohn, S.; Boeck, G.; Benning, R.; Haase, H. Systematic Studies on the Antioxidant Capacity and Volatile Compound Profile of Yellow Mealworm Larvae (*Tenebrio molitor* L.) under Different Drying Regimes. *Insects* **2022**, *13*, 166. [CrossRef]
62. Sánchez-Muros, M.J.; de Haro, C.; Sanz, A.; Trenzado, C.E.; Villareces, S.; Barroso, F.G. Nutritional evaluation of *Tenebrio molitor* meal as fishmeal substitute for tilapia (*Oreochromis niloticus*) diet. *Aquac. Nutr.* **2016**, *22*, 943–955. [CrossRef]

foods

MDPI

Review

A Holistic Approach for Ethics and Sustainability in the Food Chain: The Gateway to Oral and Systemic Health

Theodoros Varzakas [1,*] and Maria Antoniadou [2,3,*]

1 Department Food Science and Technology, University of the Peloponnese, 24100 Kalamata, Greece
2 Department of Dentistry, School of Health Sciences, National and Kapodistrian University of Athens, 11527 Athens, Greece
3 Certified Systemic Analyst Program in Systemic Management (CSAP), University of Piraeus, 18534 Piraeus, Greece
* Correspondence: t.varzakas@uop.gr (T.V.); mantonia@dent.uoa.gr (M.A.)

Abstract: Food production is a complex matter, affecting people's lives, organizations' profits, and the well-being of the whole planet, and has multifaceted ethical considerations surrounding its production, distribution, and consumption. This paper addresses the pressing need to confront ethical challenges within the food system, encompassing issues such as environmental sustainability, food security, and individual food choices for better oral and systemic health of all individuals around the globe. From agricultural practices to global trade and food waste, ethical implications are addressed across various domains, highlighting the interconnectedness of ethical decision-making in the food industry. Central themes explored include the ethical dimensions of food production methods, the impact of global trade on food ethics, and the role of individuals in making ethically informed food choices. Additionally, this paper considers the spiritual and physical significance of food, particularly through the lens of oral health as a gateway to holistic well-being. Recognizing the complexity of the food and mouth ecosystem, this paper calls for serious interventions in legislation and economics to promote ethical protocols and techniques for sustainability reasons. It emphasizes the importance of ethical considerations in food safety management systems, regulatory frameworks, and quality standards. Moreover, this paper underlines the need for a comprehensive approach to address ethical dilemmas and moral values inherent in the food industry and oral health policies, adopting the precautionary principle and ethical decision-making frameworks. This article finally aims to serve as a call to action for stakeholders across the food industry and the healthcare sector, to prioritize ethical practices, promote transparency, rearrange economic parameters, and work towards a more sustainable and equitable food system for inner and outer oral and systemic health and human sustainability for all.

Keywords: food ecosystem; food chain; food morality; ethics; oral health; systemic health; diet habits; spirituality; sustainability

1. Introduction

In contemporary society, the food ecosystem stands as a complex system involving environmental, social, economic, and ethical dimensions [1]. The production, distribution, and consumption of food are not merely utilitarian processes but rather intricate webs of interconnectedness that profoundly impact individuals, communities, and the planet at large [2]. At the heart of this multifaceted system lies ethical considerations that permeate every aspect of the food chain [3].

Human attitudes towards animals have been influenced originally by the ancient Greek philosophies addressing the formulation of terms such as ethos (ἦ θος, ἔ θος), ethics (δέον), and moral (ευδαιμονία). "Ethos" is defined as the character, sentiment, or disposition of a community or people, considered as a natural endowment. "Ethos" is a Greek word corresponding roughly to "ethics" [4]. Food ethics is not a new field, it always existed, but it appears today as a high priority, moving beyond the traditional form of ethics, as described by Monterrosa et al. [5]. Under the definition of morality [6], food morality attempts to bridge the gap between our ethical values and our food-related behaviors. This means that there are moral implications in our everyday food choices [3]. Moral considerations accompanying the production, distribution, and consumption of food can be considered as food morality [7]. The application of ethical principles in conjunction with the ethical production/manufacturing of food along with the environmental and social impacts on our diets, animal welfare, farmers, and the whole society is food morality, according to Hernandez et al. [8]. The concept of food morality encompasses issues such as climate change, resource depletion, social inequalities, and public health. A holistic approach has been considered in food research [9] on the reduction in food loss and waste [10] and on food type interdependence [11], food intake, and appetite [12]. The transformation of food environments nowadays, along with the change in the dynamics of lifestyles, implies a theory of holistic food ethics. Food ethics considers mainly animal ethics and positive and negative views regarding animal rights [13] and animal suffering [14], taking also into account environmental ethics, which analyses the sustainability of food systems [15]. Furthermore, social justice examines the nutritional side of foods along with access to them, according to Ruben et al. [16]. The adoption of vegetarian or vegan diets, consumption of organic food, or avoiding buying foods from companies adopting controversial practices means alignment of moral values with food choices [11,17]. Most people agree on the non-moral neutrality of food choices nowadays. Indeed, the relevance of food morality is greatly depicted by the heavy industrialization of food systems with negative social and environmental impacts including pollution, biodiversity loss, climate change, and labor exploitation. Model sustainability and equitability [18] are in great need today in conjunction with morality.

This study aims to provide a comprehensive examination of the ethical dimensions inherent in the modern food system, with a specific focus on the pivotal role of oral health and its interconnectedness with broader ethical considerations. Through a multidisciplinary lens encompassing ethics from Socrates to Spinoza, environmental science, oral and general health sciences, sociology, anthropology, and spirituality, our objective is to explore the area of ethical dilemmas that permeate every aspect of the food ecosystem from production to the mouth, the gateway to the whole body. From analyzing agricultural practices and production methods to investigating global trade dynamics, health economics, regulatory frameworks, and quality tools in the ethical management of the food ecosystem, we seek to uncover the moral complexities and challenges faced at each stage of the food chain. Furthermore, we aim to shed light on the often-overlooked influence of individual food choices on the ethical base of our food system, recognizing their profound impact on sustainability, social justice, and public health. We then discuss the ethical imperatives of food safety, quality standards, and the pervasive issue of food fraud. We finally examine the significance of oral health as the gateway to systemic health and the mouth as the gateway to the human body to stimulate meaningful dialogue and inspire collective action toward building a more ethical and sustainable food system that nurtures the holistic well-being of individuals and communities alike.

2. Methodology of this Review

Given the multifaceted nature of the theme and the vast body of literature available, our approach was designed to systematically sift through the variety of information to offer a comprehensive understanding of the ethical complexities surrounding food production, distribution, consumption, and sustainability. Firstly, we clearly outlined the specific ethical challenges that were about to be addressed, focusing on the multifaceted nature of the food ecosystem. These challenges included issues related to (1) morality vs. ethics, (2) cultural considerations in food ethics, (3) ethics in religious food, (4) food and spirituality issues, (5) food dynamics, (6) GMO ethics and neophobia, (7) food regulation, (8) food safety and quality standards, (9) food fraud and ethical production practices, (10) consumer rights and behavior, (11) the mouth and ethical food, (12) significance of oral health in the food chain, (13) economic implications and ethics in the prevention and/or provision of oral health, and (14) sustainability issues in the food chain up to the mouth gateway.

Then, in this review study, we employed a methodological approach that encompasses several key steps aligned with the theme of the ethical dimensions inherent in various facets of the food system. Firstly, we selected databases that cover a wide range of disciplines relevant to the topic, including PubMed, Web of Science, Scopus, and Google Scholar. Next, we developed a comprehensive search strategy using relevant keywords and phrases related to food ethics, sustainability, morality, and related concepts. We considered using a combination of controlled vocabulary (e.g., MeSH terms in PubMed). In this sense, for "sustainability of systems through food", we used the following terms: food supply/supply and distribution, food industry/economics, food preservation/methods, food handling/methods, food technology/methods, food packaging/methods, sustainable agriculture/methods, agriculture/economics, environmental sustainability, and conservation of natural resources. For "health economics", we used the following terms: "health economics", "health expenditures", "healthcare costs", "cost-benefit analysis", "cost control", "economic models", "health services accessibility", "health policy", "health insurance", and "managed care programs". Also, for "oral health" issues, we used the following terms: "oral health", "dental health services", "dental health surveys", "periodontal diseases", "tooth diseases", "dental caries", "oral health promotion", and "dental public health". Finally, we searched additional free-text terms such as "food ethics", "ethical production", "food sustainability", "food morality", "ethical consumption", "food waste ethics", "animal welfare", environmental ethics", "oral health", "general health", and "diet habits". In the end, we linked ethics with food consumption behavior.

We also defined clear inclusion/exclusion criteria based on publication date range, language, and study design, ensuring that the selected literature aligns with the objectives of this review, and extending the search from January 2000 to 31 March 2024. Following this, we conducted an initial screening of the search results based on titles and abstracts to identify potentially relevant articles, followed by a detailed review of full-text copies to assess the relevance and quality of those selected in the first place. We considered certain cases including grey literature sources such as reports and policy documents considering the nature of the theme. Additionally, we conducted manual searches of relevant journals and the reference lists of the included articles to identify additional sources. Upon data extraction, we synthesized the findings to identify common themes, patterns, and gaps in the literature. Furthermore, a quality assessment of the final included studies was conducted, and a narrative synthesis of the literature was prepared, highlighting key findings and implications.

3. Morality versus Ethics in the Food Industry

Morality is defined as the actions of people in their relationships with others [19]. This means responsibility for their actions and their engagement with others, making all co-responsible for social construction, according to Dasuky [20]. The latter author believes that ethics and morality are closely interrelated, somewhere between the passive and the dynamic or between living just for living and the longing to live well [20].

In contrast to Huxley [21], who reported on the selfishness and amorality of human nature, Hauser [22] argues about the human innate moral faculty shared by other primates, guiding and separating moral judgments from nonconscious forms. Hauser discusses a moral organ allowing for the acquirement of the moral system we wish to develop. This is in conjunction with de Waal and Aureli [23], illustrating the judgmental ability of chimpanzees regarding the consequences of certain actions in great resemblance to true moral judgments. Hauser [22] also postulates that our moral judgments are determined by a biological "moral grammar". The prohibition of murder or the promotion of reciprocity constitute basic moral principles that are shared by all cultures. On this same point,

Socrates, the ancient Greek philosopher, states that "No one does evil willingly", but this notion is a misunderstanding stemming from loose translations and misinterpretations. In Plato's work, "Protagoras" (358d) [24]), Socrates articulates a very different position. What he emphasizes is that no person commits evil willingly, nor do they choose to pursue what they believe to be evil. In this text, Socrates examines the nature of evil and good, concluding that the perception of evil is not real, as evil does not exist as an independent force in the universe. Through the distinction between evil and good, one can achieve a virtuous state.

The exploration of ethics, as discussed in the context of Socratic philosophy, can be extended to the field of food ethics. Just as Socrates examines the nature of good and evil and the conscious choice behind actions, food ethics involves considering the moral implications of food production, consumption, and distribution. In the same way that Socrates suggests that people do not knowingly choose evil [24], proponents of food ethics argue that individuals may not always be fully aware of the ethical consequences of their food choices. For example, someone might unknowingly support unethical agricultural practices or contribute to environmental degradation through dietary habits. So, by applying the Socratic principles of moral inquiry and self-awareness, individuals can engage in more conscientious decision-making regarding food. This might involve considering the environmental impact of food production, the treatment of animals in agriculture, or the social justice issues related to equal food access and distribution. In essence, just as Socrates encourages individuals to critically examine their actions and motivations [24], food ethics should encourage a similar introspection regarding the ethical dimensions of food choices and consumption patterns.

The interconnectedness between human welfare, respect for animal life, and the environment affecting our food choices is well recognized by different philosophical theories such as utilitarianism, Kantianism, natural rights, and virtue theory [3]. Ultimately, everyday choices that affect our lives and the world we share have to do with food morality and ethics. The nourishment of our bodies but also our souls with choices reflecting our deepest values and contribution to building a more just and equitable world comes from an ethical approach to food. Deeper values of justice, respect, and moral excellence might arise from an ethical reflection on food.

More specifically, utilitarian thinking refers to the Chinese philosopher Mozi, who lived between 490 and 403 BC. His work was later developed by Western thinkers such as Jeremy Bentham (1748–1832), John Stuart Mill (1806–1873), and Henry Sidgwick (1838–1900) [25] and has profound implications for the food industry's approach to safety and ethics. According to utilitarianism, the contribution of actions to the general good or overall happiness corresponds to ethical value [25]. Within the context of food morality, this perspective emphasizes the importance of assessing the utility or happiness derived from food choices and consumption patterns. The concept of "utility" or the "greatest happiness principle" serves as the foundation of moral evaluation, as actions are deemed right insofar as they contribute to the promotion of happiness, as reported by West [26]. This implies that food production, distribution, and consumption practices should prioritize the well-being and satisfaction of individuals and society. From a safety standpoint, utilitarian thinking calls for ensuring that food products are safe for consumption and free from harmful contaminants or hazards. This aligns with the notion that the promotion of happiness and well-being requires safeguarding public health and minimizing risks associated with foodborne illnesses or adulteration. Furthermore, ethical considerations in the food industry are intertwined with utilitarian principles, as decisions regarding sourcing, production methods, and marketing should aim to maximize overall happiness and minimize harm. This includes prioritizing sustainable and ethical sourcing practices, promoting fair labor conditions in the agriculture and food industries, and minimizing environmental impact throughout the supply chain. If the food industry adheres to utilitarian principles, it can strive to maximize the overall well-being and happiness of individuals while also upholding ethical standards and ensuring the safety of food products [25,26].

In addition, Kantian ethics, rooted in the philosophical reflections of Immanuel Kant, emphasizes the centrality of rationality in determining moral principles and duties [27]. Kantism asserts that ethics are grounded in the inherent rationality of human beings and the notion of duty [27]. Granja [28] posits that moral convictions necessitate freedom of choice to be valid, underscoring the importance of autonomy in ethical decision-making. Within the area of food ethics, Kantian principles find application in addressing the fair treatment of animals and upholding ethical standards in the food industry. Moreover, the rationality inherent in human nature serves as a guiding force in determining our actions, leading to opposition to practices such as animal abuse and exploitation. Kantian ethics impel individuals to recognize the intrinsic worth and dignity of sentient beings, promoting the adoption of ethical animal rights and welfare practices in food production and consumption. Aligning with Kantian principles means that individuals and industries within the food sector are compelled to uphold their moral duty to treat animals with respect and compassion, thereby guaranteeing a more ethical and human approach to food production and consumption [29]. Moreover, Kantian ethics prompt individuals to reflect on the consequences of their food choices and to consider the broader implications of their actions on both human and non-human stakeholders in the food system. Thus, Kantian principles could serve as a framework for guiding ethical behavior and decision-making in the food industry and shaping conscientious food habits that prioritize respect for all living beings.

The natural rights theory also parallels the fact that natural rights derive from the creation of certain natural laws by God, or the idea that human nature is the origin of natural laws. Therefore, human rights are considered inherent to human nature [30]. Food choice, as a decision made by every individual, involves the exercise of willpower, which is considered our divine nature, as reported by Atteshli-Theotoki [31]. This innate capacity can empower us to resist anything toxic or harmful to our health, as well as counteract any negative tendencies. It is not a matter of fighting our weaknesses but understanding them and convincing ourselves and our subconscious that we can master them and disregard them. In this direction, each person has the choice to select a diet based on the value of individual freedom and autonomy [32], and this includes the consumption of meat. Finally, individual well-being derives from freedom of choice and self-determination, which are associated with personal development and making informed decisions about one's diet [33].

Additionally, according to Cruz [34], the idea of natural rights links these rights to either divine decree or inherent human nature. This suggests that these rights are not given by human institutions. Kelsen [35] further elaborates on this concept by attributing the source of natural law to the essential nature of humanity, which serves as the ultimate authority in establishing rules and norms. Within this framework, the maintenance of a peaceful/mutual relationship with nature and all living beings assumes paramount importance, as reported by Lugo-Morin et al. [36,37]. This perspective underscores the recognition that the world and all manifestations of life are imbued with inherent value and are created by divine entities, reflecting the absolute wisdom, power, and love inherent in the universe [31]. Humanity must acknowledge and uphold the profound interconnectedness of the cosmos and all manifestations of life, which are intricately crafted by the Holy Archangels, reflecting boundless wisdom, power, and love [31]). Governed by the Circles of Possibilities inherent in each living entity, creation unfolds with divine intentionality and consciousness. Recognizing our role as co-creators, we must conscientiously control our thoughts, desires, and actions through introspection, analytical inquiry, and a commitment to truth [31]. As stewards entrusted with the care of creation, humans are further tasked with aligning their conduct with the principles of introspection, truth-seeking, and reverence for the intrinsic worth of all life forms. In the field of food ethics, this perspective underlines the imperative to honor the sanctity of nature and the inherent rights of every living being. Embracing these principles may guide our choices and behaviors towards empowering a more sustainable and ethical relationship with the natural world, thereby fulfilling our responsibility as stewards of the earth [31].

Furthermore, virtue theory lies in the development and inclusion of personal virtues and moral excellence in all aspects of human life [38]. In ancient Greece, Plato and his pupil Aristotle (384–322 BC) argued that morality has to do with leading a good life and being a morally good person [38]. Aristotle argued further that to become a good man (or a good woman), you must do what a good man does, according to Berti [39]. Aristotelian virtue ethics is a human-centered theory that relies on people and their characters and morality has more to do with the question "How ought I to be?" and does not deal with the morality of actions [40]. Virtue ethics reports that the power of striving to achieve the goal of virtue is morality and belongs to a teleological theory. However, a question arises regarding addressing ethical dilemmas where virtues compete following a viable decision-making procedure [30]. This concept of virtue ethics provides a lens through which to explore the complexities of food morality. Following this option, virtue ethics posits that morality entails the cultivation of virtuous character traits and the pursuit of virtuous goals [30]. This framework underscores the importance of striving towards ideals such as human well-being, respect for animal life, ecological sustainability, and social justice within the context of food morality. However, virtue ethics also acknowledges the inherent challenge of exploring ethical dilemmas where competing virtues pull in different directions [41]. In the field of food morality, this may manifest in situations where the pursuit of one ethical principle, such as environmental sustainability, conflicts with another, such as social equity in food access. While virtue ethics provides a robust foundation for guiding moral conduct, its lack of a concrete decision-making procedure poses a significant challenge in resolving such dilemmas [42]. Thus, applying virtue ethics to food morality necessitates careful consideration of how to balance competing virtues and reconcile conflicting ethical imperatives in the complex landscape of the food system.

Additionally, Lugo-Morin [3] urges us to integrate diverse considerations that encompass the lives of human beings, their respect for animal life, fundamental rights, environment, and social justice in creating a moral framework for food that goes beyond just eating for survival. Such a holistic approach challenges us to move beyond rigid dogmas and simplistic solutions, emphasizing the importance of fostering ties of solidarity among all who gather around the table. Indeed, the act of eating extends beyond the physical nourishment of bodies; it catalyzes nurturing relationships—most notably, our connection with the divine, whether conceived as God, the Creator, or the spiritual foundation that underpins everyone's existence. This viewpoint emphasizes the deep importance of food beyond just providing nourishment, emphasizing its ability to strengthen spiritual connections and foster community bonds, thus enriching the human journey [3].

Ethical values dominate people's lives [43–45]. Values can be defined as the sum of the positive/good properties that reflect the importance of a good person or thing, and these properties should be acquired and recommended [46,47]. There are different forms of values based on (1) material and economic issues, (2) politics, (3) social welfare (e.g., love, friendship, cooperation, and peace), (4) aesthetic conditions, (5) natural aspects of life (e.g., life, health, and nature), and (6) moral/ethical values (such as responsibility, honesty, conscientiousness, self-awareness, self-control, and dignity) [48]. The hierarchy of these values largely depends on the cultural environment [49] and may be understood differently both in terms of concepts of different values and significance [50,51]. Values in the food chain are not merely abstract concepts but are deeply embedded in the fabric of everyday life, influencing the actions and decisions of all stakeholders involved. Ethical values, as elucidated by Harris [52] and Kanungo [53], play a fundamental role in shaping the behavior and practices within the food industry. Within this framework, there exists a hierarchical structure of values, guided by rational or explicit rules set forth by the group, as noted by Landau and Osmo [54]. This hierarchy reflects the varying degrees of importance attributed to specific values within the broader system–society–culture complex, as highlighted by Jackson [50]. Particularly significant are the values that uphold the principles of human life, freedom, and justice, irrespective of contextual frames, as emphasized by Donnelly [55] and Schwartz and Bardi [56]. In the food chain, these values manifest in various forms, influencing decisions related to production methods, distribution practices, and consumer choices, thereby shaping the ethical contours of the entire ecosystem [55,56].

Finally, Spinoza's "ethics" provides us with profound insights into human nature, ethics, and the pursuit of a good life, as discussed by Curley [57]. While Spinoza's work may not directly address the food industry or diet and oral health initiatives, his philosophical principles can be applied to these areas to promote sustainability. Firstly, Spinoza emphasizes the interconnectedness of all things in the universe, thus providing us with the needed spiritual base. Applied to the food industry, this principle highlights the importance of recognizing the interdependence between food production, consumption, and environmental sustainability, as analyzed by Costa Deprá [58]. It may encourage stakeholders in the food industry to consider the broader ecological impact of their actions, promoting sustainable practices that minimize harm to the environment. Secondly, Spinoza's ethics underscore the importance of caring for oneself and others. In the context of diet habits and oral health initiatives, this translates into promoting preventive care and education to empower individuals to take responsibility for their oral health and diet, promoting well-being and sustainability, as reported by Vikram et al. [59]. Then, Spinoza posits that humans are driven by the desire to increase joy and decrease suffering. Applied to the food industry, this principle calls for practices that prioritize the well-being of all stakeholders, including consumers, producers, and the environment, according to Viles et al. [60]. Sustainable food production methods that prioritize animal welfare, reduce food waste, and promote equitable access to nutritious food contribute to maximizing joy and minimizing suffering. He also emphasizes the role of reason in guiding human behavior. In the food industry, rational decision-making involves considering the long-term consequences of production methods, supply chain practices, and consumption patterns [61]. So, by employing critical thinking and evidence-based approaches, stakeholders can make informed decisions that promote sustainability and ethical food systems. His philosophy also celebrates diversity and encourages inclusivity. Overall, applied to the food industry, this principle calls for respecting cultural food traditions, promoting diverse agricultural practices, and ensuring equitable access to nutritious food for all communities. Embracing diversity and fostering inclusive food systems is the key for stakeholders to promote sustainability and social justice [29].

4. Cultural Considerations in Food Ethics

Cultural considerations play a pivotal role in shaping food ethics, with perceptions of morally acceptable food varying significantly across different cultures, as shown by Mardian et al. [62]. What may be seen as traditional and culturally appropriate food practices in one society, such as different types of meat consumption or the eating of insects, may be met with skepticism or rejection in another [63]. These variations reflect the influence of culture and tradition on dietary norms and behaviors within each community, as discussed by Murcott [64]. Furthermore, cultural and religious diversity underlines the importance of respecting the culinary practices and food traditions of various communities, affirming individuals' rights to make decisions aligned with their cultural and religious beliefs [33].

Moreover, food aid derived from international and culturally different parts of humanity serves as a vital component of humanitarian assistance for individuals facing dire circumstances, such as immigrants and people living in war-torn regions (Tranchant et al., 2019) [65]. Designed to bolster food security and contribute to civil peace in destination countries, food aid plays a crucial role in addressing immediate needs and embeds ethical dilemmas. For example, studies examining the relationship between food aid and conflict present different conclusions [66,67]. While some research suggests that food aid may prolong armed conflicts, others argue that it reduces both the incidence and duration of conflicts [68,69]. The complexity of this relationship underscores the need for further analysis, with scholars identifying shortcomings in existing studies that raise questions about the validity of their findings [68,70–72]. Despite these challenges, it is vital to recognize that food aid serves as a lifeline in humanitarian endeavors, yet its impact hinges on meticulous assessment and continuous trial to uphold ethical standards and avert unforeseen repercussions.

5. Ethics in Religious Foods

Gillian Feeley-Harnik's seminal work, "Religion and Food: An Anthropological Perspective" by Feeley-Harnik [73], has profoundly influenced scholarly discourse on the intersection of religion and food. Feeley-Harnik's exploration emphasized the dynamic and transformative nature of food, challenging the notion of food as a static or natural symbol [73]. This perspective relates to the profound significance of food, eating, and fasting in shaping the beliefs and practices of diverse religious communities [73]. However, Feeley-Harnik's analysis also highlighted a notable gap in scholarly inquiry, as religion itself often remained unexamined and implicitly understood [74]. Building on Feeley-Harnik's insights, recent studies have begun to explore the multifaceted relationship between food and religion, offering serious perspectives on the role of food in shaping religious beliefs and practices [75,76]. According to these insights, religious food practices offer a profound glimpse into the moral, cultural, and ethical fabric of societies, often intersecting with legal and constitutional frameworks, as discussed by Pomeranz and Brownell [77]. For instance, the prohibition of certain foods, such as pork, may stem from moral or religious beliefs, necessitating careful consideration of public health, food safety, and ethical concerns, as shown by Lopez-Garcia [78]. This complex issue involves inspecting a delicate balance between upholding individual rights, cultural and religious freedoms, and the imperative to protect public health. It is important to mention that measures aimed at safeguarding public health are underpinned by robust scientific evidence regarding potential risks [78]. Moreover, in pluralistic societies, respect for cultural and religious diversity should be a guiding principle, ensuring that legal and regulatory measures are sensitive to the beliefs and practices of different communities [33]. Thus, religious freedom and conscience in many countries should be seriously considered before banning food for religious reasons [33].

Furthermore, religious dietary practices often entail periods of fasting or restriction of certain foods, deeply intertwined with spiritual and cultural traditions [79,80]. For example, the Greek Orthodox Church prescribes fasting during significant periods preceding Christmas, Easter Lent, and the Assumption, with specific guidelines on food consumption during these periods [79,80]. Similarly, adherence to Kosher and Halal dietary laws aligns with religious prescriptions, reflecting the intersection of faith and food [79,80]. There are also multiple possible versions of food having sacred significance in other civilizations. For example, Meigs [81] explained the rules regarding food among the Hua of Papua, New Guinea. The central notion has to do with nu, essence, which is transferred when touching and preparing food. Thus, the Hua have an elaborate set of rules about who can prepare food for whom, and who can eat what foods. An example is that mature, initiated males cannot eat leafy green vegetables that were picked by their real or classificatory wives.

Buddhist traditions, on the other hand, advocate for abstention from killing and the consumption of animal products, promoting a vegetarian way of eating as a means of fostering balance and spiritual well-being [82–84]. This dietary philosophy emphasizes principles of compassion, purity, and connection with the divine, underscoring the profound spiritual significance attributed to food practices, as described by Testoni et al. [85]. Similarly, within Christian traditions, the belief in humans being created in God's image underscores the importance of seeking peace and harmony with all creatures, shaping dietary choices and ethical considerations accordingly, as noted by Suzworsky [86]. Some report that this sort of behavior promotes prayer, deification, purity, and contact with the divine [85]. Hence, the assumptions that humans were created in God's image predicate Christian ideas, and some claim that peace should be sought among all creatures, especially when food preparation and consumption are involved [86].

6. Fasting and Vegetarians/Vegans

Another important argument of Christians who follow a strict vegetarian diet is fasting, which is God's first and oldest commandment. Jesus further emphasized its value when He said: "This generation shall not be brought forth except in prayer and fasting", meaning that "by fasting we fight our passions and make our human nature sensitive to His love and grace". In the religion of Christianity, and especially the Orthodox religion, about 50% of the calendar year (180–200 days) pushes believers to abstain from meat, dairy, and eggs, as reported by McPherson [87].

"Veganism", or "strict vegetarianism", is a philosophy and lifestyle that aims to avoid the use of animals for food, clothing, or any other purpose and is based on a diet of plant foods. It entails a commitment to abstain from eating products of animal origin, including meat, fish, seafood, eggs, dairy products, and honey. The majority of followers of this lifestyle extend its practice to other aspects such as clothing, toiletries (e.g., not using beeswax, lanolin, or creams with animal fats), and cosmetics (not animal-based and those not tested on animals) [88]. Overall, the spirit behind veganism is rooted in ethical concerns, environmental awareness, and a deep-rooted commitment to animal welfare. Vegans aim to minimize their impact on the environment and reduce animal exploitation by following a plant-based diet and lifestyle. Finally, the term "vegan" has become synonymous with a philosophy that supports a more mindful and conscientious approach to consumption, encouraging individuals to make choices that align with values of compassion and sustainability, hence ethics.

There is also a group of people who follow a raw food diet (raw foodism/fruitarianism) within the vegan philosophy, which has as its core principle the consumption of only fresh, unprocessed fruits and vegetables—avoiding heat-processed foods. This practice seeks to keep their diet as natural and simple as possible and pure [89].

7. Food and Spirituality

Food spirituality is defined as "An innate sense of connection that a subject can experience to and through food regarding personal and social identity, culture and ritual nature and the environment, body, and soul, the mundane and the universal", according to Michopoulou and Jauniškis [90]. Food in religious texts is associated with God's existence [73]. Spirituality affects our perceptions of everything from mundane to sacred and is overwhelmingly personal and subjective. Life and death beliefs are projected into foods. Is death the end, or does it mark the beginning of a new phase of existence, as posited by Atteshli-Theotoki [31]? Could death merely signify the separation of the physical body, leading to alignment with the psychic or the noetic realm? Atteshli-Theotoki [31] suggests that various scholars, based on her research, view the essence of the self as immortal, referring to the psychic and noetic aspects as the eternal soul. However, this perspective contradicts the findings of White Researchers, who regard the mortal aspect as not only encompassing the physical and etheric bodies but also including elements of the psychic and noetic realms [91]). According to the latter, the true self is associated with the superior spiritual body, often termed the Light Body [31]. Personal spiritual beliefs can be correlated with mental strength, stability, self-control, self-efficacy, and an improved relationship with foods [92–95]. Improvements in uncontrolled eating, emotional eating, intuitive eating, and mental and spiritual well-being have been shown following a religious program with a spiritual component, as reported by Patel et al. [94,95].

On the other hand, Ayahuasca, a spiritual ritual [96]—also known as the tea, the vine, and la purga—a brew made from the leaves of the *Psychotria viridis* shrub along with the stalks of the *Banisteriopsis caapi* vine, as reported by Savoldi et al. [97], is linked with better emotion control [98], a long-time reduction in depression and stress [99], and the potential to cure drug addiction [100].

A spiritual reconnection of the self to body, nature and society, and eating disorders is strongly discussed, but is it not a spiritual reconnection to God? Why are strong emotions like guilt, shame, well-being, and self-worth alike, or protest and anger suppression, linked with food and eating [90,101] or a feeling of powerlessness and a lack of meaning in life [102,103]? Why is overeating manifested by religious people as a divine struggle or lack of belief [104]?

Overall, religious neighborhoods and communities play a significant role in shaping food choices and practices, as discussed by Tan et al. [105]. However, merely participating in religious ceremonies does not inherently correlate with positive health outcomes [105,106]. Food consumption tends to be more influenced by personal convictions and beliefs rather than strict adherence to religious doctrines [105,106]. Therefore, fostering a deeper connection with God and the Holy Spirit, alongside a strengthening of our faith by our beliefs, appears crucial in this context. A better body–spirit connection, which naturally transfers to better intuitive eating and self-perceived body image, might also be achieved by practicing yoga [107,108].

8. Food Dynamics

Globalization, technological advances, changes in consumer preferences, environmental concerns, and socioeconomic disparities affect the dynamics of food in today's world [29]. Moreover, increased access to a wide variety of foods [109] originates from globalization and the effect of the expansion of food supply chains across different borders worldwide [110]. However, negative impacts are evident due to the vulnerability of systems to disruption, as observed during the COVID-19 pandemic with examples of shortages in some areas and surpluses in others, as reported by Moosavi et al. [111]. Preferences have been largely affected by these disruptions in food systems [112], with consumers looking for alternative food such as organic, plant-based, or locally grown. Consumption of fast and processed foods leads to obesity and diet-related diseases, as discussed and highlighted by Magano et al. [113].

Hence, the compromise of the global food system is considered true and is happening because of a wide range of factors, including the presence of chemical, physical, or microbiological hazards including pathogenic microorganisms, as shown by the WHO [114]. Other reasons involve improper handling and storage; natural toxicity; undeclared allergens; incorrect labeling; animal diseases; exceeded expiry dates; unauthorized genetic manipulation; and fraudulent practices, as elucidated by Fernandez and Paoletti [115]. But we have made this system unsafe because our vibrations of the mind, in the form of thoughts and desires materialized, attract all evil elementals present as inferior forms of life, which plague their creator, Man. It is not God and the Archangels who created mosquitoes, flies, microorganisms, and parasites causing epidemics and abominations, but Man. Man built an unconscious hell for Mankind. This is an example of the Law of Karma or Cause and Effect on a personal, but at the same time, global scale [31]. However, it is essential to recognize the role of human behavior in exacerbating all previous risks. The concept that our thoughts and desires materialize into reality, creating a negative impact on our environment and food systems, has been seriously highlighted by Atteshli-Theotoki [31]. This perspective underscores the interconnectedness of human actions and their consequences, emphasizing the importance of ethical considerations in food production, distribution, and consumption.

To address these challenges and promote a safer and more sustainable food system, stakeholders across the food industry, regulatory bodies, and policymakers must collaborate to implement comprehensive strategies [116]. This may include stringent regulations to ensure that food safety standards are met at every stage of the supply chain, investment in technology and infrastructure to enhance traceability and transparency, support for sustainable agricultural practices, and educational campaigns to raise awareness about the impacts of food choices on health and the environment [117]. Furthermore, enhancing a culture of mindfulness and ethical responsibility among consumers, producers, and policymakers can contribute to the creation of a more harmonious relationship between humanity and the food we consume, aligning with principles of sustainability, social justice, and holistic well-being, as reported by Fallah Shayan et al. [118].

9. GMOs, Ethics, and Neophobia

Global food governance needs to increase its presence to control questions raised on genetically modified organisms (GMOs), laboratory-grown meat, edible vaccines, and the environmental impact of high-tech agriculture, as discussed by Vega Rodriguez et al. [116]. The ethics or moral dimensions of the GM issue are usually presented in the recent literature as a medley of religious constructs ("playing God"), or moral dilemmas, sometimes mixed with ecological concerns, as reported by Varzakas et al. [4,80]. Playing God is still associated with religion and GMOs [119–121].

Cisgenesis and intragenesis were developed as alternatives to transgenesis. Both concepts imply that plants must only be transformed with genetic material derived from the species itself or closely related species capable of sexual hybridization. Furthermore, foreign sequences such as selection genes and vector-backbone sequences should be absent. Allowing the use of new gene combinations created by in vitro rearrangements of functional genetic elements is how intragenesis differs from cisgenesis. Several surveys show higher public acceptance of intragenic/cisgenic crops compared with transgenic crops [122]. One of the major concerns of the public about transgenic crops relates to the mixing of genetic materials between species that cannot hybridize by natural means. Hence, two transformation concepts cisgenesis and intragenesis were developed as alternatives to transgenesis [122–125].

The presentation of genetically modified produce as "Frankenfoods" [126] is observed as a political myth construct around GM food and GMOs [127]. Posting about the unnatural and artificial sources of food and its association with GMOs and GM food has been widely reported [128–134].

The issue of the unnaturalness of GM food and GMOs has been discussed from different angles such as ethics, nutrition, religion, fear, and safety among other less prominent viewpoints. They all refer to the transgenic nature of GM food and GMOs and conclude that the average consumer cannot relate to an end product that contains traits of two or more organisms that are not naturally (or sexually) compatible; therefore, the term "unnatural" and "artificial" appears in different studies and consultations, without any linkages to the scope of the inquiry of the studies, as reported by Siddiqui et al. [135].

Regarding fear, i.e., an emotion that is often expressed in food-related issues, consumer attitudes are shaped by the role of neo-phobia [128–135]. In the case of GMOs, food-related fears affect the behavior of the consumer. Studies in the field show that a lack of information and imbalanced communications constitute the driving force of consumers against GM technology [126,136–139]. And this has to do with the predominant negative feeling/emotion that is often related to different forms of fear such as fear of isolation, fear of the unknown, fear of consuming products and affecting health, and fear of speaking out on the issue. Specific fears related to GM food and GMOs that are expressed by rivals of GM technology, as reported by Uzogara [140], refer to modification of nutritive quality of foods, possible toxicity, potential allergenicity and potential antibiotic resistance from GM crops, and carcinogenicity from GM food consumption. Furthermore, several concerns refer to other aspects such as environmental contamination, accidental gene transfer to wild plants, potential formation of new viruses and toxins, monopolies in the supply chain of seeds (bio patenting of GMOs), threats to the genetic diversity of plants, and finally religious, cultural and ethical concerns, as well as fear of the unknown. Also, Laros and Steenkamp [126] discussed the frequent appearance of fear messages in the media by examining the Dutch market food market and found that Dutch consumers feel significantly more fearful of GM food than of other new food types. They assumed that there is consistent fear about GM food across society.

Moreover, neophobia still comprises a niche category among the different types of fear related to GMF; however, this has been overpassed by other fears such as allergens [141]). The display of some degree of human aversion to new foods is a trait named food neophobia, according to Cooke et al. [142]. They added that food neophobia is widespread in omnivores, warblers, rats, and chimpanzees. Knight et al. [143] argue that beliefs about the risks and benefits of the production and introduction of GMF in the food market affect consumer acceptance heavily. Moreover, attitudes towards technology are affected by the associated fears. Finally, many studies refer to "superstition", "religious fears", and "magical beliefs", factors that have been linked with negative attitudes towards GM foods [4,80,144].

However, regarding allergens, some studies show that GM crops can be used as a solution to eliminate allergens in food and fight against coeliac disease [145]. One research group developed wheat bread made from wheat flour with a very low gliadin content and the main epitopes of wheat gluten that is potentially suitable for celiac patients and other gluten-intolerant individuals, as discussed by Gil-Humanes et al. [146].

Generally, the ethical dilemmas surrounding GM food and GMOs are multifaceted and complex, encompassing a range of concerns related to human health, environmental impact, and societal values. One of the central ethical dilemmas is the perceived unnaturalness of genetically modified organisms, which raises questions about the integrity of the food supply and the potential risks posed by manipulating the genetic makeup of plants and animals, as noted by Weale [147]. This unnaturalness is often cited as a source of fear and uncertainty among consumers, who may worry about the long-term effects of consuming GM foods on their health and well-being. Additionally, ethical considerations extend to issues such as environmental sustainability and biodiversity, as the widespread adoption of GM crops could lead to unintended consequences such as the emergence of superweeds or the loss of genetic diversity in agricultural ecosystems, as shown by Ghimire et al. [148]. Furthermore, concerns about corporate control and the concentration of power in the hands of biotechnology companies raise questions about equity and justice in the food system, as reported by Fairbairn and Reisman [149]. Addressing these ethical dilemmas requires careful deliberation, transparency, and robust regulatory oversight to ensure that the benefits of GM technology outweigh the potential risks and that the interests of all stakeholders, including consumers, farmers, and the environment, are considered.

10. Food Regulation

The debate surrounding the scope of food regulation extends beyond inherently unsafe foods and is influenced by multiple factors, including considerations of public health, ethics, and regulatory efficacy [150]. Proponents of this approach argue in favor of resource efficiency, emphasizing the importance of urgency and adopting a risk-based methodology [151]. Resources can be allocated more effectively when prioritizing regulatory efforts on products with the greatest potential to harm public health. Moreover, adopting a risk-based approach allows for the implementation of stricter regulations on foods with higher associated risks [152]. This targeted strategy enables regulatory bodies to tailor their interventions to address specific threats, thereby enhancing the overall safety of the food supply chain. Such measures not only mitigate risks to public health but also optimize the allocation of regulatory resources, ensuring a more robust and responsive regulatory framework. Arguments against this topic include the complexity of the food system, prevention, and ethical considerations, as discussed by Ververis et al. [153]. The emerging risks or systemic issues in food might be overlooked if there is a limitation of regulation to inherently unsafe foods. Moreover, it should be considered that various sources might cause risks and there might not be confinement of risk to specific foods, as reported by Santeramo and Lamonaca [154].

Achieving a delicate equilibrium between safeguarding public health, upholding individual liberties, and acknowledging cultural and religious diversity is imperative when formulating food regulations for moral or religious reasons [155]. While governmental oversight of food choices plays a crucial role in public health protection, food safety enhancement, and addressing ethical concerns related to food production and consumption [156], it must be executed judiciously. Establishing rigorous standards and robust monitoring systems for food safety across production, distribution, and sale is paramount to ensure the integrity of the food supply chain, as shown by Rose et al. [157]. Between the complexities of government regulation in food and nutrition and competing interests, there is a delicate balance [158]. While it is the government's responsibility to set nutritional guidelines and standards to safeguard public health, avoidance of excessive regulation that encroaches upon personal autonomy and cultural diversity in food choices should be evident, as highlighted by Little [156]. Moreover, the regulatory landscape is often influenced by powerful commercial interests and companies, which can impede effective governance [159]. Therefore, the adaptability of government authority over food and nutrition should be looked for rather than rigidity [160]). Developing interventions that effectively address public health concerns while respecting the boundaries of government authority is essential for achieving a balanced, ethical, and evidence-based approach, as discussed by Qureshi et al. [161]. It is imperative to establish a dialogue among stakeholders, including policymakers, industry representatives, and community leaders, to devise solutions that promote both public health and individual rights, cultivating a harmonious relationship between regulation and personal freedoms. The regulation of food choices by governments is critical in public health protection, food safety adoption, and addressing ethical issues related to food production and consumption [156]. In this direction, standards and systems must be established, and food safety monitoring should be carried out through regulations for production, distribution, and sale, ensuring the non-threat of food safety, as reported by Rose et al. [157]. The basic legislative forms around the globe are mentioned in Table 1.

Table 1. Comparison of food legislation around the world [162,163] (https://food.ec.europa.eu/horizontal-topics/general-food-law_en) (accessed on 12 April 2024), https://food.ec.europa.eu/safety/biological-safety/food-hygiene_en (accessed on 12 April 2024).

Region	Legislation	Key Points
United States	Food Safety Modernization Act (FSMA)	Focuses on the prevention of foodborne illnesses
	Federal Food, Drug, and Cosmetic Act	Regulates food safety, labeling, and additives
European Union	General Food Law Regulation	Ensures food safety, traceability, and labeling
	Regulation (EC) No. 178/2002	Establishes general principles and requirements of food law
	Regulation (EU) No. 1169/2011	Deals with food information provided to consumers
	European Union Regulation (EC) No. 852/2004, 853/2004, 854/2004, 882/2004, Regulation (EU) 2017/625	Sets hygiene rules for foodstuffs in the EU
Scandinavian countries	Norway Food Act No. 124 of 2003	Regulates food safety and quality in Norway
	Sweden Food Act (2006:804).	Sets requirements for food safety and labeling in Sweden
	Finland Food Act (297/2021)	Regulates food production and safety in Finland
	Denmark Food Act (No. 46 of 2017).	Ensures food safety and quality standards in Denmark
Germany	German Food and Feed Code (LFGB)	Regulates food safety and quality in Germany
	German Food Hygiene Regulation (LMHV)	Establishes hygiene requirements for food businesses
Africa	African Union Model Law on Food Safety	Aims to harmonize food safety laws across African countries
	Tanzania Food, Drugs and Cosmetics Act, 2003	Regulates food safety, drugs, and cosmetics in Tanzania
Middle East	Gulf Cooperation Council (GCC) Food Law	Establishes food safety regulations in GCC countries
	Saudi Arabia Food Law (Royal Decree No. M/1), 30 October 2014	Regulates food safety standards and practices in Saudi Arabia
	UAE Federal Law No. 10 of 2015 on Food Safety	Focuses on ensuring food safety and quality in the UAE

Table 1. *Cont.*

Region	Legislation	Key Points
Japan	Food Sanitation Act	Regulates food safety and hygiene in Japan
Asia	ASEAN Guidelines on Food Hygiene	Provides guidelines for food safety and hygiene in ASEAN countries
	Food Safety Act (Republic of Korea) (Consolidated version of Act No. 9432 of 2009 as amended last by Act No. 18967, 10 June 2022)	Regulates food safety standards in South Korea
	Food Safety Act (Taiwan) Act Governing Food Safety and Sanitation ("Act"), last amended on 24 January 2018	Ensures food safety and quality standards in Taiwan
	Food Safety Act (Singapore) Food (Amendment) Regulations 2023, 17 April 2023	Regulates food safety and hygiene in Singapore

11. Food Safety and Quality Standards and Ethics

Food safety and quality management systems ISO 22000:2018 and ISO 9001:2015 [164] (https://www.iso.org/standard/62085.html (accessed on 12 April 2024)) discuss the issue of ethics from different perspectives. The seven principles of quality management (Figure 1) encompass various aspects such as customer focus, leadership, the engagement of people, the process approach, improvement, evidence-based decision-making, and relationship management. However, these principles may not explicitly address the ethical considerations about human beings and their conduct in the workplace. Ethical leadership and customer care are fundamental aspects that go beyond the scope of quality management principles, focusing on the moral responsibility of leaders and organizations towards their employees and customers [165]. While quality management principles emphasize efficiency and effectiveness in processes and outcomes, ethical leadership and customer care emphasize the importance of integrity, fairness, and empathy in interactions and decision-making [165]. It is then essential for organizations to integrate ethical considerations into their operations, ensuring that employees are treated ethically and customers receive fair and respectful treatment (*The Economist* 2022) [166]. However, all these principles do not deal with human beings and their ethics in a working environment applying ethical principles.

On the other hand, the quality Guru, Deming, described the PDCA (Plan-Do-Check-Act) cycle as a fundamental framework for continuous improvement [167–170]. Does that cycle not resonate with human beings and their lives? Does it not indicate that before acting ethically and applying our ethical principles, we should effectively plan, implement what has been planned, and then check and monitor accordingly? Is this not the essence of ethics? Why? Are there individuals or organizations that risk not checking monitoring and releasing their products into the market by neglecting these rules? Perhaps there are.

Under this consideration, the PDCA cycle in the food chain should then include the following (Figure 2):

Plan: In this phase, the focus is on identifying ethical goals and objectives for the food chain. This includes setting standards for sustainable food production, ethical sourcing of ingredients, fair labor practices, and minimizing environmental impact. Stakeholders collaborate to develop policies, procedures, and guidelines that align with ethical principles and regulatory requirements.

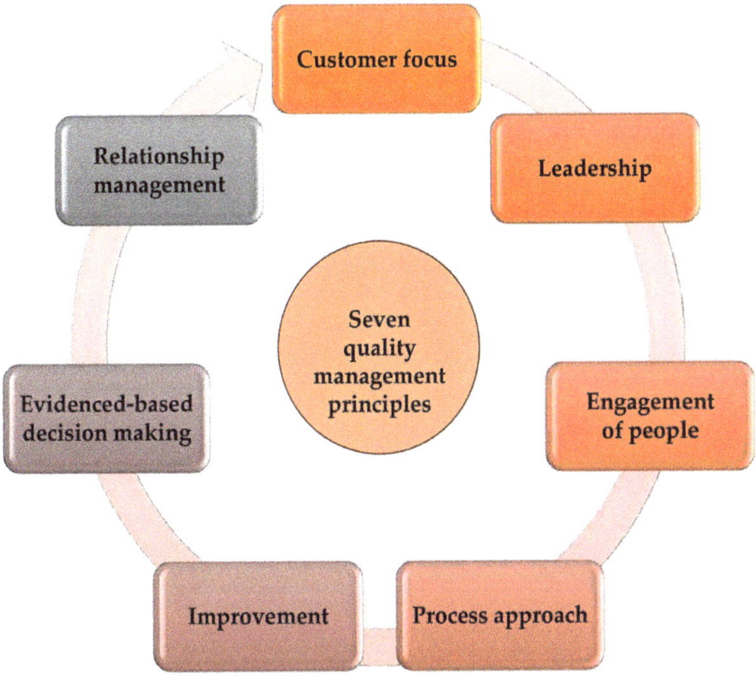

Figure 1. The seven principles of quality management according to ISO 9001:2015 [164] (https://www.iso.org/standard/62085.html (accessed on 12 April 2024)).

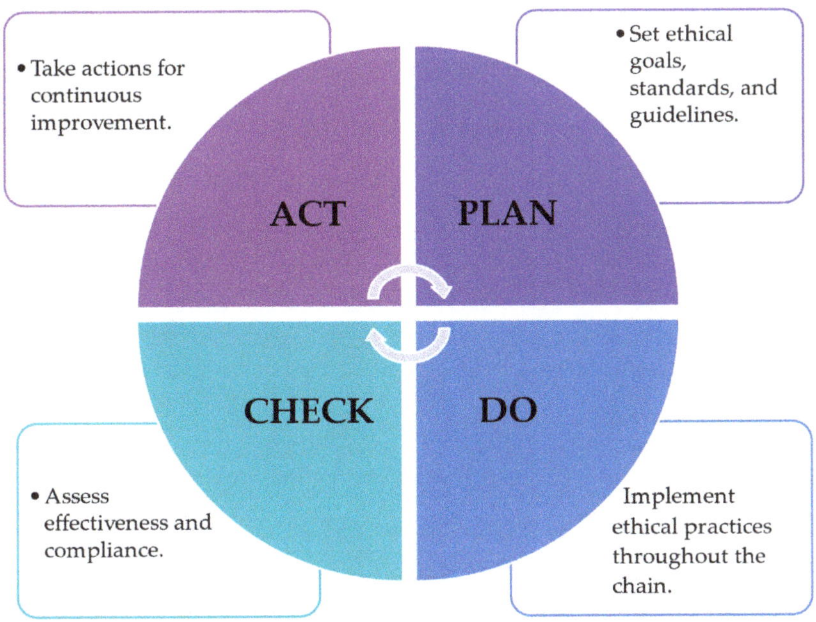

Figure 2. Description of the PCDA cycle in the food chain.

Do: Once the plan is established, it is implemented throughout the food chain. This involves executing ethical practices in food production, distribution, and consumption. Companies may adopt sustainable farming methods, ethical sourcing strategies, and transparent labeling practices. Employees are trained in ethical guidelines, and suppliers are held accountable for adhering to ethical standards.

Check: In the check phase, the effectiveness of ethical practices is assessed through monitoring, measurement, and evaluation. Key performance indicators (KPIs) are established to track progress towards ethical goals. Audits, inspections, and reviews are conducted to ensure compliance with ethical standards and identify areas for improvement. Feedback from stakeholders, including consumers, is collected and analyzed to gauge satisfaction and identify any ethical concerns.

Act: Based on the findings from the check phase, actions are taken to address any gaps or deficiencies in ethical performance. This may involve revising policies, updating procedures, providing additional training, or implementing corrective actions. Continuous communication and engagement with stakeholders are essential to drive ongoing improvement and maintain ethical integrity throughout the food chain.

Integrating risk management and risk-based thinking into food ethics is crucial for ensuring the safety, integrity, and ethicality of food production and consumption processes. By applying ISO 31000:2018 principles [171] alongside tools like SWOT analysis, stakeholders in the food industry can identify potential risks to food safety, quality, and ethical standards. This proactive approach allows for the implementation of preventive controls to mitigate risks and seize opportunities that align with ethical values.

For instance, considering the ethical implications of sourcing ingredients from suppliers with questionable labor practices or environmental sustainability could be part of the risk assessment process. By analyzing strengths, weaknesses, opportunities, and threats, food producers can identify areas where ethical concerns may arise, such as animal welfare, fair labor practices, or ecological sustainability (Leroy et al. 2022) [172]. Implementing preventive measures, such as robust supplier vetting processes or investing in sustainable farming practices, helps minimize negative impacts on ethical standards. Moreover, the concept of risk-based thinking encourages continuous improvement in ethical food practices, as discussed by Thomson [173]. By analyzing nonconformities and their potential ethical implications, organizations can take corrective actions to address underlying issues and prevent recurrence. This proactive approach fosters a culture of ethical responsibility throughout the food supply chain, promoting transparency, accountability, and trust among stakeholders.

Overall, when we incorporate risk management and risk-based thinking into food ethics, we allow for a systemic approach to identifying, assessing, and addressing ethical concerns in food production and consumption, as described by Ispas et al. [174]. If we proactively manage risks and seize opportunities that align with ethical values, we will allow the food industry to uphold its commitment to ethical standards and contribute to a more sustainable and responsible food system. Managing risk is based on the principles, framework, and process, as reported in ISO 31000:2018 (Figure 3).

Figure 4 illustrates an analysis of ethics in terms of principles, framework, and processes, as outlined in ISO 31000:2018, and connections with profit, risk, and loss [165].

Figure 5 further outlines the principles of value creation and protection (Clause 4), which are paramount in ethical considerations. These principles encompass human and cultural factors, as well as other key elements that drive continual improvement within both individuals and organizations, guided by effective management.

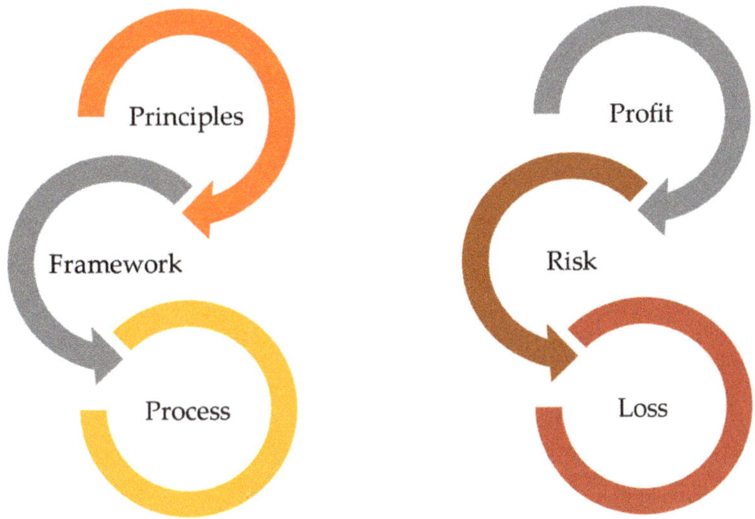

Figure 3. Overview of ISO 31000:2018 (adapted from ISO 31000:2018) [165].

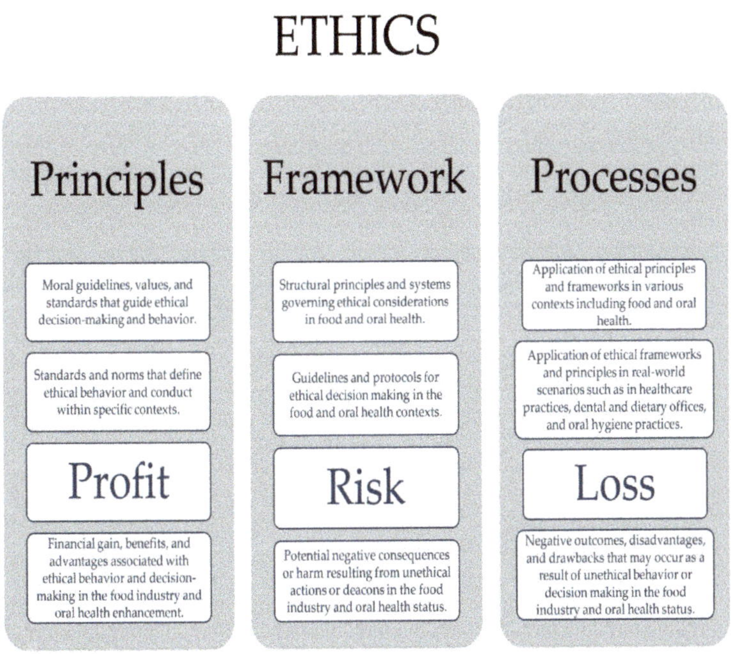

Figure 4. Analysis of ethics according to principles, framework, and processes, as outlined in ISO 31000:2018, in conjunction with profit, risk, and loss; adapted from [165].

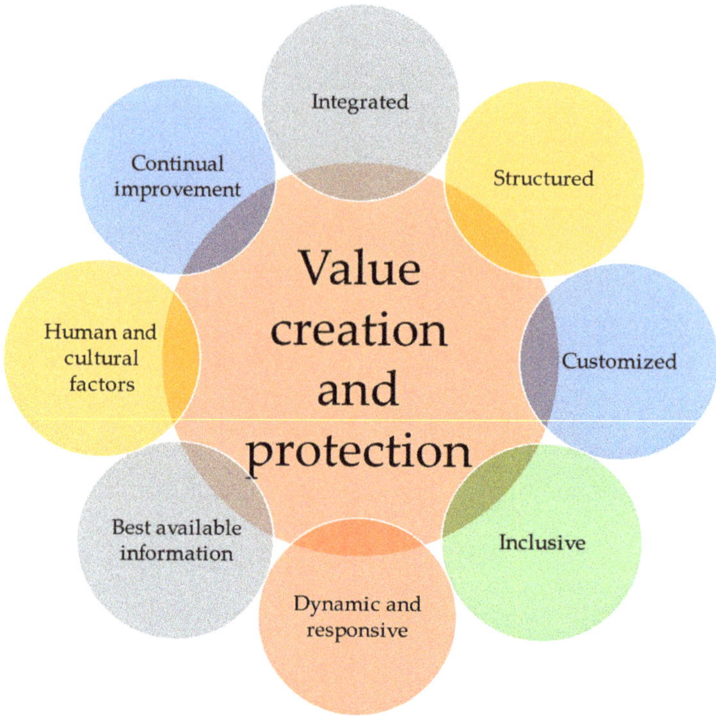

Figure 5. Principles of ISO 31000:2018 (adapted from ISO 31000:2018) [165].

A synopsis of the ISO principles in the food context is provided in Table 2.

Table 2. Ethical principles for value creation and protection in the food industry ISO 31000:2018, [165].

Ethical Principles	Application in the Food Industry
Value creation and protection	This component emphasizes the importance of creating and protecting value within the food industry. This involves not only generating profit but also ensuring that ethical considerations are prioritized to protect the well-being of consumers, workers, and the environment.
Integrated	The food industry must integrate ethical considerations into all aspects of its operations, including production, distribution, marketing, and waste management. This integration ensures that ethical values are embedded throughout the entire supply chain.
Structured	Ethical decision-making processes should be structured and systematic, guided by clear principles and guidelines. This ensures consistency and transparency in how ethical dilemmas are addressed within the food industry.
Customized	Recognizing that different contexts may require tailored ethical approaches, the food industry should customize its ethical practices to suit specific situations, regions, or cultural norms. This flexibility allows for more effective and culturally sensitive ethical decision-making.
Inclusive	Ethical practices in the food industry should be inclusive, considering the perspectives and needs of all stakeholders, including consumers, producers, workers, communities, and regulatory bodies. Inclusivity fosters collaboration and ensures that diverse voices are heard in ethical decision-making processes.
Dynamic and responsive	Ethical considerations in the food industry should be dynamic and responsive to changing circumstances, emerging issues, and stakeholder feedback. This adaptability enables the industry to address new challenges and seize opportunities for improvement.

Table 2. *Cont.*

Ethical Principles	Application in the Food Industry
Best available information	Ethical decision-making in the food industry should be informed by the best available information, including scientific research, industry standards, consumer preferences, and expert advice. This ensures that decisions are based on evidence and expertise rather than speculation or bias.
Human and cultural factors	Ethical practices in the food industry should consider the human and cultural factors that influence food consumption, production, and distribution. This includes considerations of food traditions, dietary preferences, labor rights, and social norms.
Continual improvement	The food industry should strive for continual improvement in its ethical practices, seeking to raise standards, address shortcomings, and innovate new solutions. This commitment to ongoing improvement ensures that ethical considerations remain at the forefront of industry efforts.

Figure 6 illustrates the ripple continuum of standardization, depicting how principles and values at the core of these standards influence codes of conduct and models for excellence in the food industry.

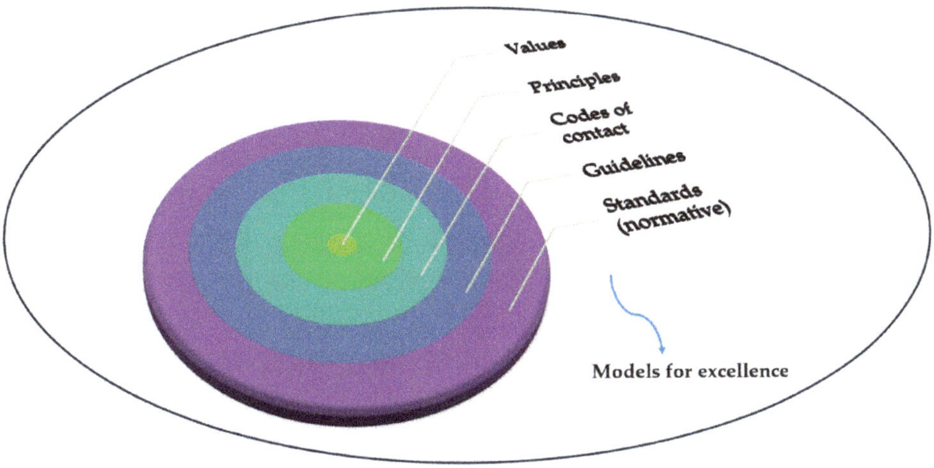

Figure 6. The ripple continuum of standardization.

Furthermore, ISO 26000:2010 provides "guidance on social responsibility", meaning how businesses and organizations can operate in a socially responsible way and refer to principles of social responsibility related to accountability, transparency, ethical behavior, respect for human rights, respect for stakeholders' interests, respect for the rule of law, and respect for international norms of behavior. The general principles of ISO 26000 are described in Figure 7.

On the other hand, food safety culture needs to be adopted effectively and requires the implementation of ethics and commitment from the side of people/personnel and management, according to Regulation (EU) 2021/382 [175], along with effective leadership and communication from all sides. In this direction, the RSC (Responsible Supply Chain) approach involves responsible sourcing, greater transparency, sustainable practices, and adherence to shared standards and values, as reported by Jacob-John et al. [176]. Amaeshi et al. [177] illustrate the role of the supply chain and the pressure in employing a socially responsible operation. Implementation can be carried out by CSR (Corporate Social Responsibility). Moreover, MacGregor [178] cites five reasons for the implementation of a CSR strategy, and these include risk mitigation, competitive marketing advantage, inter and intra-organizational demands, political–social factors, and conviction.

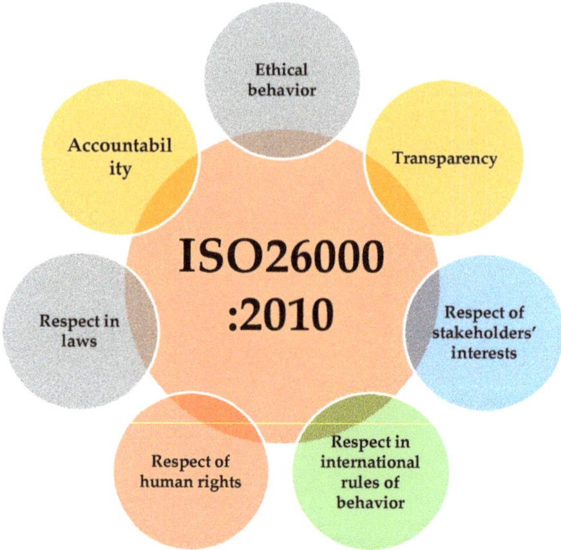

Figure 7. General principles of ISO 26000:2010; adapted from https://www.iso.org/iso-26000-social-responsibility.html, accessed on 12 April 2024.

Corporate responsibility has been broadened to include ethical management, which involves generating economic profits, satisfying the demands of different stakeholders, and ensuring business sustainability. In this context, ethical management has gained prominence as a determinant of business sustainability as it enables organizations to adhere to ethical principles, promote shared social values, and enhance the legitimacy of their operations [179].

The International Organization for Standardization (ISO) released the ISO 26000 standard in 2010 to provide international guidance on social responsibility for all types of organizations. The main objective was to implement CSR by translating the principles and key issues into practical activities and providing best practices in the area of social responsibility. ISO 26000 can be used as a benchmark for assessing the development of the expected relationships of an organization with its environment. Measurable benefits can be brought forward by strengthening stakeholder relationships, enhancing reputation, and building trust [180].

Product-related CSR assesses the management of risk for companies facing significant product recalls or losing customer confidence because of significant product quality concerns. CSR can build a positive reputation and moral capital among stakeholders, mitigating any adverse evaluation of corporate misconduct and ensuring against the potential negative consequences of any violations of stakeholder expectations. A potential product recall is the most common type of disruption with significant consequences for firms in terms of brand reputation and brand equity, as well as financial performance and value [181].

Responsible Supply Chain Management protects consumer confidence and public health while promoting a culture of transparency and ethical behavior in the food industry [176,182]. Through engagement with suppliers, employees, and local communities, RSCM enables companies to establish ethical sourcing practices, empower employees, and champion human rights throughout their supply chains. By promoting ethical labor practices and prioritizing the eradication of modern slavery, RSCM protects vulnerable workers and promotes sustainable and responsible business behavior in the food industry [176,182,183].

Research within the field of responsibility and supply chain management defines responsible behaviors in supply chains in several ways including Green Supply Chain Management, Sustainable Supply Chain Management (SSCM), and Responsible Supply Chain Management (RSCM) [176]. However, it is not only the environmental dimension but, according to Carter and Rogers [184], also the balance between the fiscal, environmental, and social dimensions of the organization. Linnenluecke and Griffiths [185] illustrate that climate change is a major problem that businesses must account for, and climate change results from anthropogenic activities leading to greenhouse gas emissions (GHG). The different perspectives on sustainable food supply are further illustrated by Smith [186] and include food safety and factors supporting the viability and diversity of the stakeholder economics and communities along with consideration of the ecological impacts. According to Smith [186], ecological aspects of food involve environmental issues such as a reduction in energy consumption and minimization of water use, whereas social aspects involve the creation and maintenance of a safe and socially acceptable working environment for all employees across the entire supply chain considering all stakeholders.

Overall, various ISO standards are relevant to ethical actions and practices in the food industry, covering areas such as social responsibility, food safety, environmental management, quality management, risk management, occupational health and safety, sustainable procurement, energy management, business continuity, information security, and anti-bribery measures. The most important ones are described in Table 3.

Table 3. ISO standards for ethical practices in the food industry.

ISO Standard	Title
ISO 26000: 2010 https://www.iso.org/iso-26000-social-responsibility.html, accessed 12 April 2024.	Guidance on social responsibility
ISO 22000:2018 https://www.iso.org/standard/65464.html accessed 12 April 2024	Food safety management systems
ISO 14001:2015 https://www.iso.org/standard/60857.html accessed 12 April 2024	Environmental management systems
ISO 9001: 2015 [164]	Quality management systems
ISO 31000:2018 [165]	Risk management
ISO 45001:2018 https://www.iso.org/standard/63787.html accessed 12 April 2024	Occupational health and safety management systems
ISO 20400:2017 https://www.iso.org/obp/ui/en/#iso:std:iso:20400:ed-1:v1:en accessed 12 April 2024	Sustainable procurement
ISO 50001:2018 https://www.iso.org/standard/69426.html accessed 12 April 2024	Energy management
ISO 22301:2019https://www.iso.org/standard/75106.html accessed 12 April 2024	Security and resilience
ISO 27001:2018 https://www.iso.org/standard/73906.html accessed 12 April 2024	Information security management system (ISMS)
ISO 37001:2016 https://www.iso.org/standard/65034.html accessed 12 April 2024	Anti-bribery management systems

12. Food Fraud

Wognum et al. [187] state that transparency is pivotal for sustainable operations within food supply chains. Better traceability and transparency within supply chains arose in 2013 with the horsemeat scandal in Europe. According to Rasul and Thapa [188], the increasing and incessant use of chemicals during farming processes by using fertilizers, insecticides, and pesticides not only affects the environment negatively with the increase in maximum residue limits (MRLs) but also the society by spreading diseases to humans and affecting aquatic life and livestock.

Spink and Moyer [189] define food fraud as all intentional acts that involve deliberate and intentional substitution, addition, tampering, misrepresentation, or false/misleading statements performed to gain economically [190–201]. This poses three types of risks—direct, indirect, and technical risks.

Rampant mislabeling is also connected with fraud. The case of New Zealand Manuka honey is well known [202,203]. Rampant use of other adulterants including formalin, urea, starch, neutralizers, detergents, sodium chloride, skim milk powder, sucrose, glucose/dextrose, and hydrogen peroxide [204,205]. It is a clear action of producers' profit through misconduct and unethical behavior versus the expense of the health of the consumer [206].

Moreover, Fassam et al. [207] reported on four supply chain drivers for food fraud including lack of trust among supply chain actors, opportunistic behaviors by supply chain partners, inadequate governance of the supply chain, and complexity of the supply chain. Niu et al. [208] further investigated the food fraud key influencing factors and their interrelationships in an emerging food market—China—by using the DEMATEL-based analytic network process (DANP) and showed that the identified key cluster was government regulation, social governance, and detection techniques. A summary of food fraud types and ways that consumers may acknowledge them is provided in Table 4.

Table 4. Possible food frauds and consumer countermeasures. Adapted from [204,205].

Food Fraud Type	Description	Consumer Countermeasures
Mislabeling	Deliberate substitution, addition, tampering, or false/misleading statements for gain	Demand transparency in labeling and certification processes, verify product authenticity, and report suspicions
Adulteration	Addition of unauthorized substances like formalin, urea, starch, etc., for economic gain	Support stringent quality control measures, seek products with reputable certifications, and report suspicions
Lack of trust in the supply chain	Distrust among supply chain actors leads to increased vulnerability to fraud	Choose products from transparent and accountable supply chains, support ethical brands, and demand traceability
Opportunistic behavior	Supply chain partners exploiting situations for personal gain	Enhance fair business practices and endorse initiatives fostering integrity and accountability
Inadequate supply chain governance	Poor oversight and control mechanisms within the supply chain, enabling fraudulent activities	Advocate for regulatory reforms and support initiatives enhancing governance and accountability
Complexity of the supply chain	Complexity of supply chain operations contributing to increased risk of fraud	Support simplified and transparent supply chain structures and favor local and short supply chains
Government regulation	Insufficient regulatory frameworks and enforcement contribute to fraud vulnerabilities	Implement stricter regulations and enforcement and support initiatives promoting regulatory compliance
Social governance	Social factors influencing fraud susceptibility within the supply chain	Promote consumer awareness and education and support initiatives fostering social responsibility and transparency
Detection techniques	Inadequate fraud detection methods and technologies allow fraud to go undetected	Invest in advanced detection technologies, support initiatives improving fraud detection, and share information

13. Consumer Behavior and Ethics

Consumers' beliefs and intentions such as health values and consciousness have been linked and interrelated with organic food purchase and consumption preferences as reported by many researchers [209–214], nutrition content [210,215,216], environmental concern [217–221], safety [222–224] and taste [225–227]. Consumer awareness about food scandals, epidemics of diseases from viruses such as Bovine Spongiform Encephalopathy or pathogenic bacteria such as *Escherichia coli* 0157 infections, and favorable weather patterns [210,228,229] has also been reported widely.

Guru et al. [230] analyzed the motivational factors critical to the purchase and consumption of organic food and extended a roadmap to the food industries for sustainable growth. They found that the purchase and consumption of organic food products are affected by issues such as being chemical-free, having no artificial ingredients, being good for the body, being healthy, and having an awareness of health.

The growing interest in fresh, healthy, and organic agriculture and food products has also been elucidated by many researchers [231–233]. Furthermore, health concerns, religious beliefs, environmental concerns, animal welfare, human rights, fair trade, and fair wages are considered by consumers in food consumption behavior [234,235]. Hence, food ethics plays an important role in understanding consumers' perceptions [236,237]. Overall, the general theory of marketing ethics developed by Hunt and Vitell [238] states that a person's deontological evaluation (DE) and teleological evaluation (TE) are related to his/her ethical judgment [235]. In this direction, DE and TE significantly affected positively perceived behavioral control and subjective norms, as reported by Ho et al. [239].

14. The Mouth—The Sacred Gateway to the Body

The phrase "the mouth is the gateway to the whole body" encapsulates a profound understanding of the interconnectedness between oral health and overall well-being, resonating throughout ancient wisdom, biblical teachings, contemporary scientific research, and holistic health principles [10,240,241]. In ancient civilizations such as Greece and Rome, this concept finds resonance in the interplay between oral health and bodily equilibrium [242–244]. Rooted in the humoral theory described by Hippocrates, oral diseases were viewed as manifestations of disharmony within the body, stemming from imbalances in bodily humor, as reported by Bujalkova et al. [245]. For ancient Greek philosophers, the mouth served not only as a site of ingestion but also as a conduit through which imbalances in diet and environment could influence systemic health [246,247]. The classification of foods based on inherent qualities and seasonal variations in dietary recommendations underscored the recognition of the mouth's pivotal role in maintaining humoral balance and preventing disease [245]. This holistic understanding suggests that oral health is not an isolated entity but rather intricately linked to the well-being of the entire body, emphasizing the importance of reviewing oral health as a cornerstone of holistic health practices across cultures and periods [12].

Seasonal dietary variations and mouth intakes, crucial for humoral balance and disease prevention, were acknowledged in ancient medical practices too [245,247]. In Hippocratic writings such as "*Airs, Waters, and Places*", seasonal changes were recognized as influencing disease patterns, highlighting the impact of environmental factors on health [246]. Consequently, dietary recommendations varied with the seasons, with specific foods prescribed to counteract seasonal imbalances, reflecting a holistic approach to health that emphasized the interconnectedness of diet, environment, and overall well-being [248–250]. The Hippocratic treatise "*On Regimen*" epitomizes the fusion of philosophy and dietetics, providing us with a balanced approach to health authored by the Hippocratic writers [251]. This work focuses on moderation and temperance in dietary habits, stressing the fundamental role of food in maintaining bodily equilibrium [251]. Central to Hippocratic philosophy is the belief that food serves as a cornerstone of well-being, resonating with the ancient Greek concept of harmony in life. Anaxagoras of Clazomenae, a pre-Socratic philosopher, contributes to this discourse through his fragmentary teachings. Curd (2007) [252] presents Anaxagoras as a thinker deeply concerned with the nature of the cosmos, yet his insights extend to the realm of nutrition. Anaxagoras proposes that food sustains life by providing nourishment to the body, aligning with his broader cosmological theory that all things are composed of infinitesimally small particles known as "nous" or mind (Curd 2015) [253]. In this context, the act of eating transcends mere sustenance; it becomes a reflection of the cosmic order, wherein the ingestion of food symbolizes the assimilation of universal principles into the individual. Pythagoras, also renowned for his mathematical and metaphysical teachings, espoused a distinct dietary philosophy [254]. Dye (1999) [255] explores the enigmatic prohibition of beans in Pythagorean doctrine, attributing it to symbolic and practical reasons. For Pythagoras, beans symbolized impurity and were associated with death rites, thereby conflicting with his belief in the transmigration of souls [256]. According to this theory, the Pythagoreans believed that all living beings, plants, animals, humans, vampires, ninja turtles, etc., shared a common soul or life force that could be reborn in different forms after death, as noted by Huffman [257]. In this context, beans may have been seen as particularly problematic, as they were believed to contain the souls of the dead. By avoiding beans, the Pythagoreans may have believed that they could avoid being contaminated by any potentially impure or malevolent spirits. It seems that beans were/are what you incarnate as if you have been a bad human. Moreover, Pythagoras suggested a diet free from animal products, viewing vegetarianism as cultivating moral and spiritual purity [258]. Through dietary restrictions, Pythagoras sought to align the physical body with the principles of harmony and ethical integrity, as discussed by Graham [259].

The philosophical inquiry into the mouth and food extends beyond individual thinkers to encompass broader cultural practices and beliefs [260]. Garnsey [261] examines the connection between food and society in classical antiquity, shedding light on how dietary customs reflected social hierarchies, religious beliefs, and agricultural practices. From the extravagant banquets of the elite to the simple fare of the common populace, food served as a marker of identity and status in ancient Greek society, as shown by Flint-Hamilton [262]. Furthermore, the philosophical contemplation of food intertwines with religious rituals and mysteries. Delatte [263] set light on the mystical significance of the "kykeon", a barley-based beverage consumed as part of the Eleusinian Mysteries. This sacred concoction symbolized nourishment for the soul through the mouth, offering initiates a transformative experience of spiritual enlightenment. The consumption of kykeon exemplifies how food transcends its material essence to become a conduit for metaphysical experiences and revelations [264]. As it seems, the mouth was considered by the ancients as "the gateway to the body", which encompasses notions of health, morality, cosmology, and spirituality that enhance the profound significance of sustenance and sustainability in shaping individual and collective well-being [265].

Central to this exploration is the passage from Matthew 15:10-1 4 [266], wherein Jesus challenges conventional understandings of purity and underscores the primacy of inner purity over external rituals, shedding light on the profound spiritual dimensions of oral health. Jesus' teachings within this passage challenge the conventional understanding of purity, urging individuals to see beyond mere external rituals and embrace the transformative power of inner purity. He elucidates that true defilement arises not from external contaminants but from the inner recesses of one's heart, emphasizing the profound link between spiritual purity, food intake, and physical well-being. This paradigm shift redirects attention to the status between one's internal disposition and external health outcomes, highlighting the holistic nature of human existence wherein spiritual and physical domains intersect [267]). Through this lens, oral health emerges not merely as a matter of hygiene but as a reflection of one's spiritual alignment with inner and outer purity. As Hawks [267] further states, this perspective invites us to recognize oral health not merely as a physical concern but as a reflection of our spiritual well-being, highlighting the transformative power of inner purity in fostering holistic health. Furthermore, Chan et al. [268] elaborate on a body–mind–spirit model in health, proposing a comprehensive approach that integrates spiritual well-being with physical and mental health. Moreover, research by Spanemberg et al. [269] and Fiorillo [270] demonstrates the profound impact of oral health on overall quality of life, highlighting its pivotal role in pursuing holistic well-being. This holistic perspective is further supported by studies on oral health-related quality of life among diverse populations, including older adults [271,272], older adults receiving home health care services [272], institutionalized residents [273,274], children and adolescents [275], individuals of varying socioeconomic status [276], and postpartum women [277].

15. Significance of Oral Health in the Food Chain

The mouth serves as a crucial gateway to the body because of its pivotal role in multiple physiological processes and its direct connection to various systemic health conditions such as digestion, respiration, pathogen defense, systemic health, and communication, as reported by Kazemi et al. [278]. First and foremost, the mouth is where the process of digestion begins. Chewing, or mastication, breaks down food into smaller particles, facilitating digestion and nutrient absorption further along the digestive tract [279]. Additionally, the mouth contains salivary glands that produce saliva, which not only moistens food to aid in swallowing but also contains enzymes that initiate the breakdown of carbohydrates (Pedersen et al. [280]). Beyond its digestive functions, the mouth is intricately connected to the respiratory system through the oral and nasal cavities. Proper breathing relies on unobstructed airflow through the mouth and nose, highlighting the mouth's role in oxygen intake and gas exchange [281]. Moreover, the mouth serves as a primary entry point for pathogens, bacteria, and foreign substances through food consumption and individual ways of function. Its warm and moist environment provides an ideal breeding ground for microorganisms, making oral hygiene crucial for preventing infections and maintaining overall health (Deo and Deshmukh 2019) [282].

Poor oral hygiene can lead to various oral health issues such as dental caries, periodontal diseases, and oral infections, which can have systemic implications, as shown by Bhatnagar [283]. Diseases of the oral cavity affect approximately 50% of the global population, which accounts for around 3.5 billion people, making them the most prevalent health condition worldwide. Among the most widespread oral diseases are untreated dental caries of both deciduous and permanent teeth, severe periodontal disease, oral candidiasis, and oral cancer (WHO, [284]). Oral health issues can have significant implications for individuals, including pain, discomfort, difficulty eating and speaking, and in severe cases, even life-threatening conditions such as oral cancer, as reported by Barranca-Enríquez and Romo-González [285]. Moreover, poor oral health can impact overall well-being, leading to social and psychological consequences, reduced quality of life, and increased healthcare costs, as highlighted by de Abreu et al. [286].

For these reasons, the connection between oral health and systemic health has been the scope of serious research investigations in the last two decades [287]). Oral health has been associated with various systemic conditions, such as cardiovascular disease, diabetes, respiratory infections, adverse pregnancy outcomes, high blood pressure, pulmonary diseases, low birth weight, Alzheimer's disease, osteoporosis, and rheumatoid arthritis [288]. Periodontal disease, characterized by inflammation and infection of the gums and supporting structures of the teeth, has been identified as a potential risk factor for the development and progression of systemic diseases such as cardiovascular disease, diabetes, and cancer [289]). Romandini et al. (2021), [290] reported that individuals with periodontal disease are 3.11 times more likely to experience mortality from stroke, 2.58 times more likely from cardiovascular disease, 1.67 times more likely from diabetes, and 1.38 times more likely from cancer. Such reports highlight the significant association between periodontal disease and various systemic conditions, indicating that oral health is closely linked to overall health outcomes [291]). This connection extends beyond dental hygiene practices to encompass broader lifestyle factors, including dietary habits. Poor dietary choices, such as high sugar intake and low nutrient consumption, can contribute to the development and progression of periodontal disease and dental caries, as reported by Dimopoulou et al. [292]. Sugary foods and beverages promote the growth of harmful bacteria in the mouth, leading to plaque formation and gum inflammation, as reported also by Antoniadou and Varzakas [138] and Pang et al. [293]. Additionally, a lack of essential nutrients, such as vitamins C and D, calcium, and antioxidants, compromises the body's ability to fight oral infections and maintain healthy gum tissue [294]). Conversely, a balanced diet rich in fruits, vegetables, lean proteins, and whole grains can support oral health by providing essential nutrients and antioxidants [295,296]. For example, foods high in vitamin C, such as citrus fruits and leafy greens, promote gum healing and reduce inflammation. Dairy products fortified with calcium and vitamin D contribute to strong teeth and bones, while lean proteins facilitate tissue repair and maintenance [138,297]. Therefore, promoting a nutritious diet as part of oral health education and preventive care initiatives is essential for reducing the incidence and severity of periodontal disease.

Moreover, dental caries, primarily caused by the consumption of free sugars, is a prevalent and costly disease with significant implications for general health and quality of life [298]. It was and still is a major public health problem globally and is the most widespread noncommunicable disease (NCD) [299,300]. It is also the most prevalent condition included in the 2015 Global Burden of Disease Study, ranking first for the decay of permanent teeth (2.3 billion people) and twelfth for deciduous teeth (560 million children) [301]. Dental caries is a leading reason for tooth extraction, which can exacerbate chronic pain and contribute to systemic infections [302]. Moreover, severe dental caries disproportionately affects vulnerable populations, including children and those in low- and middle-income countries, highlighting the need for accessible preventive and treatment measures [303]. Sugar-sweetened beverages, such as fruit-based and milk-based drinks, along with 100% fruit juices, confectioneries, cakes, biscuits, sweetened cereals, desserts, sucrose, honey, syrups, and preserves, serve as primary sources of free sugars in many countries, putting the base for the expansion of the disease [298]. To control the risk of dental caries across all stages of life, it is recommended to limit free sugar intake to less than 10% of total energy intake [292]. Ideally, reducing this intake further to less than 5% offers additional protection against dental caries [304].

If healthcare professionals emphasize the importance of dietary choices in maintaining oral hygiene and overall health, individuals can take proactive steps to safeguard against systemic conditions linked to poor oral health, such as cardiovascular disease, diabetes, and cancer [292]. This is the first step to a wide interdisciplinary approach to controlling oral diseases and other systematic ones that are reducing the sustainability of the human ecosystem [286,288,305]. Recent findings underscore the importance of maintaining good oral hygiene and seeking timely treatment for periodontal disease to reduce the risk of adverse health outcomes [306]. Moreover, they emphasize the need for interdisciplinary collaboration between dental and medical professionals to address the bidirectional relationship between oral health and systemic health effectively [307]. Early detection and management of periodontal disease may contribute to improving overall health and reducing the burden of chronic diseases globally, and the prevention of dental caries may increase quality-adjusted life expectancy [306,308,309].

All relevant data highlight the mouth's significance as a potential site for the transmission of infections and inflammation to other parts of the body [310–312]. More specifically, the study of the oral microbiome has increasingly unveiled its critical role in both oral and systemic diseases, shedding light on complex interactions within the human body [310,313]. Recent research, such as that conducted by Peng et al. [310], has emphasized the significance of oral microbiota in systemic diseases, highlighting the interconnectedness between oral health and overall well-being. So, understanding the fundamentals of the oral microbiome has become essential for oral healthcare professionals [313–315]. Moreover, investigations into bacteriophages in the oral cavity have revealed their potential roles in oral health and disease [316,317]. This microbiome complexity extends beyond the oral cavity, influencing various systemic conditions such as inflammatory bowel disease [318–320], non-alcoholic fatty liver disease [321–323], diabetes [324–326], and even neurodegenerative disorders like Alzheimer's and Parkinson's disease [327–329]. Additionally, associations have been found between the oral microbiome and conditions like depression, anxiety, and suicidal ideation [330–333] and cardiac diseases [334,335]. The implications of the oral microbiome extend to reproductive health, with evidence linking it to adverse pregnancy outcomes [336–338], while it has been reported that periodontal treatment improves prostate symptoms and lowers serum PSA in men with high PSA and chronic periodontitis [339]. Furthermore, its involvement in autoimmune diseases [340,341], autoimmune hepatitis [342], bowel disease [343], respiratory conditions [344,345], and even cancer development [346–348] underscores its systemic impact. Saliva, as a readily accessible diagnostic medium, has emerged as a valuable tool in understanding oral and systemic health [349] and body composition in early childhood [350], with ongoing research focusing on its potential for point-of-care testing and disease monitoring [351,352].

Furthermore, the mouth plays a vital role in communication and expression, serving as the primary organ for speech and articulation. The tongue, lips, and palate work together to produce sounds and convey meaning through language, facilitating social interaction and emotional expression [353]. Any disruptions or abnormalities in oral structures can impact speech clarity and communication abilities. Maintaining good oral hygiene and addressing oral health issues promptly is essential for overall well-being and systemic health [354]. Serious work has been performed also on the impact of tooth loss and edentulism on the quality of life of affected individuals [355,356]. Edentulism, the condition of being without teeth, can have profound physical, psychological, and social consequences, significantly affecting an individual's overall well-being [357]. There are multiple challenges faced by patients with edentulism, including difficulties in eating and speaking, compromised aesthetics, and decreased self-esteem. These challenges can lead to social isolation, anxiety, and depression, further exacerbating the negative impact on quality of life [138,295]. It is suggested that the needs of edentulous patients should be addressed through comprehensive oral healthcare interventions [138,357]. This includes providing access to dental prostheses, such as dentures, to restore oral function and aesthetics. Additionally, psychological support and counseling may be necessary to help patients cope with the emotional and social consequences of tooth loss. In Table 5, the significance of oral health in the food chain is presented.

Table 5. Significance of oral health in the food chain adapted from [358].

Aspect of Oral Health	Significance
Digestion	Begins in the mouth through chewing (mastication), which breaks down food into smaller particles, facilitating digestion and nutrient absorption. Saliva, produced by salivary glands, contains enzymes that initiate carbohydrate breakdown.
Respiratory system	The mouth and nasal cavities are interconnected with the respiratory system, facilitating proper breathing and oxygen intake.
Pathogen defense	The mouth acts as a primary entry point for pathogens, bacteria, and foreign substances, making oral hygiene crucial for preventing infections and maintaining overall health.
Oral health issues and systemic health	Poor oral hygiene can lead to various oral health issues such as dental caries, periodontal diseases, and oral infections, which can have systemic implications for overall health.
Interdisciplinary collaboration	Emphasizes the need for collaboration between dental and medical professionals to effectively address the bidirectional relationship between oral health and systemic health.
Oral microbiome	The oral microbiome plays a critical role in both oral and systemic diseases, influencing various conditions such as cardiovascular disease, diabetes, and even neurodegenerative disorders.
Saliva as a diagnostic medium	Saliva serves as a valuable tool for understanding oral and systemic health, with ongoing research focusing on its potential for point-of-care testing and disease monitoring.

The above synoptic report suggests that oral health stands as a fundamental component of holistic well-being, encompassing both physical and spiritual dimensions, and serves as the gateway to overall health [284,358].

16. Economic Implications and Ethics in the Prevention and/or Provision of Oral Health

The economic implications of oral health extend far beyond individual well-being to encompass broader societal and financial dimensions [359,360]. Listl et al. (2015) [361]. highlight the substantial economic burden of oral diseases, including direct treatment costs, productivity losses from missed school and workdays, and diminished quality of life. In 2015 alone, dental diseases worldwide amounted to approximately USD 545 billion in total costs, with significant direct and indirect expenses [362]. Particularly in high-income countries like the United States, oral health issues result in considerable productivity losses comparable to those associated with musculoskeletal disorders [363–366]. Moreover, oral diseases exacerbate the impacts of other conditions such as diabetes, yet effective periodontal treatment has been shown to reduce overall healthcare costs [367]. Importantly, out-of-pocket dental expenses can push economically vulnerable families towards poverty [368], while inadequate access to dental care contributes to the inappropriate use of emergency departments and physician offices [369–371].

The study conducted by Jeffcoatt et al. (2014), [372] sheds light on the significant economic benefits that could be achieved by addressing periodontal disease in patients with systemic conditions such as diabetes, cardiovascular diseases, and a history of stroke. According to their findings, treating periodontal disease could result in a substantial annual economic benefit for patients with these conditions, including a reduction in healthcare utilization by 40.2% for diabetic patients, 10.7% for patients with cardiovascular diseases, and 40.9% for patients with a history of stroke. Moreover, there could be a 67% decrease in the need for hospitalization and a 54% decrease in the need for emergency care, leading to a lighter economic burden on the healthcare system.

The implications of these findings are profound, as they suggest that investing in periodontal treatment not only improves oral health but also contributes to overall systemic health and reduces healthcare costs. Avalare Health LLC, based in Washington DC [373], projected that the expenses for periodontal treatment from 2016 to 2025 would amount to USD 7.2 billion. However, by reducing healthcare expenses related to systemic diseases by USD 70.7 billion, there could be a net gain of USD 63.5 billion. This surplus could be directed towards enhancing healthcare infrastructure, hospitals, equipment, and staffing, as well as investing in wellness initiatives and resilience-building for both patients and healthcare systems. We can imagine the amounts saved if we could provide even more preventive care for dental caries control worldwide, which is the most prevalent disease of all (Heng 2016; NIH; Kruk et al. 2018) [359,374,375].

On this issue, the WHO's Global Oral Health Status Report [376] provides a comprehensive overview of oral disease burden and serves as a reference for policymakers and stakeholders to prioritize oral health on global, regional, and national agendas. The WHO 2021 proposal [377] on oral health aims to address the global burden of dental diseases through a multifaceted approach. It emphasizes preventive measures, such as regular dental check-ups, fluoride treatments, and oral hygiene education for both caries and periodontal disease control, to reduce the prevalence of dental issues. Universal access to essential oral healthcare services is advocated for ensuring equitable access for all individuals regardless of socioeconomic status. The integration of oral health services into primary healthcare systems is promoted to deliver holistic healthcare. Community water fluoridation programs are encouraged for their effectiveness in preventing dental caries. Health promotion strategies are highlighted to raise awareness about oral health and encourage healthy behaviors. Continuous data collection and monitoring systems are also necessary to track oral health indicators and guide policy development [378]. Collaboration among governments, healthcare providers, NGOs, and other stakeholders is emphasized to implement comprehensive oral health strategies effectively. Finally, strict decision-making processes by international organizations should be followed urgently by actions globally and nationally [379].

17. Sustainability Issues in the Food Chain until the Mouth Gateway

The intersection of sustainability, oral health, and food ethics is an area of growing interest and importance. Sustainable practices in oral healthcare sectors involve not only addressing the immediate needs of patients but also considering the long-term implications for both individual health and the environment. For example, promoting preventive measures such as regular dental check-ups and oral hygiene education not only prevents oral diseases but also reduces the need for more invasive and resource-intensive treatments in the future [376]. Additionally, adopting sustainable practices in dental clinics, such as minimizing waste and using eco-friendly materials, contributes to environmental conservation efforts.

Furthermore, food ethics play a crucial role in oral health and overall well-being. A diet rich in nutritious foods not only supports oral health by providing essential nutrients for gum and tooth health but also promotes systemic health and reduces the risk of chronic diseases (Clemente-Suárez et al. 2023; Kalpe et al. 2023) [296,380]. Encouraging sustainable food choices, such as locally sourced and organic produce, not only benefits individual health but also supports ethical food production practices and reduces the carbon footprint associated with food transportation and processing (van Bussel et al. 2022) [381]. We could also enhance periodontal and oral health, enforcing research to evaluate the clinical, microbiological, and immunological effects of probiotic supplementation and other food derivatives for preventing and treating periodontal diseases (Gheisary et al. 2022) [382] and dental caries (Voidarou et al. 2022) [383]. Finally, we could suggest further research into the development and characterization of an oral microbiome transplant based on food extracts as a novel treatment approach for dental caries and periodontal disease (Nath et al. (2021)) [384].

In terms of legislation and information-sharing initiatives, there is a pressing need for comprehensive policies that recognize the intricate relationship between oral health, systemic health, and sustainability. Legislative measures should aim to incentivize preventive dental care, advocate for community water fluoridation, and implement regulations to minimize the environmental footprint of dental practices. Concurrently, information-sharing initiatives ought to focus on raising awareness about the pivotal role of oral health in overall systemic well-being, promoting sustainable oral healthcare practices in both healthcare professionals and the public, and fostering collaboration between dental and medical sectors to deliver holistic patient care (Glick et al. 2023; Fisher et al. 2023) [385,386]. Addressing oral diseases not only enhances oral health outcomes but also brings about substantial economic benefits and contributes to broader systemic health and sustainability goals (van Bussel et al. 2022) [381]. Investing in preventive oral healthcare practices, advocating for sustainable dietary choices, and implementing legislative and information-sharing initiatives is the way to collectively forge a healthier and more sustainable future for individuals, communities, and the planet. The significant economic advantages of improving oral health and ensuring equitable access to dental care underscore the critical importance of addressing oral health disparities and inequities within communities (Bhatnagar, 2021; Kruk 2021) [358,359].

Reflecting on over 2500 years since Hippocrates articulated his four pillars of medical ethics (Jhala and Jhala 2012) [387], it is evident that these principles continue to serve as fundamental guides for medical practitioners worldwide. The enduring relevance of "Primum non nocere", emphasizing the duty to avoid harm, "Beneficence", urging action in the patient's best interest, "Autonomy", recognizing patient self-determination, and "Justice", ensuring equitable healthcare, highlights their timeless significance in ethical medical practice. However, as medical knowledge and technology progress, challenges emerge in interpreting and applying these principles, particularly in areas such as nutritional support and end-of-life care (Jones, 2010) [388]. The ethical dilemma of respecting patients' wishes regarding nutrition, especially in cases of cognitive impairment or terminal illness, underscores the importance of careful ethical decision-making processes that prioritize patient well-being while acknowledging medical limitations (Cardenas 2021) [389]. Embracing a holistic approach to food consumption, individuals can honor the sanctity of life through mindful selection and consumption, acknowledging the interconnectedness of dietary choices, environmental impact, and personal well-being [390]. In such a case, the quality and quantity of what comes in through the mouth concerning the energy profile represent what is expected to come out in the form of health quality, expression, communication quality, positive choices and actions, exceptional energy performance, and ethical decisions, as there seems to be a fundamental role of sensory experiences in shaping our ethical beliefs and behaviors [391]). Overall, by honoring life through mindful oral practices and food choices, individuals not only contribute to a more ethical and sustainable food ecosystem but also nurture their health and connection with the world around them, providing tools for sustainability for the human ecosystem. Christ has urged us to be unified, echoing the sentiment of "that they may be one" (John 17:21), signifying the interconnectedness and unity we should strive for with our Creator, ourselves, and our fellow citizens.

18. Limitations of the Study

This study possesses some limitations that need consideration. Firstly, its non-systematic approach to a literature review introduces the risk of bias in the selection and interpretation of studies. Without adhering to a structured methodology, there is a possibility of overlooking key literature, leading to a potential lack of comprehensiveness in coverage. Additionally, the broad scope of the topic may hinder the depth of analysis, potentially resulting in a superficial exploration of certain complex issues. Furthermore, the inclusion of literature may be limited by factors such as availability, accessibility, and language barriers, particularly from regions with limited research infrastructure. Consequently, there is a risk of representation bias, where certain perspectives or experiences are underrepresented. Moreover, the interpretation of findings may be influenced by subjective biases, impacting the objectivity and validity of conclusions drawn. Despite these limitations, transparent reporting and cautious interpretation have mainly diminished these concerns and enhanced this study's credibility.

19. Areas for Further Research

A future perspective of this study could be conducting longitudinal studies to assess the long-term impact of legislative interventions, such as the implementation of food safety laws or sustainability initiatives, which can provide valuable insights into their effectiveness over time. Additionally, investigating the socio-cultural factors influencing food-related behaviors and ethical decision-making across diverse populations and regions can offer a more extended understanding of ethical dilemmas in food consumption and production. Furthermore, examining the role of emerging technologies, such as blockchain or artificial intelligence, in enhancing transparency and accountability within the food supply chain might contribute to more robust ethical frameworks. Exploring the intersectionality of food ethics with other disciplines, such as public health, economics, and environmental science, can also shed light on interconnected challenges and opportunities for holistic solutions across the globe. Moreover, studying the ethical implications of alternative food production methods, such as lab-grown meat or insect protein, can inform debates surrounding sustainability and animal welfare. Lastly, investigating the relationship between food ethics, spirituality, and well-being can offer novel perspectives on the ethical dimensions of food consumption and its broader implications for human flourishing. Addressing these future research directions may help scholars contribute to the development of evidence-based policies and practices that promote ethical and sustainable food systems globally.

20. Conclusions

Ethical decisions in the food chain not only contribute to preventing food fraud and ensuring the safety and integrity of the food supply but also have broader implications for global health and well-being. Societies can foster healthier communities and promote social justice if they allocate resources saved from combating food fraud and unethical practices towards addressing hunger, controlling oral diseases, and managing systemic health issues.

Moreover, ethical behavior reflects a sense of responsibility towards oneself, others, and the greater good, aligning with moral and spiritual values that emphasize compassion and care for all living beings. In essence, being ethical in the food chain encompasses not only personal integrity but also a commitment to promoting human dignity and honoring the interconnectedness of all life, thereby reflecting reverence for the divine creator.

Author Contributions: Conceptualization, T.V. and M.A.; methodology, T.V. and M.A.; formal analysis, T.V. and M.A.; investigation, T.V. and M.A.; resources, T.V. and M.A.; data curation, T.V. and M.A.; writing—original draft preparation, T.V. and M.A.; writing—review and editing, T.V. and M.A.; visualization, T.V. and M.A.; supervision, T.V.; project administration, T.V.; funding acquisition, T.V. All authors have read and agreed to the published version of the manuscript.

Funding: This research received no external funding.

Institutional Review Board Statement: Not applicable.

Informed Consent Statement: Not applicable.

Data Availability Statement: No new data were created or analyzed in this study. Data sharing does not apply to this article.

Conflicts of Interest: The authors declare no conflicts of interest.

References

1. El Bilali, H.; Strassner, C.; Ben Hassen, T. Sustainable Agri-Food Systems: Environment, Economy, Society, and Policy. *Sustainability* **2021**, *13*, 6260. [CrossRef]
2. Bindi, L.; Belliggiano, A. A Highly Condensed Social Fact: Food Citizenship, Individual Responsibility, and Social Commitment. *Sustainability* **2023**, *15*, 6881. [CrossRef]
3. Lugo-Morin, D.R. Ethical Perspectives on Food Morality: Challenges, Dilemmas and Constructs. *Food Ethic* **2024**, *9*, 9. [CrossRef]
4. Varzakas, T.; Zakynthinos, G.; Proestos, C. Chapter 10—Effect of food processing, quality, and safety with emphasis on kosher, halal, vegetarian, and GM food. In *Preparation and Processing of Religious and Cultural Foods*; Woodhead Publishing Series in Food Science, Technology and Nutrition; Ali, M.E., Nizar, N.N.A., Eds.; Elsevier: Amsterdam, The Netherlands, 2018; pp. 193–214.
5. Monterrosa, E.C.; Frongillo, E.A.; Drewnowski, A.; De Pee, S.; Vandevijvere, S. Sociocultural influences on food choices and implications for sustainable healthy diets. *Food Nutr. Bull.* **2020**, *41* (Suppl. 2), 59S–73S. [CrossRef] [PubMed]
6. Wei, Y. Human Morality: From Evolutionary to Future Perspectives. *Commun. Humanit. Res.* **2023**, *9*, 58–63. [CrossRef]
7. Askegaard, S.; Ordabayeva, N.; Chandon, P.; Cheung, T.; Chytkova, Z.; Cornil, Y.; Corus, C.; Edell, J.A.; Mathras, D.; Junghans, A.F.; et al. Moralities in food and health research. *J. Mark. Manag.* **2014**, *30*, 1800–1832. [CrossRef]
8. Hernandez, E.; Llonch, P.; Turner, P.V. Applied Animal Ethics in Industrial Food Animal Production: Exploring the Role of the Veterinarian. *Animals* **2022**, *12*, 678. [CrossRef] [PubMed]
9. Aldaco, R.; Hoehn, D.; Laso, J.; Margallo, M.; Ruiz-Salmón, J.; Cristobal, J.; Kahhat, R.; Villanueva-Rey, P.; Bala, A.; Batlle-Bayer, L.; et al. Food waste management during the COVID-19 outbreak: A holistic climate, economic and nutritional approach. *Sci. Total. Environ.* **2020**, *742*, 140524. [CrossRef] [PubMed]
10. Ni, D.; Smyth, H.; Cozzolino, D.; Gidley, M. Holistic approach to effects of foods, human physiology, and psychology on food intake and appetite (satiation & satiety). *Crit. Rev. Food Sci. Nutr.* **2022**. [CrossRef]
11. Hargreaves, S.M.; Raposo, A.; Saraiva, Z.; Puppin, R. Vegetarian diet: An overview through the perspective of quality-of-life domains. *Int. J. Environ. Res. Public Health* **2021**, *18*, 4067. [CrossRef]
12. Traverso-Yepez, M.; Hunter, K. From "Healthy Eating" to a Holistic Approach to Current Food Environments. *SAGE Open* **2016**, *6*, 2158244016665891. [CrossRef]
13. Borgdorf, L. Josh Milburn's just fodder: The ethics of feeding animals. *Food Ethics* **2024**, *9*, 10. [CrossRef]
14. Milburn, J. Relational animal ethics (and why it isn't easy). *Food Ethic* **2024**, *9*, 6. [CrossRef]
15. Varzakas, T.; Smaoui, S. Global Food Security and Sustainability Issues: The Road to 2030 from Nutrition and Sustainable Healthy Diets to Food Systems Change. *Foods* **2024**, *13*, 306. [CrossRef] [PubMed]
16. Ruben, R.; Cavatassi, R.; Lipper, L.; Smaling, E.; Winters, P. Towards food systems transformation—Five paradigm shifts for healthy, inclusive and sustainable food systems. *Food Secur.* **2021**, *13*, 1423–1430. [CrossRef] [PubMed]
17. Stein, A.; Santini, F. The sustainability of local food: A review for policy-makers. *Rev. Agric. Food Environ. Stud.* **2021**, *103*, 77–89. [CrossRef] [PubMed]
18. Carolan, M. Ethical eating as experienced by consumers and producers: When good food meets good farmers. *J. Consum. Cult.* **2022**, *22*, 103–123. [CrossRef]
19. Betancur, G.E. La ιtica y la moral: Paradojas del ser humano. *Rev. CES Psicol.* **2016**, *9*, 109–121. [CrossRef]
20. Dasuky, S. *Cuatro Versiones de la Itica y la Moral [Notas de Clase]*; Universidad Pontificia Bolivariana: Medellín, Colombia, 2010. Available online: https://www.upb.edu.co/es/home (accessed on 12 March 2024).

21. Huxley, T.H. *Evolution and Ethics*; D. Appleton and Company: New York, NY, USA, 1911.
22. Hauser, M. *Moral Minds: The Nature of Right and Wrong*; Harper Perennial: New York, NY, USA, 2006.
23. de Waal, F.B.M.; Aureli, F. Consolation, reconciliation, and a possible cognitive difference between macaques and chimpanzees. In *Reaching into Thought: The Minds of the Great Apes*; Russon, A.E., Bard, K.A., Parker, S.T., Eds.; Cambridge University Press: Cambridge, UK, 1996; pp. 80–110.
24. Shaw, J.C. *Plato's Anti-Hedonism and the Protagoras*; Cambridge University Press: Cambridge, UK, 2015; ISBN 9781107046658.
25. Lazari-Radek, K.D.; Singer, P. Origins. In *Utilitarianism: A Very Short Introduction*, online ed.; Oxford University Press: Oxford, UK, 2017. [CrossRef]
26. West, H.R. *An Introduction to Mill's Utilitarian Ethics*; Cambridge University Press: Cambridge, UK, 2003.
27. Stanford Encyclopedia of Philosophy. Kant's Moral Philosophy. First Published Monday 23 February 2004; Substantive Revision Friday 21 January 2022. Available online: https://plato.stanford.edu/entries/kant-moral/ (accessed on 25 March 2024).
28. Granja, D.M. Kant: Conciencia reflexiva y proceso humanizador. *Sociolσgica* **2004**, *19*, 213–226.
29. Perez-Cueto, F.J.A.; Olsen, A. The Multifaceted Dimensions of Food Choice and Nutrition. *Nutrients* **2020**, *12*, 502. [CrossRef]
30. Resnik, D.B. Moral Theory. In *The Ethics of Research with Human Subjects*; International Library of Ethics, Law, and the New Medicine; Springer: Cham, Switzerland, 2018; Volume 74. [CrossRef]
31. Atteshli-Theotoki, P. *The Great Voyage of the Soul. Part i & Your Wings to Fly*; The Stoa Series: Strovolos, Cyprus, 2023.
32. Botti, S.; Iyengar, S.; McGill, A. Choice freedom. *J. Consum. Psychol.* **2023**, *33*, 143–166. [CrossRef]
33. Chouraqui, J.P.; Turck, D.; Briend, A.; Darmaun, D.; Bocquet, A.; Feillet, F.; Frelut, M.L.; Girardet, J.P.; Guimber, D.; Hankard, R.; et al. Religious dietary rules and their potential nutritional and health consequences. *Int. J. Epidemiol.* **2021**, *50*, 12–26. [CrossRef] [PubMed]
34. Cruz, J.A. Derechos morales: Concepto y relevancia. *Isonomνa* **2001**, *15*, 55–79.
35. Kelsen, H. La doctrina del derecho natural y el positivismo juridico. *Acadιmica Rev. Sobre Enseρanza Derecho* **2008**, *12*, 183–198.
36. Lugo-Morin, D.R.; Frolich, L.; Magal-Royo, T. Relacion ser humano-naturaleza: Debatiendo el Desarrollo sos-tenible desde la filosofνa de la ciencia. *Eur. Sci. J.* **2015**, *11*, 1–12.
37. Lugo-Morin, D.R. Restoring the food systems resilience through the dialogue of knowledge: A case study from Mexico. *Forum Dev. Stud.* **2022**, *50*, 183–206. [CrossRef]
38. Gorichanaz, T. A Compass for What Matters: Applying Virtue Ethics to Information Behavior. *Open Inf. Sci.* **2023**, *7*, 20220151. [CrossRef]
39. Berti, E. *Las Razones de Aristσteles*; Oinos: Buenos Aires, Argentina, 2008.
40. Durant, W. *Historia de la Filosofνa: La Vida y el Pensamiento de los Mαs Grandes Filσsofos del Mundo*; Editorial Diana: Mexico City, Mexico, 1994.
41. Conroy, M.; Malik, A.Y.; Hale, C.; Weir, C.; Brockie, A.; Turner, C. Using practical wisdom to facilitate ethical decision-making: A major empirical study of phronesis in the decision narratives of doctors. *BMC Med. Ethic* **2021**, *22*, 16. [CrossRef] [PubMed]
42. Schroeder, N.B.A. Moral Dilemmas in Contemporary Virtue Ethics. A Thesis Submitted to the Graduate Faculty of the Louisiana State University and Agricultural and Mechanical College in Partial Fulfillment of the Requirements for the Degree of Master of Arts in the Department of Philosophy and Religious Studies University of New Orleans, 2008 May 2011. Available online: https://repository.lsu.edu/cgi/viewcontent.cgi?article=1640&context=gradschool_theses (accessed on 10 March 2024).
43. Smith, M.B. *Values, Self and Society: Toward a Humanist Social Psychology*; Routledge: New York, NY, USA, 2017. [CrossRef]
44. Weber, E.T. Religion, public reason, and humanism: Paul Kurtz on fallibilism and ethics. *Contemp. Pragmatism* **2008**, *5*, 131–147. [CrossRef]
45. Giannouli, V. Business Ethics in Healthcare: The Case of Greece. In *Research Anthology on Business Law, Policy, and Social Responsibility*; IGI Global: Hershey, PA, USA, 2014; pp. 115–137. [CrossRef]
46. Mayo, E. *Values: How to Bring Values to Life in Your Business*; Routledge: New York, NY, USA, 2017.
47. *Philosophical and Sociological Dictionary*; Kapopoulos Publishing: Athens, Greece, 1995; Volume 1.
48. Diamantopoulos, D. *Modern Dictionary of the Basic Concept of the Material-Technical, Spiritual and Ethical Civilization*; Patakis: Athens, Greece, 2002.
49. Mohammadi, A.; Vanaki, Z.; Memarian, R.; Fallahrafie, R.A. Islamic and Western ethical values in health services management: A comparative study. *Int. J. Nurs. Knowl.* **2019**, *30*, 239–250.
50. Jackson, T. Chapter 15—Managing managers across cultures: Different values, different ethics. In *Handbook of Research on Managing Managers*; Wil-kinson, A., Townsend, K., Suder, G., Eds.; Edward Elgar Publishing: Cheltenham, UK, 2015; pp. 283–306. [CrossRef]
51. Pletz, J. *Being Ethical*; Nova Science Publishers Inc.: Hauppauge, NY, USA, 1999.
52. Harris, J.R. Ethical values of individuals at different levels in the organizational hierarchy of a single firm. *J. Bus. Ethic* **1990**, *9*, 741–750. [CrossRef]
53. Kanungo, R.N. Ethical values of transactional and transformational leaders. *Can. J. Adm. Sci./Rev. Can. Sci. l'Administration* **2001**, *18*, 257–265. [CrossRef]
54. Landau, R.; Osmo, R. Professional and personal hierarchies of ethical principles. *Int. J. Soc. Welf.* **2003**, *12*, 42–49. [CrossRef]
55. Donnelly, J. *Universal Human Rights in Theory and Practice*; Cornell University Press: Ithaca, NY, USA, 2013.
56. Schwartz, S.H.; Bardi, A. Value hierarchies across cultures. *J. Cross-Cultural Psychol.* **2001**, *32*, 268–290. [CrossRef]

57. de Spinoza, B. *A Spinoza Reader: The Ethics and Other Works*; Curley, E., Ed.; Princeton University Press: Princeton, NJ, USA, 1994.
58. Deprá, M.C.; Dias, R.R.; Zepka, L.Q.; Jacob-Lopes, E. Building cleaner production: How to anchor sustainability in the food production chain? *Environ. Adv.* **2022**, *9*, 100295. [CrossRef]
59. Niranjan, V.R.; Kathuria, V.; Venkatraman, J.; Salve, A. Oral Health Promotion: Evidences and Strategies. In *Insights into Various Aspects of Oral Health*; Manakil, J., Ed.; IntechOpen: London, UK, 2017. [CrossRef]
60. Viles, E.; Kalemkerian, F.; Garza-Reyes, J.A.; Antony, J.; Santos, J. Theorizing the Principles of Sustainable Production in the context of Circular Economy and Industry 4.0. *Sustain. Prod. Consum.* **2022**, *33*, 1043–1058. [CrossRef]
61. Mastos, T.; Gotzamani, K. Sustainable Supply Chain Management in the Food Industry: A Conceptual Model from a Literature Review and a Case Study. *Foods* **2022**, *11*, 2295. [CrossRef] [PubMed]
62. Mardian, Y.; Shaw-Shaliba, K.; Karyana, M.; Lau, C.-Y. Sharia (Islamic Law) Perspectives of COVID-19 Vaccines. *Front. Trop. Dis.* **2021**, *2*, 788188. [CrossRef]
63. Nungesser, F.; Winter, M. *Meat and Social Change*; Springer: Berlin/Heidelberg, Germany, 2021; Volume 46, pp. 109–124. [CrossRef]
64. Murcott, A. The cultural significance of food and eating. *Proc. Nutr. Soc.* **1982**, *41*, 203–210. [CrossRef]
65. Tranchant, J.-P.; Gelli, A.; Bliznashka, L.; Diallo, A.S.; Sacko, M.; Assima, A.; Siegel, E.H.; Aurino, E.; Masset, E. The impact of food assistance on food insecure populations during conflict: Evidence from a quasi-experiment in Mali. *World Dev.* **2019**, *119*, 185–202. [CrossRef]
66. Mary, S. A replication note on humanitarian aid and violence. *Empir. Econ.* **2022**, *62*, 1465–1494. [CrossRef]
67. Koppenberg, M.; Mishra, A.K.; Hirsch, S. Food Aid and Violent Conflict: A Review of Literature. IZA DP No. 16574. November 2023. Available online: https://docs.iza.org/dp16574.pdf (accessed on 12 April 2024).
68. Nunn, N.; Qian, N. US Food Aid and Civil Conflict. *Am. Econ. Rev.* **2014**, *104*, 1630–1666. [CrossRef]
69. Mary, S.; Mishra, A.K. Humanitarian food aid and civil conflict. *World Dev.* **2020**, *126*, 104713. [CrossRef]
70. Nunn, N.; Qian, N. The Determinants of Food-Aid Provisions to Africa and the Developing World. In *African Successes Volume IV: Sustainable Growth*; Edwards, S., Johnson, S., Weil, D.N., Eds.; National Bureau of Economic Research: Cambridge, MA, USA, 2016; pp. 161–178.
71. Christian, P.; Barrett, C.B. *Revisiting the Effect of Food Aid on Conflict: A Methodological Caution*; World Bank: Washington, DC, USA, 2017.
72. Christian, P.; Barrett, C.B. Spurious Regressions and Panel IV Estimation: Revisiting the Causes of Conflict (January 1, 2021). Available online: https://ssrn.com/abstract=3950679 (accessed on 12 March 2024).
73. Feeley-Harnik, G. Religion and Food: An Anthropological Perspective[1]. *J. Am. Acad. Relig.* **1995**, *63*, 565–582. [CrossRef]
74. Casanova, J. Interdisciplinary Challenges to the Study of Religion. In *The Challenges of Religion*; Bäckström, A., Petterson, P., Eds.; Uppsala University: Uppsala, Sweden, 2010; pp. 33–37.
75. Cohen, A.B. You can learn a lot about religion from food. *Curr. Opin. Psychol.* **2020**, *40*, 1–5. [CrossRef]
76. Markowitz, F.; Avieli, N. Eating religiously: Food and faith in the 21st century. *Food Cult. Soc.* **2022**, *25*, 640–646. [CrossRef]
77. Pomeranz, J.L.; Brownell, K.D. Portion sizes and beyond—Government's legal authority to regulate food-industry practices. *N. Engl. J. Med.* **2012**, *367*, 1383–1385. [CrossRef] [PubMed]
78. Lopez-Garcia, R. Food law in Mexico: Regulatory framework and public policy strategies to address the obesity crisis in Latin America. In *International Food Law and Policy*; Steier, G., Patel, K., Eds.; Springer: Cham, Switzerland, 2016. [CrossRef]
79. Sarri, K.O.; Linardakis, M.K.; Bervanaki, F.N.; Tzanakis, N.E.; Kafatos, A.G. Greek Orthodox fasting rituals: A hidden characteristic of the Mediterranean diet of Crete. *Br. J. Nutr.* **2004**, *92*, 277–284. [CrossRef]
80. Varzakas, T.; Kandylis, P.; Dimitrellou, D.; Salamoura, C.; Zakynthinos, G.; Proestos, C. Chapter 6—Innovative and fortified food: Probiotics, prebiotics, GMOs, and superfood. In *Preparation and Processing of Religious and Cultural Foods*; Woodhead Publishing Series in Food Science, Technology and Nutrition; Ali, M.E., Nizar, N.N.A., Eds.; Elsevier: Amsterdam, The Netherlands, 2018; pp. 67–129. [CrossRef]
81. Meigs, A. *Food, Sex and Pollution: A New Guinea Religion*; Rutgers University Press: New Brunswick, NJ, USA, 1991.
82. Hopkins, E.W. The Buddhistic Rule against Eating Meat. *J. Am. Orient. Soc.* **1906**, *27*, 455. [CrossRef]
83. Sarao, K.T.S. Buddhist attitude towards meat-eating. *Tibet. J.* **2008**, *33*, 91–98.
84. Nath, J. 'God is a vegetarian': The food, health and bio-spirituality of Hare Krishna, Buddhist and Seventh-Day Adventist devotees. *Health Sociol. Rev.* **2010**, *19*, 356–368. [CrossRef]
85. Testoni, I.; Ghellar, T.; Rodelli, M.; De Cataldo, L.; Zamperini, A. Representations of death among Italian vegetarians: An ethnographic research on environment, disgust and transcendence. *Eur. J. Psychol.* **2017**, *13*, 378–395. [CrossRef] [PubMed]
86. Suzworksy, R. From the marketplace to the dinner plate: The economy, theology, and factory farming. *J. Bus. Ethics* **2001**, *29*, 177.
87. McPherson, T. The ethical basis for veganism. In *The Oxford Handbook of Food Ethics*; Oxford University Press: Oxford, UK, 2018; Chapter 10; pp. 210–215.
88. Thun, M. Veganism in a Post-Modern Society. *Crit. Reflect. A Stud. J. Contemp. Sociol. Issues* **2019**. Available online: https://ojs.leedsbeckett.ac.uk/index.php/SOC/article/view/4575/4384 (accessed on 12 April 2024).
89. Jovandaric, M. Veganism: A New Approach to Health. In *Veganism—A Fashion Trend or Food as a Medicine*; IntechOpen: London, UK, 2021; Chapter 1; pp. 1–3. [CrossRef]
90. Michopoulou, E.; Jauniškis, P. Exploring the relationship between food and spirituality: A literature review. *Int. J. Hosp. Manag.* **2020**, *87*, 102494. [CrossRef]

91. Arnett, L.D. The Soul: A Study of Past and Present Beliefs. *Am. J. Psychol.* **1904**, *15*, 121–200. [CrossRef]
92. Buser, J.K.; Parkins, R.A.; Buser, T.J. Thematic analysis of the intersection of spirituality and eating disorder symptoms. *J. Addict. Offender Couns.* **2014**, *35*, 97–113. [CrossRef]
93. Reicks, M.; Mills, J.; Henry, H. Qualitative study of spirituality in a weight loss program: Contribution to self-efficacy and locus of control. *J. Nutr. Educ. Behav.* **2004**, *36*, 13–19. [CrossRef] [PubMed]
94. Patel, R.; Lycett, D.; Coufopoulos, A.; Turner, A. A feasibility study of taste & see: A church-based programme to develop a healthy relationship with food. *Religions* **2017**, *8*, 29. [CrossRef]
95. Patel, R.; Lycett, D.; Coufopoulos, A.; Turner, A. Moving forward in their journey: Participants' experience of taste & see, a church-based programme to develop a healthy relationship with food. *Religions* **2017**, *8*, 14. [CrossRef]
96. Lafrance, A.; Loizaga-Velder, A.; Fletcher, J.; Renelli, M.; Files, N.; Tupper, K.W. Nourishing the spirit: Exploratory research on ayahuasca experiences along the continuum of recovery from eating disorders. *J. Psychoact. Drugs* **2017**, *49*, 427–435. [CrossRef]
97. Savoldi, R.; Polari, D.; Pinheiro-da-Silva, J.; Silva, P.F.; Lobao-Soares, B.; Yonamine, M.; Freire, F.A.; Luchiari, A.C. Behavioral Changes Over Time Following Ayahuasca Exposure in Zebrafish. *Front. Behav. Neurosci.* **2017**, *11*, 139. [CrossRef] [PubMed]
98. Domínguez-Clavé, E.; Soler, J.; Pascual, J.C.; Elices, M.; Franquesa, A.; Valle, M.; Alvarez, E.; Riba, J. Ayahuasca improves emotion dysregulation in a community sample and in individuals with borderline-like traits. *Psychopharmacology* **2019**, *236*, 573–580. [CrossRef] [PubMed]
99. Uthaug, M.V.; van Oorsouw, K.; Kuypers, K.P.C.; van Boxtel, M.; Broers, N.J.; Mason, N.L.; Toennes, S.W.; Riba, J.; Ramaekers, J.G. Sub-acute and long-term effects of ayahuasca on affect and cognitive thinking style and their association with ego dissolution. *Psychopharmacology* **2018**, *235*, 2979–2989. [CrossRef] [PubMed]
100. Re, T.; Palma, J.; Martins, J.; Simões, M. Transcultural perspective on consciousness: Traditional use of ayahuasca in psychotherapy in the 21st Century in Western World. *Cosm. Hist. J. Nat. Soc. Philos.* **2016**, *12*, 237–249.
101. Dell, M.L.; Josephson, A.M. Religious and spiritual factors in childhood and adolescent eating disorders and obesity. *South. Med. J.* **2007**, *100*, 628–632. [CrossRef] [PubMed]
102. Matusek, J.A.; Knudson, R.M. Rethinking recovery from eating disorders: Spiritual and political dimensions. *Qual. Health Res.* **2009**, *19*, 697–707. [CrossRef] [PubMed]
103. Rodríguez-Martín, B.C.; Gallego-Arjiz, B. Overeaters anonymous: A mutual-help fellowship for food addiction recovery. *Front. Psychol.* **2018**, *9*, 1491. [CrossRef] [PubMed]
104. Exline, J.J.; Homolka, S.J.; Harriott, V.A. Divine struggles: Links with body image concerns, binging, and compensatory behaviours around eating. *Ment. Health Relig. Cult.* **2016**, *19*, 8–22. [CrossRef]
105. Tan, M.-M.; Chan, C.K.; Reidpath, D.D. Faith, food and fettle: Is individual and neighborhood religiosity/spirituality associated with a better diet? *Religions* **2014**, *5*, 801–813. [CrossRef]
106. Tan, M.-M.; Chan, C.K.Y.; Reidpath, D.D. Religiosity, dietary habit, intake of fruit and vegetable, and vegetarian status among Seventh-Day Adventists in West Malaysia. *J. Behav. Med.* **2016**, *39*, 675–686. [CrossRef]
107. Dittmann, K.A.; Freedman, M.R. Body awareness, eating attitudes, and spiritual beliefs of women practicing yoga. *Eat. Disord.* **2009**, *17*, 273–292. [CrossRef] [PubMed]
108. Neumark-Sztainer, D.; Watts, A.W.; Rydell, S. Yoga and body image: How do young adults practicing yoga describe its impact on their body image? *Body Image* **2018**, *27*, 156–168. [CrossRef] [PubMed]
109. Gutierrez-Moya, E.; Adenso-Díaz, B.; Lozano, S. Analysis and vulnerability of the international wheat trade network. *Food Secur.* **2021**, *13*, 113–128. [CrossRef] [PubMed]
110. Nardi, V.A.M.; Auler, D.P.; Teixeira, R. Food safety in global supply chains: A literature review. *J. Food Sci.* **2020**, *85*, 883–891. [CrossRef] [PubMed]
111. Moosavi, J.; Fathollahi-Fard, A.M.; Dulebenets, M.A. Supply chain disruption during the COVID-19 pandemic: Recognising potential disruption management strategies. *Int. J. Disaster Risk Reduct.* **2022**, *75*, 102983. [CrossRef] [PubMed]
112. Kemper, J.A.; Kapetanaki, A.B.; Spotswood, F.; Roy, R.; Hassen, H.; Uzoigwe, A.G.; Fifita, 'I.M.E. Food practices adaptation: Exploring the coping strategies of low-socioeconomic status families in times of disruption. *Appetite* **2023**, *186*, 106553. [CrossRef] [PubMed]
113. Magano, N.N.; Tuorila, H.; De Kock, H.L. Food choice drivers at varying income levels in an emerging economy. *Appetite* **2023**, *189*, 107001. [CrossRef] [PubMed]
114. WHO. Food Safety. 19 May 2022. Available online: https://www.who.int/news-room/fact-sheets/detail/food-safety (accessed on 12 April 2024).
115. Fernandez, A.; Paoletti, C. What is unsafe food? Change of perspective. *Trends Food Sci. Technol.* **2021**, *109*, 725–728. [CrossRef]
116. Çakmakçı, R.; Salık, M.A.; Çakmakçı, S. Assessment and Principles of Environmentally Sustainable Food and Agriculture Systems. *Agriculture* **2023**, *13*, 1073. [CrossRef]
117. Ogiemwonyi, O.; Jan, M.T. The correlative influence of consumer ethical beliefs, environmental ethics, and moral obligation on green consumption behavior. *Resour. Conserv. Recycl. Adv.* **2023**, *19*, 200171. [CrossRef]
118. Shayan, N.F.; Mohabbati-Kalejahi, N.; Alavi, S.; Zahed, M.A. Sustainable Development Goals (SDGs) as a Framework for Corporate Social Responsibility (CSR). *Sustainability* **2022**, *14*, 1222. [CrossRef]
119. Rodríguez, A.V.; Rodríguez-Oramas, C.; Velázquez, E.S.; de la Torre, A.H.; Armendáriz, C.R.; Iruzubieta, C.C. Myths and Realities about Genetically Modified Food: A Risk-Benefit Analysis. *Appl. Sci.* **2022**, *12*, 2861. [CrossRef]

120. Scott, S.E.; Inbar, Y.; Wirz, C.D.; Brossard, D.; Rozin, P. An Overview of Attitudes Toward Genetically Engineered Food. *Annu. Rev. Nutr.* **2018**, *38*, 459–479. [CrossRef] [PubMed]
121. Rose, K.; Brossard, D.; Scheufele, D. Of Society, Nature, and Health: How Perceptions of Specific Risks and Benefits of Ge-netically Engineered Foods Shape Public Rejection. *Environ. Commun.* **2020**, *14*, 1017–1031. [CrossRef]
122. Telem, R.; Wani, S.; Singh, N.; Nandini, R.; Sadhukhan, R.; Bhattacharya, S.; Mandal, N. Cisgenics—A Sustainable Approach for Crop Improvement. *Curr. Genom.* **2013**, *14*, 468–476. [CrossRef]
123. Myskja, B.K. The Moral Difference between Intragenic and Transgenic Modification of Plants. *J. Agric. Environ. Ethic* **2006**, *19*, 225–238. [CrossRef] [PubMed]
124. Ankeny, R.; Bray, H. *Genetically Modified Food*; Oxford Handbooks Online: Oxford, UK, 2018.
125. Schouten, H.J.; Krens, F.A.; Jacobsen, E. Cisgenic plants are similar to traditionally bred plants: International regulations for genetically modified organisms should be altered to exempt cisgenesis. *EMBO Rep.* **2006**, *7*, 750–753. [CrossRef] [PubMed]
126. Laros, F.J.M.; Steenkamp, J.E.M. Importance of fear in the case of genetically modified food. *Psychol. Mark.* **2004**, *21*, 889–908. [CrossRef]
127. Lieberman, S.; Gray, T. GMOs and the Developing World: A Precautionary Interpretation of Biotechnology. *Br. J. Politics Int. Relat.* **2008**, *10*, 395–411. [CrossRef]
128. Deane-Drummond, C.; Grove-White, R.; Szerszynski, B. Genetically modified theology: The religious dimensions of public concerns about agricultural biotechnology. *Stud. Christ. Ethic* **2001**, *14*, 23–41. [CrossRef]
129. Gregorowius, D.; Lindemann-Matthies, P.; Huppenbauer, M. Ethical Discourse on the Use of Genetically Modified Crops: A Review of Academic Publications in the Fields of Ecology and Environmental Ethics. *J. Agric. Environ. Ethic* **2012**, *25*, 265–293. [CrossRef]
130. Kaptan, G.; Fischer, A.R.; Frewer, L.J. Extrapolating understanding of food risk perceptions to emerging food safety cases. *J. Risk Res.* **2018**, *21*, 996–1018. [CrossRef]
131. Mielby, H.; Sandøe, P.; Lassen, J. The role of scientific knowledge in shaping public attitudes to GM technologies. *Public Underst. Sci.* **2013**, *22*, 155–168. [CrossRef] [PubMed]
132. Robinson, E.L. Unnatural Selections: The Role of Institutional Confidence in Consumer Choice of Genetically Modified Food in the United States. Master's Thesis, Georgetown University, Washington, DC, USA, 2013. Available online: https://repository.library.georgetown.edu/bitstream/handle/10822/558604/Robinson_georgetown_0076M_12161.pdf?sequence=1&isAllowed=y (accessed on 12 April 2024).
133. Schuppli, C.A.; Molento, C.F.M.; Weary, D.M. Understanding attitudes towards the use of animals in research using an online public engagement tool. *Public Underst. Sci.* **2015**, *24*, 358–374. [CrossRef] [PubMed]
134. Wansink, B.; Tal, A.; Brumberg, A. Ingredient-based food fears and avoidance: Antecedents and antidotes. *Food Qual. Preference* **2014**, *38*, 40–48. [CrossRef]
135. Siddiqui, S.A.; Asif, Z.; Murid, M.; Fernando, I.; Adli, D.N.; Blinov, A.V.; Golik, A.B.; Nugraha, W.S.; Ibrahim, S.A.; Jafari, S.M. Consumer Social and Psychological Factors Influencing the Use of Genetically Modified Foods—A Review. *Sustainability* **2022**, *14*, 15884. [CrossRef]
136. Vilella-Vila, M.; Costa-Font, J.; Mossialos, E. Consumer involvement and acceptance of biotechnology in the European Union: A specific focus on Spain and the UK. *Int. J. Consum. Stud.* **2005**, *29*, 108–118. [CrossRef]
137. Kim, S.-H. Testing Fear of Isolation as a Causal Mechanism: Spiral of Silence and Genetically Modified (GM) Foods in South Korea. *Int. J. Public Opin. Res.* **2012**, *24*, 306–324. [CrossRef]
138. Antoniadou, M.; Varzakas, T. Breaking the vicious circle of diet, malnutrition and oral health for the independent elderly. *Crit. Rev. Food Sci. Nutr.* **2021**, *61*, 3233–3255. [CrossRef]
139. Hwang, H.; Nam, S.-J. The influence of consumers' knowledge on their responses to genetically modified foods. *GM Crops Food* **2021**, *12*, 146–157. [CrossRef]
140. Uzogara, S.G. The impact of genetic modification of human foods in the 21st century: A review. *Biotechnol. Adv.* **2000**, *18*, 179–206. [CrossRef] [PubMed]
141. Faccio, E.; Fovino, L.G.N. Food Neophobia or Distrust of Novelties? Exploring Consumers' Attitudes toward GMOs, Insects and Cultured Meat. *Appl. Sci.* **2019**, *9*, 4440. [CrossRef]
142. Cooke, R.F.; Arthington, J.D.; Staples, C.R.; Thatcher, W.W.; Lamb, G.C. Effects of supplement type on performance, reproductive, and physiological responses of Brahman-crossbred females. *J. Anim. Sci.* **2007**, *85*, 2564–2574. [CrossRef] [PubMed]
143. Knight, J.G.; Holdsworth, D.K.; Mather, D.W. GM food and neophobia: Connecting with the gatekeepers of consumer choice. *J. Sci. Food Agric.* **2008**, *88*, 739–744. [CrossRef]
144. Varzakas, T.; Tzanidis, T. Genetically Modified Foods: Risk Assessment, Legislation, Consumer Behavior, and Ethics. In *The Encyclopedia of Food and Health;* Caballero, B., Finglas, P., Toldrá, F., Eds.; Academic Press: Oxford, UK, 2016; Volume 3, pp. 204–210.
145. Gargano, D.; Appanna, R.; Santonicola, A.; De Bartolomeis, F.; Stellato, C.; Cianferoni, A.; Casolaro, V.; Iovino, P. Food Allergy and Intolerance: A Narrative Review on Nutritional Concerns. *Nutrients* **2021**, *13*, 1638. [CrossRef] [PubMed]
146. Gil-Humanes, J.; Pistón, F.; Altamirano-Fortoul, R.; Real, A.; Comino, I.; Sousa, C.; Rosell, C.M.; Barro, F. Reduced-gliadin wheat bread: Analternative to the gluten-free diet. for consumers suffering gluten-related pathologies. *PLoS ONE* **2014**, *9*, e90898. [CrossRef] [PubMed]

147. Weale, A. Ethical arguments relevant to the use of GM crops. *New Biotechnol.* **2010**, *27*, 582–587. [CrossRef] [PubMed]
148. Ghimire, B.K.; Yu, C.Y.; Kim, W.-R.; Moon, H.-S.; Lee, J.; Kim, S.H.; Chung, I.M. Assessment of Benefits and Risk of Genetically Modified Plants and Products: Current Controversies and Perspective. *Sustainability* **2023**, *15*, 1722. [CrossRef]
149. Fairbairn, M.; Reisman, E. The incumbent advantage: Corporate power in agri-food tech. *J. Peasant. Stud.* **2024**, 1–24. [CrossRef]
150. Green, J.M.; Barratt, M.J.; Kinch, M.; Gordon, J.I. Food and microbiota in the FDA regulatory framework. *Science* **2017**, *357*, 39–40. [CrossRef]
151. Van Eenennaam, A.L.; Wells, K.D.; Murray, J.D. Proposed U.S. regulation of geneedited food animals is not fit for purpose. *NPJ Sci. Food* **2019**, *3*, 3. [CrossRef]
152. Zwietering, M. Risk assessment and risk management for safe foods: Assessment needs inclusion of variability and uncertainty, management needs discrete decisions. *Int. J. Food Microbiol.* **2015**, *213*, 118–123. [CrossRef] [PubMed]
153. Ververis, E.; Ackerl, R.; Azzollini, D.; Colombo, P.A.; de Sesmaisons, A.; Dumas, C.; Fernandez-Dumont, A.; da Costa, L.F.; Germini, A.; Goumperis, T.; et al. Novel foods in the European Union: Scientific requirements and challenges of the risk assessment process by the European Food Safety Authority. *Food Res. Int.* **2020**, *137*, 109515. [CrossRef] [PubMed]
154. Santeramo, F.G.; Carlucci, D.; De Devitiis, B.; Seccia, A.; Stasi, A.; Viscecchia, R.; Nardone, G. Emerging trends in European food, diets and food industry. *Food Res. Int.* **2018**, *104*, 39–47. [CrossRef]
155. Kearney, J. Food consumption trends and drivers. *Philos. Trans. R. Soc. Lond. Ser. B Biol. Sci.* **2010**, *365*, 2793–2807. [CrossRef]
156. Little, D. Governments as regulators. In *A New Social Ontology of Government*; Foundations of Government and Public Administration; Palgrave Pivot: Cham, Switzerland, 2020. [CrossRef]
157. Rose, N.; Reeve, B.; Charlton, K. Barriers and enablers for healthy food systems and environments: The role of local governments. *Curr. Nutr. Rep.* **2022**, *11*, 82–93. [CrossRef]
158. Mozaffarian, D.; Angell, S.Y.; Lang, T.; Rivera, J.A. Role of government policy in nutrition—Barriers to and opportunities for healthier eating. *BMJ* **2018**, *361*, k2426. [CrossRef] [PubMed]
159. Kjaer, P.F.; Vetterlein, A. Regulatory governance: Rules, resistance and responsibility. *Contemp. Politics* **2018**, *24*, 497–506. [CrossRef]
160. Canfield, M.C.; Duncan, J.; Claeys, P. Reconfiguring food systems governance: The UNFSS and the battle over authority and legitimacy. *Development* **2021**, *64*, 181–191. [CrossRef]
161. Qureshi, M.E.; Dixon, J.; Wood, M. Public policies for improving food and nutrition security at different scales. *Food Secur.* **2015**, *7*, 393–403. [CrossRef]
162. Available online: https://food.ec.europa.eu/horizontal-topics/general-food-law_en (accessed on 12 April 2024).
163. Available online: https://food.ec.europa.eu/safety/biological-safety/food-hygiene_en (accessed on 12 April 2024).
164. Available online: https://www.iso.org/standard/62085.html (accessed on 12 April 2024).
165. Malik, M.; Mahmood, F.; Sarwar, N.; Obaid, A.; Memon, M.A.; Khaskheli, A. Ethical leadership: Exploring bottom-line mentality and trust perceptions of employees on middle-level managers. *Curr. Psychol.* **2023**, *42*, 16602–16617. [CrossRef]
166. The Economist. Leadership Ethics: Building Trust and Integrity in the Workplace. Last Updated: 22 October 2023, 09:14:00 PM IST. Available online: https://economictimes.indiatimes.com/jobs/c-suite/leadership-ethics-building-trust-and-integrity-in-the-workplace/articleshow/104631305.cms?utm_source=contentofinterest&utm_medium=text&utm_campaign=cppst (accessed on 14 April 2014).
167. Gartner, W.B.; Naughton, M.J.; Deming, W.E.; Gitlow, H.W.; Gitlow, S.J.; Mann, N.; Scherkenbach, W.W.; Walton, M. The Deming Theory of Management Reviewed Works: A Review of Out of the Crisis by W. Edwards Deming; Deming Guide to Achieving Quality and Competitive Position by Howard W. Gitlow, Shelly J. Gitlow; The Keys to Excellence: The Story of the Deming Philosophy by Nancy Mann; The Deming Route to Quality and Productivity: Roadmaps and Roadblocks by William W. Scherkenbach; The Deming Management Method by Mary Walton. *Acad. Manag. Rev.* **1988**, *13*, 138–142. [CrossRef]
168. Lee, J.C.; Neonaki, M.; Alexopoulos, A.; Varzakas, T. Case Studies of Small-Medium Food Enterprises around the World: Major Constraints and Benefits from the Implementation of Food Safety Management Systems. *Foods* **2023**, *12*, 3218. [CrossRef] [PubMed]
169. Daraba, A.; Varzakas, T.; Lee, J. Implementation of FSMS and Correlation with Microbiological Criteria, Systems Thinking, and Food Safety Culture. Food Safety Magazine. Column Process Control. 2022. Available online: https://digitaledition.food-safety.com/december-2022-january-2023/column-process-control/ (accessed on 15 April 2024).
170. Lee, J.C.; Daraba, A.; Voidarou, C.; Rozos, G.; El Enshasy, H.A.; Varzakas, T. Implementation of Food Safety Management Systems along with Other Management Tools (HAZOP, FMEA, Ishikawa, Pareto). The Case Study of *Listeria monocytogenes* and Correlation with Microbiological Criteria. *Foods* **2021**, *10*, 2169. [CrossRef] [PubMed]
171. *ISO 31000:2018*; Risk Management—Guidelines. ISO: Geneva, Switzerland, 2018. Available online: https://www.iso.org/iso-31000-risk-management.html (accessed on 15 April 2024).
172. Leroy, F.; Abraini, F.; Beal, T.; Dominguez-Salas, P.; Gregorini, P.; Manzano, P.; Rowntree, J.; van Vliet, S. Animal board invited review: Animal source foods in healthy, sustainable, and ethical diets—An argument against drastic limitation of livestock in the food system. *Animal* **2022**, *16*, 100457. [CrossRef] [PubMed]

173. Thompson, P. Food and Agricultural Biotechnology: Incorporating Ethical Considerations. Prepared for the Canadian Biotechnology Advisory Committee. Project Steering Committee on the Regulation of Genetically Modified Foods. October 2000. Available online: https://www.iatp.org/sites/default/files/Food_and_Agricultural_Biotechnology_Incorporat.htm (accessed on 12 March 2024).
174. Ispas, L.; Mironeasa, C.; Silvestri, A. Risk-Based Approach in the Implementation of Integrated Management Systems: A Systematic Literature Review. *Sustainability* **2023**, *15*, 10251. [CrossRef]
175. Regulation (EU) 2021/382. Available online: https://eur-lex.europa.eu/eli/reg/2021/382/oj (accessed on 12 March 2024).
176. Jacob-John, J.; Veerapa, N.K.; Eller, C. Responsible Food Supply Chain Management: Cases of Irresponsible Behaviour and Food Fraud. In *Corporate Social Responsibility in Rising Economies, CSR, Sustainability, Ethics & Governance*; Mitra, N., Schmidpeter, R., Eds.; Springer: Cham, Switzerland, 2020. [CrossRef]
177. Amaeshi, K.M.; Osuji, O.K.; Nnodim, P. Corporate Social Responsibility in Supply Chains of Global Brands: A Boundaryless Responsibility? Clarifications, Exceptions and Implications. *J. Bus. Ethic* **2007**, *81*, 223–234. [CrossRef]
178. Macgregor, S.F.J.; Fontrodona, J. *Exploring the Fit between CSR and Innovation*; Working Paper WP 759; University of Navarra: Pamplona, Spain, 2008.
179. Choi, J.; Roh, T.; Lee, J.-H. The pitfalls of corporate social irresponsibility: Hypocrisy of family firms in South Korea. *J. Clean. Prod.* **2024**, *435*, 140557. [CrossRef]
180. Abu Ayyash, A.H.; Saadon, M.S.I.; Nordin, N.; Othman, M.R. The impact of implementing ISO 26000 on Firms' performance: The mediating role of Corporate Governance. *Int. J. Prof. Bus. Rev.* **2023**, *8*, e01042. [CrossRef]
181. Gong, R. Does Rivals' product recall increase Firm's product related corporate social Responsibility? *J. Bus. Res.* **2024**, *170*, 114343. [CrossRef]
182. Costa, R.; Pittia, P. *Food Ethics Education*; Integrating Food Science and Engineering Knowledge Into the Food Chain; Series Editor Kristberg Kristbergsson; Springer: Cham, Switzerland, 2018. [CrossRef]
183. Davies, J.; Ollus, N. Labour exploitation as corporate crime and harm: Outsourcing responsibility in food production and cleaning services supply chains. *Crime Law Soc. Chang.* **2019**, *72*, 87–106. [CrossRef]
184. Carter, C.R.; Rogers, D.S. A framework of sustainable supply chain management: Moving toward new theory. *Int. J. Phys. Distrib. Logist. Manag.* **2008**, *38*, 360–387. [CrossRef]
185. Linnenluecke, M.K.; Griffiths, A. (Eds.) Chapter 2: The threat of climate change. In *The Climate Resilient Organization: Adaptation and Resilience to Climate Change and Weather Extremes*; Edward Elgar Publishing: Cheltenham, UK, 2015.
186. Smith, B.G. Developing sustainable food supply chains. *Philos. Trans. R. Soc. B Biol. Sci.* **2008**, *363*, 849–861. [CrossRef] [PubMed]
187. Wognum, P.N.; Bremmers, H.; Trienekens, J.H.; Van Der Vorst, J.G.; Bloemhof, J.M. Systems for sustain-ability and transparency of food supply chains—Current status and challenges. *Adv. Eng. Inform.* **2011**, *25*, 65–76. [CrossRef]
188. Rasul, G.; Thapa, G.B. Sustainability of ecological and conventional agricultural systems in Bangladesh: An as-sessment based on environmental, economic and social perspectives. *Agric. Syst.* **2004**, *79*, 327–351. [CrossRef]
189. Spink, J.; Moyer, D.C. Defining the public health threat of food fraud. *J. Food Sci.* **2011**, *76*, R157–R163. [CrossRef] [PubMed]
190. Yang, Y.; Huisman, W.; Hettinga, K.; Liu, N.; Heck, J.; Schrijver, G.; Gaiardoni, L.; van Ruth, S. Fraud vulnerability in the Dutch milk supply chain: Assessments of farmers, processors and retailers. *Food Control* **2018**, *95*, 308–317. [CrossRef]
191. Everstine, K.; Spink, J.; Kennedy, S. Economically motivated adulteration (EMA) of food: Common characteristics of EMA incidents. *J. Food Prot.* **2013**, *76*, 723–735. [CrossRef] [PubMed]
192. McElwee, G.; Smith, R.; Lever, J. Illegal activity in the UK halal (sheep) supply chain: Towards greater understanding. *Food Policy* **2017**, *69*, 166–175. [CrossRef]
193. van der Meulen, B. Is current EU food safety law geared up for fighting food fraud? *J. Consum. Prot. Food Saf.* **2015**, *10*, 19–23. [CrossRef]
194. Liu, C.-Y. Institutional isomorphism and food fraud: A longitudinal study of the mislabeling of rice in Taiwan. *J. Agric. Environ. Ethic* **2016**, *29*, 607–630. [CrossRef]
195. Nöhle, U. Food fraud, food crime oder kalter kaffee? *J. Consum. Prot. Food Saf.* **2017**, *12*, 197–199. [CrossRef]
196. Curll, J. The significance of food fraud in Australia. *Aust. Bus. Law Rev.* **2015**, *43*, 270–302. Available online: https://www.researchgate.net/publication/281366422 (accessed on 4 April 2024).
197. Manning, L.; Soon, J.M. Developing systems to control food adulteration. *Food Policy* **2014**, *49*, 23–32. [CrossRef]
198. Bouzembrak, Y.; Steen, B.; Neslo, R.; Linge, J.; Mojtahed, V.; Marvin, H. Development of food fraud media monitoring system based on text mining. *Food Control* **2018**, *93*, 283–296. [CrossRef]
199. Zhang, W.; Xue, J. Economically motivated food fraud and adulteration in China: An analysis based on 1553 media reports. *Food Control* **2016**, *67*, 192–198. [CrossRef]
200. Schaefer, K.A.; Scheitrum, D.; Nes, K. International sourcing decisions in the wake of a food scandal. *Food Policy* **2018**, *81*, 48–57. [CrossRef]
201. Rocchi, B.; Romano, D.; Sadiddin, A.; Stefani, G. Assessing the economy-wide impact of food fraud: A SAM-based counterfactual approach. *Agribusiness* **2020**, *36*, 167–191. [CrossRef]
202. Akkoc, R. Popular with Celebrities but Could that Manuka Honey in Your Cupboard be Fake? *The Telegraph*. 30 June 2014. Available online: https://www.telegraph.co.uk/news/health/news/10935779/Popular-with-celebrities-but-could-that-manuka-honey-in-your-cupboard-be-fake.html (accessed on 4 April 2024).

203. Leak, J. Food Fraud Buzz over Fake Manuka Honey. 2013. Available online: https://www.theaustralian.com.au/subscribe/news/1/?sourceCode=TAWEB_WRE170_a_GGL&dest=https://www.theaustralian.com.au/news/world/food-fraud-buzz-over-fake-manuka-honey/news-story/e58d5d067d615b20c71bd04864f4397c&memtype=anonymous&mode=premium&v21=GROUPA-Segment-2-NOSCORE&V21spcbehaviour=append (accessed on 12 April 2024).
204. Singuluri, H.; Sukumaran, M. Milk adulteration in Hyderabad, India—A comparative study on the levels of different adulterants present in milk. *J. Chromatogr. Sep. Tech.* **2014**, *5*, 1. [CrossRef]
205. Agriopoulou, S.; Tarapoulouzi, M.; Varzakas, T. *Chemometrics and Authenticity of Foods of Plant Origin*; CRC Press: Boca Raton, FL, USA; Taylor and Francis Group: Abingdon, UK, 2023; ISBN 9781032199450.
206. Bhaskaran, M. White Poison: Why Drinking Milk Could Prove Fatal for You. 2015. Available online: https://indianexpress.com/article/lifestyle/health/white-poison-why-drinking-milk-could-prove-fatal-for-this-generation/ (accessed on 12 April 2024).
207. Fassam, L.; Dani, S.; Hills, M. Supply chain food crime & fraud: A systematic literature review of food criminality. In Proceedings of the 20th International Symposium on Logistics (ISL 2015): Reflections on Supply Chain Research and Practice, Bologna, Italy, 5–8 July 2015.
208. Niu, L.; Chen, M.; Chen, X.; Wu, L.; Tsai, F.-S. Enterprise Food Fraud in China: Key Factors Identification From Social Co-governance Perspective. *Front. Public Health* **2021**, *9*, 752112. [CrossRef]
209. Arora, A.; Rani, N.; Devi, C.; Gupta, S. Factors affecting consumer purchase intentions of organic food through fuzzy AHP. *Int. J. Qual. Reliab. Manag.* **2022**, *39*, 1085–1103. [CrossRef]
210. Basha, M.B.; Lal, D. Indian consumers' attitudes towards purchasing organically produced foods: An empirical study. *J. Clean. Prod.* **2019**, *215*, 99–111. [CrossRef]
211. Kamboj, S.; Matharu, M.; Gupta, M. Examining consumer purchase intention towards organic food: An empirical study. *Clean. Responsible Consum.* **2023**, *9*, 100121. [CrossRef]
212. Nagaraj, S. Role of consumer health consciousness, food safety and attitude on organic food purchase in emerging market: A serial mediation model. *J. Retail. Consum. Serv.* **2021**, *59*, 102423. [CrossRef]
213. Teng, C.C.; Lu, C.H. Organic food consumption in Taiwan: Motives, involvement, and purchase intention under the moderating role of uncertainty. *Appetite* **2016**, *105*, 95–105. [CrossRef] [PubMed]
214. Yang, Q.; Al Mamun, A.; Naznen, F.; Siyu, L.; Makhbul, Z.K.M. Modelling the significance of health values, beliefs and norms on the intention to consume and the consumption of organic foods. *Heliyon* **2023**, *9*, e17487. [CrossRef] [PubMed]
215. Singh, A.; Verma, P. Factors influencing Indian consumers' actual buying behaviour towards organic food products. *J. Clean. Prod.* **2017**, *167*, 473–483. [CrossRef]
216. Talwar, S.; Jabeen, F.; Tandon, A.; Sakashita, M.; Dhir, A. What drives willingness to purchase and stated buying behavior toward organic food? A Stimulus–Organism–Behavior–Consequence (SOBC) perspective. *J. Clean. Prod.* **2021**, *293*, 125882. [CrossRef]
217. Kirmani, M.D.; Shahzad, N.; Ahmad, A.; Uddin, S.F.; Ayyub, S.; Adil, M. Socio-environmental considerations and organic food consumption: An empirical investigation of the attitude of Indian consumers. *Food Qual. Prefer.* **2022**, *100*, 104604. [CrossRef]
218. Magnusson, M.K.; Arvola, A.; Hursti, U.K.K.; Åberg, L.; Sjödén, P.O. Choice of organic foods is related to perceived consequences for human health and to environmentally friendly behaviour. *Appetite* **2003**, *40*, 109–117. [CrossRef]
219. Molinillo, S.; Vidal-Branco, M.; Japutra, A. Understanding the drivers of organic foods purchasing of millennials: Evidence from Brazil and Spain. *J. Retail. Consum. Serv.* **2020**, *52*, 101926. [CrossRef]
220. Nautiyal, S.; Lal, C. Product knowledge as a facilitator of organic purchase intention in emerging markets: Empirical evidence from India. *J. Clean. Prod.* **2022**, *372*, 133782. [CrossRef]
221. Tandon, A.; Dhir, A.; Kaur, P.; Kushwah, S.; Salo, J. Why do people buy organic food? The moderating role of environmental concerns and trust. *J. Retail. Consum. Serv.* **2020**, *57*, 102247. [CrossRef]
222. Chekima, B.; Igau, A.; Wafa SA WS, K.; Chekima, K. Narrowing the gap: Factors driving organic food consumption. *J. Clean. Prod.* **2017**, *166*, 1438–1447. [CrossRef]
223. Parashar, S.; Singh, S.; Sood, G. Examining the role of health consciousness, environmental awareness and intention on purchase of organic food: A moderated model of attitude. *J. Clean. Prod.* **2023**, *386*, 135553. [CrossRef]
224. Rana, J.; Paul, J. Consumer behavior and purchase intention for organic food: A review and research agenda. *J. Retail. Consum. Serv.* **2017**, *38*, 157–165. [CrossRef]
225. Danner, H.; Menapace, L. Using online comments to explore consumer beliefs regarding organic food in German-speaking countries and United States. *Food Qual. Prefer.* **2020**, *83*, 103912. [CrossRef]
226. Matharu, G.K.; von der Heidt, T.; Sorwar, G.; Sivapalan, A. What motivates young Indian consumers to buy organic food? *J. Int. Consum. Mark.* **2022**, *34*, 497–516. [CrossRef]
227. Sadiq, M.A.; Rajeswari, B.; Ansari, L.; Kirmani, M.D. The role of food eating values and exploratory behavior traits in predicting intention to consume organic foods: An extended planned behaviour approach. *J. Retail. Consum. Serv.* **2021**, *59*, 102352. [CrossRef]
228. Fortune Business Insights. Organic Foods Market Size. Share and Industry Analysis. 2020. Available online: https://www.fortunebusinessinsights.com/industry-reports/organic-foods-market-101470 (accessed on 11 April 2024).
229. USDA. India: Indian Organic Foods Market Report. 2020. Available online: https://apps.fas.usda.gov/newgainapi/api/Report/DownloadReportByFileName?fileName=Indian%20Organic%20Foods%20Market%20Report_New%20Delhi_India_06-08-2020 (accessed on 12 April 2024).

230. Guru, S.; Thanki, S.; Thakkar, J.J. Analysing Factors Driving Purchase and Consumption Behaviour Towards Organic Food. *Vis. J. Bus. Perspect.* **2024**. [CrossRef]
231. Katt, F.; Meixner, O. A Systematic Review of Drivers Influencing Consumer Willingness to Pay for Organic Food. *Trends Food Sci. Technol.* **2020**, *100*, 374–388. [CrossRef]
232. Wu, Y.; Wang, L.; Patras, A. Safety, Nutrition, and Health Aspects of Organic Food. In *Organic Farming for Sustainable Agriculture*; Springer: Cham, Switzerland, 2016; pp. 89–106.
233. Zámková, M.; Rojík, S.; Prokop, M.; Činčalová, S.; Stolín, R. National Labelling System of Organic Agriculture and Food Products—How Familiar Are Czech Consumers with the National Organic Agri-Food Brand? *Agriculture* **2024**, *14*, 100. [CrossRef]
234. Carrigan, M.; Szmigin, I.; Wright, J. Shopping for a better world? An interpretive study of the potential for ethical consumption within the older market. *J. Consum. Mark.* **2004**, *21*, 401–417. [CrossRef]
235. Hassan, S.M.; Rahman, Z. The evolving passage of consumer ethics research: A systematic literature review. *Int. J. Emerg. Mark.* **2023**, *18*, 3043–3064. [CrossRef]
236. Ghvanidze, S.; Velikova, N.; Dodd, T.H.; Oldewage-Theron, W. Consumers' environmental and ethical consciousness and the use of the related food products information: The role of perceived consumer effectiveness. *Appetite* **2016**, *107*, 311–322. [CrossRef]
237. Batat, W. Consumers' perceptions of food ethics in luxury dining. *J. Serv. Mark.* **2022**, *36*, 754–766. [CrossRef]
238. Hunt, S.D.; Vitell, S. A General Theory of Marketing Ethics. *J. Macromarketing* **1986**, *6*, 5–16. [CrossRef]
239. Ho, Y.-H.; Alam, S.S.; Ahsan, M.N.; Lin, C.-Y. Consumers' intention toward buying ethically produced products in Bangladesh. *Int. J. Emerg. Mark.* **2023**, *18*, 5798–5817. [CrossRef]
240. Pearson, H.H. The mouth is the gateway to the body. *J. Dent. Que* **1971**, *8*, 8–9.
241. Iyer, P. Oral Cavity is the Gateway to the Body: Role of Oral Health Professionals: A Narrative Review. *J. Calif. Dent. Assoc.* **2023**, *51*, 2193372. [CrossRef]
242. Prioreschi, P. *A History of Medicine: Roman Medicine*; Edwin Mellen Press: Lewiston, NY, USA, 1998; p. 360. ISBN 978-1-888456-03-5.
243. Jacques, J.; Allies, N. Wine and Medicine in Ancient Greece. In *Greek Medicine from Hippocrates to Galen: Selected Papers*; Van der Eijk, P., Ed.; Brill: Leiden, The Netherlands; Boston, MA, USA, 2012; pp. 173–194. Available online: www.jstor.org/stable/10.1163/j.ctt1w76vxr.15 (accessed on 18 December 2023).
244. Gritzalis, K.C.; Karamanou, M.; Androutsos, G. Gout in the writings of eminent ancient Greek and Byzantine physicians. *Acta Med. Hist. Adriat.* **2011**, *9*, 83–88.
245. Bujalkova, M.; Straka, S.; Jureckova, A. Hippocrates' humoral pathology in nowaday's reflections. *Bratisl. Lek. Listy* **2001**, *102*, 489–492. [PubMed]
246. Skiadas, P.K.; Lascaratos, J.G. Dietetics in ancient Greek philosophy: Plato's concepts of healthy diet. *Eur. J. Clin. Nutr.* **2001**, *55*, 532–537. [CrossRef]
247. Wilkins, J.; Harvey, F.D.; Dobson, M.J. *Food in Antiquity*; University of Exeter Press: Exeter, UK, 1995; p. 345. ISBN 0-85989-418-5.
248. Ancient Greek Medicine. World History Encyclopedia. Available online: https://www.worldhistory.org/Greek_Medicine/ (accessed on 12 March 2024).
249. Davison, G.; Kehaya, C.; Wyn Jones, A. Nutritional and Physical Activity Interventions to Improve Immunity. *Am. J. Lifestyle Med.* **2014**, *10*, 152–169. [CrossRef] [PubMed]
250. Macdiarmid, J.I. Seasonality and dietary requirements: Will eating seasonal food contribute to health and environmental sustainability? *Proc. Nutr. Soc.* **2014**, *73*, 368–375. [CrossRef]
251. Bartoš, H. *Philosophy and Dietetics in the Hippocratic on Regimen: A Delicate Balance of Health*; Hippocrates, Ed.; Brill: Boston, MA, USA, 2015.
252. Curd, P. *Anaxagoras of Clazomenae: Fragments and Testimonia*; University of Toronto Press: Toronto, ON, Canada, 2007.
253. Curd, P. Anaxagoras. In *Stanford Encyclopedia of Philosophy*, Winter 2015 ed.; Zalta, E.N., Ed.; Stanford University: Stanford, CA, USA, 2015; Available online: https://plato.stanford.edu/entries/anaxagoras/ (accessed on 12 April 2024).
254. Luchte, J. Wandering Souls: The Doctrine of Transmigration in Pythagorean Philosophy. Philosophical Writings. Published with Minor Variations by Bloomsbury as Pythagoras and the Doctrine of Transmigration: Wandering Souls in 2009. Available online: https://luchte.wordpress.com/wandering-souls-the-doctrine-of-transmigration-in-pythagorean-philosophy/ (accessed on 4 April 2024).
255. Dye. Explaining Pythagorean Abstinence from Beans. 1999. Available online: https://periodicos.unb.br/index.php/archai/article/view/30363 (accessed on 12 March 2024).
256. Nutton, V. Healers and the healing act in Classical Greece. *Eur. Rev.* **1999**, *7*, 27–35. [CrossRef]
257. Huffman, C. Pythagoras. In *Stanford Encyclopedia of Philosophy*, Summer 2014 ed.; Zalta, E.N., Ed.; Stanford University: Stanford, CA, USA, 2014. Available online: https://plato.stanford.edu/archives/sum2014/entries/pythagoras/ (accessed on 12 March 2024).
258. Dalby, A. *Food in the Ancient World from A to Z*; Routledge: London, UK; New York, NY, USA, 2003.
259. Graham, D.W. *The Texts of Early Greek Philosophy: The Complete Fragments and Selected Testimonies of the Major Presocratics*; Cambridge University Press: Cambridge, UK, 2010.
260. Edelstein, L. The Relation of Ancient Philosophy to Medicine. In *Ancient Medicine*; Temkin, O., Temkin, C.L., Eds.; Johns Hopkins University Press: Baltimore, MD, USA, 1967; pp. 349–366.
261. Garnsey, P. *Food and Society in Classical Antiquity*; Cambridge University Press: Cambridge, UK, 1999.

262. Flint-Hamilton, K.B. Legumes in Ancient Greece and Rome: Food, Medicine, or Poison? *Hesperia: J. Am. Sch. Class. Stud. Athens* **1999**, *68*, 371–385. [CrossRef]
263. Delatte, A. *Le Cycéon, Breuvage Habitual des Mystères d'Éleusis*; Belles Lettres: Paris, France, 1955.
264. Dombrowski, D.A. *The Philosophy of Vegetarianism*; University of Massachusetts Press: Amherst, MA, USA, 1984.
265. Kirk, G.S.; Raven, J.E.; Schofield, M. *The Presocratic Philosophers*; Cambridge University Press: Cambridge, UK, 1984.
266. Matthew 15:10-14. Bible. Available online: https://www.biblegateway.com/passage/?search=Matthew%2015:1-10&version=NIV (accessed on 12 March 2024).
267. Hawks, S. Spiritual Wellness, Holistic Health, and the Practice of Health Education. *Am. J. Health Educ.* **2004**, *35*, 11–18. [CrossRef]
268. Chan, C.; Ho, P.S.Y.; Chow, E. A Body-Mind-Spirit Model in Health. *Soc. Work Health Care* **2001**, *34*, 261–282. [CrossRef]
269. Spanemberg, J.C.; Cardoso, J.A.; Slob, E.M.G.B.; López-López, J. Quality of life related to oral health and its impact in adults. *J. Stomatol. Oral Maxillofac. Surg.* **2019**, *120*, 234–239. [CrossRef]
270. Fiorillo, L. Oral Health: The First Step to Well-Being. *Medicina* **2019**, *55*, 676. [CrossRef] [PubMed]
271. Kossioni, A.E.; Hajto-Bryk, J.; Janssens, B.; Maggi, S.; Marchini, L.; McKenna, G.; Müller, F.; Petrovic, M.; Roller-Wirnsberger, R.E.; Schimmel, M.; et al. Practical Guidelines for Physicians in Promoting Oral Health in Frail Older Adults. *J. Am. Med. Dir. Assoc.* **2018**, *19*, 1039–1046. [CrossRef] [PubMed]
272. Henni, S.H.; Skudutyte-Rysstad, R.; Ansteinsson, V.; Hellesø, R.; Hovden, E.A.S. Oral health and oral health-related quality of life among older adults receiving home health care services: A scoping review. *Gerodontology* **2023**, *40*, 161–171. [CrossRef] [PubMed]
273. Wong, F.M.F.; Ng, Y.T.Y.; Leung, W.K. Oral Health and Its Associated Factors Among Older Institutionalized Residents—A Systematic Review. *Int. J. Environ. Res. Public Health* **2019**, *16*, 4132. [CrossRef] [PubMed]
274. Takeuchi, N.; Sawada, N.; Ekuni, D.; Morita, M. Association between oral condition and subjective psychological well-being among older adults attending a university hospital dental clinic: A cross-sectional study. *PLoS ONE* **2023**, *18*, e0295078. [CrossRef] [PubMed]
275. Thomson, W.M.; Broder, H.L. Oral–Health–Related Quality of Life in Children and Adolescents. *Pediatr. Clin. N. Am.* **2018**, *65*, 1073–1084. [CrossRef] [PubMed]
276. Chaffee, B.W.; Rodrigues, P.H.; Kramer, P.F.; Vítolo, M.R.; Feldens, C.A. Oral health-related quality-of-life scores differ by socioeconomic status and caries experience. *Community Dent. Oral Epidemiol.* **2017**, *45*, 216–224. [CrossRef] [PubMed]
277. Meyling, M.M.G.; Frieling, M.E.; Vervoort, J.P.M.; de Jong, E.I.F.; Jansen, D.E.M.C. Health problems experienced by people during the first year postpartum: A systematic review. *Eur. J. Midwifery* **2023**, *7*, 1–20. [CrossRef]
278. Kazemi, S.; Savabi, G.; Khazaei, S.; Savabi, O.; Esmaillzadeh, A.; Keshteli, A.H.; Adibi, P. Association between food intake and oral health in elderly: SEPAHAN systematic review no. 8. *Dent. Res. J.* **2011**, *8* (Suppl. 1), S15–S20.
279. Sensoy, I. A review on the food digestion in the digestive tract and the used in vitro models. *Curr. Res. Food Sci.* **2021**, *4*, 308–319. [CrossRef]
280. Pedersen, A.; Sørensen, C.; Proctor, G.; Carpenter, G. Salivary functions in mastication, taste and textural perception, swallowing and initial digestion. *Oral Dis.* **2018**, *24*, 1399–1416. [CrossRef] [PubMed]
281. Lizal, F.; Elcner, J.; Jedelsky, J.; Maly, M.; Jicha, M.; Farkas, Á.; Belka, M.; Rehak, Z.; Adam, J.; Brinek, A.; et al. The effect of oral and nasal breathing on the deposition of inhaled particles in upper and tracheobronchial airways. *J. Aerosol Sci.* **2020**, *150*, 105649. [CrossRef] [PubMed]
282. Deo, P.; Deshmukh, R. Oral microbiome: Unveiling the fundamentals. *J. Oral Maxillofac. Pathol.* **2019**, *23*, 122–128. [CrossRef] [PubMed]
283. Bhatnagar, D.M. Oral health: A gateway to overall health. *Contemp. Clin. Dent.* **2021**, *12*, 211–212. [CrossRef] [PubMed]
284. WHO's Global Oral Health Status Report. Available online: https://www.who.int/publications/i/item/9789240061484 (accessed on 12 March 2024).
285. Barranca-Enríquez, A.; Romo-González, T. Your health is in your mouth: A comprehensive view to promote general wellness. *Front. Oral Health* **2022**, *3*, 971223. [CrossRef] [PubMed]
286. de Abreu, M.H.N.G.; Cruz, A.J.S.; Borges-Oliveira, A.C.; Martins, R.d.C.; Mattos, F.d.F. Perspectives on Social and Environmental Determinants of Oral Health. *Int. J. Environ. Res. Public Health* **2021**, *18*, 13429. [CrossRef] [PubMed]
287. Nazir, M.; Izhar, F.; Akhtar, K.; Almas, K. Dentists' awareness about the link between oral and systemic health. *J. Fam. Community Med.* **2019**, *26*, 206–212. [CrossRef] [PubMed]
288. Lee, Y.-H.; Chung, S.W.; Auh, Q.-S.; Hong, S.-J.; Lee, Y.-A.; Jung, J.; Lee, G.-J.; Park, H.J.; Shin, S.-I.; Hong, J.-Y. Progress in oral microbiome related to oral and systemic diseases: An update. *Diagnostics* **2021**, *11*, 1283. [CrossRef] [PubMed]
289. Kapila, Y.L. Oral health's inextricable connection to systemic health: Special populations bring to bear multimodal relationships and factors connecting periodontal disease to systemic diseases and conditions. *Periodontology* **2021**, *87*, 11–16. [CrossRef]
290. Romandini, M.; Baima, G.; Antonoglou, G.; Bueno, J.; Figuero, E.; Sanz, M. Periodontitis, Edentulism, and Risk of Mortality: A Systematic Review with Meta-analyses. *J. Dent. Res.* **2021**, *100*, 37–49. [CrossRef]
291. Schwarz, C.; Hajdu, A.I.; Dumitrescu, R.; Sava-Rosianu, R.; Bolchis, V.; Anusca, D.; Hanghicel, A.; Fratila, A.D.; Oancea, R.; Jumanca, D.; et al. Link between Oral Health, Periodontal Disease, Smoking, and Systemic Diseases in Romanian Patients. *Healthcare* **2023**, *11*, 2354. [CrossRef]
292. Dimopoulou, M.; Antoniadou, M.; Amargianitakis, M.; Gortzi, O.; Androutsos, O.; Varzakas, T. Nutritional Factors Associated with Dental Caries across the Lifespan: A Review. *Appl. Sci.* **2023**, *13*, 13254. [CrossRef]

293. Pang, L.; Zhi, Q.; Jian, W.; Liu, Z.; Lin, H. The Oral Microbiome Impacts the Link between Sugar Consumption and Caries: A Preliminary Study. *Nutrients* **2022**, *14*, 3693. [CrossRef]
294. Shetty, A.S.; Shenoy, R.; Bajaj, P.D.; Rao, A.; Ks, A.; Pai, M.; Br, A.; Jodalli, P. Role of nutritional supplements on oral health in adults—A systematic review. *F1000Research* **2023**, *12*, 492. [CrossRef]
295. Antoniadou, M.; Varzakas, T. Diet and Oral Health Coaching Methods and Models for the Independent Elderly. *Appl. Sci.* **2020**, *10*, 4021. [CrossRef]
296. Clemente-Suárez, V.J.; Beltrán-Velasco, A.I.; Redondo-Flórez, L.; Martín-Rodríguez, A.; Tornero-Aguilera, J.F. Global Impacts of Western Diet and Its Effects on Metabolism and Health: A Narrative Review. *Nutrients* **2023**, *15*, 2749. [CrossRef]
297. Peponis, M.; Antoniadou, M.; Pappa, E.; Rahiotis, C.; Varzakas, T. Vitamin D and Vitamin D Receptor Polymorphisms Relationship to Risk Level of Dental Caries. *Appl. Sci.* **2023**, *13*, 6014. [CrossRef]
298. Feldens, C.A.; Pinheiro, L.L.; Cury, J.A.; Mendonça, F.; Groisman, M.; Costa, R.A.H.; Pereira, H.C.; Vieira, A.R. Added Sugar and Oral Health: A Position Paper of the Brazilian Academy of Dentistry. *Front. Oral Health* **2022**, *3*, 869112. [CrossRef]
299. Nicklisch, N.; Oelze, V.M.; Schierz, O.; Meller, H.; Alt, K.W. A Healthier Smile in the Past? Dental Caries and Diet in Early Neolithic Farming Communities from Central Germany. *Nutrients* **2022**, *14*, 1831. [CrossRef]
300. WHO. Oral Health. Available online: https://www.who.int/data/gho/data/themes/topics/topic-details/GHO/oral-health#:~:text=The%20World%20Health%20Assembly%20approved,in%20universal%20health%20coverage%20programs (accessed on 12 March 2024).
301. WHO. Sugars and Dental Caries. 2017. Available online: https://www.who.int/news-room/fact-sheets/detail/sugars-and-dental-caries#:~:text=Teeth%20affected%20by%20caries%20are,infection%20or%20adverse%20growth%20patterns (accessed on 12 March 2024).
302. Sabharwal, A.; Stellrecht, E.; Scannapieco, F.A. Associations between dental caries and systemic diseases: A scoping review. *BMC Oral Health* **2021**, *21*, 472. [CrossRef]
303. Karnaki, P.; Katsas, K.; Diamantis, D.V.; Riza, E.; Rosen, M.S.; Antoniadou, M.; Gil-Salmerón, A.; Grabovac, I.; Linou, A. Dental Health, Caries Perception and Sense of Discrimination among Migrants and Refugees in Europe: Results from the Mig-HealthCare Project. *Appl. Sci.* **2022**, *12*, 9294. [CrossRef]
304. Jain, N.; Dutt, U.; Radenkov, I.; Jain, S. WHO's global oral health status report 2022: Actions, discussion and implementation. *Oral Dis.* **2023**, *30*, 73–79. [CrossRef]
305. Giordano-Kelhoffer, B.; Lorca, C.; Llanes, J.M.; Rábano, A.; del Ser, T.; Serra, A.; Gallart-Palau, X. Oral microbiota, its equilibrium and implications in the pathophysiology of human diseases: A systematic review. *Biomedicines* **2022**, *10*, 1803. [CrossRef]
306. Matsuyama, Y.; Tsakos, G.; Listl, S.; Aida, J.; Watt, R. Impact of Dental Diseases on Quality-Adjusted Life Expectancy in US Adults. *J. Dent. Res.* **2019**, *98*, 510–516. [CrossRef]
307. Bell, V.; Rodrigues, A.R.; Antoniadou, M.; Peponis, M.; Varzakas, T.; Fernandes, T. An Update on Drug–Nutrient Interactions and Dental Decay in Older Adults. *Nutrients* **2023**, *15*, 4900. [CrossRef]
308. Aida, J.; Takeuchi, K.; Furuta, M.; Ito, K.; Kabasawa, Y.; Tsakos, G. Burden of Oral Diseases and Access to Oral Care in an Ageing Society. *Int. Dent. J.* **2022**, *72*, S5–S11. [CrossRef]
309. van der Putten, G.-J.; de Baat, C. An Overview of Systemic Health Factors Related to Rapid Oral Health Deterioration among Older People. *J. Clin. Med.* **2023**, *12*, 4306. [CrossRef]
310. Peng, X.; Cheng, L.; You, Y.; Tang, C.; Ren, B.; Li, Y.; Xu, X.; Zhou, X. Oral microbiota in human systematic diseases. *Int. J. Oral Sci.* **2022**, *14*, 14. [CrossRef]
311. Radaic, A.; Kapila, Y.L. The oralome and its dysbiosis: New insights into oral microbiome-host interactions. *Comput. Struct. Biotechnol. J.* **2021**, *19*, 1335–1360. [CrossRef]
312. Kleinstein, S.; Nelson, K.; Freire, M. Inflammatory Networks Linking Oral Microbiome with Systemic Health and Disease. *J. Dent. Res.* **2020**, *99*, 1131–1139. [CrossRef]
313. Sedghi, L.; DiMassa, V.; Harrington, A.; Lynch, S.V.; Kapila, Y.L. The oral microbiome: Role of key organisms and complex networks in oral health and disease. *Periodontology* **2021**, *87*, 107–131. [CrossRef] [PubMed]
314. Kilian, M.; Chapple, I.L.C.; Hannig, M.; Marsh, P.D.; Meuric, V.; Pedersen, A.M.L.; Tonetti, M.S.; Wade, W.G.; Zaura, E. The oral microbiome—An update for oral healthcare professionals. *Br. Dent. J.* **2016**, *221*, 657–666. [CrossRef] [PubMed]
315. Schamarek, I.; Anders, L.; Chakaroun, R.M.; Kovacs, P.; Rohde-Zimmermann, K. The role of the oral microbiome in obesity and metabolic disease: Potential systemic implications and effects on taste perception. *Nutr. J.* **2023**, *22*, 28. [CrossRef] [PubMed]
316. Edlund, A.; Santiago-Rodriguez, T.M.; Boehm, T.K.; Pride, D.T. Bacteriophage and their potential roles in the human oral cavity. *J. Oral Microbiol.* **2015**, *7*, 27423. [CrossRef] [PubMed]
317. Parras-Moltó, M.; López-Bueno, A. Methods for enrichment and sequencing of oral viral assemblages: Saliva, oral mucosa, and dental plaque viromes. *Methods Mol. Biol.* **2018**, *1838*, 143–161. [CrossRef]
318. Abdelbary, M.M.H.; Hatting, M.; Bott, A.; Dahlhausen, A.; Keller, D.; Trautwein, C.; Conrads, G. The oral-gut axis: Salivary and fecal microbiome dysbiosis in patients with inflammatory bowel disease. *Front. Cell. Infect. Microbiol.* **2022**, *12*, 1010853. [CrossRef]
319. Kitamoto, S.; Nagao-Kitamoto, H.; Hein, R.; Schmidt, T.M.; Kamada, N. The Bacterial Connection between the Oral Cavity and the Gut Diseases. *J. Dent. Res.* **2020**, *99*, 1021–1029. [CrossRef]

320. Atarashi, K.; Suda, W.; Luo, C.; Kawaguchi, T.; Motoo, I.; Narushima, S.; Kiguchi, Y.; Yasuma, K.; Watanabe, E.; Tanoue, T.; et al. Ectopic colonization of oral bacteria in the intestine drives TH cell induction and inflammation. *Science* **2017**, *358*, 359–365. [CrossRef]
321. Kuraji, R.; Sekino, S.; Kapila, Y.; Numabe, Y. Periodontal disease–related nonalcoholic fatty liver disease and nonalcoholic steatohepatitis: An emerging concept of oral-liver axis. *Periodontology* **2021**, *87*, 204–240. [CrossRef]
322. Yoneda, M.; Naka, S.; Nakano, K.; Wada, K.; Endo, H.; Mawatari, H.; Imajo, K.; Nomura, R.; Hokamura, K.; Ono, M.; et al. Involvement of a periodontal pathogen, Porphyromonas gingivalis on the pathogenesis of non-alcoholic fatty liver disease. *BMC Gastroenterol.* **2012**, *12*, 16. [CrossRef]
323. Abe, K.; Fujita, M.; Hayashi, M.; Okai, K.; Takahashi, A.; Ohira, H. Gut and oral microbiota in autoimmune liver disease. *Fukushima J. Med. Sci.* **2020**, *65*, 71–75. [CrossRef] [PubMed]
324. Xiao, E.; Mattos, M.; Vieira, G.H.A.; Chen, S.; Corrêa, J.D.; Wu, Y.; Albiero, M.L.; Bittinger, K.; Graves, D.T. Diabetes enhances IL-17 expression and alters the oral microbiome to increase its pathogenicity. *Cell Host Microbe* **2017**, *22*, 120–128.e4. [CrossRef] [PubMed]
325. Matsha, T.; Prince, Y.; Davids, S.; Chikte, U.; Erasmus, R.; Kengne, A.; Davison, G. Oral microbiome signatures in diabetes mellitus and periodontal disease. *J. Dent. Res.* **2020**, *99*, 658–665. [CrossRef]
326. Negrini, T.d.C.; Carlos, I.Z.; Duque, C.; Caiaffa, K.S.; Arthur, R.A. Interplay among the oral microbiome, oral cavity conditions, the host immune response, diabetes mellitus, and its associated-risk factors—An overview. *Front. Oral Health* **2021**, *2*, 697428. [CrossRef]
327. Fleury, V.; Zekeridou, A.; Lazarevic, V.; Gaïa, N.; Giannopoulou, C.; Genton, L.; Cancela, J.; Girard, M.; Goldstein, R.; Bally, J.F.; et al. Oral dysbiosis and inflammation in Parkinson's disease. *J. Park. Dis.* **2021**, *11*, 619–631. [CrossRef] [PubMed]
328. Fan, Z.; Li, Z.; Zhao, S.; Chen, Y.; Su, Y.; Peng, G.; Luo, B. Salivary Aβ1–42 may be a quick-tested biomarker for clinical use in Alzheimer's disease: A meta-analysis. *J. Neurol.* **2022**, *270*, 1945–1954. [CrossRef]
329. Liu, S.; Dashper, S.G.; Zhao, R. Association between oral bacteria and Alzheimer's disease: A systematic review and meta-analysis. *J. Alzheimer's Dis.* **2022**, *91*, 129–150. [CrossRef]
330. Li, C.; Chen, Y.; Wen, Y.; Jia, Y.; Cheng, S.; Liu, L.; Zhang, H.; Pan, C.; Zhang, J.; Zhang, Z.; et al. A genetic association study reveals the relationship between the oral microbiome and anxiety and depression symptoms. *Front. Psychiatry* **2022**, *13*, 960756. [CrossRef] [PubMed]
331. Simpson, C.A.; Adler, C.; du Plessis, M.R.; Landau, E.R.; Dashper, S.G.; Reynolds, E.C.; Schwartz, O.S.; Simmons, J.G. Oral microbiome composition, but not diversity, is associated with adolescent anxiety and depression symptoms. *Physiol. Behav.* **2020**, *226*, 113126. [CrossRef]
332. Wingfield, B.; Lapsley, C.; McDowell, A.; Miliotis, G.; McLafferty, M.; O'neill, S.M.; Coleman, S.; McGinnity, T.M.; Bjourson, A.J.; Murray, E.K. Variations in the oral microbiome are associated with depression in young adults. *Sci. Rep.* **2021**, *11*, 1–9. [CrossRef]
333. Ahrens, A.P.; Sanchez-Padilla, D.E.; Drew, J.C.; Oli, M.W.; Roesch, L.F.W.; Triplett, E.W. Saliva microbiome, dietary, and genetic markers are associated with suicidal ideation in university students. *Sci. Rep.* **2022**, *12*, 14306. [CrossRef] [PubMed]
334. Blekkenhorst, L.C.; Bondonno, N.P.; Liu, A.H.; Ward, N.C.; Prince, R.L.; Lewis, J.R.; Devine, A.; Croft, K.D.; Hodgson, J.M.; Bondonno, C.P. Nitrate, the oral microbiome, and cardiovascular health: A systematic literature review of human and animal studies. *Am. J. Clin. Nutr.* **2018**, *107*, 504–522. [CrossRef] [PubMed]
335. Zhang, J.; Xie, M.; Huang, X.; Chen, G.; Yin, Y.; Lu, X.; Feng, G.; Yu, R.; Chen, L. The effects of porphyromonas gingivalis on atherosclerosis-related cells. *Front. Immunol.* **2021**, *12*, 766560. [CrossRef] [PubMed]
336. Cobb, C.M.; Kelly, P.J.; Williams, K.B.; Babbar, S.; Angolkar, M.; Derman, R. The oral microbiome and adverse pregnancy outcomes. *Int. J. Women's Health* **2017**, *ume 9*, 551–559. [CrossRef] [PubMed]
337. Balan, P.; Chong, Y.S.; Umashankar, S.; Swarup, S.; Loke, W.M.; Lopez, V.; He, H.G.; Seneviratne, C.J. Keystone species in pregnancy gingivitis: A snapshot of oral microbiome during pregnancy and postpartum period. *Front. Microbiol.* **2018**, *9*, 2360. [CrossRef] [PubMed]
338. Ye, C.; You, M.; Huang, P.; Xia, Z.; Radaic, A.; Tang, J.; Wu, W.; Wu, Y.; Kapila, Y. Clinical study showing a lower abundance of Neisseria in the oral microbiome aligns with low birth weight pregnancy outcomes. *Clin. Oral Investig.* **2022**, *26*, 2465–2478. [CrossRef] [PubMed]
339. Bissada, N.A.N.F. Periodontal treatment improves prostate symptoms and lowers serum PSA in men with high PSA and chronic periodontitis. *Dentistry* **2015**, *5*, 1. [CrossRef]
340. Cheng, Z.; Do, T.; Mankia, K.; Meade, J.; Hunt, L.; Clerehugh, V.; Speirs, A.; Tugnait, A.; Emery, P.; Devine, D. Dysbiosis in the oral microbiomes of anti-CCP positive individuals at risk of developing rheumatoid arthritis. *Ann. Rheum. Dis.* **2021**, *80*, 162–168. [CrossRef] [PubMed]
341. Corrêa, J.D.; Calderaro, D.C.; Ferreira, G.A.; Mendonça, S.M.S.; Fernandes, G.R.; Xiao, E.; Teixeira, A.L.; Leys, E.J.; Graves, D.T.; Silva, T.A. Subgingival microbiota dysbiosis in systemic lupus erythematosus: Association with periodontal status. *Microbiome* **2017**, *5*, 34. [CrossRef]
342. Rao, B.; Lou, J.; Lu, H.; Liang, H.; Li, J.; Zhou, H.; Fan, Y.; Zhang, H.; Sun, Y.; Zou, Y.; et al. Oral microbiome characteristics in patients with autoimmune hepatitis. *Front. Cell. Infect. Microbiol.* **2021**, *11*, 656674. [CrossRef]
343. Qi, Y.; Wu, H.-M.; Yang, Z.; Zhou, Y.-F.; Jin, L.; Yang, M.-F.; Wang, F.-Y. New insights into the role of oral microbiota dysbiosis in the pathogenesis of inflammatory bowel disease. *Dig. Dis. Sci.* **2022**, *67*, 42–55. [CrossRef] [PubMed]

344. Mammen, M.J.; Sethi, S. COPD and the microbiome. *Respirology* **2016**, *21*, 590–599. [CrossRef] [PubMed]
345. Pathak, J.L.; Yan, Y.; Zhang, Q.; Wang, L.; Ge, L. The role of oral microbiome in respiratory health and diseases. *Respir. Med.* **2021**, *185*, 106475. [CrossRef] [PubMed]
346. Peters, B.A.; Wu, J.; Pei, Z.; Yang, L.; Purdue, M.P.; Freedman, N.D.; Jacobs, E.J.; Gapstur, S.M.; Hayes, R.B.; Ahn, J. Oral microbiome composition reflects prospective risk for esophageal cancers. *Cancer Res.* **2017**, *77*, 6777–6787. [CrossRef] [PubMed]
347. Yuan, S.; Fang, C.; Leng, W.-D.; Wu, L.; Li, B.-H.; Wang, X.-H.; Hu, H.; Zeng, X.-T. Oral microbiota in the oral-genitourinary axis: Identifying periodontitis as a potential risk of genitourinary cancers. *Mil. Med. Res.* **2021**, *8*, 54. [CrossRef]
348. Gaba, F.I.; González, R.C.; Martínez, R.G. The Role of oral fusobacterium nucleatum in female breast cancer: A systematic review and meta-analysis. *Int. J. Dent.* **2022**, *2022*, 1876275. [CrossRef]
349. Tutuncu, E.E.; Ozgur, D.; Karamese, M. Saliva samples for detection of SARS-CoV-2 in mildly symptomatic and asymptomatic patients. *J. Med. Virol.* **2021**, *93*, 2932–2937. [CrossRef]
350. Coker, M.O.; Lebeaux, R.M.; Hoen, A.G.; Moroishi, Y.; Gilbert-Diamond, D.; Dade, E.F.; Palys, T.J.; Madan, J.C.; Karagas, M.R. Metagenomic analysis reveals associations between salivary microbiota and body composition in early childhood. *Sci. Rep.* **2022**, *12*, 13075. [CrossRef] [PubMed]
351. Dawes, C.; Wong, D. Role of saliva and salivary diagnostics in the advancement of oral health. *J. Dent. Res.* **2019**, *98*, 133–141. [CrossRef]
352. Landry, V.; Coburn, P.; Kost, K.; Liu, X.; Li-Jessen, N.Y.K. Diagnostic accuracy of liquid biomarkers in airway diseases: Toward point-of-care applications. *Front. Med.* **2022**, *9*, 855250. [CrossRef]
353. Perkell, J.S. Movement goals and feedback and feedforward control mechanisms in speech production. *J. Neurolinguistics* **2012**, *25*, 382–407. [CrossRef]
354. Slavkin, H.C.; A Dubois, P.; Kleinman, D.V.; Fuccillo, R. Science-Informed Health Policies for Oral and Systemic Health. *J. Health Leadersh.* **2023**, *ume 15*, 43–57. [CrossRef]
355. Gkavela, G.; Kossioni, A.; Lyrakos, G.; Karkazis, H.; Volikas, K. Oral health related quality of life in older people: Preliminary validation of the Greek version of the Geriatric Oral Health Assessment Index (GOHAI). *Eur. Geriatr. Med.* **2015**, *6*, 245–250. [CrossRef]
356. Paredes-Rodriguez, V.; Torrijos-Gomez, G.; Gonzalez-Serrano, J.; Lopez-Pintor-Munoz, R.; Lopez-Bermejo, M.; Hernandez-Vallejo, G. Quality of life and oral health in elderly. *J. Clin. Exp. Dent.* **2016**, *8*, e590–e596. [CrossRef]
357. Kossioni, A.; Bellou, O. Eating habits in older people in Greece: The role of age, dental status and chewing difficulties. *Arch. Gerontol. Geriatr.* **2011**, *52*, 197–201. [CrossRef] [PubMed]
358. Alpert, P.T. Oral health: The oral-systemic health connection. *Home Health Care Manag. Pract.* **2017**, *29*, 56–59. [CrossRef]
359. E Kruk, M.; Gage, A.D.; Arsenault, C.; Jordan, K.; Leslie, H.H.; Roder-DeWan, S.; Adeyi, O.; Barker, P.; Daelmans, B.; Doubova, S.V.; et al. High-quality health systems in the Sustainable Development Goals era: Time for a revolution. *Lancet Glob. Health* **2018**, *6*, e1196–e1252, Erratum in *Lancet Glob. Health* **2018**, *6*, e1162; Erratum in *Lancet Glob. Health* **2021**, *9*, e1067. [CrossRef] [PubMed]
360. Section 1: Effect of Oral Health on the Community, Overall Well-Being, and the Economy. Chapter 1: Status of Knowledge, Practice, and Perspectives. In *Oral Health in America: Advances and Challenges [Internet]*; National Institute of Dental and Craniofacial Research(US): Bethesda, MD, USA, 2021. Available online: https://www.ncbi.nlm.nih.gov/books/NBK578297/ (accessed on 12 March 2024).
361. Listl, S.; Galloway, J.; Mossey, P.; Marcenes, W. Global economic impact of dental diseases. *J. Dent. Res.* **2015**, *94*, 1355–1361. [CrossRef]
362. Righolt, A.; Jevdjevic, M.; Marcenes, W.; Listl, S. Global-, regional-, and country-level economic impacts of dental diseases in 2015. *J. Dent. Res.* **2018**, *97*, 501–507. [CrossRef]
363. Australian Research Centre for Population Oral Health, The University of Adelaide, South Australia. Productivity losses from dental problems. *Aust. Dent. J.* **2012**, *57*, 393. [CrossRef]
364. Guarnizo-Herreño, C.C.; Wehby, G.L. Children's dental health, school performance, and psychosocial well-being. *J. Pediatr.* **2012**, *161*, 1153–1159. [CrossRef]
365. Hayes, A.; Azarpazhooh, A.; Dempster, L.; Ravaghi, V.; Quiñonez, C. Time loss due to dental problems and treatment in the Canadian population: Analysis of a nationwide cross-sectional survey. *BMC Oral Health* **2013**, *13*, 17. [CrossRef]
366. Singhal, S.; Correa, R.; Quiñonez, C. The impact of dental treatment on employment outcomes: A systematic review. *Health Policy* **2013**, *109*, 88–96. [CrossRef] [PubMed]
367. Nasseh, K.; Vujicic, M.; Glick, M. The relationship between periodontal interventions and healthcare costs and utilization. Evidence from an integrated dental, medical, and pharmacy commercial claims database. *Health Econ.* **2017**, *26*, 519–527. [CrossRef]
368. Bernabé, E.; Masood, M.; Vujicic, M. The impact of out-of-pocket payments for dental care on household finances in low and middle income countries. *BMC Public Health* **2017**, *17*, 109. [CrossRef] [PubMed]
369. Allareddy, V.; Rampa, S.; Lee, M.K.; Allareddy, V.; Nalliah, R.P. Hospital-based emergency department visits involving dental conditions: Profile and predictors of poor outcomes and resource utilization. *J. Am. Dent. Assoc.* **2014**, *145*, 331–337. [CrossRef]
370. Vujicic, M.; Nasseh, K. A Decade in Dental Care Utilization among Adults and Children (2001–2010). *Health Serv. Res.* **2014**, *49*, 460–480. [CrossRef]

371. Singh, A.; Peres, M.; Watt, R. The relationship between income and oral health: A critical review. *J. Dent. Res.* **2019**, *98*, 853–860. [CrossRef] [PubMed]
372. Jeffcoat, M.K.; Jeffcoat, R.L.; Gladowski, P.A.; Bramson, J.B.; Blum, J.J. Impact of periodontal therapy on general health: Evidence from insurance data for five systemic conditions. *Am. J. Prev. Med.* **2014**, *47*, 166–174. [CrossRef]
373. Avalare Health LLC to Pacific Dental Services Foundation: Evaluation of Cost Savings Associated with Periodontal Disease Treatment Benefit. Washington, DC, USA. Available online: https://assets.senate.mn/committees/2021-2022/3095_Committee_on_Health_and_Human_Services_Finance_and_Policy/Avalere%20Health%202016%20analysis%20of%20MC%20savings%20for%20perio%20in%203%20diseases.pdf (accessed on 14 March 2024).
374. Botelho, J.; Machado, V.; Leira, Y.; Proença, L.; Chambrone, L.; Mendes, J.J. Economic burden of periodontitis in the United States and Europe: An updated estimation. *J. Periodontol.* **2022**, *93*, 373–379. [CrossRef]
375. NHI. National Institute of Dental and Craniofacial Research. Dental Caries (Tooth Decay). Available online: https://www.nidcr.nih.gov/research/data-statistics/dental-caries (accessed on 12 March 2024).
376. WHO. Global Oral Health Status Report: Towards Universal Health Coverage for Oral Health by 2030: Regional Summary of the Eastern Mediterranean Region (4 March 2024). Available online: https://www.who.int/publications/i/item/9789240070806 (accessed on 4 April 2024).
377. WHO. Oral Health. 2021 Proposal. Available online: https://www.who.int/data/gho/data/themes/topics/topic-details/GHO/oral-health (accessed on 12 March 2024).
378. Varenne, B.; Fox, C. The Role of Research in the WHO Oral Health Resolution. *JDR Clin. Transl. Res.* **2021**, *6*, 112–114. [CrossRef] [PubMed]
379. Lamster, I.B. The 2021 WHO Resolution on Oral Health. *Int. Dent. J.* **2021**, *71*, 279–280. [CrossRef] [PubMed]
380. Kalpe, S.; Mathur, A.; Kharat, P. How fad diets may jeopardize your oral well-being: The hidden consequences. *Hum. Nutr. Metab.* **2023**, *33*, 200214. [CrossRef]
381. van Bussel, L.; Kuijsten, A.; Mars, M.; van 't Veer, P. Consumers' perceptions on food-related sustainability: A systematic review. *J. Clean. Prod.* **2022**, *341*, 130904. [CrossRef]
382. Gheisary, Z.; Mahmood, R.; Shivanantham, A.H.; Liu, J.; Lieffers, J.R.L.; Papagerakis, P.; Papagerakis, S. The clinical, microbiological, and immunological effects of probiotic supplementation on prevention and treatment of periodontal diseases: A systematic review and meta-analysis. *Nutrients* **2022**, *14*, 1036. [CrossRef] [PubMed]
383. Voidarou, C.; Antoniadou, M.; Rozos, G.; Alexopoulos, A.; Giorgi, E.; Tzora, A.; Skoufos, I.; Varzakas, T.; Bezirtzoglou, E. An In Vitro Study of Different Types of Greek Honey as Potential Natural Antimicrobials against Dental Caries and Other Oral Pathogenic Microorganisms. Case Study Simulation of Oral Cavity Conditions. *Appl. Sci.* **2021**, *11*, 6318. [CrossRef]
384. Nath, S.; Zilm, P.; Jamieson, L.; Kapellas, K.; Goswami, N.; Ketagoda, K.; Weyrich, L.S. Development and characterization of an oral microbiome transplant among Australians for the treatment of dental caries and periodontal disease: A study protocol. *PLoS ONE* **2021**, *16*, e0260433. [CrossRef] [PubMed]
385. Glick, M.; Urquhart, O.; Bhosale, A.S.; Carrasco-Labra, A.; Edelson, J.T. A unified voice to drive global improvements in oral health. *BMC Glob. Public Health* **2023**, *1*, 19. [CrossRef]
386. Fisher, J.; Berman, R.; Buse, K.; Doll, B.; Glick, M.; Metzl, J.; Touger-Decker, R. Achieving Oral Health for All through Public Health Approaches, Interprofessional, and Transdisciplinary Education. *NAM Perspect.* **2023**, *13*. [CrossRef]
387. Jhala, K.N.; Jhala, C. The Hippocratic oath: A comparative analysis of the ancient text's relevance to American and Indian modern medicine. *Indian J. Pathol. Microbiol.* **2012**, *55*, 279–282. [CrossRef]
388. Jones, B.J. Ethics and artificial nutrition towards the end of life. *Clin. Med.* **2010**, *10*, 607–610. [CrossRef]
389. Cardenas, D. Ethical issues and dilemmas in artificial nutrition and hydration. *Clin. Nutr. ESPEN* **2021**, *41*, 23–29. [CrossRef] [PubMed]
390. Thiermann, U.B.; Sheate, W.R.; Vercammen, A. Practice Matters: Pro-environmental Motivations and Diet-Related Impact Vary With Meditation Experience. *Front. Psychol.* **2020**, *11*, 584353. [CrossRef] [PubMed]
391. Rozin, P.; Haidt, J.; Fincher, K. From oral to moral. *Science* **2009**, *323*, 1179–1180. [CrossRef]

foods MDPI

Review

Global Food Security and Sustainability Issues: The Road to 2030 from Nutrition and Sustainable Healthy Diets to Food Systems Change

Theodoros Varzakas [1,*] and Slim Smaoui [2]

1 Department of Food Science and Technology, University of the Peloponnese, Antikalamos, 24100 Kalamata, Greece
2 Laboratory of Microbial, Enzymatic Biotechnology, and Biomolecules (LBMEB), Center of Biotechnology of Sfax, University of Sfax-Tunisia, Sfax 3029, Tunisia; slim.smaoui@cbs.rnrt.tn
* Correspondence: t.varzakas@uop.gr; Tel.: +30-2721045279

Abstract: The accomplishment of food/nutrition security for all across sustainable food systems (SFS) is tied to the Sustainable Development Goals (SDGs). SFS is connected to all SDGs via the traditional framework of social inclusion, economic development, environmental safety, inclusivity, and the development of sustainable food systems. We suggest that, for the world to achieve sustainable development, a shift to SFS is necessary to guarantee food/nutrition security for all, while operating within planetary boundaries to protect ecosystems and adapt to and mitigate climate change. Therefore, there is a requirement for original approaches that implement systemic and more participatory methods to engage with a wider range of food system stakeholders. However, the lack of skills and tools regarding novel methodologies for food system transformation is a key obstacle to the deployment of such approaches in practice. In the first part of this review, a summary of some challenges that occur in the governance of food system transformation is given. Through a case study of plant-based proteins and their biological and chemical modification as diets shift towards alternative proteins, we demonstrate that resource-efficient food systems and food waste, through system transformation, are useful in understanding both (i) how food system transformation has ensued and (ii) how the required transformation is prohibited. Finally, we discuss the implications of food system transformation in terms of nutrition and sustainable healthy diets, which are needed to achieve changes in food safety systems in the future. The linkage of food and the environment is evident, focusing on nutrition and sustainable healthy diets. This cannot be accomplished without system change and research towards new foods and, more specifically, new proteins such as plant-based ones and their biological and chemical modification.

Keywords: governance; food safety; food system transformation; sustainability transitions

1. Introduction

Owing to the pressures provoked by the present, allied, global food systems leading to health/environmental degradation, challenges to redevelop them to be more sustainable are progressively emerging across the world. Above all, the tendency is to change from individualized agendas to cooperative strategies that can successfully promote the authentic transformation of food systems to be more sustainable. In this sense, according to the European Commission (2020) [1], food system transformation is required in order to shift towards a more sustainable and healthy diet, ensuring holistic food and nutrition security. Hence, a more thorough and comprehensive understanding of the different components of the current food systems and their interactions is required for the maximum co-benefits.

Future-proof sustainable food systems with a focus on health and inclusion are a key focus of the European Commission (EC). The Farm to Fork strategy and the European Green Deal policy are important tools related to the success of these food systems. Food

2030, an EU research and innovation policy framework, supports the transition towards maintainable, innovative, and comprehensive food systems that respect planetary boundaries. Human health, the climate, the planet, and communities will all benefit from the implementation of these systems.

The United Nations Sustainable Development Goals (UN SDGs) [2] and the European Green Deal [3] are considered essential to mitigate the anthropogenic climate change (CC) crisis. They are synergetic since they endorse maintainable agrifood systems and the preservation of the environment [4].

The Sustainable Development Goals (SDGs), approved by the UN in 2015, include several worldwide goals focused on accomplishing a maintainable future for all by 2030 [5]. It should be noted that the change in food systems is crucial due to their unsustainable nature. Climate change, resource scarcity, effluence and waste, environmental degradation, biodiversity damage, human development, undernourishment, and diet-connected non-communicable diseases are all drivers heavily affecting this change.

Food 2030 requires that the entire food system is linked together, associating multiple sectors from farm to fork, i.e., from primary production and food processing to retailing and distribution, food services, and consumption.

All stakeholders should be involved in this process, engaging science–policy–society (consumers). In this way, research and innovation policy will be improved, aiming at coherence and stability, and research and innovation funding and investment will be increased. Hence, the consumer should be an integral part of this process. Finally, the role of innovative technologies should be supported and initiated, along with new approaches and business models, accompanied by social, institutional, and governance innovation relevant to food system change (Figure 1).

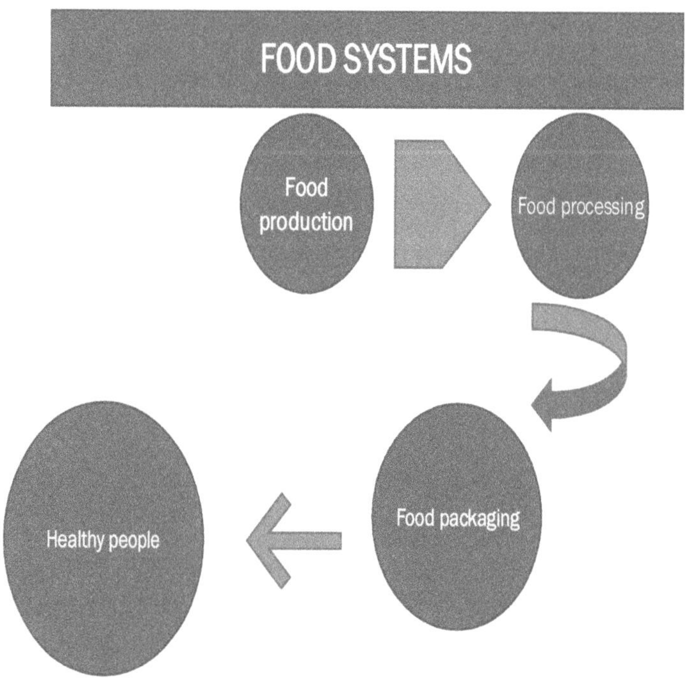

Figure 1. Research and innovation for future-proofing of food systems (adapted from European Commission, 2020).

The co-benefits of these four thematic priorities are presented in the following.

1.1. Nutrition for Maintainable and Healthy Diets

It is important to tackle important issues such as malnutrition and obesity, the support of healthy ageing, the development of novel protein substitutes towards plant-based diets, the improvement of food authenticity/traceability, the support of the cultivation and consumption of overlooked harvests for nutrition and resilience, and the support of the shift towards sustainable healthy diets in Europe and Africa.

Additional expansions and applications of EU food directives and food safety policies (Food Safety—European Commission (europa.eu)) and the Expert Group on Public Health—European Commission (europa.eu) (https://ec.europa.eu/health/non_communicable_diseases/steeringgroup_promotionprevention_en and https://ec.europa.eu/jrc/en/health-knowledge-gateway) (accessed on 28 December 2023) are relevant to Sustainable Development Goals 2, 3, 8, and 10 [1].

1.2. Food Systems Supporting a Healthy Planet

Water, soil, land, and sea should be managed dependably, thus making them available in the future. Smarter food systems are the priority of Food 2030. Hence, they will be better aligned with climate change, and, in this way, will help to preserve the environment. In this direction, environmental risks will be limited and the flow of greenhouse gases into the atmosphere will be reduced [1]. The priority is to devise and operate environmentally friendly and resilient food systems that boost biodiversity, fostering sustainable and healthy agriculture and aquaculture.

1.3. Circularity and Resource Efficiency

The aim of circularity and resource efficiency is to use more efficient and greener industrial processes and logistics in order to reduce food, water, and energy waste. This can be achieved by using unavoidable biomass and waste resources. Another solution is the provision of local food on demand for short supply chains.

Circularity can be defined by the use of maintainable, resource-effective food systems that can manage the 1.3 billion tons of food lost and wasted each year. This could be achieved by zero food waste (FW) policies, the efficient recycling of food waste, the biodegradation of food packaging, limiting microplastics, and responding to the increased demand for more local and healthy food [1].

1.4. Innovation and Empowering Communities

The development of an ecosystem supporting new business models and solutions for society is the aim of the fourth Food 2030 priority.

The achievement of this goal will help to link urban/rural/coastal economies and establish communities across the EU. Closer linkages and partnerships among industry and society will help to create new jobs, decrease prices, and enhance sustainability. Key challenges in this direction include governance innovation, social innovation through citizens' involvement, citizens' engagement in food science and policy, a farm to fork economy with a focus on social innovation, and the development of data-driven food and nutrition systems with the goal of meeting societal needs [1].

1.5. Impact of Climate Change on Food Security

Biodiversity is an essential source of food. In this context, an awareness of species disappearance is necessary, caused by factors such as pollution, pests, and food and medicine control. As an illustration, between 1996 and 2003, the precipitation in parts of equatorial East Africa provoked flooding and reductions in crops and agricultural yields [6]. Consequently, climate change has a direct impact on food production and distribution [7]. Firstly, an increase in the incidence of pests and diseases has been observed, and a loss of biodiversity and a decline in ecosystem functioning has been noted. Secondly, the accessibility of water for crops and fish production and a sea-level rise has been observed [8]. The impacts include the loss of life and food security of millions of people in disaster-prone areas. Through extreme weather, CC will disturb food security and crop yields too. By 2050, it is projected that agricultural yields in Africa alone could decline by >30% [9].

On the other hand, food preparation, processing, acquisition, distribution, and consumption are impacted by CC [10], which influences plant and animal growth, water cycles, biodiversity and nutrient cycling, and the ways in which these are managed for agricultural practices and food production [11]. In addition, CC could amend suitable cultivation zones with a wide range of crops.

CC also influences on income-earning balances, which could affect the ability to buy food, and a changing climate or climate extremes may affect the availability of certain food products. For example, in Tunisia and Egypt, there have been augmented prices for basic foodstuffs [12].

CC has augmented the genetic erosion of landraces and threatens wild species, including crops' wild relatives [13]. As a result, the existing varieties could be lost as farmers replace them with other landraces and improved varieties that are better adapted to the new conditions.

In this review, a summary of some challenges that occur in the governance of food system transformation is firstly given. Through a case study of plant-based proteins and their biological and chemical modification due to the dietary shift towards alternative proteins, we demonstrate that resource-efficient food systems and FW on system transformation is useful in understanding both (i) how food system transformation has ensued and (ii) how the required transformation is prohibited. Finally, we discuss the implications for governing food system transformations in terms of the nutrition and sustainable healthy diets that are needed to implement changes in the food safety systems of the future (Figure 2).

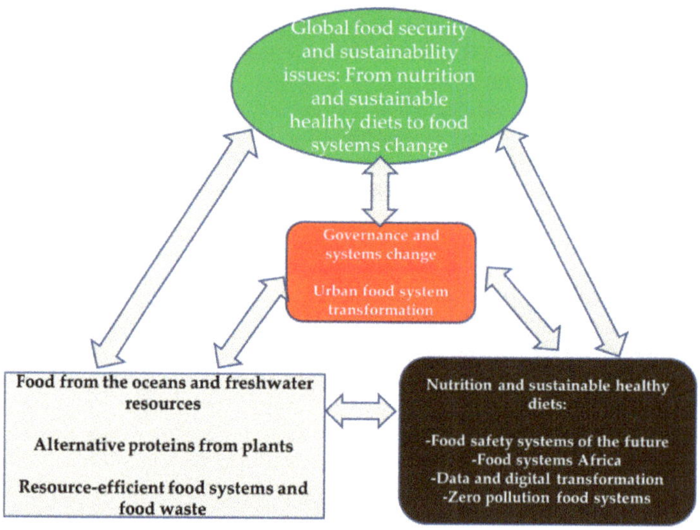

Figure 2. Flowchart underlying governance, food system transformation, and sustainability transitions.

2. Governance and System Change

The first pathway for action is governance and system change. Food systems need to be resistant towards global challenges such as CC, but this will depend upon the successful development, integration, and implementation of policies and strategies, such as the Green Deal and Farm to Fork strategy. The latter will set out an R&I policy and this will lead to effective food system transformation [1].

In this context, the protection and restoration of natural ecosystems, along with the sustainable use of resources, can lead to improvements in human health. Benefits in this direction include climate impartiality and adaption to climate variation. Biodiversity increases and natural resource exploitation will be useful in order to apply farm to fork food policies and nutrition security. In order to achieve these benefits, the consideration of the interactive, socioeconomic/demographic drivers of change is necessary for the redesign of sustainable rural, coastal, peri-urban, and urban areas. It is noteworthy that zero pollution effects should be considered, as noted in the 11th pathway, as well as a fair and democratic environment [1].

An example of this is the 'European SUStainable Food and Nutrition Security' (SUSFANS) project (2015–2019), with the objective of building a framework, evidence base, and analytical tools for EU-wide food policies and their association with consumer diets. Finally, their implications for nutrition and public health in the EU have been outlined in SUSFANS PUBLICATIONS | SUSFANS [14].

According to Rutten et al. [15], there should be a correlation between improvements in European diets and sustainable food systems. Metrics, models, and navigation tools regarding sustainable food and nutrition security have been employed.

SUSFANS developed metrics, identified and analyzed drivers, integrated data and modeling, and formulated foresight for EU sustainable food and nutrition security (FNS). It developed an analytical toolbox integrating new and improved micro-level models of nutrient intake with habitual dietary patterns and the preferences of consumers, along with health impacts [16,17]. Connections among the agri-food/fish sectors with the energy sector, factor markets, labor supply, and health sector and intercontinental employment were assessed by the Computable General Equilibrium (CGE) model MAGNET. However, partial equilibrium (PE) economic and biophysical models in the forest/crop/livestock sectors were analyzed by the Global Biosphere Management Model (GLOBIOM). In the latter, the CC impacts on global agriculture and food availability were discussed thoroughly. In addition, EPIC, the biophysical crop growth model, provided management system-specific weather and soil information.

Food production seems to be affected by CC and plant variability effects. The physiology of crops seems to be affected by the gradual increase in temperatures throughout many regions, hence leading to production and quality limitations [18] Moreover, the capacity of food production is reduced, leading to major economic losses derived from unpredictable cycles of drought and excess humidity in crops [19–22]. This might favor the appearance of new variants of pests and diseases, which will be uncontrollable and aggressive [23].

Jimenez et al. [24] exploited the possibility of using native microbiota as a practical alternative to converse pliability on harvests. Anthropogenic emissions of greenhouse gases (GHGs) affect the development and lifestyles of society, leading to CC [25]. These effects cause irreversible impacts [26]. The Intergovernmental Panel on Climate Change (IPCC), in its sixth assessment report (AR6) in 2018, estimated that, since pre-industrial times (1850–1900), global warming of approximately 1.1 °C was caused by human activities by 2021 [27].

A good solution to the food crisis in many areas of the world is represented by changes in agricultural, forestry, or livestock processes. During 2007–2016, these activities represented around 13% of global carbon dioxide (CO_2) emissions, 44% of methane (CH_4), and 82% of nitrous oxide (N_2O) [28]. Food security is heavily affected by extreme temperatures since many crops are destroyed. Hence, areas of cultivation are being reduced, along with effects on livestock in tropical areas.

The three most significant aspects for changes in precipitation in the future account for (a) an increase in precipitation intensity [29]; (b) hurricanes and tropical storms, causing an increase in the frequency and speed of winds [30]; (c) changes in the start and end of the rainy season in tropical areas [27]. These implications will undoubtedly affect crop cycles, agricultural production, and later post-harvest stages, hence affecting food systems.

In order to reduce GHG emissions, bio-based products should be manufactured using eco-efficient and robust technologies for the processing of biomass and waste in biorefineries [31]. These bio-based production methods and new processes should generate new biomolecules for biostimulation, biocontrol, and fertilization (e.g., struvite, integrated biochar–compost), leading to innovation in crop protection [32]. The bioeconomy and agroecology could also constitute a synergistic solution for the further development of these new climate-smart agricultural systems, as reported by Faucon et al. [33]. They reported on the ecological role of crop diversification, waste recycling, and biomass transformation for agroecological development. Finally, they discussed the holistic approach of the combination of agroecology and the bioeconomy for sustainable agricultural systems.

3. Urban Food System Transformation

In order to understand urban food system transformation, it is essential to consider science-based multi-actor governance processes [1].

Nowadays, urban areas accommodate >50% of the world's populace [34], with an estimated increase of over 70% by 2050 [35,36]. Considering that food consumption in cities is centrally linked to 79% of all produced food [37], the changing demand for food is linked to the urbanization of food [30]. This will of course affect rural areas and agricultural supply chains [38,39].

The understanding of how food is manufactured and consumed comprises one of the main aspects of urbanized justifiable expansion and food security but also affects rural areas, in relation to CC and socioeconomic inequalities [40]. Moreover, the globalization of the 1980s led to the increased disconnection of cities from food [41]. Hence, food systems are managed at the national level, since urban policies and regulations do not often pay significant attention [42,43].

Food security is a major urban problem in developed and high-income countries and around 50 million urban dwellers were found to be food-insecure in 2015, in North America and Europe [44]. This now includes food accessibility.

Cities rely on external markets and long food chains, hence being vulnerable to supply chain shocks, including CC [45] or pandemics [46].

Recently, [47] stated that the key actors towards more sustainable food systems, despite the lack of a clear mandate, are city governments (and territorial communities) [48].

Food system transformation can be defined as 'a process of major and key change in the food system structural, functional and relational issues leading to more equitable relationships and more benign patterns of interactions and outcomes' [49,50].

Enhanced participatory governance structures using a multi-actor approach can be achieved with cities playing a pivotal role, according to Mattioni et al. [51]. National governments, due to their capacity to invest resources in the food system infrastructure, should promote food system transformation beyond local areas to create cohesion [52].

Eight projects contribute to the Food 2030 priority of nutrition and sustainable and healthy diets, nine projects contribute to the Food 2030 priority of the climate and environment, and twelve projects contribute to the Food 2030 priority of innovation through empowering communities.

A useful framework based on place-based solutions, the connection of food with the climate and community, and the circularity and diversity of approaches is the recently developed Client-Led Information System Creation (CLIC) framework [53]. CLIC stands for 'conceptual framework for integrated food policies and intervention design' and is conceptualized by four pillars:

1. co-benefits across social, environmental, and economic objectives;
2. linkages between rural and urban areas;
3. the inclusion of all stakeholders and their knowledge;
4. connectivity between food and other policy priorities (e.g., Food 2030).

4. Food from Ocean and Freshwater Resources

A key factor for European and global food and nutrition security is seafood production through harvesting (fisheries) and farming (aquaculture). Primary food production systems contributing to food and nutrition security by 2030 comprise sustainable fisheries and aquaculture [54].

Europeans consume roughly twice as much as they produce [55] and most imports come from Asian countries.

By 2030, aquaculture could enhance seafood production and deliver close to two thirds of the global seafood demand [56]. However, this necessitates development in sustainable and less impactful ways, including freshwater aquaculture, which is decisive for noncoastal countries, as reported in the Blue Growth Strategy.

Sustainably farmed seafood production requires overcoming obstacles such as a lack of knowledge of the elementary biology/ecology of fish and shellfish, sickness prevention, and management. Moreover, there is a need to control weak governance structures for fisheries management and build on new technology uptake by the fisheries sector, as well as the consumer acceptance of farmed seafood. Hence, better risk assessment and management in seafood systems will be required [1].

Approximately three billion people are supplied with fish, with an average per capita animal protein intake of 20%, accompanying various crucial micronutrients. Around 10–12% of the world's population depends on blue foods for their livelihoods [57].

Operational costs for aquaculture producers, seafood processors, and fishermen come from energy and raw material price increases, according to Rahman et al. [58]. Prices are reaching EUR 1 per liter in several EU nations, with the industry claiming profits from EU vessel operations of up to EUR 0.60 per liter [59,60].

The worldwide transition to a sustainable agri-food system will be supported by the EU. This will include the sustainable management of fish and seafood resources and the control of ocean governance, marine cooperation, and coastal management. Illegal, unreported, and unregulated fisheries will face a zero tolerance policy. The governance of the agriculture and fishery industries can enhance the cycle of sustainable development for food and nutrition security, as reported by [61–64]. This will affect global food security and development.

5. Alternative Proteins and Dietary Shift with a Focus on Biological and Chemical Modification of Plant-Based Proteins as a New Sustainable Solution

Changes in ecosystem services have resulted from heavy industrialization and agricultural intensification (excessive use of fertilizers and pesticides), which have radically changed the N and P cycles, indispensable for plant growth. Hence, if water is polluted, this will affect soil productivity [65].

The application of a reasonable reduction in animal-based calorific consumption can reduce emissions from agricultural production, as stated by the Commission's Communication, 'A Clean Planet for All' [66].

The Intergovernmental Panel on Climate Change (IPCC) [67] stated that composed diets (plant-/animal-based foods) can achieve climate change adaptation and mitigation and this can be beneficial for human health [68]. Moreover, one fifth of the change requested to limit warming to <2 °C could arise from dietary shifts.

Milford et al. [69] suggested the indirect interrelationship of consumers' preferences and consumption habits. Castellani et al. [70] showed that dietary shifts towards less animal-based food, resulting in a decrease in the environmental impact of food consumption, will be affected by marketing issues. Consumers' choices and food production will affect the linkage between the food supply and diets [1]. Hence, research on the improvement of the processing of alternative proteins is imperative, taking into account biotechnology pathways that provide nutritional and sensorial food quality coupled with environmental sustainability.

Regarding plant-based proteins, a focus on biological and chemical modification is described, since it is a means to improve them and contribute to their sustainability and full utilization.

Proteins offer a range of amino acids that are essential in preserving human health, and, in food technology, they provide some functional properties, e.g., as stabilizing, emulsifying, thickening, gelling, foaming, and binding agents [71–73]. Biochemically, proteins' capability to retain their functional traits relies on their intrinsic structure, their shape/configuration, and how they interrelate with food constituents [74]. Nowadays, the processing and exploitation of plant proteins have attracted universal attention and numerous scientific investigations are focused on the improvement of the application of plant proteins in the pharmaceutical and food industries through modification techniques. The latter could increase their techno-functional aspects, bioavailability, bioactivity, and digestibility characteristics [75–77].

Biological and conventional food processing changes to food proteins are easy to implement and have been extensively employed. Current chemical protein strategies for site-selective modification can be mastered by adding new, efficient moieties or the exclusion of components from the protein structure. Numerous chemical and biological modification approaches will be discussed in this section, with examples contributed precisely in the plant-based protein context. Martinez-Alvarenga et al. [78] produced glycoproteins with maltodextrins (MD) attached per whey protein isolate (WPI). Their solubility was augmented at the pI by combining them with MD. These authors demonstrated that the increase in the foaming capability and foam stability of WPI after glycation with maltodextrin was linked. In the same way, emulsification was improved and presented monomodal performance due to the developed steric revulsion [79]. In this study, the authors demonstrated the potential of the electrospinning technique in the modification of pea protein isolate and as a pretreatment to facilitate the production of Maillard conjugates with improved functionality. In another study, oat protein was glycated with dextran and β-glucan, which enhanced its solubility and emulsification potential via alterations in its secondary structure, leading to the random coiling of the protein conjugate [80]. The conjugated product displayed elevated viscosity, attributed to the functional properties of oat β-glucan and its concentration. In the Meng et al. study, it was reported that the Maillard reaction enhanced the solubility and emulsion capacity of rice dreg protein (RDP) [81]. It was reported that the modified protein conjugates had improved immunomodulatory properties when examined using a cyclophosphamide-induced immunodeficiency animal model. To improve the thermal stability of canola protein isolate, Maillard glycation was effectively applied by Pirestani et al. [82]. Through Maillard reactions in aqueous solutions, Xue et al. [83] revealed that buckwheat protein isolates reacted with dextran to produce conjugates and this reaction was enhanced with ultrasound. Structurally, more random coils and less α-helices in the conjugates were generated. The conjugates presented comparable solubility behavior and improved emulsification and thermal stability. The functional properties of food proteins can thus be increased by several modifications.

Amongst all chemical modifications, phosphorylation has been established as an effective method of enhancing the functional properties of proteins like potato protein isolate (PPI). At pH 8, phosphorylated products (PP-PPI) with sodium trimetaphosphate (STMP) were marked by the maximum oil absorption and foam capacity and emulsion activity. At pH 10.5, WHC was superior to the native ones [84]. Similarly, Sánchez-Reséndiz et al. [85] demonstrated the improved emulsifying activity of peanut and soy protein isolates after phosphorylation using STMP. In this study, peanut protein isolates' phosphorylation changed greatly the values of the emulsifying activity (+6.6) and in vitro protein digestibility (1%). Remarkably, in soybean, all functional properties were improved, excluding the water solubility index (WSI) and foam activity (FA). The thermal aggregation and viscoelasticity of rice glutelin (RG) were also enhanced after STMP phosphate modification, as stated in the study of Wang et al. [86]. These authors concluded that the generation of protein aggregates was credited to the interactions between proteins, comprising covalent and non-covalent interactions. The particle size distribution, intrinsic fluorescence emission spectra, surface hydrophobicity, and rheological behavior results indicated that three steps were realized: (i) the RG was unfolded; (ii) the unfolded proteins created oxidized free SH groups to form disulfide bonds and hydrophobic interactions; and (iii) a 3D system structure was formed. The modification of the secondary structures through heat treatment presented an increase in β-sheets and a decrease in α-helix content, leading to the generation of protein aggregates [86].

Protein acylation between the amino acid residues of proteins and anhydrides is one type of chemical modification that serves to increase the function and physicochemical properties of food proteins. Zhao et al. [87] found that the addition of succinic anhydride to oat proteins apparently improved their gel properties compared to the unmodified oat protein. After acylation, the gelling property of rapeseed protein achieved additional functionality, as assessed by Chen et al. [88]. Heightened thermal stability, solubility, and emulsifying properties were also reported due to succinylation-induced changes in the secondary structure of male date palm pollen protein concentrate [89].

In contrast to acylation and alkylation, involving chemicals, deamidation can be achieved under moderate conditions and without additional molecules. Since legumes/cereals have very high proportions of Gln and Asn, deamidation can be a suitable tool to apply them in the food industry using several approaches (such as alkali, acid, and enzyme). For instance, by using acids such as acetic acid, citric acid, and tartaric acid, the bitterness of wheat gluten hydrolysates was masked [90]. To improve the solubility of rice bran protein, Guan et al. [91] used alkaline deamidation (pH 12 and 120 °C for 15–30 min). For enzymatic deamidation, it was reported that GTase was the most investigated enzyme in the deamidation of plant-based proteins. Through GTase deamidation, Hadidi et al. [92] improved the WHC, solubility, and emulsifying and foaming properties of *Oenothera biennis* L. protein; in the same way, a techno-functionality improvement in pea protein isolates was observed when GTase deamidation was utilized [93]. Industrially, it should be noted that only deamidase isolated from *Chryseobacterium proteolyticum* has been applied [94].

Biological modification, linking enzymatic and fermentation approaches, is an alternative modification technique that is eco-friendly, less energy-consuming, and free from the production of toxic by-products. Regarding the enzymatic approach, the class of the employed enzyme plays a central role in the final features of the modified plant proteins, considering that their molecular changes translate into different specific cleavage sites [94]. For example, tryptic hydrolysis, especially at pH 4, resulted in the formation of oat protein peptides with an improved, homogenous foam structure, a rapid foaming ability, and a highly viscoelastic interfacial film [95]. In addition, pepsin can generate hydrophobic peptides with high hydrophobicity and surface-active properties [96]. These authors observed the improved WHC and OHC of pea protein-enriched flour after pepsin hydrolysis. Sun et al. [97] prepared enzymatically modified walnut dregs (CPMP), based on protease, and indicated that the generated product showed interesting emulsifying and foaming properties. On the other hand, fermentation has been used as a biological tool for plant-based protein modification. In this sense, various starter cultures have been used for the fermentation of plant proteins, such as lactic acid bacteria (LAB), mold, yeast, and *Bacillus* strains. Fermentation, using Lactobacilli strains, was recorded to enhance soy protein's solubility, WHC and OHC, and foaming properties [98,99]. Similar trends were reported by Klupsaite et al. [100], who used *Pediococcus pentosaceus* KTU05-9 in the solid state fermentation of lupin protein. Interestingly, fermentation can significantly decrease the beany and bitter off-flavors of different plant-based proteins [101]. These authors used eight LAB starins, and *Lactobacillus brevis* had strong potential to improve the techno-functional properties of lupin protein. A summary of the above chemical and biological modifications is given in Table 1.

Table 1. A summary of the chemical and biological modification approaches for plant-based proteins.

Modification Strategy	Methods/Reactions	Source of Plant Protein	Techno-Functional Properties	References
Chemical	Deamidation—alkaline	Evening primrose (*Oenothera biennis* L.)	- Improvement in functional traits - Edible protein was generated	[92]
	Deamidation—glutaminase	Pea	- Improvement in solubility - Improvement in techno-functionality characteristics of pea protein isolate - Increase of beany flavor, grittiness and lumpiness.	[93]
	Acylation and additional transglutaminase catalysis	Rapeseed	- Improvement in gelation properties - Rapeseed protein isolate (RPI) had good thermal stability, gel strength, apparent viscosity, and surface roughness	[88]
	Glycation by electrospun fiber-assisted drying	Pea	- The emulsion stability and solubility of pea protein hydrolysate were improved	[79]
	Glycosylation (microwave-assisted wetting)	Rice	- The solubility was increased - With Maillard reaction, the emulsifying capacity was enhanced - Good immunomodulatory properties of rice dreg protein	[81]
	STMP phosphorylation	Rice glutelin	- The turbidity of phosphorylated rice glutelin (PPRG) was enhanced - The viscoelasticity, phosphorylation, and thermal aggregation of rice gluten were improved	[86]
	Deamidation—alcalase hydrolysis	Wheat	- The bitterness was masked	[102]
	Acylation and glycation	Rapeseed	- Improvement in gelation properties - Improvement in water absorption capacity and textural properties	[103]
	STMP phosphorylation	Soybean and peanut	- Improvement in emulsifying activity - Improvement in in vitro protein digestibility >1%	[85]
	Deamidation	Rice bran protein	- The water solubility was enhanced (=90%) at pH 12 and 120 °C for 15–30 min - Thermal property was preserved	[91]
	Glycosylation (wet)	Canola	- Improvement in viscosity - Improvement in physical structure	[82]
	Glycosylation (ultrasound-assisted drying)	Buckwheat	- Improvement in surfactant capacity - Ultrasonication enhanced the functional properties - Improvement in emulsion stability and solubility	[83]
	Deamidation—proteax/glutaminase SD-C100S	Wheat	- The bitterness was masked	[76]
	Glycation	Oat	- Improvement in emulsification ability and solubility	[72]
	Glycation	Whey	- Improvement in foaming properties and protein functionality	[70]
	Phosphorylation with sodium trimetaphosphate (STMP)	Potato	- Phosphorylation is pH-dependent: (pH 5.2) -Improvement in content of all amino acids of potato protein isolate (PP-PPI) - At pH 10.5, decrease in all content - At pH 8: improvement in oil absorption capacity and emulsion activity and ↗ foam capacity - At pH 10.5: improvement in water absorption capacity	[84]

Table 1. *Cont.*

Modification Strategy	Methods/Reactions	Source of Plant Protein	Techno-Functional Properties	References
Biological	Enzymatic: TGase (transglutaminase)	Peanut	- Improvement in the emulsifying activity index (EAI), emulsifying stability index (ESI) - Improvement in gelation and oil-binding properties - Decrease in protein solubility	[104]
	Enzymatic: proteolytic enzymes	Pea	- Improvement in protein solubility at pH 4.5 at all times during hydrolysis - Improvement in foaming with trypsin hydrolysates and emulsifying capacity	[105]
	Enzymatic: papain and pepsin	Pea	- Improvement in WHC and OHC - Decrease in foaming properties and emulsifying properties	[96]
	Enzymatic: pectin methyl esterase	Pea	- Improvement in the degree of esterification - Improvement in solubility	[106]
	Enzymatic by complex proteases	Walnut	- Improvement in solubility - Improvement in water holding capacity - Improvement in emulsifiability and emulsion stability - Decrease in oiliness - No change in foaming features	[97]
	Enzymatic: TGase	Coconut	- Improvement in the mechanical and barrier properties of films based on modified coconut protein	[107]
	Enzymatic: chymotrypsin and protease	Quinoa and amaranth	- Improvement in antioxidant, antimicrobial, and antihemolytic properties	[108]
	Fermentation: *Lactobacilli* strains and *Staphylococcus xylosus*	Lupin	- Improvement in foaming properties and emulsifying properties - At pH 4: no change in solubility - At pH 7, a decrease in solubility and bitterness	[101]
	Fermentation: *Lactobacillus plantarum* strains	Soy	- Improvement in surface hydrophobicity - Emergence of β-strand structure	[99]
	Enzymatic glycosylation: TGase	Black soybean	- Improvement in solubility, rheological properties, and emulsification	[97]
	Fermentation: *Pediococcus pentosaceus* KTU05-9	Lupine	- Improvement in solublity and functional properties, at pH 8.0 - Decrease in bread hardness, chewiness, and resilience - Improvement in bread springiness	[94]
	Fermentation: *Lactobacillus helveticus*	Soy	- Decrease in the beany and bitter off-flavors	[98]
	Fermentation: *Bacillus licheniformis*	Peanut	- Improvement in nutritional properties - Improvement in antioxidant potential	[109]

6. Resource-Efficient Food Systems and Food Waste

Accelerated action to reduce food loss and waste represents United Nations Sustainable Development Goal 12.3, with the target of halving FW by 2030 [110]. "A reduction in food quantity and quality" is the definition of food loss and waste [111] and refers to "food lost or wasted in the part of food chains leading to edible products going to human consumption" [112]. Fish provides 20% of the average per capita intake of animal protein, making fisheries central to achieving food security [113] in order to feed more than 3.3 billion people globally [114]. All food losses take place along the food supply chain (FSC) and the retail level is included through FW [1]. Two separate indexes, the Food Loss Index (FLI) and the Food Waste Index (FWI), have been reported by the FAO and the United Nations

Environment Program (UNEP). It is projected by the FLI that 14% of food produced is lost from post-harvest without retail [115,116].

Drivers of FW are differentiated as below.

Specific food products are represented by generic and systemic approaches [117]. The highest FW generation (46%) comes from the consumption stage, succeeded by principal production and processing/production at 25 and 24%, respectively.

Regarding FLW, according to the World Food Programme [118], hunger is a common paradox as it leads to food insecurity. In 2019 alone, the EPA estimated [119] that the food retail, food service, and residential sectors accounted for approximately 66 million tons of wasted food, with most of this waste (about 60%) directed to landfills. The EPA estimated that, in 2018 in the U.S., 24 percent of the amount was landfilled and 22 percent of the amount was combusted with energy recovery. Moreover, they reported that landfills and combustion facilities were overloaded with more food than any other single material [120].

Reducing wasted food saves resources such as land, water, energy, and labor; reduces greenhouse gas emissions (the majority of greenhouse gas emissions from wasted food result from production, transport, processing, and distribution); and reduces methane from landfills. It is necessary to consider that when wasted food enters landfill, the nutrients in the food never return to the soil. The EPA estimates that 58% of landfill methane emissions to the atmosphere come from wasted food [121]. Nutrients can be returned to the soil by composting, hence supporting a circular economy (CE).

In 2021, the EPA released the first of two reports in a series on the environmental impacts of wasted food. Part 1 was titled From Farm to Kitchen: The Environmental Impacts of U.S. Food Waste (https://www.epa.gov/land-research/farm-kitchen-environmental-impacts-us-food-waste [120]) (accessed on 28 December 2023), and Part 2 was titled From Field to Bin: The Environmental Impacts of U.S. Food Waste Management Pathways [121], released in 2023.

The most important themes in discussing FW nowadays are as follows: (1) the anaerobic digestion of FW for CE conception; (2) FW systems and life cycle valuations for CE; (3) bio-based CE methods; (4) consumer performance and approaches towards CE; (5) food supply chains/FW in a CE; (6) material flow analysis and sustainability; (7) challenges, policies, and practices involved in achieving circularity; and (8) CE and outlines of consumption [122].

The CE is designed to substitute traditional linear supply chains with systems in which materials are recycled within creation systems, grounded on the principle of "waste = food" [123,124], reinforcing the transition from recycling to upcycling [125]. This refers to any waste transformation process transforming waste into higher-value products by using them as input for other products. Hence, the CE seeks to transform one person's waste into another person's resources [126], stimulating radical innovation and integrating human activities into ecosystems [127]. International policy makers often discuss the need to shift towards a CE [128]. However, in order to allow the CE to shift towards sustainability, several actors need to be engaged [118], including society and consumers [129].

A food use hierarchy should be employed focusing on prevention, followed by the redistribution and reprocessing of surplus food to people in need, the production of animal feed, and recycling and disposal, as shown in Figure 3 [130].

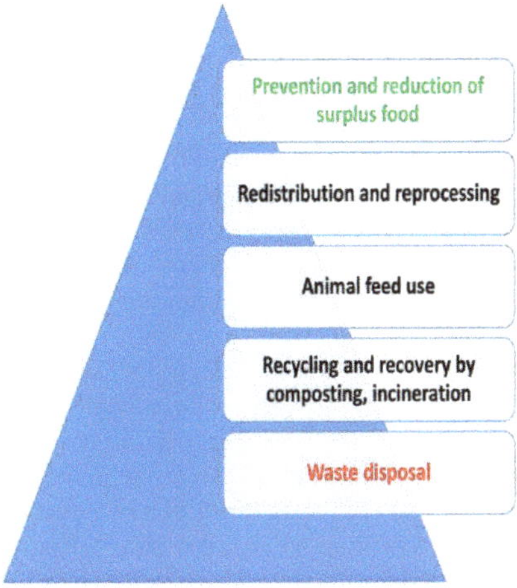

Figure 3. Sequence of management of food surplus, by-products, and FW deterrence: 10 strategies. Adapted from [130].

A Malaysian model has been proposed analyzing the factors affecting FW [131]. Poor food management practices and gender are important issues that affect FW, reinforced by the consumer behavior concept. The projected presented rice waste in the CE model, which was well accepted by the public. Moreover, they mentioned people's readiness to pay a certain sum of money to process their FW.

A study conducted in Daegu, South Korea in 2019–2020, collecting FW from 218 households, showed an average daily contribution of FW of 0.73 kg per household, with the equivalent greenhouse gas emissions of 0.71 CO_2, a water footprint of 0.46 m^3, and economic losses of KRW 3855.93 [132].

7. The Microbiome

Relevant EU policies in this direction involve energy security and climate change, climate mitigation [133], and adaption to CC, referring to the new bioeconomy strategy [134].

Beneficial symbiosis provided by certain microbiomes can lead to higher yields and nutritious food. The Soil Health and Food mission [135] requires sustainable and circular management, the use of natural resources, and soil health improvements.

A possible agronomic management measure for crops is the exploitation and exploration of the microbiota in agroecosystems in symbiosis with crop plants. Multiple benefits are underlined by this innovative strategy, such as coping with biotic and abiotic stresses—for instance, heat stress [136,137], drought [138,139], waterlogging [140,141], and the attack of pests [142] and diseases [143,144].

8. Nutrition and Sustainable Healthy Diets

Major challenges in global and EU food systems involve all aspects of nutrition, including malnutrition (undernutrition, over-nutrition, and micronutrient deficiencies), as well as environmental issues such as climate change and resource scarcity, in connection with urbanization and food poverty.

According to IPES FOOD [145], approximately 2 billion people are overweight or obese, 2 billion are affected by micronutrient deficiencies, and 800 million people are undernourished.

A leading non-communicable disease (NCD) hazard feature and a principal cause of obesity is an unmaintainable and unnatural diet, with an augmented demand for livestock products, along with the consumption of (ultra)processed food that are high in calories, with no nutrients [146].

The reduction of obesity rates over the last few decades has not been accomplished by any EU country to a significant extent [147]. In Europe, at present, children and the poorest individuals are severely affected, with more than half of the adult population being overweight or obese [1].

The adoption of long-term healthy and sustainable diets and the reduction of non-communicable diseases (NCDs) are representative innovative solutions to achieve optimal health and well-being. A more thorough understanding of the relation between lifestyle (including nutrition and alcohol), geographical (national/regional and rural/urban zone), and environmental factors is required, as well as assessing biological parameters and the risk of NCDs.

Consumer choice is affected by consumer behavior, and dietary behavior is not only impacted from birth, but also by food environments, policies, gender, and nutrition labeling [1]. Food choices, lifestyle, motivation, and decision-making are heavily affected by human neurobiological pathways. Hence, people should modify their dietary models and make healthier selections. In this direction, the combination of data from different domains and artificial intelligence needs to be employed towards the study of consumer behavior and dietary intake, as well as their impact on health and environmental sustainability. In addition, the optimization of nutritional, structural, and functional food properties from new raw materials or organic ones with a low impact on the environment and in accordance with new style food consumption preferences should be implemented [1].

A One Health approach should be adopted, showing the links among the health/well-being of people–animals–plants–environment, that sustains their existence [148], optimizing the health of people, animals, and ecosystems [149]. This includes nutritious food [150].

Threats to "life on Earth" have been addressed by the Manhattan Principles [151]. They introduced 12 maintainable methods to evade and avoid epidemic or epizootic diseases [151,152]. Some actions require holistic expertise [153–157]. In this context, potential regional differences regarding One Health research priorities were observed by [158]. Potential regional gaps and differences in One Health research priorities were highlighted, emphasizing the surveillance versus policy activities in One Health.

Global food systems should be able to provide access to diets that allow adolescents to grow and thrive and not to fail [159]. Access to nutritious, safe, affordable, and sustainable foods for all adolescents should be critical [160]. Dietary intake should be improved for most adolescents around the world and not be based on energy-dense, nutrient-poor processed and ultra-processed foods [159,161] with low intake of fruits, vegetables, and whole grains. Another reason is that ultra-processed foods have a higher environmental impact and affect the sustainability of dietary patterns [162].

Despite this, sustainable adolescent nutrition should cover the large gaps in dietary intervention, policy, and programming that still remain [163]. Their complex food environments should be explored further, investigating their decisions regarding what to eat, where to eat, and where to buy food [164]. Adolescent policy and practice initiatives should be adopted and implemented effectively [165,166].

Moreover, the basis of all policy design should be explicitly adolescent-centered and designed to co-construct knowledge [167–170].

Nutritional life cycle assessment (n-LCA) can be used to measure environmental impacts against the nutritional value of food levels (i.e., production systems, food items, and diets/food supply). It can use nutrient or health metrics in the impact assessment phase, as carried out in the CONE-LCA framework, reported by Stylianou et al. [171].

An illustration of milk and dairy production in the food economy system and the food chain is represented by the attitudes of a group of young Polish consumers towards selected features of dairy products [172]. Consumer opinion surveys include key elements of improving the food market and assessing consumers' approaches to current issues related to access to high-quality food. Gaworski et al.'s [172] investigation aimed to determine the attitudes of young Polish consumers about dairy products. The objective was linked to the evaluation of selected features of dairy products and their packaging and the assessment of regional products and novelties in dairy production. These consumers pointed out the importance of the quality/taste of dairy products and the minor role of packaging. Additionally, most respondents claimed that they did not notice to the biodegradability of the packaging. When asked about regional dairy products, respondents paid great attention to their value, resulting from natural methods of production, without preservatives. However, a small number of young respondents showed knowledge of the idea of dairy production 'from grass to glass', which would indicate inadequate interest in innovative solutions in the dairy sector.

Nowadays, the most comprehensive assessment of nutrient metrics in LCA is the recent FAO paper by McLaren [173], which establishes a basic foundation and high-level overview. Other papers on LCA include [174–177].

The nutritional and environmental contributions of selected production practices can be measured by farmers, who receive higher prices for foods with a stronger sustainability profile [178,179].

Green et al. [180] also highlighted the inclusion of capping, weighting, energy standardization, across-the-board versus group-specific aspects, dietary- and/or context-specific aspects, validation, disqualifying nutrients (e.g., saturated fat), reference amounts, processing quality, the selection of nutrients/ingredients, interpretation, and data quality. The largest impact was attributed to energy standardization and dietary specificities when assessing nutrient indices in isolation, as well as capping and disqualifying nutrients. Spearman rank correlations and Wilcoxon signed rank p-values were also calculated.

9. Food Safety Systems of the Future

To become a climate-neutral continent by 2050, the EU has promoted the Green Deal, with the Farm to Fork and the Biodiversity strategies as its foundations [181]. The goal of the Farm to Fork strategy aims to transition to a maintainable food system [182]. The Biodiversity strategy involves numerous proposed policy actions with the goal of lower greenhouse gas emissions, biodiversity conservation, reduced pesticide use, and augmented consumer empowerment [183]. The Farm to Fork strategy predicts the creation of a more justifiable food system by applying restrictions on fertilizer and pesticide use and placing at least 25% of the EU's agricultural land under organic farming by 2030 [182]. The Farm to Fork strategy and the Biodiversity strategy aim to enhance European biodiversity levels by increasing the agricultural land amount under high-diversity landscape features to at least 10% [182]. However, there are concerns regarding whether organic production and certification are adequate to preserve and expand biodiversity and to achieve the climate targets as defined in the UN Sustainable Development Goals. Therefore, holistic approaches that go beyond organic production, which also address the role of consumers in food labeling, are suggested [184]. In this context, the ecological food system framework initiative was introduced, with the goal of making the EU food system sustainable and integrating sustainability into all food-related policies [182]. While progress has been made in quantifying product-specific environmental impacts, major challenges in how to track and communicate these influences continue. For instance, there is a lack of EU-wide monitoring frameworks to appraise progress towards food sustainability objectives. An initiative for green claims was launched to substantiate green product credentials against a regular methodology to evaluate their impact on the environment.

An integral part of food and nutrition security (FNS) that is very important and significant for health and a sustainable environment is food safety. The Rapid Alert System of Food and Feed (RASFF) and legislation from farm to fork focus on ensuring a high level of food safety and animal health and welfare and plant health in Europe. The European Food Safety Authority (EFSA), responsible for scientific advice, is developing new risk assessment methodologies for emerging new foods and existing ones. Despite the strong focus on food safety, there is limited evidence of an integrated systemic approach to FNS, leading to the decentralization of food policy [185,186]. Food safety is part of a sustainable food system, minimizing the risk of the transmission of toxins or pathogens through the food system and minimizing the use of antibiotics, pesticides, and other substances of concern. Complex interactions due to CC with a number of food safety hazards lead to increased risks of foodborne illnesses and affect safe and nutritious food for millions of people around the globe, and these should be considered and addressed by innovative global food security systems. Food security and livelihoods could be at risk despite the lack of connectivity of the current COVID-19 pandemic to food safety in the EU. Proactivity regarding biological hazards and, in the case of the food system, those that emerge throughout the food chain remains imperative. Food safety risks might also be increased by changes in food and farming systems. According to the recently published FAO document on CC and food safety [187–191], the understanding of CC and novel approaches and applications should be furthered. Emerging technologies in various areas of the food chain should be further strengthened among all actors and stakeholders. This will lead to innovation and provide solutions to address food chain challenges in association with consumers.

10. Food Systems in Africa

Meta-analyses suggest that mobile devices disseminating agricultural information in sub-Saharan Africa and India have improved yields by 4% and the adoption of agrochemical inputs has increased by 22% [192].

Better nutrition performance of African farming systems should be sought, thus strengthening the link between agro-biodiversity, aquaculture systems, and food diversity. Technological, food safety, social, and gender issues should be taken into account regarding local food systems. These will include sustainable post-harvest technologies and bio-based packaging approaches, for the reduction of food waste. The diversity of diets and improvement of food identity will be derived from small farmers and processors, benefitting rural areas. This means that a focus on food supplies for local urban markets and high-value global markets is essential. Diet-related, non-communicable diseases and persistent undernutrition are some of the most commonly reported nutritional imbalances in both Europe and Africa. The UN predicts that the global population will increase from 7 billion to >9 billion by 2050 [193], of which the majority will reside in Africa. Nutrition performance needs to be linked with sustainable agricultural systems, thus strengthening the agro-biodiversity of resilient cropping systems. This will lead to a more balanced, healthy diet in order to satisfy population growth and address challenges associated with enhanced climate change. Furthermore, the development of resource-efficient, resilient food value chains should deliver sufficient, safe, affordable, and nutritious food to local consumers and for high-value global markets. Africa's wealth of local varieties, food intelligence, and healthy diets, including plant-based proteins, should be exploited and explored, as they are currently not considered and not reaching the market.

In the framework of the SDGs, the EU–Africa RI Partnership on FNSSA proposals describe how projects can contribute to

- sustainable, healthy African diets based on sustainable and secure food systems (comparable to the Mediterranean diet);
- the sustainable growth of food chain operators (SMEs) in rural areas in Africa, along with the involvement of small farmers (including aquafarmers);
- novel food products, tools, and processes applicable in Africa, addressing food safety issues across the entire food value chain;
- the implementation of nutritional recommendations leading to significant reductions in childhood malnutrition in Africa;
- the founding of the EU–Africa Research and Innovation Partnership on FNSSA and influence at a local level;
- pilot modernization activities for the benefit of African and European consumers.

InnoFoodAfrica [194] focuses on nutrition, with the overall objective of improving the nutrition and well-being of African people via the analysis of diets, growing a wider variety of crops, and developing new food products and ingredients and new food processing technologies. This is complemented by the use of crop side streams to produce biodegradable composite materials for packaging and other end uses.

To obtain a deeper understanding of the nutritional situation in several African cities, a survey was conducted on sub-Saharan diets. Through the survey results, nutrient gaps were analyzed and the authors created recommendations to close these gaps. Here, the Food 2030 priorities become very tangible: affordability needs to be a key element in relation to nutrient availability, ensuring that healthy diets are affordable. Some investigations have been conducted that indicate how the aquaculture and fishery industries are managed in Africa as a means of guaranteeing food security and economic growth [195–198].

11. Data and Digital Transformation

The economy and society are being transformed by digital technologies. Data are growing and expected to increase significantly by the year 2030. Newly evolved technologies such as artificial intelligence and new smart-connected objects (e.g., farm machinery, robotics, home appliances, wearables) have led to this data increase, along with data processing and analysis [1,199].

Enormous benefits for citizens, farmers, food businesses, researchers, and society have been brought by this data-driven innovation.

Digital technologies have entered the whole food supply chain and allow us to address techno-economic challenges in the agro-food sector. Digitalization in agricultural and food systems, introducing innovative tools and systems, can help in the achievement of the 2030 United Nations Sustainable Development Goals.

In Sridhar et al. [200], the application of digitalization in the agriculture and food sectors has been reviewed. Artificial intelligence, precision farming, and big data analytics have been employed and applied in agriculture. Environmental, social, and economic sustainability would be the end result of the integration of these techniques. A 23% reduction in costs and a nearly 5% decrease in the volume of medium-level waste have resulted from the application of these digital technologies in agriculture.

The amalgamation of these smart systems and technologies in the agricultural and food sectors might affect their productivity and output, since they allow the highly effective storage, transmission, and retrieval of electronic data.

All sustainable goals will be affected by the application of digitalization in the agro-food sector; however, the main pillars are related to SDG 2, which centers on eradicating hunger and achieving food security, and SDG 1 (no poverty), focused on improving access to markets, financial services, and valuable agronomic information. The final outcome will be the enhancement of the livelihoods of smallholder farmers and marginalized communities. Real-time monitoring and traceability, thereby reducing foodborne illnesses and making food safer, in the context of SDG 3 (good health and well-being), will also be beneficial due to the application of digital technologies.

SDG 6 (clean water and sanitation and improved water resource management and conservation) will be satisfied by the use of digital technologies in agriculture. SDG 8 (decent work and economic growth), which requires the creation of employment opportunities within the agro-food tech sector and the promotion of sustainable economic growth will also be affected by the application of digitalization.

SDG 9 (industry, innovation, and infrastructure) will also benefit from innovative agro-food technologies, through improving the production and distribution infrastructure.

SDG 12 (responsible consumption and production), focusing on resource-efficient digitalized farming practices, will be affected, hence reducing food waste and enhancing traceability in supply chains.

Digitalization will also affect climate resilience and mitigation effects, hence supporting SDG 13 (climate action). SDG 14 (life below water) will also be affected due to more sustainable marine ecosystems [200].

The optimization of seafood farming practices can be achieved by digitalization, thereby reducing the strain on marine resources and aiding the conservation of aquatic life.

Furthermore, sustainable land use and agriculture, as achieved with the use of digital technologies, aligns with SDG 15 (life on land). The minimization of the use of harmful agrochemicals, the reduction of soil erosion, and the encouragement of sustainable land management, affecting the conservation of terrestrial ecosystems and biodiversity, can be achieved by data analytics.

Industry 4.0 is a new era with biofortification. The enhancement of the nutritional content of crops will be carried out, directly aligning with SDG 2.2 and attaining SDG 2.3 through the improvement of agricultural productivity. Ensuring sustainable food production systems will also contribute to SDG 2.4. Sensors could also be used along with hydroponics systems and vertical farming [201].

Artificial intelligence (AI) connects three subfields: machine learning, deep learning, and neural networks. Algorithms for statistical predictions and inferences, such as image recognition, are employed in machine learning [202]. The connection of information arises from deep learning via the use of various datasets or nodes. Speech recognition is accomplished by deep learning. Neural networks, which leverage statistical models and algorithms, can predict data and recognize patterns [203]. The prediction of weather conditions and soil patterns essential for the growth of crops can be achieved by employing these technologies.

Big data involve the five Vs, namely volume, variety, velocity, veracity, and value [204]. Volatility, validity, visualization, variability, vulnerability, visibility, and vagueness are also considered [205]. Robots, satellite imaging, remote sensing, and geospatial data are being used the most in understanding crops and performing land mapping and soil management [206–212].

Several barriers exist that limit the potential of data-driven innovation.

Lack of data governance: missing organizational approaches and structures.

Imbalances in market power: a large amount of data is accumulated by a small number of players (e.g., food retailers, data aggregators like Google or Facebook, equipment manufacturers, ERP platforms). The implementation of new approaches (towards SMEs, farmers, and citizens) will tackle the oligopolistic characteristics of today's data economy [213,214].

Empowerment of individuals to exercise their 'data rights': the generation of large amounts of data might lead to risks of discrimination, unfair practices, and lock-in effects. GDPR and ePrivacy access at a granular level should be granted [215].

- **Low availability of data:** the scarcity of data available for innovative reuse and low awareness of benefits of data sharing. Some of these data could be part of a 'data commons' for EU food systems.
- **Low awareness of the potential of data-driven innovation and low skills and low uptake:** in EU food systems, cost-effective solutions at the farm and post-farm level should be tested and demonstrated [216]. The path to digitalization is not clear and transparent to many of them [217], which negatively affects the uptake of data-driven innovations. There is a delay in digitalization among SMEs in general and the difference between SMEs and large firms is shown in the 2020 EU SME strategy. Only 17% of SMEs have employed digital technologies in their businesses, whereas large companies represent a higher percentage of 54% [218].
- **Lack of data infrastructure and technologies and lack of cybersecurity:** cloud services, allowing secure, energy-efficient, affordable, and high-quality data processing processes, will aid the digital transformation process.

12. Zero Pollution Food Systems

CC is threatening global crop production. A major cause of global yield losses is the rising temperature [219–221], especially in less developed and warm areas such as sub-Saharan Africa and Latin America [222]. Exposure and vulnerability to CC can be reduced by strengthening the adaptations of agricultural systems [223]. Cultivar shifts and agronomic management practices to enhance adaptation have been employed [224–227]. Cultivars (with higher heat tolerance) and management (e.g., irrigation) adaptations could reduce yield losses due to warming by ~5% in the mid-21st century [227]. Innovative adaptation pathways might include soil-based strategies [223,227] to address climate risks in agricultural systems.

Improving soil health and resilience is fundamental for sustainable food production; in particular, the role of soil in maintaining or improving global crop productivity under climate warming has been identified and quantified by Deng et al. [228].

Since the 1990s, the push towards decarbonization in the chemical industries (e.g., cosmetic, pharmaceutics, and food) coupled with that of the energy sector is imperative. As proposed by de Boer and Van Ittersum [229], the bioeconomy aims to reduce food and organic waste, using biomass as a source of energy as well as feed and then upcycling organic by-products and waste (manures, agri-food wastes, etc.). The reduction of nitrogen (N) inputs and the development of agroecological practices is a means to stabilize or decrease agriculture GHG emissions, as carried out by several countries [230,231].

13. Conclusions

The importance of science–policy–society interfaces (SPSIs) is outlined by the transformation of food systems by connecting science, policy, and society. On the other hand, research and innovation might be a strategic driver in the transformation to more sustainable food systems and thus characterizes a key planned area in the Farm to Fork strategy and EU Green Deal. In this way, all stakeholders should not only be involved in discussing the interpretation of evidence but also in the process of determining the action pathways to pursue towards quality, security, and safety. In addition, capitalization and underwriting for the development of skills and tools for research and innovation among food system stakeholders to achieve food system transformation is vital in supporting this transformation.

Referring to this gap, the current review describes tools to improve the capabilities among food system stakeholders in terms of understanding both how food system transformation has ensued and how the required transformation is restricted. Moreover, an extensive discussion is presented on the implications for food system transformation in terms of nutrition and sustainable healthy diets. The latter are needed to achieve changes in the food safety systems of the future. The linkage of food and the environment is evident, and the focus on nutrition and a sustainable healthy diet is well established.

Governance and system change, food security and safety, system transformation, and sustainability transitions have been well analyzed and underlined. All have their significance and the end result is manifested in how consumers will interact with these systems and how their lives will be changed by the consumption of a more nutritious diet. Of course, it is crucial that the environment is not further affected negatively. In this direction, a dietary shift towards alternative proteins and, more specifically, plant-based proteins is imperative. Hence, the technical aspect of their biological and chemical modification has been critically analyzed. This paper will be of value to all readers, whether they work with food, health, or in the food economy system, enabling them to apply and integrate the tools for transformation and adopt more inclusive, transdisciplinary, and systemic approaches to the demanding challenges that we face today.

Author Contributions: Both authors contributed equally. All authors have read and agreed to the published version of the manuscript.

Funding: This research received no external funding.

Institutional Review Board Statement: Not applicable.

Informed Consent Statement: Not applicable.

Data Availability Statement: No new data were created or analyzed in this study. Data sharing is not applicable to this article.

Conflicts of Interest: The authors declare no conflicts of interest.

References

1. European Commission Directorate-General for Research and Innovation. *Food 2030. Pathways for Action. Research and Innovation Policy as a Driver for Sustainable, Healthy and Inclusive Food Systems*; European Commission Directorate-General for Research and Innovation: Brussel, Belgium, 2020.
2. United Nations. The Sustainable Development Goals Report 2022. 2022. Available online: https://unstats.un.org/sdgs/report/2022/The-Sustainable-Development-Goals-Report-2022.pdf (accessed on 28 December 2023).
3. European Commission. The European Green Deal: Striving to Be the First Climate-Neutral Continent. 2021. Available online: https://commission.europa.eu/strategy-and-policy/priorities-2019-2024/european-green-deal_en (accessed on 28 December 2023).
4. Kuc-Czarnecka, M.; Markowicz, I.; Sompolska-Rzechuła, A. SDGs implementation, their synergies, and trade-offs in EU countries–Sensitivity analysis-based approach. *Ecol. Indic.* **2023**, *146*, 109888. [CrossRef]
5. Mondejar, M.E.; Avtar, R.; Diaz, H.L.B.; Dubey, R.K.; Esteban, J.; Gomez-Morales, A.; Hallam, B.; Mbungu, N.T.; Okolo, C.C.; Prasad, K.A.; et al. Digitalization to achieve sustainable development goals: Steps towards a Smart Green Planet. *Sci. Total Environ.* **2021**, *794*, 148539. [CrossRef]
6. Funk, W.C.; Blouin, M.S.; Corn, P.S.; Maxell, B.A.; Pilliod, D.S.; Amish, S.; Allendorf, F.W. Population structure of Columbia spotted frogs (*Rana luteiventris*) is strongly affected by the landscape. *Mol. Ecol.* **2005**, *14*, 483–496. [CrossRef] [PubMed]
7. Meybeck, A.; Laval, E.; Lévesque, R.; Parent, G. Food security and nutrition in the age of climate change. In Proceedings of the International Symposium Organized by the Government of Québec in Collaboration with FAO, Québec City, QC, Canada, 24–27 September 2017; p. 132.
8. Intergovernmental Panel on Climate Change. *Climate Change and Biodiversity*; IPCC Technical Paper V; Intergovernmental Panel on Climate Change: Geneva, Switzerland, 2002; Volume 24, p. 77.
9. Juma, S.G.; Kelonye, F. Projected rainfall and temperature changes over Bungoma county in western Kenya by the year 2050 based precis modeling system. *Ethiop. J. Environ. Stud. Manag.* **2016**, *9*, 625–640. [CrossRef]
10. Wheeler, T.; Von Braun, J. Climate change impacts on global food security. *Science* **2013**, *341*, 508–513. [PubMed]
11. Yadav, S.S.; Redden, R.J.; Hatfield, J.L.; Lotze-Campen, H.; Hall, A.E. *Crop Adaptation to Climate Change*; John Wiley & Sons: Hoboken, NJ, USA, 2011.
12. Sasson, A. Food security for Africa: An urgent global challenge. *Agric. Food Secur.* **2012**, *1*, 1–16. [CrossRef]
13. Jarvis, A.; Lane, A.; Hijmans, R.J. The effect of climate change on crop wild relatives. *Agric. Ecosyst. Environ.* **2008**, *126*, 13–23. [CrossRef]
14. Available online: https://www.susfans.eu/portfolio (accessed on 28 December 2023).
15. Rutten, M.; Achterbosch, T.; de Boer, I.; Cuaresma, J.C.; Geleijnse, M.; Havlík, P.; Heckelei, T.; Ingram, J.; Marette, S.; van Meijl, H.; et al. Metrics, models and foresight for European sustainable food and nutrition security: The vision of the SUSFANS project. *Agric. Syst.* **2018**, *163*, 45–57. [CrossRef]
16. Acharya, T.; Fanzo, J.; Gustafson, D.; Ingram, J.; Schneeman, B.; Allen, L.; Boote, K.; Drewnowski, A.; Ewert, F.; Hall, S.; et al. *Assessing Sustainable Nutrition Security: The Role of Food Systems*; The International Life Sciences Institute, Research Foundation, Center for Integrated Modeling of Sustainable Agriculture and Nutrition: Washington, DC, USA, 2014; Available online: https://catalogue.unccd.int/455_CIMSANS-Sustainable-Nutrition-Security-FINAL%2030may2014-4.pdf (accessed on 28 December 2023).
17. Ingram, J.S.I.; Porter, J.R. Plant science and the food security agenda. *Nat. Plants* **2015**, *1*, 15173. [CrossRef]
18. Kiełkowska, A. Allium cepa root meristem cells under osmotic (sorbitol) and salt (NaCl) stress in vitro. *Acta Bot. Croat.* **2017**, *76*, 146–153. [CrossRef]
19. Grüter, R.; Trachsel, T.; Laube, P.; Jaisli, I. Expected global suitability of coffee, cashew and avocado due to climate change. *PLoS ONE* **2022**, *17*, e0261976. [CrossRef] [PubMed]
20. Bunn, C.; Läderach, P.; Rivera, O.O.; Kirschke, D. A bitter cup: Climate change profile of global production of Arabica and Robusta coffee. *Clim. Chang.* **2015**, *129*, 89–101. [CrossRef]
21. Läderach, P.; Martinez-Valle, A.; Schroth, G.; Castro, N. Predicting the future climatic suitability for cocoa farming of the world's leading producer countries, Ghana and Côte d'Ivoire. *Clim. Chang.* **2013**, *119*, 841–854. [CrossRef]
22. Zhang, T.; Van der Wiel, K.; Wei, T.; Liu, Y.; Cui, X.; Yang, X. Increased wheat price spikes and larger economic inequality with 2 °C global warming. *One Earth* **2022**, *5*, 907–916. [CrossRef]
23. Berge, J.B.; Ricroch, A.E. Emergence of minor pests becoming major pests in GE cotton in China: What are the reasons? What are the alternatives practices to this change of status? *GM Crops* **2010**, *1*, 214–219. [CrossRef]
24. Jiménez, O.R.; Bornemann, A.C.; Medina, Y.E.; Romero, K.; Bravo, J.R. Prospects of biological inputs as a measure for reducing crop losses caused by climate change effects. *J. Agric. Food Res.* **2023**, *14*, 100689.

25. IPCC. *Cambio Climático 2007: Informe de Síntesis Contribución de los Grupos de Trabajo, I.; II y III al Cuarto Informe de Evaluación del Grupo Intergubernamental de Expertos Sobre el Cambio Climático [Equipo de Redacción Principal in:, R.K. Pachauri, A. Reisinger (Eds.), (Directores de la Publicación)]*; IPCC: Geneva, Switzerland, 2007; p. 104. Available online: https://www.ipcc.ch/site/assets/uploads/2018/02/ar4_syr_sp.pdf (accessed on 28 December 2023).
26. IPCC. *Cambio Climático 2014: Informe de Síntesis Contribución de los Grupos de Trabajo, I.; II y III al Quinto Informe de Evaluación del Grupo Intergubernamental de Expertos Sobre el Cambio Climático [Equipo Principal de Redacción in:, R.K. Pachauri, L.A. Meyer (Eds.)]*; IPCC: Geneva, Switzerland, 2014; p. 157. Available online: https://www.ipcc.ch/site/assets/uploads/2018/02/SYR_AR5_FINAL_full_es.pdf (accessed on 28 December 2023).
27. IPCC. Summary for policymakers. In *Climate Change 2021: The Physical Science Basis Contribution of Working Group I to the Sixth Assessment Report of the Intergovernmental Panel on Climate Change*; Masson-Delmotte, P., Zhai, A., Pirani, S.L., Péan, C.C., Berger, S., Caud, N., Chen, Y., Goldfarb, L., Gomis, M.I., Huang, M., et al., Eds.; Cambridge University Press: Cambridge, UK; New York, NY, USA, 2021; pp. 3–32. Available online: https://www.ipcc.ch/report/ar6/wg1/downloads/report/IPCC_AR6_WGI_SPM.pdf (accessed on 28 December 2023).
28. IPCC. Resumen para responsables de políticas. In *El Cambio Climático y la Tierra*; Shukla, P.R., Skea, J., Buendia, E.C., Masson-Delmotte, V., Pörtner, H.-O., Roberts, D.C., Zhai, P., Slade, R., Connors, S., van Diemen, R., et al., Eds.; Informe especial del IPCC sobre el cambio climático, la desertificación, la degradación de las tierras, la gestión sostenible de las tierras, la seguridad alimentaria y los flujos de gases de efecto invernadero en los ecosistemas terrestres; IPCC: Geneva, Switzerland, 2019; En prensa; Available online: https://www.ipcc.ch/site/assets/uploads/sites/4/2020/06/SRCCL_SPM_es.pdf (accessed on 28 December 2023).
29. US Environmental Protection Agency (EPA). Climate Change Indicators: Heavy Precipitation. 2022. Available online: https://www.epa.gov/climate-indicators/climate-change-indicators-heavy-precipitation (accessed on 28 December 2023).
30. USGCRP. *Climate Science Special Report: Fourth National Climate Assessment Volume I*; Wuebbles, D.J., Fahey, D.W., Hibbard, K.A., Dokken, D.J., Stewart, B.C., Maycock, T.K., Eds.; U.S. Global Change Research Program: Washington, DC, USA, 2017; p. 470. Available online: https://science2017.globalchange.gov/chapter/7/ (accessed on 28 December 2023).
31. Jonkman, J.; Kanellopoulos, A.; Bloemhof, J.M. Designing an eco-efficient biomass-based supply chain using a multi-actor optimisation model. *J. Clean. Prod.* **2019**, *210*, 1065–1075. [CrossRef]
32. Zin, M.M.T.; Kim, D.J. Simultaneous recovery of phosphorus and nitrogen from sewage sludge ash and food wastewater as struvite by Mg-biochar. *J. Hazard. Mater.* **2021**, *403*, 123704.
33. Faucon, M.P.; Aussenac, T.; Debref, R.; Firmin, S.; Houben, D.; Marraccini, E.; Sauvée, L.; Trinsoutrot-Gattin, I.; Gloaguen, R. Combining agroecology and bioeconomy to meet the societal challenges of agriculture. *Plant Soil* **2023**, *492*, 61–78. [CrossRef]
34. UN Habitat, 2019. Several Definitions of what Constitutes a City or an Urban Area Are in Use, with Significant Variations between Countries. Available online: https://unhabitat.org/sites/default/files/2020/06/city_definition_what_is_a_city.pdf (accessed on 28 December 2023).
35. World Economic Forum. *Project MainStream. Urban Biocycles*; World Economic Forum: Cologny, Switzerland, 2017.
36. FAO. *Urban Food Action*; FAO: Rome, Italy, 2019.
37. Available online: https://urban.jrc.ec.europa.eu/thefutureofcities/what-is-a-city#the-chapter (accessed on 28 December 2023).
38. Brussels, European Commission. DG research and Innovation. Available online: https://op.europa.eu/en/web/eu-law-and-publications/publication-detail/-/publication/ca8ffeda-99bb-11ea-aac4-01aa75ed71a1 (accessed on 28 December 2023).
39. IPES-Food. 2017. What Makes Urban Food Policy Happen? Insights from Five Case Studies. International Panel of Experts on Sustainable Food Systems. Available online: www.ipes-food.org (accessed on 28 December 2023).
40. Doernberg, A.; Horn, P.; Zasada, I.; Piorr, A. Urban food policies in German city regions: An overview of key players and policy instruments. *Food Policy* **2019**, *89*, 101782. [CrossRef]
41. FAO—Food and Agriculture Organizaion of the United Nations. Food, Agriculture and Cities: Challenges of Food and Nutrition Security, Agriculture and Ecosystem Management in an Urbanizing World, FAO Food for the Cities Multi-Disciplinary Initiative Position Paper. 2011. Available online: http://www.fao.org/fileadmin/templates/FCIT/PDF/FoodAgriCities_Oct2011.pdf (accessed on 19 December 2023).
42. MUFPP—Milan Urban Food Policy Pact. Milan Urban Food Policy Pact and Framework for Action. 2015. Available online: https://www.milanurbanfoodpolicypact.org/ (accessed on 28 December 2023).
43. C40 Cities Report. The Future of Urban Consumption in a 1.5 °C World. 2019. Available online: https://www.c40knowledgehub.org/s/article/The-future-of-urban-consumption-in-a-1-5-C-world?language=en_US (accessed on 28 December 2023).
44. FAO. *Integrating Food into Urban Planning*; UCL Press: London, UK, 2018; ISBN 9781787353763.
45. Halliday, J.; Platenkamp, L.; Nicolarea, Y. A Menu of Actions to Shape Urban Food Environments for Improved Nutrition, Global Alliance for Improved Nutrition (GAIN), MUFPP and RUA. 2019. Available online: https://www.gainhealth.org/resources/reports-and-publications/menu-actions-shape-urban-food-environments-improved-nutrition (accessed on 28 December 2023).
46. IPES-Food. *COVID-19 and the Crisis in Food Systems: Symptoms, Causes, and Potential Solutions*; IPES-Food: Brussels, Belgium, 2020.
47. Piorr, A.; Zasada, I.; Doernberg, A.; Zoll, F.; Ramme, W. Research for AGRI Committee—Urban and Peri-Urban Agriculture in the EU, Brussels, 2018. Available online: http://www.europarl.europa.eu/RegData/etudes/STUD/2018/617468/IPOL_STU(2018)617468_EN.pdf (accessed on 19 December 2023).

48. Battersby, J. Food system transformation in the absence of food system planning: The case of supermarket and shopping mall retail expansion in Cape Town, South Africa. *Built Environ.* **2017**, *43*, 417–430. [CrossRef]
49. European Commission Directorate-General for Research and Innovation. *Urban Food System Transformation in the Context of Food 2030—Current Practice and Outlook towards 2030*; European Commission Directorate-General for Research and Innovation: Brussel, Belgium, 2023.
50. Sonnino, R.; Milbourne, P. Food system transformation: A progressive place-based approach. *Local Environ.* **2022**, *27*, 915–926. [CrossRef]
51. Mattioni, D.; Milbourne, P.; Sonnino, R. Destabilizing the food regime "from within": Tools and strategies used by urban food policy actors. *Environ. Innov. Soc. Transit.* **2022**, *44*, 48–59. [CrossRef]
52. Sonnino, R. City-To-City Cooperation for More Democratic and Inclusive Food Systems: What Has Been Achieved and What Remains to Be Done. Cidob Opinion. November 2021. Available online: https://www.cidob.org/es/publicaciones/serie_de_publicacion/opinion_cidob/2021/city_to_city_cooperation_for_more_democratic_and_inclusive_food_systems_what_has_been_achieved_and_what_remains_to_be_done (accessed on 28 December 2023).
53. Edwards, F.; Sonnino, R.; Lopez-Cifuentes, M. Report on Facilitators of and Barriers to the Development and Implementation of Evidence-Based and Integrated Food Policies and Planning Frameworks. 2023. Available online: https://openresearch.surrey.ac.uk/esploro/outputs/99783469502346 (accessed on 28 December 2023). [CrossRef]
54. EC Website. Food from the Oceans. Scientific Advice in the Area of Food and Biomass from the Oceans. Research and Innovation—European Commission (europa.eu). Available online: https://commission.europa.eu/research-and-innovation_en (accessed on 28 December 2023).
55. Eurostat and Eumofa. The EU Fish Market. 2018. Available online: https://oceans-and-fisheries.ec.europa.eu/system/files/2019-02/2019-02-18-eu-fish-market-publication_en.pdf (accessed on 28 December 2023).
56. World Bank; FAO. *Fish to 2030. Prospects for Fisheries and Aquaculture*; World Bank: Washington, DC, USA; FAO: Rome, Italy, 2014.
57. Tidd, A.N.; Rousseau, Y.; Ojea, E.; Watson, R.A.; Blanchard, J.L. Food security challenged by declining efficiencies of artisanal fishing fleets: A global country-level analysis. *Glob. Food Sec.* **2022**, *32*, 100598. [CrossRef]
58. Rahman, M.S.; Huang, W.C.; Toiba, H.; Efani, A. Does adaptation to climate change promote household food security? Insights from Indonesian fishermen. *Int. J. Sustain. Dev. World Ecol.* **2022**, *29*, 611–624. [CrossRef]
59. Syddall, V.M.; Fisher, K.; Thrush, S. Collaboration a solution for small island developing states to address food security and economic development in the face of climate change. *Ocean Coast. Manag.* **2022**, *221*, 106132. [CrossRef]
60. Alsaleh, M. The role of the fishery industry in the shift towards sustainable food security: A critical study of blue food. *Environ. Sci. Pollut. Res.* **2023**, *30*, 105575–105594. [CrossRef]
61. Collins, A.M. Empowerment, rights, and global food governance: Gender in the UN Committee for World Food Security. *Globalizations* **2022**, *19*, 220–237. [CrossRef]
62. Jones, E.C.; Ong, C.; Haynes, J. Disaster-related food security and past general governance strategies in a worldwide sample. *Weather Clim. Soc.* **2022**, *14*, 3–18. [CrossRef]
63. Adelle, C.; Dekeyser, K. Food and sustainable development governance. In *Handbook on the Governance of Sustainable Development*; Edward Elgar Publishing: Cheltenham, UK, 2022; pp. 14–29. [CrossRef]
64. Leeuwis, C.; Boogaard, B.K.; Atta-Krah, K. How food systems change (or not): Governance implications for system transformation processes. *Food Secur.* **2021**, *13*, 761–780. [CrossRef]
65. EEA. *The European Environment—State and Outlook 2020 Knowledge for Transition to a Sustainable Europe*; EEA: København, Denmark, 2019.
66. European Commission. *In-Depth Analysis in Support of the Commission Communication COM(2018) 773—A Clean Planet for All—A European Long-Term Strategic Vision for a Prosperous, Modern, Competitive and Climate Neutral Economy*; European Commission: Brussels, Belgium, 2018.
67. IPCC. Summary for policymakers. In *Climate Change and Land: An IPCC Special Report on Climate Change, Desertification, Land Degradation, Sustainable Land Management, Food Security, and Greenhouse Gas Fluxes in Terrestrial Ecosystems*; IPCC: Geneva, Switzerland, 2019.
68. de Coninck, H.; Revi, A.; Babiker, M.; Bertoldi, P.; Buckeridge, M.; Cartwright, A.; Dong, W.; Ford, J.; Fuss, S.; Hourcade, J.C.; et al. Strengthening and implementing the global response. In *Global Warming of 1.5 °C. An IPCC Special Report on the Impacts of Global Warming of 1.5 °C Above Pre-Industrial Levels and Related Global Greenhouse Gas Emission Pathways, in the Context of Strengthening the Global Response to the Threat of Climate Change, Sustainable Development, and Efforts to Eradicate Poverty*; Intergovernmental Panel on Climate Change: Geneva, Switzerland, 2018.
69. Milford, A.B.; Le Mouel, C.; Bodirsky, B.L.; Rolinski, S. Drivers of meat consumption. *Appetite* **2019**, *141*, 104313. [CrossRef] [PubMed]
70. Castellani, V.; Fusi, A.; Sala, S. *Consumer Footprint. Basket of Products Indicator on Food, EUR 28764 EN*; Publications Office of the European Union: Luxembourg, 2017; ISBN 978-92-7973194-5.
71. Tan, M.; Nawaz, M.A.; Buckow, R. Functional and food application of plant proteins—A review. *Food Rev. Int.* **2023**, *39*, 2428–2456. [CrossRef]
72. Kaur, T.; Pal, A.; Kumar, S.; Saharan, V. Protein Chemistry and Gelation. In *Advances in Food Chemistry: Food Components, Processing and Preservation*; Springer Nature: Singapore, 2022; pp. 161–207.

73. Day, L.; Cakebread, J.A.; Loveday, S.M. Food proteins from animals and plants: Differences in the nutritional and functional properties. *Trends Food Sci. Technol.* **2022**, *119*, 428–442. [CrossRef]
74. Lai, H.; Peng, X.; Li, L.; Zhu, D.; Xiao, P. Novel monomers for photopolymer networks. *Prog. Polym. Sci.* **2022**, *128*, 101529. [CrossRef]
75. Dhiman, A.; Thakur, K.; Parmar, V.; Sharma, S.; Sharma, R.; Kaur, G.; Suhag, R. New insights into tailoring physicochemical and techno-functional properties of plant proteins using conventional and emerging technologies. *J. Food Meas. Charact.* **2023**, *17*, 3845–3873.
76. Bou, R.; Navarro-Vozmediano, P.; Domínguez, R.; López-Gómez, M.; Pinent, M.; Ribas-Agustí, A.; Benedito, J.J.; Lorenzo, J.M.; Terra, X.; García-Pérez, J.V.; et al. Application of emerging technologies to obtain legume protein isolates with improved techno-functional properties and health effects. *Compr. Rev. Food Sci. Food Saf.* **2022**, *21*, 2200–2232. [CrossRef]
77. Nartea, A.; Kuhalskaya, A.; Fanesi, B.; Orhotohwo, O.L.; Susek, K.; Rocchetti, L.; Di Vittori, V.; Bitocchi, E.; Pacetti, D.; Papa, R. Legume byproducts as ingredients for food applications: Preparation, nutrition, bioactivity, and techno-functional properties. *Compr. Rev. Food Sci. Food Saf.* **2023**, *22*, 1953–1985.
78. Martinez-Alvarenga, M.S.; Martinez-Rodriguez, E.Y.; Garcia-Amezquita, L.E.; Olivas, G.I.; Zamudio-Flores, P.B.; Acosta-Muniz, C.H.; Sepulveda, D.R. Effect of Maillard reaction conditions on the degree of glycation and functional properties of whey protein isolate–Maltodextrin conjugates. *Food Hydrocoll.* **2014**, *38*, 110–118. [CrossRef]
79. Kutzli, I.; Beljo, D.; Gibis, M.; Baier, S.K.; Weiss, J. Effect of maltodextrin dextrose equivalent on electrospinnability and glycation reaction of blends with pea protein isolate. *Food Biophys.* **2020**, *15*, 206–215. [CrossRef]
80. Zhang, B.; Guo, X.; Zhu, K.; Peng, W.; Zhou, H. Improvement of emulsifying properties of oat protein isolate–dextran conjugates by glycation. *Carbohydr. Polym.* **2015**, *127*, 168–175.
81. Meng, X.; Li, T.; Song, T.; Chen, C.; Venkitasamy, C.; Pan, Z.; Zhang, H. Solubility, structural properties, and immunomodulatory activities of rice dreg protein modified with sodium alginate under microwave heating. *Food Sci. Nutr.* **2019**, *7*, 2556–2564.
82. Pirestani, S.; Nasirpour, A.; Keramat, J.; Desobry, S.; Jasniewski, J. Effect of glycosylation with gum Arabic by Maillard reaction in a liquid system on the emulsifying properties of canola protein isolate. *Carbohydr. Polym.* **2017**, *157*, 1620–1627. [PubMed]
83. Xue, F.; Wu, Z.; Tong, J.; Zheng, J.; Li, C. Effect of combination of high-intensity ultrasound treatment and dextran glycosylation on structural and interfacial properties of buckwheat protein isolates. *Biosci. Biotechnol. Biochem.* **2017**, *81*, 1891–1898. [PubMed]
84. Miedzianka, J.; Pęksa, A. Effect of pH on phosphorylation of potato protein isolate. *Food Chem.* **2013**, *138*, 2321–2326. [CrossRef] [PubMed]
85. Sánchez-Reséndiz, A.; Rodríguez-Barrientos, S.; Rodríguez-Rodríguez, J.; Barba-Dávila, B.; Serna-Saldívar, S.O.; Chuck-Hernández, C. Phosphoesterification of soybean and peanut proteins with sodium trimetaphosphate (STMP): Changes in structure to improve functionality for food applications. *Food Chem.* **2018**, *260*, 299–305. [PubMed]
86. Wang, Y.R.; Yang, Q.; Fan, J.L.; Zhang, B.; Chen, H.Q. The effects of phosphorylation modification on the structure, interactions and rheological properties of rice glutelin during heat treatment. *Food Chem.* **2019**, *297*, 124978. [PubMed]
87. Zhao, C.B.; Zhang, H.; Xu, X.Y.; Cao, Y.; Zheng, M.Z.; Liu, J.S.; Wu, F. Effect of acetylation and succinylation on physicochemical properties and structural characteristics of oat protein isolate. *Process Biochem.* **2017**, *57*, 117–123. [CrossRef]
88. Chen, C.; Zhang, C.; Zhang, R.; Ju, X.; He, R.; Wang, Z. Enzyme-catalyzed acylation improves gel properties of rapeseed protein isolate. *J. Sci. Food Agric.* **2020**, *100*, 4182–4189.
89. Sebii, H.; Karra, S.; Bchir, B.; Nhouchi, Z.; Ghribi, A.M.; Karoui, R.; Besbes, S. Effect of succinylation on the secondary structures, surface, and thermal properties of date palm pollen protein concentrate. *J. Food Sci. Technol.* **2021**, *58*, 632–640. [CrossRef]
90. Liu, B.Y.; Zhu, K.X.; Guo, X.N.; Peng, W.; Zhou, H.M. Effect of deamidation-induced modification on umami and bitter taste of wheat gluten hydrolysates. *J. Sci. Food Agric.* **2017**, *97*, 3181–3188. [CrossRef]
91. Guan, J.; Takai, R.; Toraya, K.; Ogawa, T.; Muramoto, K.; Mohri, S.; Cho, S.J. Effects of alkaline deamidation on the chemical properties of rice bran protein. *Food Sci. Technol. Res.* **2017**, *23*, 697–704.
92. Hadidi, M.; Ibarz, A.; Pouramin, S. Optimization of extraction and deamidation of edible protein from evening primrose (*Oenothera biennis* L.) oil processing by-products and its effect on structural and techno-functional properties. *Food Chem.* **2021**, *334*, 127613. [CrossRef] [PubMed]
93. Talema, A. Causes, negative effects, and preventive methods of water pollution in Ethiopia. *Qual. Assur. Saf. Crops Foods* **2023**, *15*, 129–139. [CrossRef]
94. Nasrabadi, M.N.; Doost, A.S.; Mezzenga, R. Modification approaches of plant-based proteins to improve their techno-functionality and use in food products. *Food Hydrocoll.* **2021**, *118*, 106789. [CrossRef]
95. Brückner-Gühmann, M.; Heiden-Hecht, T.; Sözer, N.; Drusch, S. Foaming characteristics of oat protein and modification by partial hydrolysis. *Eur. Food Res. Technol.* **2018**, *244*, 2095–2106. [CrossRef]
96. Konieczny, D.; Stone, A.K.; Korber, D.R.; Nickerson, M.T.; Tanaka, T. Physicochemical properties of enzymatically modified pea protein-enriched flour treated by different enzymes to varying levels of hydrolysis. *Cereal Chem.* **2020**, *97*, 326–338. [CrossRef]
97. Sun, Q.; Ma, Z.F.; Zhang, H.; Ma, S.; Kong, L. Structural characteristics and functional properties of walnut glutelin as hydrolyzed: Effect of enzymatic modification. *Int. J. Food Prop.* **2019**, *22*, 265–279. [CrossRef]
98. Meinlschmidt, P.; Ueberham, E.; Lehmann, J.; Schweiggert-Weisz, U.; Eisner, P. Immunoreactivity, sensory and physicochemical properties of fermented soy protein isolate. *Food Chem.* **2016**, *205*, 229–238.

99. Rui, X.; Huang, J.; Xing, G.; Zhang, Q.; Li, W.; Dong, M. Changes in soy protein immunoglobulin E reactivity, protein degradation, and conformation through fermentation with Lactobacillus plantarum strains. *LWT* **2019**, *99*, 156–165.
100. Klupsaite, D.; Juodeikiene, G.; Zadeike, D.; Bartkiene, E.; Maknickiene, Z.; Liutkute, G. The influence of lactic acid fermentation on functional properties of narrow-leaved lupine protein as functional additive for higher value wheat bread. *LWT* **2017**, *75*, 180–186. [CrossRef]
101. Schlegel, K.; Leidigkeit, A.; Eisner, P.; Schweiggert-Weisz, U. Technofunctional and sensory properties of fermented lupin protein isolates. *Foods* **2019**, *8*, 678. [CrossRef] [PubMed]
102. He, W.; Yang, R.; Zhao, W. Effect of acid deamidation-alcalase hydrolysis induced modification on functional and bitter-masking properties of wheat gluten hydrolysates. *Food Chem.* **2019**, *277*, 655–663. [CrossRef] [PubMed]
103. Wang, Z.; Zhang, C.; Zhang, T.; Ju, X.; He, R. Effects of acylation and glycation treatments on physicochemical and gelation properties of rapeseed protein isolate. *RSC Adv.* **2018**, *8*, 40395–40406.
104. Zhang, S.B.; Wang, X.H.; Li, X.; Yan, D.Q. Effects of Tween 20 and transglutaminase modifications on the functional properties of peanut proteins. *J. Am. Oil Chem. Soc.* **2020**, *97*, 93–103.
105. Arteaga, V.G.; Guardia, M.A.; Muranyi, I.; Eisner, P.; Schweiggert-Weisz, U. Effect of enzymatic hydrolysis on molecular weight distribution, techno-functional properties and sensory perception of pea protein isolates. *Innov. Food Sci. Emerg. Technol.* **2020**, *65*, 102449.
106. Pillai, P.K.; Morales-Contreras, B.E.; Wicker, L.; Nickerson, M.T. Effect of enzyme de-esterified pectin on the electrostatic complexation with pea protein isolate under different mixing conditions. *Food Chem.* **2020**, *305*, 125433. [CrossRef]
107. Sorde, K.L.; Ananthanarayan, L. Effect of transglutaminase treatment on properties of coconut protein-guar gum composite film. *LWT* **2019**, *115*, 108422. [CrossRef]
108. Mudgil, P.; Omar, L.S.; Kamal, H.; Kilari, B.P.; Maqsood, S. Multi-functional bioactive properties of intact and enzymatically hydrolysed quinoa and amaranth proteins. *LWT* **2019**, *110*, 207–213.
109. Yang, X.; Teng, D.; Wang, X.; Guan, Q.; Mao, R.; Hao, Y.; Wang, J. Enhancement of nutritional and antioxidant properties of peanut meal by bio-modification with Bacillus licheniformis. *Appl. Biochem. Biotechnol.* **2016**, *180*, 1227–1242.
110. FAO. *State of World Fisheries and Aquaculture 2022: Towards Blue Transformation*; Food and Agriculture Organization: Rome, Italy, 2022; pp. 1–236.
111. Definitional Framework of Food Loss. Global Initiative on Food Loss and Waste Reduction. 2014. Available online: http://www.fao.org/fileadmin/user_upload/save-food/PDF/FLW_Definition_and_Scope_2014.pdf (accessed on 23 December 2023).
112. Food and Agriculture Organization. *FAO State of Food Security and Nutrition in the World 2020: Transforming Food Systems for Affordable. Healthy Diets*; Food and Agriculture Organization: Rome, Italy, 2020; pp. 1–283.
113. Kruijssen, F.; Tedesco, I.; Ward, A.; Pincus, L.; Love, D.; Thorne-Lyman, A.L. Loss and waste in fish value chains: A review of the evidence from low and middle-income countries. *Glob. Food Secur.* **2020**, *26*, 100434. [CrossRef]
114. Peter, S.M.J.; Blomberg, S.P.; Holden, M.H.; Hoffman, L.C.; Tibbetts, I.R. Is Mushy Tuna Syndrome a Growing Problem for the Tuna Industry? *Foods* **2023**, *12*, 3590. [CrossRef] [PubMed]
115. FAO. *The State of Food and Agriculture*; FAO: Rome, Italy, 2019.
116. United States Environmental Protection Agency. Available online: https://www.epa.gov/sustainable-management-food/sustainable-management-food-basics (accessed on 15 December 2023).
117. REFRESH H2020 Project. Available online: https://eu-refresh.org/results.html (accessed on 28 December 2023).
118. World Food Programme. *Hunger Statistics*; World Food Programme: Rome, Italy, 2016.
119. United States Environmental Protection Agency. *Quantifying Methane Emissions from Landfilled Food Waste (PDF)*; United States Environmental Protection Agency: Washington, DC, USA, 2023.
120. From Farm to Kitchen: The Environmental Impacts of U.S. Food Waste. Available online: https://www.epa.gov/system/files/documents/2021-11/from-farm-to-kitchen-the-environmental-impacts-of-u.s.-food-waste_508-tagged.pdf (accessed on 28 December 2023).
121. The Part 2, From Field to Bin: The Environmental Impacts of U.S. Food Waste Management Pathways. Available online: https://www.epa.gov/system/files/documents/2023-10/part2_wf-pathways_report_formatted_no-appendices_508-compliant.pdf (accessed on 28 December 2023).
122. Tamasiga, P.; Miri, T.; Onyeaka, H.; Hart, A. Food Waste and Circular Economy: Challenges and Opportunities. *Sustainability* **2022**, *14*, 9896. [CrossRef]
123. Wang, T.W.; Zhang, M.L.; Sun, N.N.; Chen, H.H.; Zhang, C.X.; Wang, Q.; Zhang, W.Z. Analysis of the economically motivated food adulteration in China based on 6477 events from 2000 to 2020. *Qual. Assur. Saf. Crops Foods* **2023**, *15*, 222–231. [CrossRef]
124. Moreno, M.; de los Rios, C.; Rowe, Z.; Charnley, F. A Conceptual Framework for Circular Design. *Sustainability* **2016**, *8*, 937.
125. Kenny, S.T.; Runic, J.N.; Kaminsky, W.; Woods, T.; Babu, R.P.; Keely, C.M.; Blau, W.; O'Connor, K.E. Up-cycling of PET (polyethylene terephthalate) to the biodegradable plastic PHA (polyhydroxyalkanoate). *Environ. Sci. Ttechnol.* **2008**, *42*, 7696–7701. [CrossRef] [PubMed]
126. Borrello, M.; Lombardi, A.; Pascucci, S.; Cembalo, L. The Seven Challenges for Transitioning into a Bio-Based Circular Economy in the Agri-Food Sector. *Recent Pat. Food Nutr. Agric.* **2016**, *8*, 39–47. [PubMed]

127. Ellen MacArthur Foundation. *Towards the Circular Economy, Report—Economic and Business Rationale for an Accelerated Transition*; Ellen MacArthur Foundation: Cowes, UK, 2012; Volume 1.
128. European Union (EU). European Commission, Communication from the Commission to the European Parliament, the Council, the European Economic and Social Committee and the Committee of the Regions, Closing the Loop—An EU Action Plan for the Circular Economy, COM(2015) 614/2. 2015. Available online: http://eur-lex.europa.eu/legal-content/EN/TXT/?uri=CELEX:52015DC0614 (accessed on 17 January 2023).
129. Mohajeri, S.; Harsej, F.; Sadeghpour, M.; Nia, J.K. Integrated reverse supply chain model for food waste based on industry 4.0 revolutions: A case study of producing the household waste recycling machine. *Qual. Assur. Saf. Crops Foods* **2021**, *13*, 70–83. [CrossRef]
130. United Nations. *EU Platform on Food Losses and Food Waste, Draft Key Recommendations for Action to Prevent and Reduce Food Waste*; United Nations: New York, NY, USA, 2019.
131. Jamaludin, H.; Elmaky, H.S.E.; Sulaiman, S. The future of food waste: Application of circular economy. *Energy Nexus* **2022**, *7*, 100098. [CrossRef]
132. Adelodun, B.; Kim, S.H.; Odey, G.; Choi, K.-S. Assessment of environmental and economic aspects of household food waste using a new Environmental-Economic Footprint (EN-EC) index: A case study of Daegu, South Korea. *Sci. Total Environ.* **2021**, *776*, 145928. [CrossRef]
133. 2050 Long-Term Strategy. Available online: https://climate.ec.europa.eu/eu-action/climate-strategies-targets/2050-long-term-strategy_en (accessed on 28 December 2023).
134. A New Bioeconomy Strategy for a Sustainable Europe. Available online: https://knowledge4policy.ec.europa.eu/publication/new-bioeconomy-strategy-sustainable-europe_en (accessed on 28 December 2023).
135. Available online: https://research-and-innovation.ec.europa.eu/system/files/2020-04/he_missions_soil_quiz_eng.pdf (accessed on 28 December 2023).
136. Franzoni, G.; Cocetta, G.; Prinsi, B.; Ferrante, A.; Espen, L. Biostimulants on crops: Their impact under abiotic stress conditions. *Horticulture* **2022**, *8*, 189. [CrossRef]
137. Cocetta, G.; Landoni, M.; Pilu, R.; Repiso, C.; Nolasco, J.; Alajarin, M.; Ugena, M.L.; Levy, C.C.B.; Scatolino, G.; Villa, D.; et al. Priming treatments with biostimulants to cope the short-term heat stress response: A transcriptomic profile evaluation. *Plants* **2022**, *11*, 113. [CrossRef] [PubMed]
138. Grover, M.; Ali, S.Z.; Sandhya, V.; Rasul, A.; Venkateswarlu, B. Role of microorganisms in adaptation of agriculture crops to abiotic stresses. *World J. Microbiol. Biotechnol.* **2011**, *27*, 1231–1240. [CrossRef]
139. Jacomassi, L.M.; Viveiros, J.O.; Oliveira, M.P.; Momesso, L.; de Siqueira, G.F.; Crusciol, C.A.C. A seaweed extract-based biostimulant mitigates drought stress in sugarcane. *Front. Plant Sci.* **2022**, *13*, 865291. [CrossRef] [PubMed]
140. Ali, S.; Moon, Y.S.; Hamayun, M.; Khan, M.A.; Bibi, K.; Lee, I.J. Pragmatic role of microbial plant biostimulants in abiotic stress relief in crop plants. *J. Plant Interact.* **2022**, *17*, 705–718. [CrossRef]
141. Hasanuzzaman, M.; Parvin, K.; Bardhan, K.K.; Nahar, K.; Anee, T.I.; Masud, A.A.C.; Fotopoulos, V. Biostimulants for the regulation of reactive oxygen species metabolism in plants under abiotic stress. *Cells* **2021**, *10*, 2537. [CrossRef]
142. Melo, A.; Ariza, P.; Lissbrant, S.; Tofiño, A. Evaluation of agrochemicals and bioinputs for sustainable bean management on the Caribbean coast of Colombia. *Agron. Colomb.* **2015**, *33*, 203–2011. [CrossRef]
143. Shinwari, Z.K.; Tanveer, F.; Iqrar, I. Role of microbes in plant health, disease management, and abiotic stress management. In *Microbiome in Plant Health and Disease*; Springer: Singapore, 2019; Volume 11. [CrossRef]
144. Pereira, R.V.; Filgueiras, C.C.; Dória, J.; Pe, M.F.G.V.; Willett, D.S. The effects of biostimulants on induced plant defense. *Front. Agron.* **2021**, *3*, 630596. [CrossRef]
145. IPES Food. *International Panel of Experts on Sustainable Food Systems, Towards a Common Food Policy for the European Union*; IPES-Food: Brussels, Belgium, 2019.
146. Scientific Advice Mechanism. Scoping Review Report: Towards an EU Sustainable Food System. 2019. Available online: https://ec.europa.eu/research/sam/pdf/meetings/hlg_sam_032019_scoping_report_sustainable-food.pdf#view=fit&pagemode=none (accessed on 28 December 2023).
147. Ng, M.; Fleming, T.; Robinson, M.; Thomson, B.; Graetz, N.; Margono, C.; Mullany, E.C.; Biryukov, S.; Abbafati, C.; Abera, S.F.; et al. Global, regional, and national prevalence of overweight and obesity in children and adults during 1980–2013: A systematic analysis for the Global Burden of Disease Study 2013. *Lancet* **2014**, *384*, 766–781. [CrossRef]
148. Zinsstag, J.; Schelling, E.; Waltner-Toews, D.; Tanner, M. From "one medicine" to "one health" and systemic approaches to health and well-being. *Prevent. Vet. Med.* **2011**, *101*, 148–156. [CrossRef]
149. Adisasmito, W.B.; Almuhairi, S.; Behravesh, C.B.; Bilivogui, P.; Bukachi, S.A.; Casas, N.; Becerra, N.C.; Charron, D.F.; Chaudhary, A.; Zanella, J.R.C. One Health: A new definition for a sustainable and healthy future. *PLoS Pathog.* **2022**, *18*, e1010537. [CrossRef]
150. Villanueva-Cabezas, J.P.; Rajkhowa, A.; Campbell, A.J. One Health needs a vision beyond zoonoses. *Transbound. Emerg. Dis.* **2020**, *67*, 2271–2273. [CrossRef]
151. Gruetzmacher, K.; Karesh, W.B.; Amuasi, J.H.; Arshad, A.; Farlow, A.; Gabrysch, S.; Jetzkowitz, J.; Lieberman, S.; Palmer, C.; Winkler, A.S. The Berlin principles on One Health–bridging global health and conservation. *Sci. Total Environ.* **2021**, *764*, 142919. [CrossRef] [PubMed]
152. Zinsstag, J. *Convergence of Ecohealth and One Health*; Springer: Berlin/Heidelberg, Germany, 2012. [CrossRef]

153. Binot, A.; Duboz, R.; Promburom, P.; Phimpraphai, W.; Cappelle, J.; Lajaunie, C.; Goutard, F.L.; Pinyopummintr, T.; Figui, M.; Roger, F.L. A framework to promote collective action within the One Health community of practice: Using participatory modelling to enable interdisciplinary, cross-sectoral and multi-level integration. *One Health* **2015**, *1*, 44–48. [CrossRef]
154. Lerner, H.; Berg, C. The concept of health in One Health and some practical implications for research and education: What is One Health? *Infect. Ecol. Epidemiol.* **2015**, *5*, 25300. [CrossRef] [PubMed]
155. Tarazona, A.M.; Ceballos, M.C.; Broom, D.M. Human relationships with domestic and other animals: One Health, one welfare, one biology. *Animals* **2020**, *10*, 43. [CrossRef]
156. Kingsley, P.; Taylor, E. One Health: Competing perspectives in an emerging field. *Parasitology* **2017**, *144*, 7–14. [CrossRef]
157. European Food Safety Authority; European Centre for Disease Prevention and Control. The European Union One Health 2018 zoonoses report. *EFSA J.* **2019**, *17*, e05926. [CrossRef]
158. Ahmed, T.; Tahir, M.F.; Boden, L.; Kingston, T. Future directions for One Health research: Regional and sectoral gaps. *One Health* **2023**, *17*, 100584.
159. Kupka, R.; Siekmans, K.; Beal, T. The diets of children: Overview of available data for children and adolescents. *Glob. Food Secur.* **2020**, *27*, 100442.
160. Raza, A.; Fox, E.L.; Morris, S.S.; Kupka, R.; Timmer, A.; Dalmiya, N.; Fanzo, J. Conceptual framework of food systems for children and adolescents. *Glob. Food Secur.* **2020**, *27*, 100436. [CrossRef]
161. Neufeld, L.M.; Andrade, E.B.; Ballonoff Suleiman, A.; Barker, M.; Beal, T.; Blum, L.S.; Demmler, K.M.; Dogra, S.; Hardy-Johnson, P.; Lahiri, A.; et al. Food choice in transition: Adolescent autonomy, agency, and the food environment. *Lancet* **2022**, *399*, 185–197. [CrossRef] [PubMed]
162. Anastasiou, K.; Baker, P.; Hadjikakou, M.; Hendrie, G.A.; Lawrence, M. A conceptual framework for understanding the environmental impacts of ultra-processed foods and implications for sustainable food systems. *J. Clean. Prod.* **2022**, *368*, 133155. [CrossRef]
163. Lelijveld, N.; Wrottesley, S.V.; Aburmishan, D.; Hemler, E.C.; Fentahun, N.; Ali, Z.; Menezes, R.; Roche, M.; Sharma, D.; Wazny, K.; et al. Research priorities for nutrition of school-aged children and adolescents in low- and middle-income countries. *PLoS ONE* **2023**, *18*, e0280510. [CrossRef] [PubMed]
164. Hawkes, C.; Fox, E.; Downs, S.M.; Fanzo, J.; Neve, K. Child-centered food systems: Reorienting food systems towards healthy diets for children. *Glob. Food Secur.* **2020**, *27*, 100414. [CrossRef]
165. Third, A.; Lala, G.; Moody, L.; Theakstone, G. Children's views on digital health in the Global South: Perspectives from crossnational, creative and participatory workshops. In *Creative Approaches to Health Education*; Lupton, D., Leahy, D., Eds.; Routledge: Oxfordshire, UK, 2021; pp. 173–189.
166. Fleming, C.A.; De Oliveira, J.D.; Hockey, K.; Lala, G.; Schmied, V.; Theakstone, G.; Third, A. *Food and Me. How Adolescents Experience Nutrition across the World. A Companion Report to the State of the World's Children*; UNICEF: Sydney, Australia, 2019; p. 2020.
167. Hunner-Kreisel, C.; März, S. Children and participation: Mapping social inequalities within concepts of well-being in qualitative research. *Child Indic. Res.* **2019**, *12*, 425–442.
168. United Nations. *UN Convention on the Rights of the Child (UNCRC). O_ce of the United Nations High Commissioner for Human Rights*; United Nations: New York, NY, USA, 1989.
169. Browne, S.; Barron, C.; Staines, A.; Sweeney, M.R. Participatory approaches to understand dietary behaviours of adolescents in the secondary school setting. *Nutrients* **2020**, *12*, 3761. [CrossRef]
170. Fleming, C.A.K.; Sharma, D.; Brunacci, K.; Chandra, S.; Lala, G.; Munn, L. Amanda Third. Fix my food: An urgent call to action from adolescents on how they experience and want to see change in their food systems. *J. Hum. Nutr. Diet.* **2023**, *36*, 2295–2309. [CrossRef]
171. Stylianou, K.S.; Heller, M.C.; Fulgoni, V.L.; Ernstoff, A.S.; Keoleian, G.A.; Jolliet, O. A life cycle assessment framework combining nutritional and environmental health impacts of diet: A case study on milk. *Int. J. Life Cycle Assess.* **2016**, *21*, 734–746.
172. Gaworski, M.; Borowski, P.F.; Zajkowska, M. Attitudes of a group of young Polish consumers towards selected features of dairy products. *Agron. Res.* **2021**, *19*, 1023–1038.
173. McLaren, S. Integration of Environment and Nutrition in Life Cycle Assessment of Food Items: Opportunities and Challenges. FAO. 2021. Available online: https://www.fao.org/documents/card/en/c/cb8054en/?utm_source=twitter&utm_medium=social+media&utm_campaign=faoknowledge (accessed on 28 December 2023).
174. Bianchi, M.; Strid, A.; Winkvist, A.; Lindroos, A.-K.; Sonesson, U.; Hallstrom, E. Systematic evaluation of nutrition indicators for use within food LCA studies. *Sustainability* **2020**, *12*, 8992. [CrossRef]
175. Green, A.; Nemecek, T.; Chaudhary, A.; Mathys, A. Assessing nutritional, health, and environmental sustainability dimensions of agri-food production. *Glob. Food Secur.* **2020**, *26*, 100406. [CrossRef]
176. McAuliffe, G.A.; Takahashi, T.; Beal, T.; Huppertz, T.; Leroy, F.; Buttriss, J.; Collins, A.L.; Drewnowski, A.; McLaren, S.J.; Ortenzi, F.; et al. Protein quality as a complementary functional unit in life cycle assessment (LCA). *Int. J. Life Cycle Assess.* **2023**, *28*, 146–155. [CrossRef] [PubMed]
177. McAuliffe, G.A.; Takahashi, T.; Lee, M.R.F. Applications of nutritional functional units in commodity-level life cycle assessment (LCA) of agri-food systems. *Int. J. Life Cycle Assess.* **2019**, *25*, 208–221. [CrossRef] [PubMed]

178. Madanayake, N.H.; Hossain, A.; Adassooriya, N.M. Nanobiotechnology for agricultural sustainability, and food and environmental safety. *Qual. Assur. Saf. Crops Foods* **2021**, *13*, 20–36. [CrossRef]
179. Castro, S.M.; Inαcio, R.S.; Alexandre, E.M.C.; Fidalgo, L.G.; Pereira, S.; Quaresma, P.; Freitas, P.; Teixeira, P.; Pintado, M.; Gomes, A.M.; et al. Environmental footprint of emerging technologies, regulatory and legislative issues. In *Innovative Technologies for Food Preservation: Inactivation of Spoilage and Pathogenic Microorganisms*; Barba, F.J., Sant, A.S., Orlien, V., Koubaa, M., Eds.; Elsevier: Amsterdam, The Netherlands, 2018; pp. 255–276. [CrossRef]
180. Green, A.; Nemecek, T.; Mathys, A. A proposed framework to develop nutrient profiling algorithms for assessments of sustainable food: The metrics and their assumptions matter. *Int. J. Life Cycle Assess.* **2023**, *28*, 1326–1347. [CrossRef]
181. Bazzan, G.; Daugbjerg, C.; Tosun, J. Attaining policy integration through the integration of new policy instruments: The case of the Farm to Fork Strategy. *Appl. Econ. Perspect. Policy* **2023**, *45*, 803–818.
182. Schulze, C.; Matzdorf, B.; Rommel, J.; Czajkowski, M.; García-Llorente, M.; Gutiérrez-Briceño, I.; Zawadzki, W. Between farms and forks: Food industry perspectives on the future of EU food labelling. *Ecol. Econ.* **2024**, *217*, 108066.
183. Schebesta, H.; Candel, J.J. Game-changing potential of the EU's Farm to Fork Strategy. *Nat. Food* **2020**, *1*, 586–588. [CrossRef]
184. Riccaboni, A.; Neri, E.; Trovarelli, F.; Pulselli, R.M. Sustainability-oriented research and innovation in 'farm to fork' value chains. *Curr. Opin. Food Sci.* **2021**, *42*, 102–112.
185. European Commission. *Food Systems Research and Innovation Investment Gap Study: Policy Report, 2023*; European Commission: Brussels, Belgium, 2023.
186. World Health Organisation. *WHO Estimates of the Global Burden of Foodborne Diseases: Foodborne Disease Burden Epidemiology Reference Group 2007–2015*; WHO Library Cataloguing-in-Publication Data, Printed in Switzerland; World Health Organisation: Geneva, Switzerland, 2015; ISBN 9789241565165.
187. FAO. 'Climate Change: Unpacking the Burden on Food Safety', Food Safety and Quality Series No 8, Rome, 2020. Available online: http://www.fao.org/3/ca8185en/CA8185EN.pdf (accessed on 28 December 2023).
188. Lee, J.C.; Neonaki, M.; Alexopoulos, A.; Varzakas, T. Case Studies of Small-Medium Food Enterprises around theWorld: Major Constraints and Benefits from the Implementation of Food Safety Management Systems. *Foods* **2023**, *12*, 3218. [CrossRef]
189. Alli, I.; Okorie, N. Food safety and developing countries: The benefits of technology transfer. *J. Food Saf.* **2015**, *35*, 312–330.
190. Grace, D. ILSI. Food safety in Developing Countries: An Overview. 2015. Available online: https://cgspace.cgiar.org/server/api/core/bitstreams/ac307a2d-4fbd-40bc-8781-68fa5040600c/content (accessed on 14 January 2024).
191. World Health Organization (WHO). The WHO's Website Includes Information on Food Safety, Global Foodborne Diseases, and International Food Safety Standards. Available online: https://www.who.int/foodsafety/en/ (accessed on 26 December 2022).
192. Fabregas, R.; Kremer, M.; Schilbach, F. Realizing the potential of digital development: The case of agricultural advice. *Science* **2019**, *366*, 6471. [CrossRef]
193. Available online: http://www.un.org/en/development/desa/news/population/2015-report.html (accessed on 28 December 2023).
194. Locally-Driven Co-Development of Plant-Based Value Chains Towards More Sustainable African Food System with Healthier Diets and Export Potential | InnoFoodAfrica | Project | Fact sheet | H2020 | CORDIS | European Commission (europa.eu). Available online: https://cordis.europa.eu/project/id/862170 (accessed on 28 December 2023).
195. Iitembu, J.A.; Gabriel, N.N.; Tjipute, M.; Asino, H.; Hamukwaya, J. The governance of aquaculture in Namibia as a vehicle for food security and economic growth. In *Food Security for African Smallholder Farmers*; Springer: Singapore, 2022; pp. 391–403. [CrossRef]
196. Ahwireng, A.K. Small Pelagic Fish for Food: Governance and Performance of Small Pelagic Fish Value Chains for Food Security and Nutrition in Ghana. 2022. Available online: https://hdl.handle.net/11245.1/a2938d0d-feeb-4775-8e0a-d480810d49fb (accessed on 13 January 2022).
197. Anser, M.K.; Osabohien, R.; Olonade, O.; Karakara, A.A.; Olalekan, I.B.; Ashraf, J.; Igbinoba, A. Impact of ICT adoption and governance interaction on food security in West Africa. *Sustainability* **2021**, *13*, 5570. [CrossRef]
198. Haysom, G. Integrating food sensitive planning and urban design into urban governance actions. *Urban Forum* **2021**, *32*, 289–310. [CrossRef]
199. Available online: https://www.gartner.com/smarterwithgartner/2017-the-year-that-data-and-analytics-go-mainstream (accessed on 28 December 2023).
200. Sridhar, A.; Ponnuchamy, M.; Kumar, P.S.; Kapoor, A.; Vo, D.V.N.; Rangasamy, G. Digitalization of the agro-food sector for achieving sustainable development goals: A review. *Sustain. Food Technol.* **2023**, *1*, 783. [CrossRef]
201. Bouis, H.E.; Saltzman, A. Improving nutrition through biofortification: A review of evidence from HarvestPlus, 2003 through 2016. *Global Food Secur.* **2017**, *12*, 49–58. [CrossRef]
202. Smith, M.J. Getting value from artificial intelligence in agriculture. *Anim. Prod. Sci.* **2020**, *60*, 46–54. [CrossRef]
203. Jain, D.K.; Shamsolmoali, P.; Sehdev, P. Extended deep neural network for facial emotion recognition. *Pattern Recognit. Lett.* **2019**, *120*, 69–74. [CrossRef]
204. Liu, J.; Li, T.; Xie, P.; Du, S.; Teng, F.; Yang, X. Urban big data fusion based on deep learning: An overview. *Inf. Fusion* **2020**, *53*, 123–133. [CrossRef]
205. Storey, V.C.; Song, I.Y. Big data technologies and Management: What conceptual modeling can do. *Data Knowl. Eng.* **2017**, *108*, 50–67.

206. Barrett, B.; Nitze, I.; Green, S.; Cawkwell, F. Assessment of multi-temporal, multi-sensor radar and ancillary spatial data for grasslands monitoring in Ireland using machine learning approaches. *Remote Sens. Environ.* **2014**, *152*, 109–124. [CrossRef]
207. Van Arkel, Z.; Kaleita, A.L. Identifying sampling locations for field-scale soil moisture estimation using K-means clustering. *Water Resour. Res.* **2014**, *50*, 7050–7057.
208. Monteiro, A.L.; de Freitas Souza, M.; Lins, H.A.; Teofilo, T.M.d.S.; Barros Júnior, A.P.; Silva, D.V.; Mendonça, V. A new alternative to determine weed control in agricultural systems based on artificial neural networks (ANNs). *Field Crops Res.* **2021**, *263*, 108075. [CrossRef]
209. Schnase, J.L.; Duffy, D.Q.; Tamkin, G.S.; Nadeau, D.; Thompson, J.H.; Grieg, C.M.; McInerney, M.A.; Webster, W.P. MERRA analytic services: Meeting the big data challenges of climate science through cloud-enabled climate analytics-as-a-service. *Comput. Environ. Urban Syst.* **2017**, *61*, 198–211. [CrossRef]
210. Marcot, B.G.; Holthausen, R.S.; Raphael, M.G.; Rowland, M.M.; Wisdom, M.J. Using Bayesian belief networks to evaluate fish and wildlife population viability under land management alternatives from an environmental impact statement. *For. Ecol. Manag.* **2001**, *153*, 29–42.
211. Becker-Reshef, I.; Justice, C.; Sullivan, M.; Vermote, E.; Tucker, C.; Anyamba, A.; Small, J.; Pak, E.; Masuoka, E.; Schmaltz, J.; et al. Monitoring global croplands with coarse resolution earth observations: The Global Agriculture Monitoring (GLAM) project. *Remote Sens.* **2010**, *2*, 1589–1609.
212. Nativi, S.; Mazzetti, P.; Santoro, M.; Papeschi, F.; Craglia, M.; Ochiai, O. Big data challenges in building the global earth observation system of systems. *Environ. Model. Softw.* **2015**, *68*, 1–26.
213. Available online: https://ec.europa.eu/jrc/en/publication/semicircular-flow-data-economy (accessed on 28 December 2023).
214. European Commission, DG COMP. *Competition Policy for the Digital Era*; European Commission, DG COMP: Brussel, Belgium, 2019.
215. Horizon 2020 Decode and LEDGER Projects. Available online: https://mydata.org/ (accessed on 28 December 2023).
216. EIB. *Feeding Future Generations: How Finance Can Boost Innovation in Agri-Food*; EIB: Luxembourg, 2019; p. 12.
217. S3 Food-Project, Inventory of the Challenges and Needs of the Food Sector with Regard to the Digitalisation in the Value Chain and the Related Technologies. 2020. Available online: https://s3platform.jrc.ec.europa.eu/en/w/digitalisation-and-new-technologies-in-agri-food-and-s3p-agrifood-s-international-b2b-meetings (accessed on 28 December 2023).
218. 2018 Report by Digital Innovation Hubs Working Group; See Also COM(2020) 103 'An SME Strategy for a Sustainable and Digital Europe'. Available online: https://data.consilium.europa.eu/doc/document/ST-6783-2020-INIT/en/pdf (accessed on 28 December 2023).
219. Lobell, D.B.; Asseng, S. Comparing estimates of climate change impacts from process-based and statistical crop models. *Environ. Res. Lett.* **2017**, *12*, 015001. [CrossRef]
220. Rosenzweig, C.; Elliott, J.; Deryng, D.; Ruane, A.C.; Müller, C.; Arneth, A.; Boote, K.J.; Folberth, C.; Glotter, M.; Khabarov, N.; et al. Assessing agricultural risks of climate change in the 21st century in a global gridded crop model intercomparison. *Proc. Natl. Acad. Sci. USA* **2014**, *111*, 3268–3273. [CrossRef]
221. Zhao, C.; Liu, B.; Piao, S.; Wang, X.; Lobell, D.B.; Huang, Y.; Huang, M.; Yao, Y.; Bassu, S.; Ciais, P.; et al. Temperature increase reduces global yields of major crops in four independent estimates. *Proc. Natl. Acad. Sci. USA* **2017**, *114*, 9326–9331. [CrossRef]
222. Ortiz-Bobea, A.; Ault, T.R.; Carrillo, C.M.; Chambers, R.G.; Lobell, D.B. Anthropogenic climate change has slowed global agricultural productivity growth. *Nat. Clim. Chang.* **2021**, *11*, 306–312. [CrossRef]
223. IPCC. *Climate Change 2022 Impacts Adaptation and Vulnerability Summary for Policy Makers*; IPCC: Geneva, Switzerland, 2022.
224. Challinor, A.J.; Watson, J.; Lobell, D.B.; Howden, S.M.; Smith, D.R.; Chhetri, N. A meta-analysis of crop yield under climate change and adaptation. *Nat. Clim. Chang.* **2014**, *4*, 287–291. [CrossRef]
225. Peng, B.; Guan, K.Y.; Tang, J.Y.; Ainsworth, E.A.; Asseng, S.; Bernacchi, C.J.; Cooper, M.; Delucia, E.H.; Elliott, J.W.; Ewert, F.; et al. Towards a multiscale crop modelling framework for climate change adaptation assessment. *Nat. Plants* **2020**, *6*, 338–348. [CrossRef] [PubMed]
226. Sloat, L.L.; Davis, S.J.; Gerber, J.S.; Moore, F.C.; Ray, D.K.; West, P.C.; Mueller, N.D. Climate adaptation by crop migration. *Nat. Commun.* **2020**, *11*, 1243. [CrossRef] [PubMed]
227. Hasegawa, T.; Wakatsuki, H.; Ju, H.; Vyas, S.; Nelson, G.C.; Farrell, A.; Deryng, D.; Meza, F.; Makowski, D. A global dataset for the projected impacts of climate change on four major crops. *Sci. Data* **2022**, *9*, 58. [CrossRef] [PubMed]
228. Deng, X.; Huang, Y.; Yuan, W.; Zhang, W.; Ciais, P.; Dong, W.; Smith, P.; Qin, Z. Building soil to reduce climate change impacts on global crop yield. *Sci. Total Environ.* **2023**, *903*, 166711.
229. De Boer, I.J.; van Ittersum, M.K. *Circularity in Agricultural Production*; Wageningen University & Research: Wageningen, The Netherlands, 2018.

230. Garnier, J.; Le Noë, J.; Marescaux, A.; Sanz-Cobena, A.; Lassaletta, L.; Silvestre, M.; Billen, G. Long-term changes in greenhouse gas emissions from French agriculture and livestock (1852–2014): From traditional agriculture to conventional intensive systems. *Sci. Total Environ.* **2019**, *660*, 1486–1501. [CrossRef]
231. Mrówczyńska-Kamińska, A.; Bajan, B.; Pawłowski, K.P.; Genstwa, N.; Zmyślona, J. Greenhouse gas emissions intensity of food production systems and its determinants. *PLoS ONE* **2021**, *16*, e0250995.

foods

MDPI

Review

Innovative Applications of *Tenebrio molitor* Larvae in Food Product Development: A Comprehensive Review

Konstantina Kotsou [1], Theodoros Chatzimitakos [1], Vassilis Athanasiadis [1], Eleni Bozinou [1], Christos G. Athanassiou [2] and Stavros I. Lalas [1,*]

1 Department of Food Science and Nutrition, University of Thessaly, Terma N. Temponera Str., 43100 Karditsa, Greece; kkotsou@agr.uth.gr (K.K.); tchatzimitakos@uth.gr (T.C.); vaathanasiadis@uth.gr (V.A.); empozinou@uth.gr (E.B.)

2 Laboratory of Entomology and Agricultural Zoology, Department of Agriculture, Crop Production and Rural Environment, School of Agricultural Sciences, University of Thessaly, Phytokou Str., 38446 Volos, Greece; athanassiou@uth.gr

* Correspondence: slalas@uth.gr

Abstract: The utilization of alternative and sustainable food sources has garnered significant interest as a means to address the challenges of food security and environmental sustainability. *Tenebrio molitor* larvae, commonly known as mealworms, have emerged as a promising candidate in this context, as they are a rich source of nutrients and can be reared with relatively low resource input. This review article presents an in-depth analysis of the diverse range of food products developed using *T. molitor* larvae and the distinctive properties they bestow on these products. The review encompasses an exploration of the nutritional composition of the larvae, emphasizing their rich protein content, balanced amino acid profile, fatty acids with health benefits, vitamins, and minerals. It delves into how these attributes have been harnessed to enhance the nutritional value of a variety of food items, ranging from protein-rich snacks and energy bars to pasta, bakery goods, etc. Each of these applications is discussed with regard to how *T. molitor* larvae contribute to the nutritional content and sensory characteristics of the final product. Furthermore, this review sheds light on the innovative techniques and processing methods employed to incorporate *T. molitor* larvae into different food matrices. It addresses challenges related to taste, texture, and appearance that have been encountered and the strategies devised to overcome related problems. Overall, this comprehensive review elucidates the diverse food products that have been developed utilizing *T. molitor* larvae as a key ingredient. Highlighting the nutritional, sensory, and sustainability aspects of these products, this review offers valuable insights to harness the potential of this alternative protein source to meet the evolving needs of modern food systems.

Keywords: Tenebrionidae; edible insects; food additive; food products; nutritional enrichment

1. Introduction

The exploration of alternative food sources for human consumption has garnered significant attention due to several compelling factors, with most of them revolving around overpopulation and the high nutritional value of these novel food sources. This ongoing issue of overpopulation, which has been a topic of concern in the research community since 1997 [1], naturally correlates with an increasing demand for conventional protein sources, namely, meat and fish [2–5]. The expansion of livestock farming, driven by this escalating demand, poses the risk of encroaching on previously uncultivated land, exacerbating the issue of land use, considering that 75% of arable land is currently dedicated to livestock activities [6,7]. Moreover, the livestock sector is a major contributor, accounting for 14.5% of total anthropogenic greenhouse gas emissions [8], raising concerns about the environmental impact of this sector, which is likely to escalate with an increase in livestock units [9]. In the context of fish as a key source of protein, numerous fish species are endangered worldwide as a consequence of overfishing, pollution, coastal development, climate change and other anthropogenic actions [10]. An example is

Chondrichthyes. The *Chondrichthyes* species is in danger, since 32.6% (391 out of 1199 species) are threatened with complete extinction due to overfishing [11]. This overexploitation of marine resources has resulted in the degradation of several marine habitats [11–13]. Given these ecological and sustainability concerns surrounding conventional protein sources, the scientific community has shifted its focus toward identifying alternative, sustainable sources of protein. Among these, insects have emerged as one of the most promising options [14–18].

Insects exhibit distinct advantages from economic and ecological standpoints. They require considerably less space, minimal water (relying primarily on moisture), and reduced feed to complete their biological life cycle compared to traditional protein sources [19–22]. However, it is noteworthy that the concept of insects as a food source is not entirely novel. Recent archaeological evidence from Tanzania suggests that insects played a crucial role in the nutrition of early humans, dating back 1.8 million years [23]. Additionally, historical depictions, such as those found in the Artamila caves in northern Spain (3000–9000 BC), portray the collection of bee nests and larvae, further highlighting the historical significance of insect consumption [24]. Despite this historical practice, insect consumption has declined significantly over time and is now primarily narrowed to traditional practices in certain regions like Southeast Asia and Australia [25,26]. In Europe, insect consumption remains relatively rare, largely due to "neophobia" surrounding insect consumption in any form [27–30]. Nevertheless, as of 2021, three insect species have received approval for safe human consumption by the European Parliament and Council [31]. In particular, on 12 November 2021 they authorized the placing on the market of frozen, dried, and powdered forms of *Locusta migratoria* Linnaeus (Orthoptera: Acrididae) [32], on 1 June 2021 the placing on the market of dried *Tenebrio molitor* Linnaeus (Coleoptera: Tenebrionidae) larvae as a novel food [33] and, most recently, on 5 January 2023, the frozen, paste, dried and powder forms of *Alphitobius diaperinus* Panzer (Coleoptera: Tenebrionidae) larvae (lesser mealworm) [34].

One of the approved species, *Tenebrio molitor* Linnaeus (Coleoptera: Tenebrionidae), is particularly noteworthy for its potential to be more readily accepted by consumers, especially in Europe [35–38]. For instance, Europeans tend to find the larvae of *T. molitor* more appealing than alternatives like cockroaches or ants [39]. Moreover, the rearing of *T. molitor* shows great promise in the rapidly expanding, global, edible insect market, where it is already being widely produced as feed for fish and poultry [40–43].

A comparative analysis of the rearing of *T. molitor* and *Alphitobius diaperinus* Panzer (Coleoptera: Tenebrionidae) reveals several advantages of the former. *T. molitor* larvae complete their development and growth in a shorter time frame, boasting a higher survival rate and greater weight compared to *A. diaperinus* [44]. Specifically, the survival rate of *T. molitor* larvae is as high as 98.8%, while that of *A. diaperinus* ranges from 88 to 95%. Additionally, the weight of *T. molitor* larvae is substantially greater, compared to the weight of *A. diaperinus* larvae [45–47]. This disparity is expected, given that *T. molitor* is larger than *A. diaperinus* at all stages of development [48,49]. When comparing the rearing of *T. molitor* to that of *Locusta migratoria* Linnaeus (Orthoptera: Acrididae), it becomes evident that *T. molitor* rearing is both faster and more straightforward. The development of *T. molitor* larvae is completed in just 2–3 months, whereas *L. migratoria* larvae require 2–4 months for their development [50–52]. This efficiency in rearing further highlights the potential of *T. molitor* as a viable and sustainable protein source for human consumption, particularly in regions where insect consumption is gaining recognition as an environmentally responsible dietary choice.

The pursuit of alternative food sources, driven by concerns for overpopulation and the environmental impact of traditional protein sources, has led to the exploration of insects as a sustainable protein option. *T. molitor*, in particular, holds promise due to its appeal to European consumers, efficient rearing, and favorable growth characteristics compared to other insect species. This comprehensive review aims to provide a thorough exploration of the diverse applications of *T. molitor* larvae, in the realm of food product development. It endeavors to elucidate the multifaceted potential of *T. molitor* larvae as a sustainable and nutritious protein source for human consumption. By comprehensively reviewing the existing research and applications of *T. molitor* larvae in food product development, this review aims to provide valuable insights for researchers, food industry professionals, and policymakers. Furthermore, it underscores the need for continued exploration and development in this field to advance the sustainable food options required to meet the demands of a growing global population while mitigating environmental concerns.

2. Nutritional Value of *T. molitor*

As is frequently mentioned, their high nutritional value is the reason that led researchers to extensively study the larvae of *T. molitor*. This species is, to this day, the main subject of research since it is characterized by its content of essential nutrients, including crude protein, essential amino acids (EAAs), fat, and essential fatty acids [24,53–57]. The quantification of the crude protein of *T. molitor* and its flour (dried and finely ground larvae) has received special attention. Due to this, several analytical methods have been used, such as the combustion (Dumas) method [56], Randall method [35], Kjeldahl method [58], elemental analysis method [59], Bradford method [60], and precipitation [61], yielding a variety of results, ranging from 36.8–75.1%. This variation can be attributed to the different rearing conditions of each *T. molitor* larvae. At this point, it is worth emphasizing that conventional animal, protein-rich foods, such as beef, chicken, and tuna, contain 21.4, 19.4 [62], and 22.7% crude protein [63], respectively, values that can easily be considered low compared to the protein content of *T. molitor* larvae. In Table 1 the crude protein values contained in *T. molitor* larvae according to various surveys are presented.

Table 1. Nutritional value of *T. molitor* larvae according to the crude protein quantification method.

Composition of Dry Weight (%)			
Crude Protein	**Crude Fat**	**Ash**	**Ref.**
36.8	26.0	~1.0	[60]
46.4	32.2	2.9	[35]
47.0	29.6	2.6	[56]
51.0	ne *	ne	[59]
60.2	19.1	4.2	[58]
75.1	ne	ne	[61]

* ne: not examined.

Histidine (His), isoleucine (Iso), leucine (Leu), lysine (Lys), methionine (Met), phenylalanine (Phe), threonine (Thr), tryptophan (Trp), and valine (Val) are nine amino acids that cannot be synthesized by mammals and they must be obtained through food consumption. Therefore, they are called dietarily essential, indispensable nutrients or essential amino acids (EAAs) [64]. *T. molitor* larvae contain large amounts of Leu, Val, and Lys but low amounts of His, Met, and Trp EAAs, with Trp rarely occurring in these insects [55]. Similarly, to crude protein, EAA concentrations are influenced by the rearing conditions of the *T. molitor* larvae, temperature, humidity, feed substrate and rearing region [65,66]. For instance, Lys quantities in *T. molitor* larvae can range from 1.6 to 5.8%, Thr quantities can range from 1.3 to 4.3%, Met quantities can range from 0.6 to 2.2% and Trp quantities can range from 0.02 to 1.9% [35,42–45]. More details for the quantities of the EAAs are presented in Table 2.

Table 2. Amino acid composition (mg/g of dry weight) of *T. molitor* larvae.

His	Iso	Leu	Lys	Met	Phe	Thr	Trp	Val	Ref.
0.8	1.3	2.2	1.6	0.6	1.3	1.3	0.3	2.2	[67]
1.5	3.6	3.4	2.9	0.7	1.6	1.8	nd *	2.4	[57]
2.6	2.8	4.8	1.8	1.4	1.4	2.9	1.9	4.0	[43]
2.8	6.5	6.2	5.3	1.2	3.2	3.3	0.02	4.5	[68]
3.1	4.0	7.3	5.8	2.2	1.8	4.3	0.7	5.3	[69]

* nd: not detected; His: histidine; Iso: isoleucine; Leu: leucine; Lys: lysine; Met: methionine; Phe: phenylalanine; Thr: threonine; Trp: tryptophane; Val: valine.

The second most plentiful nutrient in the composition of *T. molitor* larvae is crude fat [70], which exhibits variability in values depending on the content of other nutrients and the processing method [55]. Predominantly, these larvae contain about 27% fat, whereas the percentage was found to range from 19.1 to 32.2% [56–60]. In Table 1, a detailed presentation of the crude fat values, with the corresponding crude protein content, is shown. In addition to EAAs, in *T. molitor* larvae, essential fatty acids (EFAs), such as linoleic acid (ω-6 group) and alpha-linolenic acid (ω-3 group) [71], are also detected [72]. Humans cannot produce these EFAs and thus must be obtained through the diet they are based on. The fatty acid composition of *T. molitor* larvae is presented in Table 3. As far as linolenic acid (C18:2 ω-6) is concerned, it is present in high amounts in *T. molitor* larvae, ranging from 11.5 to 48.1%. In contrast, as far as the content of alpha-linolenic acid (C18:3 ω-3) in larvae is concerned, it covers a range of values, from 0.2 to 2.3%. Furthermore, considerable quantities of fatty acids such as myristic acid (C14:0), palmitic acid (C16:0), palmitoleic acid (C16:1), and ω-9 oleic acid (C18:1) are also detected [38,42,45–47].

Table 3. Fatty acid composition (%) of *T. molitor* larvae.

C14:0	C16:0	C16:1	C18:1	C18:2 (ω-6)	C20:0	C18:3 (ω-3)	∑ SFA	∑ UFA	Ref.
2.1	18.8	0.5	28.6	48.1	0.6	0.2	22.5	77.5	[58]
2.1	17.2	1.9	43.8	29.4	nr *	2.3	21.0	79.0	[62]
2.3	21.4	0.1	39.1	27.3	1.4	0.1	31.6	nr	[65]
4.3	21.1	1.9	52.9	11.5	0.5	0.2	33.2	nr	[66]
5.0	19.1	1.7	49.9	18.1	0.1	0.4	28.6	nr	[67]

* nr: not referred.

Crude ash constitutes one more important aspect of *T. molitor* larvae, which is tested in all studies investigating the nutritional value of *T. molitor* larvae. Crude ash refers to the total of all mineral elements such as phosphorus (P), calcium (Ca), sodium (Na), and magnesium (Mg) [73]. The crude ash as presented in Table 1 ranges from about 1.0 to 4.2% [56–60] and is also affected by the composition of the other nutrients, crude protein, and crude fat. Regarding, the minerals contained in *T. molitor* larvae were identified and quantified in a number of studies. In particular, the P content of larvae has recorded values up to about 1.0% [74], Ca content up to 0.5% [58], Na content up to 0.4% [57], Mg content up to 1.6% [58] and potassium (K) content up to about 1.0% [57]. In addition, *T. molitor* larvae contain up to 100.02 mg/kg of iron (Fe), up to 117.4 mg/kg, of zinc (Zn), and 20.0 mg/kg of copper (Cu) [74].

3. *Tenebrio molitor* and Its Derivatives as Innovative Products for Human Consumption

As stated in the introduction, humans have been consuming insects in their daily diet since ancient times [23]. However, it is presumed that, at that time, insects were mainly eaten raw. In the modern world, the habit of eating insects, known as entomophagy, was first described in 1975 by Meyer-Rochow, in which edible insects were proposed as a future dietary solution [75]. Nowadays, in Western countries the predominant consumption way of *T. molitor* larvae is in the form of 'flour' [76]. The preparation method of this flour has been relatively similar in all investigations to date. First, the larvae are dried either using lyophilization (e.g., lyophilizer for 48 h at a temperature of −96 °C and vacuum) [77] or an oven (e.g., air circulation oven at 40 °C for 48 h) [78] and then pulverized until they are in powder form. Consequently, most of the products obtained from mealworm larvae are bakery products, ranging from bread to pastries [79].

Moreover, *T. molitor* larvae are also well known for their high oil and fatty acid content [55]. Various methods and solvents carry out the extraction of the oil. For instance, the use of petroleum ether as solvent and extraction using a Soxhlet device at 50 °C for 6 h and evaporation with a rotary evaporator has been proposed [80]. Another method is the use of supercritical carbon dioxide (CO_2) extraction, where for maximum degreasing

(95%) pressure at 400/250 bar, a temperature of 45 °C and an extraction time of 105 min was applied [81]. In addition, extraction with hydroethanolic solvent (20:80, *v*/*v*) under continuous stirring (150 rpm) for 3 days has been proposed [78]. Alternatively, extraction using stirring at 40 °C and evaporation of the solvent with a rotary evaporator has also been proposed [60].

3.1. Bakery Goods

3.1.1. Bread

Kim et al. [82] pioneered the incorporation of flour from *T. molitor* larvae for bread production, marking a significant milestone in the exploration of alternative protein-rich ingredients for bakery products. Their study involved the creation of three separate bread formulations, each containing varying proportions of *T. molitor* larvae powder (2, 4, and 6%). A comprehensive evaluation of key parameters, including fermentation characteristics, pH level, total acidity, specific volume, color characteristics, and sensory evaluations were performed, all of which were compared to a control bread, produced with conventional flour. The findings of this study revealed exciting insights into the impact of *T. molitor* larval flour on bread properties. Initially, as the concentration of insect flour in the bread increased, the amount of fermentation swelling decreased; this may be due to the fact that *T. molitor* larvae are considered gluten-free [83]. Along with the increase in insect flour in bread, the pH levels also increased, which may be due to the fact that the pH of the insect ranges from 6.4 to 7.5 [84]. As expected, with the increase in insect flour in bread, acidity levels decreased, which possibly resulted from the previous fact. A further interesting result is that the color characteristics of the bread sample remained almost unaffected, even though *T. molitor* larvae flour has a dark color owing to its composition. After a sensory evaluation (using a 9-point scale system), the bread with the addition of 2% *T. molitor* larvae flour recorded the highest scores in appearance, taste, and overall acceptability, indicating the promising potential of this flour as a protein-rich alternative ingredient in bakery products.

A few years later, breads were prepared with the addition of *T. molitor* larvae flour by Roncolini et al. [85]. Higher percentages of *T. molitor* larval flour, namely 5 and 10%, were incorporated in this later research. This research extended its focus to the analysis of the nutritional value of bread (protein, fat, EAAs, and EFAs) and the effect of the addition of *T. molitor* larval flour on the antimicrobial activity of bread products. According to the results, the bread with wheat flour contained 8.9% crude protein and as the addition of *T. molitor* larvae powder increased, the protein percentage also increased, finally reaching 11.6%, in the sample with 10% addition of insect flour, up to 30%. The same effects were shown for the fat content. In the control sample, bread with wheat flour, the fat percentage was 0.1%, while for the sample with 10% insect flour from *T. molitor* larvae, the fat percentage was 1.1%, showing a ten-fold increase. Finally, the ash content was also increased significantly from 0.5 to 0.6% in the above-mentioned samples. It is therefore shown that the addition of nutritional ingredients to conventional foods may enhance their nutritional value without being influenced by the preparation process, i.e., baking. In addition, the content of EAAs and EFAs was also studied, whereby by increasing the content of insect flour from *T. molitor* larvae, these important nutrients were also enhanced. Worth mentioning is the evaluation of the presence of endospores in the baked bread samples, along with the occurrence of the following human pathogens: *Clostridium perfringens* and *Bacillus cereus*. The occurrence of spores in all samples containing *T. molitor* flour was <1 while in the control sample was 0.38, and all prepared bread samples were subjected to enumeration of these two human pathogens and showed safe and viable counts below 1 log colony forming unit (CFU)/g [86]. Regarding sensory characteristics, color was evaluated where, unlike the previous study by Kim et al. [82] in which the color remained unaffected, the color of bread in this case progressively deepened with the increasing proportion of insect-derived flour. This observation likely stemmed from the notably higher percentages employed in this experiment compared to the previous study. Finally, in terms of consumption preference, white flour bread proved to be the most preferable, with the study conclusion being that, so far, the addition of more than 2% insect flour is not as preferred by consumers.

A expansion of the previous investigations was carried out by study Kuenper et al. [87], in which baked bread with a wider range of *T. molitor* larval flour concentrations, namely 0, 5, 10, and 15%, was examined. A combination of the characteristics studied in the previous studies was examined, leading to similar outcomes. In more detail, nutritional characteristics (protein, fat, carbohydrates, and fiber), antimicrobial activity, and consumer preference for the products were also explored in the present study. According to the results, increasing the content of *T. molitor* flour led to an increase in crude protein, fat, and fiber while simultaneously decreasing carbohydrates. The crude protein in bread made with conventional flour was recorded at 11.3 g while the highest value was recorded in bread with a 15% *T. molitor* larvae content, which was a 75.2% increase compared to the control sample. Fat increased from 2.2 to 4.8 g and fiber from 7.5 to 7.7 g, among the control sample and that containing 15% insect flour in both cases. Meanwhile, the carbohydrate content decreased as the content of *T. molitor* larval meal increased, which is expected since the carbohydrate content of insects is reduced due to their high crude protein content. Regarding the antimicrobial activity of the samples, the resistance of the bread to yeasts and molds was studied after 7 days. The results are highly promising, since no significant differences between the samples were found in terms of shelf life. Finally, according to the consumers, the bread with 5% *T. molitor* larval flour added was the most preferable despite being darker than the control one. This fact is greatly encouraging considering that, until now, consumers have not preferred the addition of insect flour to exceed 2%.

3.1.2. Biscuits

Biscuits are the second most consumed bakery product after bread [88]. For this reason, *T. molitor* flour was also used to enhance these nutrient-poor products. The first attempt was made by Zielińska et al. [89] by preparing three types of biscuits enriched with *T. molitor* flour and the control biscuit with conventional flour. The control biscuit consisted of 300 g flour (M0), while the three versions receiving *T. molitor* flour contained 15 g (M15), 20 g (M20), and 30 g (M30), with the last sample being the darkest. As mentioned in the fortified bread, the higher the content of *T. molitor* flour, the higher the crude protein content of the biscuits. In M0, 9.1% crude protein content was recorded while in M30, it was 13.5%, i.e., a 48.3% increased content. However, the ash content was found to be greater in sample M20, with a value of 0.7%, while the highest antioxidant activity was found in sample M10, about 1 mM TE. Mixed results were obtained between the samples since the increase in *T. molitor* flour in the samples did not lead to a similar increase in nutrients in the biscuit samples. Nevertheless, it is worth noting that samples M15, M20, and M30 were significantly enriched compared to M0.

Biscuits with percentages of *T. molitor* larval flour, in order to enrich their nutritional value, were also prepared by Xie et al. [90]. Specifically, four biscuits with different percentages of *T. molitor* flour of 5% (M5), 10% (M10), 15% (M15), and 20% (M20) were prepared and compared with an insect-free control sample (M0). Concerning the nutritional value of the different biscuits in terms of protein, lipid, and dietary fiber, the higher the level of substitution of conventional flour with *T. molitor* flour, the higher the content of the above-mentioned nutrients in the samples. Sample M0 showed an equal crude protein content as in the aforementioned study, i.e., 9.1% [67], while the highest protein content was recorded in M20, which reached 16.0%. Moreover, the fat and dietary fiber content of M20 showed an increase of 20.5 and 21.7% compared to those of M0. Regarding organoleptic characteristics, the score for sensory properties showed no significant difference up to the 15% substitution level; however, in terms of preference, the biscuit with a 5% *T. molitor* flour content prevailed. Therefore, it was proven again that consumers do not prefer products containing over 5% *T. molitor* flour.

The latest study on the nutritional quality of biscuits enriched with *T. molitor* flour was conducted by Cozmuta et al. [91], where biscuits were prepared with 0, 10, 15, and 20% replacement of conventional flour with *T. molitor* flour. According to consumer preference, a significant result was recorded as the biscuit enriched with 15% *T. molitor* flour was selected as the most preferred biscuit, and it was the first time that consumers showed preference for such a high percentage of *T. molitor* flour in a food product. Comparing the conventional biscuit with the fortified biscuit, the nutritional value was significantly increased. Specifically, crude protein increased from 29.1 to 31.3%, crude fat increased from 46.4 to 47.8%, and ash increased from 1.4 to 1.6%. In addition, all individual minerals were identified and quantified, where in particular the addition of 15% *T. molitor* flour resulted in enhanced nutritional value in mineral elements. In detail, Na, K, Ca, Mg, P, Cu, Zn, Mn, Fe and Li showed an increase in content of about 2.1, about 7.1, 6.4, 20.6, 0.1, 39.5, 50.1, 13.2, 25.3 and 13.3%, respectively, proving once again that *T. molitor* larval flour can contribute to enhancing otherwise-nutrient-poor food products, such as bread and biscuits [92].

3.1.3. Cracker

Further study of the addition of *T. molitor* larvae flour to bread has not been carried out and, thus, the first attempt to enrich another amylaceous product, cracker, was made by Djouadi et al. [93]. In this study, crackers incorporating varying proportions of *T. molitor* larvae flour (0, 2, 4, 6, 10, 15, and 20%) were prepared (see Figure 1), where 0% was used as a control sample as in previous studies. The color of the crackers darkened greatly as the *T. molitor* larvae flour content increased, and therefore, consumers showed no attraction to crackers containing 10% or more *T. molitor* larvae flour. Hence, the researchers evaluated only the cracker with 6% insect flour and compared it to the control cracker. Regarding the nutritional value of the two crackers, 0 and 6% insect flour, the protein content scored values from 9.7 to 13.9%, respectively, and the ash content of the crackers scored values from 1.9 to 2.2%. Moreover, besides the increase in total ash content, the content of constituent minerals such as Na, K, Ca, Mg, P, Fe, Cu, and Zn in the cracker with 6% *T. molitor* flour increased strongly. Delving deeper into the study of the nutritional value of the 6% cracker, it was found that its polyphenol content also showed a statistically significant ($p < 0.05$) increase compared to the control sample, which was enhanced by 148.8%. Polyphenols are known for their strong antioxidant properties [94]. Therefore, it is believed that the high polyphenol content may have enhanced the antioxidant activity of the cracker, since through this study using two different methods (2,2-diphenyl-1-picrylhydrazyl (DPPH) radical scavenging activity and the reducing power of ferric ion), the cracker with the *T. molitor* larvae flour showed a significantly higher antioxidant activity than the control sample.

Figure 1. Crackers containing 0 (control) to 20% *T. molitor* flour (% *w*/*w*) from (**right**) to (**left**) [93]. Reprinted/adapted with permission from Ref. [93]. 2022, MDPI.

3.1.4. Bars

An investigation in which whole *T. molitor* larvae were used to produce a snack (bars) was carried out in Poland [95]. The research team developed four kinds of bars: bars in the absence of edible insects (serving as control), bars containing 7.4% *w*/*w* whole mealworms (they were visible to consumers), bars containing 7.4% *w*/*w* ground mealworms, and bars containing 7.4% *w*/*w* ground crickets. This study was conducted to assess consumer acceptance of snack bars enriched with edible insects and whether preference is influenced by the appearance of a food rather than its composition.

The findings of this study revealed intriguing consumer preferences. Notably, the control bars, which served as the control, were favored over the ground insect-enriched counterparts. Moreover, participants noted that the appearance of bars containing crickets was less appealing compared to the other products, highlighting the significance of visual appeal in consumer choices. The study underscored that deliciousness emerged as the primary determinant factor of the acceptability of insect-infused bars as a food source. According to the results, the bar with *T. molitor* larvae flour was much more preferable than the one with whole larvae, and in fact, the study underlines that in terms of appearance and taste, the control bars and those containing non-visibly embedded *T. molitor* larvae do not show significant differences. Additionally, the odor of the product emerged as a statistically significant indicator of quality, further emphasizing the importance of sensory aspects in consumer preferences. Considering the results from the odor of the samples, the control sample had the most attractive odor, followed by the sample with no obvious edible insects and the sample with obvious edible insects. As far as the samples with *T. molitor* larvae were concerned, the appearance of the larvae seemed to have influenced the odor of the food product to a large extent. Collectively, this research highlights that in terms of appearance and taste, control bars and those containing non-visibly integrated edible insects do not exhibit significant differences. This suggests that it is the phenomenon of neophobia, the reluctance to embrace new or unfamiliar foods like insects, that instills suspicion in people regarding insect consumption. These findings are invaluable in understanding the nuanced factors that influence consumer acceptance of insect-based food products. The proximate composition of the different baked products containing *T. molitor* flour is shown in Table 4 below.

Table 4. Proximate composition of the various bakery products made with the addition of *T. molitor* flour.

Bakery Products	Percentage (%) of *T. molitor* Flour Addition	Protein Content (g/100 g)	Fat Content (g/100 g)	Ash Content (g/100 g)	Ref.
Bread	0	8.9	0.1	0.5	[85]
	5	10.5	0.5	0.5	
	10	11.6	1.1	0.6	
	0	9.6	ne *	ne	[87]
	5	12.6	ne	ne	
	10	13.2	ne	ne	
	15	13.7	ne	ne	
Biscuits	0	9.1	27.0	0.3	[89]
	15	13.5	27.2	0.6	
	20	11.9	26.9	0.7	
	30	10.8	28.5	0.4	
	0	9.1	18.5	2.0	[90]
	5	10.3	19.4	2.1	
	10	13.0	19.7	2.3	
	15	14.2	20.5	2.3	
	20	16.0	22.3	2.4	
	0	29.1	46.4	1.4	[91]
	15	31.3	47.8	1.6	
Cracker	0	9.7	12.7	1.9	[93]
	6	13.9	11.1	2.2	

* ne: not examined.

3.2. *Meat*

3.2.1. Burgers

T. molitor, aside from the bakery products, was also employed in meat-based products. To this end, distinct burger formulations were prepared at the University of Liège [96]. Specifically, a beef burger consisting of unflavored ground beef, a lentil-based variant (consisting of 95% green lentils), a burger with 45% green lentils and 50% *T. molitor*, and a burger containing 45% ground beef and 50% *T. molitor* were produced. In all cases, a 5% fraction of each burger consisted of a flavoring mix (containing onions, carrots, tomato paste, garlic, salt and pepper). These burgers were prepared in order to examine the degree of insect neophobia in Western societies. In the analysis of the participants' ratings, the gender of the participant was found to play a significant role. As far as the taste of the burger was concerned, all participants favored the traditional beef burger, ranking it first. However, it is worth noting that the male participants preferred burger preparations containing beef or *T. molitor* more than the fully vegetable ones. At the same time, women showed a greater preference for conventional beef burgers only. In addition, odor did not seem to be significantly affected by the composition of the samples according to both genders. One notable outcome was the positive inclination observed among the participants regarding the integration of insects into their future dietary habits. Impressively, 70% of the respondents expressed confidence in the prospective inclusion of insects within their diets.

3.2.2. Sausages

Kim et al. [97], being the first researchers to do so, decided to prepare emulsion sausages which also contained various types of *T. molitor* flour. Specifically, they prepared a conventional pork sausage containing 60% lean pork, 20% fat, and 20% ice (control sausage). At the same time, those containing insects were prepared by replacing 10% of the lean pork with flour of *T. molitor* larvae. Essentially, three different insect sausages were prepared, since three different *T. molitor* flours were used. Specifically, untreated *T. molitor* larvae flour, defatted *T. molitor* larvae flour, and defatted and acid-hydrolyzed *T. molitor* larvae flour were employed. Results indicated that the increased treatment of *T. molitor* flour resulted in an increased protein content of the final sample. In other words, a higher percentage of proteins was found in the sausage with defatted and acid-hydrolyzed flour from *T. molitor* larvae; the value increased from 22.6 to 31.3%. In addition to protein, minerals contained in the final samples were also evaluated. The results of these showed varying results. The addition of *T. molitor* larval flour, regardless of the treatment it received, increased the content of all the minerals tested (P, K, Ca, Mg, Na, Zn, Fe, Cu, and Mn). Nevertheless, the sample containing defatted and acid-hydrolyzed flour from *T. molitor* larvae presented the lowest mineral content compared to the other two samples containing *T. molitor*. The highest content of Ca (13.9 mg/100 g), Na (913.7 mg/100 g), Fe (9.9 mg/100 g), Cu (0.3 mg/100 g), and Mn (0.4 mg/100 g) was recorded in the sample with 10% *T. molitor* larvae flour that no additional treatment was applied to. Simultaneously, a higher content of P (202.1 mg/100 g), K (325.2 mg/100 g), Mg (47.7 mg/100 g), and Zn (2.5 mg/100 g) was recorded in the sample with 10% defatted *T. molitor* larvae flour. At this point it should be noted that the control sample showed the following mineral contents (P—145.5 mg/100 g, K—245.7 mg/100 g, Ca—9.3 mg/100 g, Mg—20.9 mg/100 g, Na—799.5 mg/100 g, Zn—1.2 mg/100 g, Fe—3.8 mg/100 g, Cu—0.04 mg/100 g and Mn—0.1 mg/100 g). Thus, it is proven that a nutrient product can be enhanced even with the addition of 10% of insect flour. Furthermore, it is proven that the addition of *T. molitor* larvae flour to a food product can achieve different outcomes depending on the treatment it has received. Hence, depending on the nutrients insisted upon to be enhanced in a product, the flour used should receive a corresponding treatment.

Frankfurt sausages are known worldwide due to their inclusion in the well-known product "hot dog" [98]. Choi et al. [99] attempted to enrich this sausage with insect flour in various proportions in order to evaluate its nutritional value and consumer preference. This substitution with insect flour is also carried out in order to reduce the negative environmental footprint caused by the production of the pork [8,100] from which the Frankfurt sausage is made. As control samples, the sausages were prepared with 50% pork ham, 25% back fat, and 25% ice. The other samples were prepared as follows: T1, 45% pork ham + 5% *T. molitor* larvae flour; T3, 35% pork ham + 15% *T. molitor* larvae flour; T4, 30% pork ham + 20% *T. molitor* larvae flour; T5, 25% pork ham + 25% *T. molitor* larvae flour; and T6, 20% pork ham + 30% *T. molitor* larvae flour. The composition of back fat and ice in each treatment was not different from the control. As occurred in previous studies, increasing the amount of *T. molitor* flour in the samples resulted in an enhanced protein content of the samples. Specifically, the crude protein content was 10.0% in the control sample, 13.9% in T1, 14.3% in T2, 15.6% in T3, 17.5% in T4, 21.1% in T5 and 24.0% in T6. Likewise, a similar increase was observed in the total mineral content (ash) as the content was recorded at 2.1% in the control sample, 2.3% in T1, 2.3% in T2, 2.3% in T3, 2.6% in T4, and about 2.9% in T5 and T6. Hence, the use of *T. molitor* larvae flour is not necessarily limited to the preparation of bakery products, since it may also enrich the nutritional value of further foodstuffs. According to the sensory evaluation, sausages with *T. molitor* were poorer in color, flavor, off-flavor, and juiciness scores. However, consumers indicated that up to the T2 sample, there was no noticeable difference from the control sample. Consequently, another widely consumed food can be enriched and rendered eco-friendlier without affecting consumer preferences.

3.3. Sauces

In Asia, one of the main protein foods is soy sauce [101], which apart from its high crude protein content, contains all the EAAs such as His, Met, which imparts the bitter flavor and Lys, which imparts the sweetness [102,103]. In an attempt to enhance the nutritional value of this sauce, six different variations of the sauce were prepared by Cho et al. [104]. Specifically, the samples were as follows: S8: soy sauce with soybean meju:koji:roasted rice flour = 8:1:1; S6: soy sauce with soybean meju:koji:roasted rice flour = 6:2:2; M8: raw insect sauce with mealworm meju:koji:roasted rice flour = 8:1:1; M6: raw insect sauce with mealworm meju:koji:roasted rice flour = 6:2:2; DM8: defatted insect sauce with defatted mealworm meju:koji:roasted rice flour = 8:1:1; DM6: defatted insect sauce with defatted mealworm meju:koji:roasted rice flour = 6:2:2; and conventional soy sauce. The fermentation period extended to 20 days at a controlled temperature of 25 °C. After the nutritional value assessment, consumer acceptance was examined. With respect to the results of EAAs, the composition of the soy sauce seemed to influence their quantity in a different way than presented in previous studies. This means that the addition of *T. molitor* larvae to the samples did not seem to enhance the content of all EAAs. To illustrate, the arginine content was found to be highest in samples S6 (225.4 mg/100 g) and S8 (100.8 mg/100 g) and the Phe content was highest in sample S6 (292.0 mg/100 g). Concerning the other EAAs, the addition of *T. molitor* larvae greatly favored their content, with sample M6 leading the way. Moreover, the defatting of the samples seemed to have negatively affected the *T. molitor* samples, possibly due to the longer series of experimental treatments the larvae received. Thereby, it was concluded that *T. molitor* larvae can be added to a range of foods, which can be either in solid or liquid form. However, it is worth mentioning that the total nitrogen content, a critical parameter, in samples S8, DM6, and DM8 ranged between 1.06 and 1.19% after 20 days of fermentation, exceeding the reference limit of 0.70% total nitrogen set by the FAO/WHO *Codex Alimentarius* international standard for soy sauce. Considering some organoleptic characteristics, it was reported that sauces with *T. molitor* larvae had a yellowish color but pH items did not seem to be influenced by the composition of the sauce. Of great interest was that consumer preference for the samples was studied on a comprehensive 7-point scale and showed no significant differences between the insect and soy sauce. Consequently, insect sauce appeared similar to soy sauce, confirming its potential to be marketed in liquid spice form. This mitigates consumer rejection due to its shape and texture characteristics, ensuring wider acceptance of this product.

Due to the popularity of soy sauce, a second attempt was made to produce enriched soy sauce with flour from *T. molitor* larvae. Specifically, Lee and Kim [105] used four different combinations of materials to prepare the soy sauces. The materials used were ground soybean, mixed ground soybean with heated okara, defatted *T. molitor* larvae powder, and both heated okara and defatted *T. molitor* larvae. From these materials, five sauces were prepared: one high-salt soy sauce with ground soybean (sample 1), one low-salt soy sauce with ground soybean (sample 2), one low-salt soy sauce with mixed ground soybean with heated okara (sample 3), one low-salt soy sauce with defatted *T. molitor* larvae powder (sample 4), and finally, one low-salt soy sauce with both heated okara and defatted *T. molitor* larvae (sample 5). The sauces were analyzed for their content of free amino acids, their antioxidant capacity, and their sensory characteristics. Regarding the different free amino acids, all 20 [106] were detected in the soy sauces, with sample 4 having the highest content of free amino acids (28.6 mg/L), followed by samples 5, 1, 2, and 3 with contents of 25.7, 11.5, 9.6, and 2.1 mg/L, respectively. Furthermore, with regard to amino acids, the most abundant amino acid in these soy sauces was glutamic acid (593–5600 mg/L), with samples 4 and 5 recording the highest values of 4830 and 5600 mg/L, respectively. Therefore, it has been initially confirmed that the enrichment of existing foods with *T. molitor* flour enhances their nutritional value, and more specifically, the addition of nymph powder increased the glutamic acid content through fermentation, as previously reported by Yoo et al. [107]. In addition to the high content of amino acids, samples 4 and 5 also showed a significant increase in their antioxidant capacity. Finally, perhaps the most significant result is the preference of consumers for the sauces. Sample 5 seems to have made the strongest impression among the other samples, as it ranked first in terms of taste and overall preference. Therefore, for the second time, it has been demonstrated that flour from *T. molitor* larvae is suitable for producing more nutritionally enriched soy sauces, with a low salt content, in fact.

3.4. *Dairy Products*

3.4.1. Ice Cream

Ice cream is a frozen dairy food product consisting of cream or butter fat, milk, sugar, and flavors [108,109]. As a favorite dessert for all age groups of people of all genders, hundreds of flavors have been invented, with vanilla, chocolate, and strawberry being the most popular. Ice creams are considered fattening desserts with no particular nutritional value; hence, an effort was made to enrich the nutritional value of ice cream [110]. To this end, a distinctive product—strawberry–cranberry ice cream containing *T. molitor* larvae, was developed [111]. In more detail, four products were developed, including a control sample (HCOo). The first was with the addition of *T. molitor* larvae (HT), the second with a mixture of *T. molitor* larvae and chia (HTC), and the last with the addition of a mixture of larvae and quinoa (HTQ). The ingredients were weighed, mixed, and pasteurized at 68 °C for 30 min. The addition of the strawberry–cranberry pulp, seeds, and larvae necessitated an aging process at 4 °C for 24 h, after which the mixture was beaten for 10 min and frozen at −4 °C. The addition of an insect led to an increase in crude protein, vitamins, and minerals in the product, resulting in a significant increase in the nutritional value of the ice cream. In detail, the highest percentage of protein was found in the HT sample, 11.8%, and the following protein richest ice cream was the HTC sample, 11.4%, i.e., 26.2 and 21.4% higher than the control sample, respectively. The most nutritionally enhanced samples were HT and HTC, with the latter showing the highest nutritional value. Namely, the content of Ca, P, Na, K, Mg, and Fe in HTC was 332.0, 224.5, 232.8, 408.2, 50.0, and 1.6 mg/100 g, respectively, while the Ca, P, Na, K, Mg, and Fe content of HT was 265.4, 185.1, 215.3, 403.9, 32.1, and 1.5 mg/100 g, respectively. In addition to the increase in minerals, there was a great increase in vitamins, where the contents of vitamins A, D, E, K B1, B2, B3, B9, and B12 in HTC were 1.9, 949.2, 9.7, 997.0, 27.3, 33.4, 102.9, 40.0 and 4.7 mg/100 g, respectively. At the same time, the contents of HTC in vitamins A, D, E, K B1, B2, B3, B9, B12 were 1.0, 584.3, 5.9, 0.0, 22.6, 33.7, 86.7, 36.8 and 4.4 mg/100 g, respectively, while the contents of the same vitamins in the control sample were 0.9, 479.7, 4.8, 0.0, 21.5, 30.6, 74.6, 17.5 and 4.4 mg/100 g, respectively. A further application of *T. molitor* larvae flour is demonstrated in the present research, paving the way for other foods in which *T. molitor* flour can be incorporated and enhance its nutritional value.

3.4.2. Milk

Milk is a common household food product with unique functional, nutritional, and sensory properties [112,113]. Recent research from Maastricht University confirmed that the protein of *T. molitor* larvae matches the gold standard of milk protein due to its complete profile of EAAs and digestion [27], and for this reason, Tello et al. [114] decided to prepare an innovative product, i.e., milk from *T. molitor* larvae, crafted to mitigate the environmental impact associated with conventional bovine's milk production. The insect milk created consisted of 1.19% crude protein, 5.76% fat, and <1% carbohydrate. The impact of the insect milk prototype was 0.76 kg CO_2 equivalents per kg FPCM in global warming potential, while bovine milk accounts for 1.6 gigatons CO_2 equivalents per year [115]. However, this milk is in a pilot stage and individual processes need to be implemented in order to prepare a final product and make it edible and probably marketable. Nevertheless, it has paved the way for a still-widely consumable product that will greatly reduce the negative impact of bovine milk production.

3.4.3. Spreadable Cheese

The most recent research conducted for the preparation of enriched food with *T. molitor* larvae was carried out in the current year, 2023. A hybrid product of spreadable cheese was produced using nine different combinations of three different ingredients: milk protein concentrate, faba bean flours, and *T. molitor* flour, which together constituted 7.1% of the formula [116]. The samples had the following proportions of the ingredients: milk protein concentrate, faba bean flours, and *T. molitor* flour—100/0/0 (control sample)—C1, 50/50/0—C2, 0/50/50—C3, 50/0/50—C4, 75/0/25—C5, 75/25/0—C6, 25/50/25—C7, 25/25/50—C8, and 50/25/25—C9 (see Figure 2).

Figure 2. Image of the nine spreadable cheese analogues used in the experiments, labeled from C1 to C9 [116]. Reprinted/adapted with permission from Ref. [116]. 2023, MDPI.

These nine samples were evaluated both in terms of their nutritional value and their sensory characteristics. Regarding the protein profile of the samples, it appears that the addition of two protein ingredients, *T. molitor* flour and faba bean flour [117], did not enhance the protein content of the spreadable cheese analogues. Specifically, the protein contents recorded in samples C1–C9 were 5.2%, 4.8%, 4.2%, 4.6%, 4.9%, 5.0%, 4.5%, 4.4%, and 4.7%, respectively. Additionally, it was evident that the immediately higher protein value was recorded in sample C6, 3.9% lower than the control sample. As expected, the sample with the highest protein content also had the lowest fat content. Specifically, sample C1 (high protein sample) had a fat content of 30.1%, while sample C3 (low protein sample) had a fat content of 31.9%. Furthermore, in terms of sensory characteristics, samples C2, C7, and C9 had the best flavor as they exhibited a more characteristic cheesy taste compared to the other samples. The most promising result is that the samples containing *T. molitor* flour showed a more preferable flavor profile compared to the samples that did not contain *T. molitor* flour.

3.5. Oil

A survey on the suitability of *T. molitor* larvae oil for consumption or as an ingredient in various foods was carried out by Son et al. [118]. The oil from larvae was presented as a light-yellow liquid, like commonly used edible oils. The peroxide value of commercialized edible oils was estimated at less than 2 mequivalents/kg oil [119] while *T. molitor* oil showed a much higher value, of the order of 3.5 mequivalents/kg oil. In order to examine the samples for their stability, secondary oxidation markers (TBARS) were determined. The lower the value of TBARS, the higher the stability of the oils [120], and in this case the oil showed a value of 1.8 mg MDA/kg oil. Furthermore, the tocopherol (vitamin E) content of the oil was studied, which is a natural antioxidant that prevents lipid oxidation by free radicals [121]. In terms of animal food sources, chicken and pork contained 2.7–4.2 and 9.5–100.0 mg/kg of total vitamin E, respectively [122,123]. Meanwhile, olive oil and soybean oil ranged from 80.0 to 1360.0 mg/kg of vitamin E [124]. In accordance with the present study, the oil from *T. molitor* larvae was rich in vitamin E, with a content of 144.3 mg/kg oil; therefore, it was concluded that it can be easily compared with the main consumed oils. Rigorous safety assessments, including a comprehensive toxicological profile on rats over 28 days, revealed significant reductions in cholesterol and glucose levels without adverse hematological effects or mortality. This unequivocally establishes the safety and nutritional superiority of insect oils. Intriguingly, after a one-year storage period, the oil was found to be stable, highlighting its potential as a novel food ingredient suitable for commercial distribution.

3.6. *Tenebrio molitor* as Animal Feed to Enhance Meat for Human Consumption

In addition to the production of food products fortified with *T. molitor* larvae flour, a further way of consuming insect-fortified conventional foods exist. In particular, this alternative route for consuming insects exists indirectly through the consumption of meat or fish raised either exclusively or for a large part of their lives with insects. One such case is the study by Gasco et al. [125], in which they bred rabbits subjected to diets enriched with fat from *T. molitor* and other insects to ascertain the advantages that a rabbit diet can provide on the final nutritional value. Three samples were tested: the control sample and rabbits fed with a diet containing 50% fat from *T. molitor* and rabbits fed with a diet containing 100% fat from *T. molitor*. The final product was examined for its crude protein content and this revealed that all the samples had the same protein content. In addition, the EFA content was not affected by the diet of the samples. Nevertheless, the rabbit meat fed with diets containing *T. molitor* fat was less susceptible to oxidation. Moreover, a sensory test was carried out by young individuals who consumed meat at least twice a week. According to their preferences, they did not perceive any difference between the rabbit meats, which means that organoleptically there would be no difference between the 'classic' meats and the meat that was fed with insects. However, the meat of rabbits reared with insects was not nutritionally enhanced so it cannot be considered enriched or extra beneficial for consumption.

4. Conclusions and Future Perspectives

T. molitor larvae are increasingly gaining ground in the food sector as they are no longer studied only for their nutritional value and their value as animal food, but also ways of introducing them into the human diet are being investigated. A growing amount of research is being carried out in order to develop new tasty and nutritious products prepared with *T. molitor* larvae, especially in bakery products or products containing flour, such as burgers. There are three reasons why flour is the main product of *T. molitor* larvae used for food preparation. Firstly, the production of flour from *T. molitor* larvae is exceptionally simple and economical; secondly, the flour can be added to a variety of foods, increasing their nutritional characteristics (proteins, EAAs, EFAs, and vitamins); and finally, they are largely accepted in foods by consumers. This means that neophobia in Europe is slowly disappearing in terms of insect consumption and people are starting to accept insects as a new and innovative food, either directly or indirectly, given that the use of insects as animal feed has been ongoing for years. This fact is very encouraging as new enhanced nutritional foods can now be produced to promote a more sustainable way of production. These foodstuffs are not required to be completely novel, since the enhancement of conventional foodstuffs with flour from *T. molitor* larvae has been proven to result in new and advanced products. These products may differ in color from conventional ones, as mentioned above. Still, once a preferable rate of insect flour addition/replacement by consumers has been established, products can be created more efficiently. Finally, as stated above, flour is not the only product that can be used containing *T. molitor* larvae, as it turned out that oil containing them can also be a safe and nutritional choice either for direct consumption or as an additive or substitute for oils in various food products. Moreover, the oil is an undetectable ingredient in food, which will promote rapid consumer acceptance. Overall, the development of food products that are either based on or enhanced with *T. molitor* larvae has begun, and the current studies are paving the way for new avenues to be explored, so as to revolutionize the food market.

Author Contributions: Conceptualization, S.I.L., C.G.A. and K.K.; writing—original draft preparation, K.K. and T.C.; writing—review and editing, K.K., T.C., V.A., E.B., C.G.A. and S.I.L.; visualization, K.K.; supervision, C.G.A. and S.I.L. All authors have read and agreed to the published version of the manuscript.

Funding: This research received no external funding.

Data Availability Statement: Data is contained within the article.

Conflicts of Interest: The authors declare no conflict of interest.

References

1. Wallace, B. The Multidimensional Nature of Population/Environmental Problems. *Polit. Life Sci.* **1997**, *16*, 224–226. [CrossRef]
2. Béné, C.; Barange, M.; Subasinghe, R.; Pinstrup-Andersen, P.; Merino, G.; Hemre, G.I.; Williams, M. Feeding 9 Billion by 2050–Putting Fish Back on the Menu. *Food Secur.* **2015**, *7*, 261–274. [CrossRef]
3. Boyd, C.E.; McNevin, A.A.; Davis, R.P. The Contribution of Fisheries and Aquaculture to the Global Protein Supply. *Food Secur.* **2022**, *14*, 805–827. [CrossRef]
4. Van Bavel, J. The World Population Explosion: Causes, Backgrounds and -Projections for the Future. *Facts Views Vis. ObGyn* **2013**, *5*, 281–291.
5. Henchion, M.; Hayes, M.; Mullen, A.M.; Fenelon, M.; Tiwari, B. Future Protein Supply and Demand: Strategies and Factors Influencing a Sustainable Equilibrium. *Foods* **2017**, *6*, 53. [CrossRef]
6. Foley, J.A.; Ramankutty, N.; Brauman, K.A.; Cassidy, E.S.; Gerber, J.S.; Johnston, M.; Mueller, N.D.; O'Connell, C.; Ray, D.K.; West, P.C.; et al. Solutions for a Cultivated Planet. *Nature* **2011**, *478*, 337–342. [CrossRef]
7. Le Mouël, C.; Forslund, A. How Can We Feed the World in 2050? A Review of the Responses from Global Scenario Studies. *Eur. Rev. Agric. Econ.* **2017**, *44*, 541–591. [CrossRef]
8. Caro, D. *Greenhouse Gas and Livestock Emissions and Climate Change*; Elsevier: Amsterdam, The Netherlands, 2018; Volume 1, ISBN 9780128126882.
9. Van Der Zijpp, A.; Wilke, P.; Carsan, S. *Sustainable Livestock Intensification the Role of Livestock in Developing Communities: Enhancing Multifunctionality*; UJ Press: Johannesburg, South Africa, 2008; pp. 123–150.
10. Copping, A.E.; Hemery, L.G.; Viehman, H.; Seitz, A.C.; Staines, G.J.; Hasselman, D.J. Are Fish in Danger? A Review of Environmental Effects of Marine Renewable Energy on Fishes. *Biol. Conserv.* **2021**, *262*, 109297. [CrossRef]
11. Dulvy, N.K.; Pacoureau, N.; Rigby, C.L.; Pollom, R.A.; Jabado, R.W.; Ebert, D.A.; Finucci, B.; Pollock, C.M.; Cheok, J.; Derrick, D.H.; et al. Overfishing Drives over One-Third of All Sharks and Rays toward a Global Extinction Crisis. *Curr. Biol.* **2021**, *31*, 4773–4787.e8. [CrossRef]

12. Coleman, F.C.; Williams, S.L. Overexploiting Marine Ecosystem Engineers: Potential Consequences for Biodiversity. *Trends Ecol. Evol.* **2002**, *17*, 40–44. [CrossRef]
13. Lachs, L.; Oñate-Casado, J. *Fisheries and Tourism: Social, Economic, and Ecological Trade-Offs in Coral Reef Systems*; Springer: Berlin/Heidelberg, Germany, 2020; ISBN 9783030203887.
14. Patel, S.; Suleria, H.A.R.; Rauf, A. Edible Insects as Innovative Foods: Nutritional and Functional Assessments. *Trends Food Sci. Technol.* **2019**, *86*, 352–359. [CrossRef]
15. Egas-Montenegro, E.; Ordo, R. International Journal of Gastronomy and Food Science Edible Insects: A Food Alternative for the Sustainable Development of the Planet. *Int. J. Gastron. Food Sci.* **2021**, *23*, 100304. [CrossRef]
16. Guin, R.P.F.; Florença, S.G.; Boustani, N.M.; Chuck-hern, C.; Sari, M.M.; Ferreira, M.; Costa, C.A.; Cardoso, A.P.; Tarcea, M.; Correia, P.M.R.; et al. Are Consumers Aware of Sustainability Aspects Related to Edible Insects? Results from a Study Involving 14 Countries. *Sustainability* **2022**, *14*, 14125. [CrossRef]
17. Florença, S.G.; Guin, R.P.F.; Gonçalves, F.J.A.; Jo, M.; Ferreira, M.; Costa, C.A.; Correia, P.M.R.; Cardoso, A.P.; Campos, S. The Motivations for Consumption of Edible Insects: A Systematic Review. *Foods* **2022**, *11*, 3643. [CrossRef]
18. Ha, N.I.; Mun, S.K.; Im, S.B.; Jang, H.Y.; Jeong, H.G.; Kang, K.Y.; Park, K.W.; Seo, K.S.; Ban, S.E.; Kim, K.J.; et al. Changes in Functionality of *Tenebrio molitor* Larvae Fermented by *Cordyceps militaris* Mycelia. *Foods* **2022**, *11*, 2477. [CrossRef]
19. Premalatha, M.; Abbasi, T.; Abbasi, T.; Abbasi, S.A. Energy-Efficient Food Production to Reduce Global Warming and Ecodegradation: The Use of Edible Insects. *Renew. Sustain. Energy Rev.* **2011**, *15*, 4357–4360. [CrossRef]
20. Krongdang, S.; Phokasem, P.; Venkatachalam, K.; Charoenphun, N. Edible Insects in Thailand: An Overview of Status, Properties, Processing, and Utilization in the Food Industry. *Foods* **2023**, *12*, 2162. [CrossRef]
21. Siddiqui, S.A.; Shah, M.A.; Centoducati, G. Prospects of Edible Insects as Sustainable Protein for Food and Feed—A Review. *J. Insects Food Feed.* **2023**, *1*, 4588. [CrossRef]
22. van Huis, A. Edible Insects Contributing to Food Security? *Agric. Food Secur.* **2015**, *4*, 1–9. [CrossRef]
23. Van Huis, A. Edible Insects Are the Future? *Proc. Nutr. Soc.* **2016**, *75*, 294–305. [CrossRef]
24. Kouřimská, L.; Adámková, A. Nutritional and Sensory Quality of Edible Insects. *NFS J.* **2016**, *4*, 22–26. [CrossRef]
25. Ojeda-Avila, T.; Woods, H.A.; Raguso, R.A. Effects of Dietary Variation on Growth, Composition, and Maturation of *Manduca sexta* (Sphingidae: Lepidoptera). *J. Insect Physiol.* **2003**, *49*, 293–306. [CrossRef]
26. Bjørge, J.D.; Overgaard, J.; Malte, H.; Gianotten, N.; Heckmann, L.H. Role of Temperature on Growth and Metabolic Rate in the Tenebrionid Beetles *Alphitobius Diaperinus* and *Tenebrio molitor*. *J. Insect Physiol.* **2018**, *107*, 89–96. [CrossRef]
27. Mancini, S.; Moruzzo, R.; Riccioli, F.; Paci, G. European Consumers' Readiness to Adopt Insects as Food. A Review. *Food Res. Int.* **2019**, *122*, 661–678. [CrossRef]
28. Piha, S.; Pohjanheimo, T.; Lähteenmäki-Uutela, A.; Křečková, Z.; Otterbring, T. The Effects of Consumer Knowledge on the Willingness to Buy Insect Food: An Exploratory Cross-Regional Study in Northern and Central Europe. *Food Qual. Prefer.* **2018**, *70*, 1–10. [CrossRef]
29. Moruzzo, R.; Mancini, S.; Boncinelli, F.; Riccioli, F. Exploring the Acceptance of Entomophagy: A Survey of Italian Consumers. *Insects* **2021**, *12*, 123. [CrossRef]
30. van Huis, A.; Rumpold, B. Strategies to Convince Consumers to Eat Insects? A Review. *Food Qual. Prefer.* **2023**, *110*, 104927. [CrossRef]
31. Żuk-Gołaszewska, K.; Gałęcki, R.; Obremski, K.; Smetana, S.; Figiel, S.; Gołaszewski, J. Edible Insect Farming in the Context of the EU Regulations and Marketing—An Overview. *Insects* **2022**, *13*, 446. [CrossRef]
32. Commission Implementing Regulation (EU) 2021/1975 of 12 November 2021 Authorising the Placing on the Market of Frozen, Dried and Powder Forms of *Locusta Migratoria* as a Novel Food under Regulation (EU) 2015/2283 of the European Parliament and of the Council and Amending Commission Implementing Regulation (EU) 2017/2470. *Off. J. Eur. Union L402* **2021**, *1975*, 10.
33. Commission Implementing Regulation (EU) 2021/882 of 1 June 2021 Authorising the Placing on the Market of Dried *Tenebrio molitor* Larva as a Novel Food under Regulation (EU) 2015/2283 of the European Parliament and of the Council, and Amending Commission Implementing Regulation (EU) 2017/2470. *Off. J. Eur. Union L194* **2021**, *882*, 16–20.
34. Commission Implementing Regulation (EU) 2023/58 of 5 January 2023 Authorising the Placing on the Market of the Frozen, Paste, Dried and Powder Forms of *Alphitobius Diaperinus* Larvae (Lesser Mealworm) as a Novel Food and Amending Implementing Regulation (EU) 2017/2470. *Off. J. Eur. Union L5* **2023**, *58*, 10–15.
35. Moruzzo, R.; Riccioli, F.; Espinosa Diaz, S.; Secci, C.; Poli, G.; Mancini, S. Mealworm (*Tenebrio molitor*): Potential and Challenges to Promote Circular Economy. *Animals* **2021**, *11*, 2568. [CrossRef]
36. Bordiean, A.; Krzyżaniak, M.; Stolarski, M.J.; Czachorowski, S.; Peni, D. Will Yellow Mealworm Become a Source of Safe Proteins for Europe? *Agriculture* **2020**, *10*, 233. [CrossRef]
37. Tzompa-Sosa, D.A.; Moruzzo, R.; Mancini, S.; Schouteten, J.J.; Liu, A.; Li, J.; Sogari, G. Consumers' Acceptance toward Whole and Processed Mealworms: A Cross-Country Study in Belgium, China, Italy, Mexico, and the US. *PLoS ONE* **2023**, *18*, e0279530. [CrossRef]
38. Gkinali, A.-A.; Matsakidou, A.; Vasileiou, E.; Paraskevopoulou, A. Potentiality of *Tenebrio molitor* Larva-Based Ingredients for the Food Industry: A Review. *Trends Food Sci. Technol.* **2022**, *119*, 495–507. [CrossRef]
39. Fischer, A.R.H.; Steenbekkers, L.P.A. All Insects Are Equal, but Some Insects Are More Equal than Others. *Br. Food J.* **2018**, *120*, 852–863. [CrossRef]

40. Mancuso, T.; Pippinato, L.; Gasco, L. The European Insects Sector and Its Role in the Provision of Green Proteins in Feed Supply. *Qual. Access Success* **2019**, *20*, 374–381.
41. Sogari, G.; Amato, M.; Biasato, I.; Chiesa, S.; Gasco, L. The Potential Role of Insects as Feed: A Multi-Perspective Review. *Animals* **2019**, *9*, 119. [CrossRef]
42. Thévenot, A.; Rivera, J.L.; Wilfart, A.; Maillard, F.; Hassouna, M.; Senga-Kiesse, T.; Le Féon, S.; Aubin, J. Mealworm Meal for Animal Feed: Environmental Assessment and Sensitivity Analysis to Guide Future Prospects. *J. Clean. Prod.* **2018**, *170*, 1260–1267. [CrossRef]
43. Selaledi, L.; Mbajiorgu, C.A.; Mabelebele, M. The Use of Yellow Mealworm (*T. Molitor*) as Alternative Source of Protein in Poultry Diets: A Review. *Trop. Anim. Health Prod.* **2020**, *52*, 7–16. [CrossRef]
44. van Broekhoven, S. *Quality and Safety Aspects of Mealworms as Human Food*; Wageningen University and Research: Wageningen, The Netherlands, 2008; Volume 12, ISBN 9789462575714.
45. Toviho, O.A. Nutrient Composition and Growth of Yellow Mealworm (*Tenebrio molitor*) at Different Ages and Stages of the Life Cycle. *Agriculture* **2022**, *12*, 1924. [CrossRef]
46. Hosen, M.; Khan, A.R.; Hossain, M. Growth and Development of the Lesser Mealworm, *Alphitobius Diaperinus* (Panzer) (Coleoptera: Tenebrionidae) on Cereal Flours. *Pak. J. Biol. Sci.* **2004**, *7*, 1505–1508. [CrossRef]
47. Rumbos, C.I.; Bliamplias, D.; Gourgouta, M.; Michail, V.; Athanassiou, C.G. Rearing *Tenebrio molitor* and *Alphitobius diaperinus* larvae on seed cleaning process byproducts. *Insects* **2021**, *12*, 293. [CrossRef]
48. Rumbos, C.I.; Karapanagiotidis, I.T.; Mente, E.; Athanassiou, C.G. The Lesser Mealworm *Alphitobius Diaperinus*: A Noxious Pest or a Promising Nutrient Source? *Rev. Aquac.* **2019**, *11*, 1418–1437. [CrossRef]
49. Mariod, A.A.; Mirghani, M.E.S.; Hussein, I. Tenebrio molitor Mealworm. In *Unconventional Oilseeds and Oil Sources*; Mariod, A.A., Mirghani, M.E.S., Hussein, I., Eds.; Academic Press: San Diego, CA, USA, 2017; pp. 331–336. [CrossRef]
50. Gomaa, K.F.S. Evaluation of Five Pearl Millet Ecotypes Susceptibility to the Nymphal Instars of Migratory Locust. *Trop. Drylands* **2017**, *1*, 57–63. [CrossRef]
51. Rumbos, C.I.; Karapanagiotidis, I.T.; Mente, E.; Psofakis, P.; Athanassiou, C.G. Evaluation of Various Commodities for the Development of the Yellow Mealworm, *Tenebrio molitor*. *Sci. Rep.* **2020**, *10*, 11224. [CrossRef]
52. Dreyer, M.; Hörtenhuber, S.; Zollitsch, W.; Jäger, H.; Schaden, L.M.; Gronauer, A.; Kral, I. Environmental Life Cycle Assessment of Yellow Mealworm (*Tenebrio molitor*) Production for Human Consumption in Austria—A Comparison of Mealworm and Broiler as Protein Source. *Int. J. Life Cycle Assess.* **2021**, *26*, 2232–2247. [CrossRef]
53. Zielińska, E.; Baraniak, B.; Karaś, M.; Rybczyńska, K.; Jakubczyk, A. Selected Species of Edible Insects as a Source of Nutrient Composition. *Food Res. Int.* **2015**, *77*, 460–466. [CrossRef]
54. Jantzen da Silva Lucas, A.; Menegon de Oliveira, L.; da Rocha, M.; Prentice, C. Edible Insects: An Alternative of Nutritional, Functional and Bioactive Compounds. *Food Chem.* **2020**, *311*, 126022. [CrossRef] [PubMed]
55. Hong, J.; Han, T.; Kim, Y.Y. Mealworm (*Tenebrio molitor* Larvae) as an Alternative Protein Source for Monogastric Animal: A Review. *Animals* **2020**, *10*, 2068. [CrossRef]
56. Benzertiha, A.; Kierończyk, B.; Kołodziejski, P.; Pruszyńska–Oszmałek, E.; Rawski, M.; Józefiak, D.; Józefiak, A. *Tenebrio molitor* and *Zophobas Morio* Full-Fat Meals as Functional Feed Additives Affect Broiler Chickens' Growth Performance and Immune System Traits. *Poult. Sci.* **2020**, *99*, 196–206. [CrossRef]
57. Ravzanaadii, N.; Kim, S.-H.; Choi, W.-H.; Hong, S.-J.; Kim, N.-J. Nutritional Value of Mealworm, *Tenebrio molitor* as Food Source. *Int. J. Ind. Entomol.* **2012**, *25*, 93–98. [CrossRef]
58. Heidari-Parsa, S.; Imani, S.; Fathipour, Y.; Kheiri, F.; Chamani, M. Determination of Yellow Mealworm (*Tenebrio molitor*) Nutritional Value as an Animal and Human Food Supplementation. *Arthropods* **2018**, *7*, 94–102.
59. Boulos, S.; Tännler, A.; Nyström, L. Nitrogen-to-Protein Conversion Factors for Edible Insects on the Swiss Market: *T. molitor*, *A. domesticus*, and *L. migratoria*. *Front. Nutr.* **2020**, *7*, 89. [CrossRef]
60. Kotsou, K.; Chatzimitakos, T.; Athanasiadis, V.; Bozinou, E.; Rumbos, C.I.; Athanassiou, C.G.; Lalas, S.I. Enhancing the Nutritional Profile of *Tenebrio molitor* Using the Leaves of *Moringa oleifera*. *Foods* **2023**, *12*, 2612. [CrossRef]
61. Gkinali, A.; Matsakidou, A.; Paraskevopoulou, A. Characterization of *Tenebrio molitor* Larvae Protein Preparations Obtained by Different Extraction Approaches. *Foods* **2022**, *11*, 3852. [CrossRef]
62. Siulapwa, N.; Mwambungu, A.; Lungu, E.; Sichilima, W. Nutritional Value of Four Common Edible Insects in Zambia. *Int. J. Sci. Res.* **2012**, *3*, 2319–7064.
63. Aberoumand, A.; Baesi, F. The Nutritional Quality and Contents of Heavy Elements Due to Thermal Processing and Storage in Canned *Thunnus tonggol* Fish Change Compared to Fresh Fish. *Food Sci. Nutr.* **2023**, *11*, 3588–3600. [CrossRef]
64. Jensen, I.J.; Bodin, N.; Govinden, R.; Elvevoll, E.O. Marine Capture Fisheries from Western Indian Ocean: An Excellent Source of Proteins and Essential Amino Acids. *Foods* **2023**, *12*, 1015. [CrossRef] [PubMed]
65. Adámková, A.; Mlček, J.; Adámek, M.; Borkovcová, M.; Bednářová, M.; Hlobilová, V.; Knížková, I.; Juríková, T. *Tenebrio molitor* (Coleoptera: Tenebrionidae)—Optimization of Rearing Conditions to Obtain Desired Nutritional Values. *J. Insect Sci.* **2020**, *20*, 24. [CrossRef]
66. Machona, O.; Matongorere, M.; Chidzwondo, F.; Mangoyi, R. Evaluation of Nutritional Content of the Larvae of *Tenebrio molitor*, and Formulation of Broiler Stockfeed. *Entomol. Appl. Sci. Lett.* **2022**, *9*, 48–56. [CrossRef]

67. Wu, R.A.; Ding, Q.; Yin, L.; Chi, X.; Sun, N.; He, R.; Luo, L.; Ma, H.; Li, Z. Comparison of the Nutritional Value of Mysore Thorn Borer (*Anoplophora chinensis*) and Mealworm Larva (*Tenebrio molitor*): Amino Acid, Fatty Acid, and Element Profiles. *Food Chem.* **2020**, *323*, 126818. [CrossRef]
68. Ao, X.; Yoo, J.S.; Wu, Z.L.; Kim, I.H. Can Dried Mealworm (*Tenebrio molitor*) Larvae Replace Fish Meal in Weaned Pigs? *Livest. Sci.* **2020**, *239*, 104103. [CrossRef]
69. Yoo, J.S.; Cho, K.H.; Hong, J.S.; Jang, H.S.; Chung, Y.H. Nutrient Ileal Digestibility Evaluation of Dried *Tenebrio molitor*. *Asian Australas. J. Anim. Sci.* **2018**, *32*, 387. [CrossRef]
70. Dreassi, E.; Cito, A.; Zanfini, A.; Materozzi, L.; Botta, M.; Francardi, V. Dietary Fatty Acids Influence the Growth and Fatty Acid Composition of the Yellow Mealworm *Tenebrio molitor* (Coleoptera: Tenebrionidae). *Lipids* **2017**, *52*, 285–294. [CrossRef]
71. Di Pasquale, M.G. The Essentials of Essential Fatty Acids. *J. Diet. Suppl.* **2009**, *6*, 143–161. [CrossRef]
72. Lawal, K.G.; Kavle, R.R.; Akanbi, T.O.; Mirosa, M.; Agyei, D. Enrichment in Specific Fatty Acids Profile of *Tenebrio molitor* and *Hermetia illucens* Larvae through Feeding. *Futur. Foods* **2021**, *3*, 100016. [CrossRef]
73. Ali, M.Y.; Sina, A.A.I.; Khandker, S.S.; Neesa, L.; Tanvir, E.M.; Kabir, A.; Khalil, M.I.; Gan, S.H. Nutritional Composition and Bioactive Compounds in Tomatoes and Their Impact on Human Health and Disease: A Review. *Foods* **2021**, *10*, 45. [CrossRef]
74. Ghosh, S.; Lee, S.M.; Jung, C.; Meyer-Rochow, V.B. Nutritional Composition of Five Commercial Edible Insects in South Korea. *J. Asia Pac. Entomol.* **2017**, *20*, 686–694. [CrossRef]
75. Meyer-Rochow, V.B. Can Insects Help to Ease the Problem of World Food Shortage? *Search* **1975**, *6*, 261–262.
76. Errico, S.; Spagnoletta, A.; Verardi, A.; Moliterni, S.; Dimatteo, S.; Sangiorgio, P. *Tenebrio molitor* as a Source of Interesting Natural Compounds, Their Recovery Processes, Biological Effects, and Safety Aspects. *Compr. Rev. Food Sci. Food Saf.* **2022**, *21*, 148–197. [CrossRef]
77. Sete da Cruz, R.M.; da Silva, C.; da Silva, E.A.; Hegel, P.; Barão, C.E.; Cardozo-Filho, L. Composition and Oxidative Stability of Oils Extracted from *Zophobas morio* and *Tenebrio molitor* Using Pressurized N-Propane. *J. Supercrit. Fluids* **2022**, *181*, 105504. [CrossRef]
78. Alves, A.V.; Freitas de Lima, F.; Granzotti da Silva, T.; de Oliveira, V.S.; Kassuya, C.A.L.; Sanjinez-Argandoña, E.J. Safety Evaluation of the Oils Extracted from Edible Insects (*Tenebrio molitor* and *Pachymerus Nucleorum*) as Novel Food for Humans. *Regul. Toxicol. Pharmacol.* **2019**, *102*, 90–94. [CrossRef]
79. Gantner, M.; Kr, K.; Piotrowska, A.; Sionek, B.; Sadowska, A. Adding Mealworm (*Tenebrio molitor* L.) Powder to Wheat Bread: Effects on Physicochemical, Sensory and Microbiological Qualities of the End-Product. *Molecules* **2022**, *27*, 6155. [CrossRef]
80. Wu, R.A.; Ding, Q.; Lu, H.; Tan, H.; Sun, N.; Wang, K.; He, R.; Luo, L.; Ma, H.; Li, Z. Caspase 3-Mediated Cytotoxicity of Mealworm Larvae (*Tenebrio molitor*) Oil Extract against Human Hepatocellular Carcinoma and Colorectal Adenocarcinoma. *J. Ethnopharmacol.* **2020**, *250*, 112438. [CrossRef]
81. Purschke, B.; Stegmann, T.; Schreiner, M.; Jäger, H. Pilot-scale Supercritical CO_2 Extraction of Edible Insect Oil from *Tenebrio molitor* L. Larvae—Influence of Extraction Conditions on Kinetics, Defatting Performance and Compositional Properties. *Eur. J. Lipid Sci. Technol.* **2017**, *119*, 1600134. [CrossRef]
82. Kim, Y.M. Quality characteristics of white bread with *Tenebrio molitor* linne powder. *Korean J. Food Nutr.* **2017**, *30*, 1164–1175. [CrossRef]
83. Mancini, S.; Fratini, F.; Tuccinardi, T.; Degl'Innocenti, C.; Paci, G. *Tenebrio molitor* Reared on Different Substrates: Is It Gluten Free? *Food Control* **2020**, *110*, 20–23. [CrossRef]
84. Harrison, J.F. Insect Acid-Base Physiology. *Annu. Rev. Entomol.* **2001**, *46*, 221–250. [CrossRef]
85. Roncolini, A.; Milanović, V.; Cardinali, F.; Osimani, A.; Garofalo, C.; Sabbatini, R.; Clementi, F.; Pasquini, M.; Mozzon, M.; Foligni, R.; et al. Protein Fortification with Mealworm (*Tenebrio molitor* L.) Powder: Effect on Textural, Microbiological, Nutritional and Sensory Features of Bread. *PLoS ONE* **2019**, *14*, e0211747. [CrossRef] [PubMed]
86. Commission Regulation (EC). N0.1441/2007 Amending Regulation (EC) No 2073/2005 on Microbiological Criteria for Foodstuffs. *Off. J. Eur. Union* **2007**, *322*, 12–29.
87. Khuenpet, K.; Pakasap, C.; Vatthanakul, S.; Kitthawee, S. Effect of Larval—Stage Mealworm (*Tenebrio molitor*) Powder on Qualities of Bread. *Int. J. Agric. Technol.* **2020**, *16*, 283–296.
88. Ahmad, S.; Naz, A.; Usman, M.; Amjad, A.; Pasha, I.; Farooq, U. Impediment Effect of Chemical Agents (Additives) on Gluten Development in Cookie Dough. *J. Food Sci. Technol.* **2022**, *59*, 1396–1406. [CrossRef]
89. Zielińska, E.; Pankiewicz, U. Nutritional, Physiochemical, and Antioxidative Characteristics of Shortcake Biscuits Enriched with *Tenebrio molitor* Flour. *Molecules* **2020**, *25*, 5629. [CrossRef]
90. Xie, X.; Yuan, Z.; Fu, K.; An, J.; Deng, L. Effect of Partial Substitution of Flour with Mealworm (*Tenebrio molitor* L.) Powder on Dough and Biscuit Properties. *Foods* **2022**, *11*, 2156. [CrossRef]
91. Mihaly Cozmuta, A.; Uivarasan, A.; Peter, A.; Nicula, C.; Kovacs, D.E.; Mihaly Cozmuta, L. Yellow Mealworm (*Tenebrio molitor*) Powder Promotes a High Bioaccessible Protein Fraction and Low Glycaemic Index in Biscuits. *Nutrients* **2023**, *15*, 997. [CrossRef]
92. Biltoft-Jensen, A.; Matthiessen, J.; Ygil, K.H.; Christensen, T. Defining Energy-Dense, Nutrient-Poor Food and Drinks and Estimating the Amount of Discretionary Energy. *Nutrients* **2022**, *14*, 1477. [CrossRef]
93. Djouadi, A.; Sales, J.R.; Carvalho, M.O.; Raymundo, A. Development of Healthy Protein-Rich Crackers Using *Tenebrio molitor* Flour. *Foods* **2022**, *11*, 702. [CrossRef]
94. Tsao, R. Chemistry and Biochemistry of Dietary Polyphenols. *Nutrients* **2010**, *2*, 1231–1246. [CrossRef]

95. Bartkowicz, J.; Babicz-Zielińska, E. Acceptance of Bars with Edible Insects by a Selected Group of Students from Tri-City, Poland. *Czech J. Food Sci.* **2020**, *38*, 192–197. [CrossRef]
96. Caparros Megido, R.; Gierts, C.; Blecker, C.; Brostaux, Y.; Haubruge, É.; Alabi, T.; Francis, F. Consumer Acceptance of Insect-Based Alternative Meat Products in Western Countries. *Food Qual. Prefer.* **2016**, *52*, 237–243. [CrossRef]
97. Kim, H.W.; Setyabrata, D.; Lee, Y.J.; Jones, O.G.; Kim, Y.H.B. Pre-Treated Mealworm Larvae and Silkworm Pupae as a Novel Protein Ingredient in Emulsion Sausages. *Innov. Food Sci. Emerg. Technol.* **2016**, *38*, 116–123. [CrossRef]
98. Hirschman, C. The Contributions of Immigrants to American Culture. *Daedalus* **2013**, *142*, 26–47. [CrossRef]
99. Choi, Y.S.; Kim, T.K.; Choi, H.D.; Park, J.D.; Sung, J.M.; Jeon, K.H.; Paik, H.D.; Kim, Y.B. Optimization of Replacing Pork Meat with Yellow Worm (*Tenebrio molitor* L.) for Frankfurters. *Korean J. Food Sci. Anim. Resour.* **2017**, *37*, 617–625. [CrossRef]
100. Grossi, G.; Goglio, P.; Vitali, A.; Williams, A.G. Livestock and Climate Change: Impact of Livestock on Climate and Mitigation Strategies. *Anim. Front.* **2019**, *9*, 69–76. [CrossRef] [PubMed]
101. Woo, K.-L.; Lee, S.-C.; Jang, D.-K. Quality Characteristics of Soy Sauces Containing Shiitake Mushroom (*Lentinus edodes*). *Appl. Biol. Chem.* **2003**, *46*, 220–224.
102. Nelson, G.; Chandrashekar, J.; Hoon, M.A.; Feng, L.; Zhao, G.; Ryba, N.J.; Zuker, C.S. An Amino-Acid Taste Receptor. *Nature* **2002**, *416*, 199–202. [CrossRef] [PubMed]
103. Choi, J.M.; Lee, C.B.; Kim, H.S. Quality Characteristics of Soy Sauces by Various Manufacturing Methods. *Culin. Sci. Hosp. Res.* **2016**, *22*, 57–65. [CrossRef]
104. Cho, J.H.; Zhao, H.L.; Kim, J.S.; Kim, S.H.; Chung, C.H. Characteristics of Fermented Seasoning Sauces Using *Tenebrio molitor* Larvae. *Innov. Food Sci. Emerg. Technol.* **2018**, *45*, 186–195. [CrossRef]
105. Lee, H.; Kim, Y. Development of a Low-Salt Soy Sauce with Enhanced Flavor and Functionality with Soy Residue and *Tenebrio molitor* Larvae Powder. *J. Food Process. Preserv.* **2022**, *46*, e16546. [CrossRef]
106. Doig, A.J. Frozen, but No Accident—Why the 20 Standard Amino Acids Were Selected. *FEBS J.* **2017**, *284*, 1296–1305. [CrossRef]
107. Yoo, J.; Hwang, J.-S.; Goo, T.-W.; Yun, E.-Y. Comparative Analysis of Nutritional and Harmful Components in Korean and Chinese Mealworms (*Tenebrio molitor*). *J. Korean Soc. Food Sci. Nutr.* **2013**, *42*, 249–254. [CrossRef]
108. Alvarez, V.B. Ice Cream and Related Products. In *The Sensory Evaluation of Dairy Products*; Clark, S., Costello, M., Drake, M., Bodyfelt, F., Eds.; Springer US: New York, NY, USA, 2009; pp. 271–331, ISBN 978-0-387-77408-4.
109. Alvarez, V. Ice Cream and Frozen Desserts. In *Ullmann's Encyclopedia of Industrial Chemistry*; Wiley Online Library: Hoboken, NJ, USA, 2023; pp. 281–344, ISBN 978-3-031-30018-9.
110. Balaji, P.; Shalini, N.; Santhasheela, M.; Vidhyavathi, A. Consumer Choice of Ice Creams: A Binary Logit Model of Analysis. *Madras Agric. J.* **2021**, *108*, 1–6. [CrossRef]
111. Hernández Toxqui, A.G.; Ramírez Ramírez, J.; Pino Moreno, J.M.; Talamantes Gómez, J.M.; Angeles Campos, S.C.; Ramírez Orejel, J.C. Development of Nutraceutical Ice Creams Using Flour Yellow Worm Larvae (*Tenebrio molitor*), Chia (*Salvia Hispanica*), and Quinoa (*Chenopodium Quinoa*). *Front. Vet. Sci.* **2021**, *8*, 629180. [CrossRef]
112. Bus, A.; Worsley, A. Consumers' Sensory and Nutritional Perceptions of Three Types of Milk. *Public Health Nutr.* **2003**, *6*, 201–208. [CrossRef]
113. Smith, T.; Campbell, R.; Drake, M. Sensory Properties of Milk Protein Ingredients. In *Advanced Dairy Chemistry: Volume 1B: Proteins: Applied Aspects*; Springer: Berlin/Heidelberg, Germany, 2016; pp. 197–223, ISBN 978-1-4939-2799-9.
114. Tello, A.; Aganovic, K.; Parniakov, O.; Carter, A.; Heinz, V.; Smetana, S. Product Development and Environmental Impact of an Insect-Based Milk Alternative. *Futur. Foods* **2021**, *4*, 100080. [CrossRef]
115. Opio, C. *The Global Livestock Environmental Assessment Model*; Food and Agriculture Organization of the United Nations (FAO): Rome, Italy, 2017; pp. 22–26.
116. Garcia-Fontanals, L.; Llorente, R.; Valderrama, J.; Bravo, S.; Talens, C. Hybrid Spreadable Cheese Analogues with Faba Bean and Mealworm (*Tenebrio molitor*) Flours: Optimisation Using Desirability-Based Mixture Design. *Foods* **2023**, *12*, 1522. [CrossRef]
117. Martineau-Côté, D.; Achouri, A.; Karboune, S.; L'Hocine, L. Faba Bean: An Untapped Source of Quality Plant Proteins and Bioactives. *Nutrients* **2022**, *14*, 1541. [CrossRef]
118. Son, Y.J.; Choi, S.Y.; Hwang, I.K.; Nho, C.W.; Kim, S.H. Could Defatted Mealworm (*Tenebrio molitor*) and Mealworm Oil Be Used as Food Ingredients? *Foods* **2020**, *9*, 40. [CrossRef]
119. Kamal-Eldin, A. Effect of Fatty Acids and Tocopherols on the Oxidative Stability of Vegetable Oils. *Eur. J. Lipid Sci. Technol.* **2006**, *108*, 1051–1061. [CrossRef]
120. Kalompatsios, D.; Athanasiadis, V.; Chatzimitakos, T.; Palaiogiannis, D.; Lalas, S.I.; Makris, D.P. Sustainable Exploitation of Waste Orange Peels: Enrichment of Commercial Seed Oils and the Effect on Their Oxidative Stability. *Waste* **2023**, *1*, 761–774. [CrossRef]
121. Rizvi, S.; Raza, S.T.; Ahmed, F.; Ahmad, A.; Abbas, S.; Mahdi, F. The Role of Vitamin E in Human Health and Some Diseases. *Sultan Qaboos Univ. Med. J.* **2014**, *14*, 157–165.
122. Maraschiello, C.; Sárraga, C.; García Regueiro, J.A. Glutathione Peroxidase Activity, TBARS, and α-Tocopherol in Meat from Chickens Fed Different Diets. *J. Agric. Food Chem.* **1999**, *47*, 867–872. [CrossRef]
123. Surai, P.F.; Sparks, N.H.C. Tissue-Specific Fatty Acid and α-Tocopherol Profiles in Male Chickens Depending on Dietary Tuna Oil and Vitamin E Provision. *Poult. Sci.* **2000**, *79*, 1132–1142. [CrossRef]

124. Okogeri, O.; Tasioula-Margari, M. Changes Occurring in Phenolic Compounds and α-Tocopherol of Virgin Olive Oil during Storage. *J. Agric. Food Chem.* **2002**, *50*, 1077–1080. [CrossRef]
125. Gasco, L.; Dabbou, S.; Gai, F.; Brugiapaglia, A.; Schiavone, A.; Birolo, M.; Xiccato, G.; Trocino, A. Quality and Consumer Acceptance of Meat from Rabbits Fed Diets in Which Soybean Oil Is Replaced with Black Soldier Fly and Yellow Mealworm Fats. *Animals* **2019**, *9*, 629. [CrossRef]

foods

MDPI

Review

Green Solvents: Emerging Alternatives for Carotenoid Extraction from Fruit and Vegetable By-Products

Adriana Viñas-Ospino [1], Daniel López-Malo [2], María José Esteve [1,*], Ana Frígola [1] and Jesús Blesa [1]

[1] Nutrition and Food Chemistry, University of Valencia, Avenida Vicent Andrés Estellés, s/n, 46100 Burjassot, Spain

[2] Department of Biomedical Sciences, Faculty of Health Sciences, European University of Valencia, Paseo de La Alameda, 7, 46010 Valencia, Spain

* Correspondence: maria.jose.esteve@uv.es

Abstract: Carotenoids have important implications for human health and the food industry due to their antioxidant and functional properties. Their extraction is a crucial step for being able to concentrate them and potentially include them in food products. Traditionally, the extraction of carotenoids is performed using organic solvents that have toxicological effects. Developing greener solvents and techniques for extracting high-value compounds is one of the principles of green chemistry and a challenge for the food industry. This review will analyze the use of green solvents, namely, vegetable oils, supercritical fluids, deep eutectic solvents, ionic liquids, and limonene, combined with nonconventional techniques (ultrasound-assisted extraction and microwave), for carotenoid extraction from fruit and vegetable by-products as upcoming alternatives to organic solvents. Recent developments in the isolation of carotenoids from green solvents and their inclusion in food products will also be discussed. The use of green solvents offers significant advantages in extracting carotenoids, both by decreasing the downstream process of solvent elimination, and the fact that the carotenoids can be included directly in food products without posing a risk to human health.

Keywords: green extractions; green solvents; carotenoids; fruit by-products; vegetable by-products

1. Introduction

Carotenoids are natural pigments produced as secondary metabolites in fruits, vegetables, algae, fungi, and bacteria. These compounds are lipophilic isoprenoids associated with a wide range of health benefits and antioxidant properties [1,2]. Considerable interest has developed over the past few years in the extraction of carotenoids from wastes and agro-industrial by-products as an alternative to synthetic carotenoids [3–5]. Additionally, the extraction of bioactive compounds is gaining importance in response to consumer inclination and demand for more natural products containing biological properties with health benefits [6–8]. As a result, one of the goals for the food industry is to enrich their products for human consumption with substances that have functional activities that can reduce the risk of many chronic diseases. The first step in producing functional food products is the extraction of target compounds.

Carotenoids tend to oxidize, which limits their ability to withstand exposure to heat, light, acids, and long extraction times; due to their heterogeneous polarity and hydrophobic nature, their extraction is difficult and usually carried out using organic solvents [9]. The conventional extraction methods such as Soxhlet, maceration, and hydrodistillation normally use organic solvents such as hexane, acetone, methanol, ethanol, and their combinations [10,11]. These hazardous solvents are associated with harmful effects on the environment and human health [7]. In order to eliminate these weaknesses, it has been suggested that new, green, and sustainable options must be used. A viable alternative is solvent-free extraction, because it reduces the use of a large volume of solvents and, in consequence, the associated costs. However, solvents are almost unavoidable due to their crucial roles in dissolving solids, transferring mass and heat, decreasing viscosity, and in the separation and purification process [12]. One of the fundamental steps in the green extraction process is solvent elimination from the extract. Moreover, it is time- and energy-consuming when the process has not been optimized and, if the boiling time of the solvent is high, the target compound can be damaged [13]. There are two ways to solve this problem: to use solvents that can be removed after the extraction process is over, or to use safe-for-consumption solvents that can be present in the final extract formulation [14]. However, even though a solvent is safe for consumption, product recovery and solvent recycling is required for large-scale application and to ensure the sustainability of the process in the long term [15].

In recent years, the extraction methods for carotenoids have attracted great interest, and the improvement of green chemistry and the use of green solvents, as an eco-friendly option to replace organic solvents, has developed quickly [16]. Furthermore, the use of green extraction techniques conforms to the global guidelines indicated by the Food and Agriculture Organization (FAO) and the EU environmental policy and legislation for the period 2010–2050 for the reduction of hazardous solvents in industry [17,18].

Considering agro-industrial by-products as an alternative source of bioactive compounds is one of the key points for the development of the "zero waste" industry. Fruit and vegetable wastes are usually undervalued, though it is well known that those by-products often contain valuable compounds in theirs peels, pulps, and seeds [19]. These by-products contain a variety of bioactive compounds, such as polyphenols, carotenoids, ascorbic acid, essential oils, and dietary fiber, which have considerable health benefits [2,20–22]. Lycopene is the most prevalent and well-known carotenoid, accounting for more than 85% of total carotenoids. For example, tomato peel contains five times the amount of lycopene compared to the pulp [4]. Considering this, the extraction of carotenoids from a variety of fruit and vegetable by-products, which are emerging as extraordinarily rich sources of bioactive compounds, should be explored and widely employed in the food industry.

To date, other reviews have been published about various aspects of carotenoid extraction using nonconventional techniques and diverse sources of vegetables, fruits, plants, and algae [1,2,9,23]. However, the extraction of carotenoids using green solvents on vegetable by-products has not been analyzed. This review aims to discuss, compare, and provide information about the use of novel green solvents combined with green extraction techniques for the extraction of carotenoids from fruit and vegetable by-products. The following information regarding green solvents will be analyzed from the point of view of their recovery, the isolation of the extracted compound, their possible toxicological effects, and application in food products. The information in this review can contribute to a more sustainable utilization of fruit and vegetable by-products and the production of functional food products enriched with carotenoids extracted using green solvents.

2. Carotenoids: General Aspects

2.1. Chemistry and Classification

Carotenoids are lipophilic isoprenoids and natural pigments present in the chromoplasts and chloroplasts of plants and various algae, fungi, and bacteria [9]. They are responsible for the color in fruits, vegetables, flowers, leaves, fish, and crustaceans, and their colors can vary between yellow, orange, red, and green [2,24]. Their chemical structure consists of a long system of conjugated double bonds that acts as a chromophore, which is responsible for its color and photoprotective properties. There are more than 600 known carotenoids and they have been divided into two major groups on the basis of function: (i) carotenes, oxygen-free hydrocarbon carotenoids; and (ii) xanthophylls, oxygenated derivatives of hydrocarbon carotenoids [2,25]. Figure 1 shows examples of the two groups of carotenoids. The xanthophylls are synthesized within the plastids and, due to their different polarity, they are chromatographically separated from other carotenes. In this group are included lutein, zeaxanthin, neoxanthin, violaxanthin, and α-cryptoxanthin. In contrast, carotenes are orange photosynthetic pigments, and they cooperate in photosynthesis due to their ability of transmitting light energy. In this second group are included α-carotene, β-carotene, and lycopene [26,27].

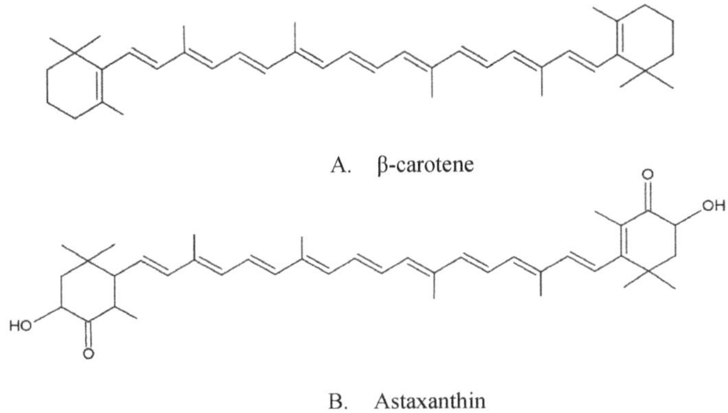

Figure 1. Carotenoid molecular structure: (**A**) example from carotene group; (**B**) example from xanthophyll group.

Carotenoids can be free or esterified with fatty acids; however, esterification does not modify the chromophore capacity of carotenoids [1]. This conjugated polyene chromophore can determine light absorption and light properties. The molecule of β-carotene has eleven conjugated double bonds forming the chromophore, which has the ability to absorb light in a visible range of the electromagnetic spectrum between 400 and 500 nm [25,26,28]. Colors in food are provided by carotenoid pigments; for example, tomatoes are red because of lycopene, and the yellow and orange colors in fruits and vegetables are due to the presence of β-carotene and α-carotene, while dark green vegetables are rich in lutein and zeaxanthin [29]. The structures of 563 carotenoids have been identified and only 50 of them can be metabolized by humans; α-carotene, β-carotene, lutein, and lycopene constitute 90% of the carotenoids included in the human diet and they are responsible for functional properties in health [25,30,31].

2.2. Biological Functions

In addition to their important applications as coloring agents and preventing oxidation in many types of food, carotenoids are also recognized for playing a vital role in human health [3]. Carotenoids are crucial for their antioxidant activity, intercellular communication, and immune system activity. It has been demonstrated that carotenoid-rich diets are associated with a lower incidence of cancer, cardiovascular disease, age-related macular degeneration, and cataract formation [28]. With respect to the effects on vision health, lutein and zeaxanthin play a crucial role in protecting photoreceptor cells from light-generated oxygen radicals and they are fundamental in the prevention of age-related macular degeneration. Intake of 15–60 mg of β-carotene, 6–20 mg of lutein, and 2–5 mg of zeaxanthin per day are recommended for preventing macular degeneration, and 15 mg of lutein per week for cataract patients [9].

Pro-vitamin A carotenoids, such as α-carotene, β-carotene, and β-cryptoxanthin, have many health benefits by providing a source of vitamin A for the growth, development, and proper functioning of the immune system and vision. Carotenoids also have a positive effect in cardiovascular health, and the efficiency of lycopene in the protection of cardiovascular disease has been well studied. In one study, the consumption of 15 mg per day of lycopene by hypercholesterolemic patients decreased systolic and diastolic blood pressure [9]. The recommended levels for lycopene intake are 5–7 mg for healthy patients, but in cases of disease, the amount of lycopene is increased to 35–75 mg. Studies have shown that lower levels of lycopene in serum are related to the incidence of microangiopathy [32]. Thus, a diet rich in carotenoid-containing fruits and vegetables must be considered to avoid cardiovascular disease.

There are other benefits of including carotenoids in the human diet, such as the positive effects on different kinds of cancer, decreasing the risk of osteoporosis, and playing a key role in the immune system. However, the antioxidant properties and protection from photooxidation are crucial for plants and humans. Xanthophylls, α-carotene, β-carotene, and cryptoxanthin, present in human serum and tissues, have been found to be efficient singlet oxygen quenchers, inhibiting DNA and cell membrane damage [25,33]. The biological functions of carotenoids in human health have encouraged industry to search for natural sources and environmentally friendly techniques for their extraction. The importance of choosing the optimal extraction conditions without degradation and alteration of their bioactive properties is a crucial issue for the food industry.

3. Green Solvents: Alternatives for Organic Solvents

The principles of green extraction dictate that it should be based on a process that reduces energy consumption, includes alternative solvents, uses renewable and natural products, and can guarantee a safe and high-quality extraction of bioactive compounds [34]. In traditional extraction and separation processes, huge quantities of organic solvents are used. Furthermore, the final extract may contain leftover solvents, impurities from the original material, or denatured chemicals as a result of the harsh extraction conditions [11]. In this way, an innovative process using green solvents and green techniques to avoid the degradation of the target compound is of great interest. There are several examples of green solvents, and they can be classified in diverse ways. Figure 2 shows the different solvents used for carotenoid extraction and their classification as green or not. Supercritical fluids (SCFs), vegetable oils, ionic liquids (ILs), natural deep eutectic solvents (NADESs), and limonene are some green solvents used for carotenoid extraction. Table 1 shows their advantages and disadvantages when used for green extraction compared with organic solvents. The characteristics, properties, optimal conditions, and most recent developments in the process used for the isolation of carotenoids from the green solvents in various different fruit and vegetable by-products are then explained and summarized in Table 2.

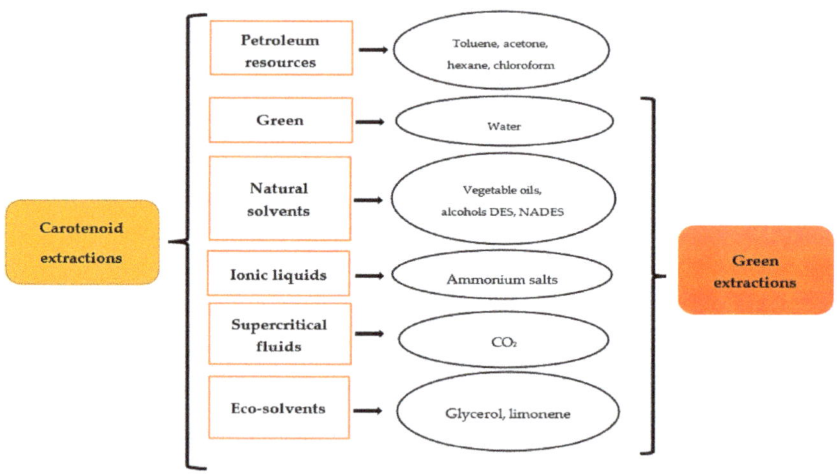

Figure 2. Alternative solvents for carotenoid green extraction (DES: deep eutectic solvents, NaDES: natural deep eutectic solvent).

Table 1. Advantages and disadvantages of green solvents for carotenoid extraction.

Solvent	Advantages	Disadvantages	Environmental Impact	Cost
Supercritical fluids (e.g., CO_2)	Provide carotenoids with high purity. Little time needed. Non-explosive, nontoxic, and inexpensive technique. Avoids thermal degradation of carotenoids.	High pressures needed. Limitations for samples containing large amounts of water. Require a large investment at the beginning.	Low	Medium
Ionic liquids (e.g., ammonium salts)	High biocompatibility with several compounds. Recyclability. Good chemical and electrochemical properties (volatility, stability, nonflammability, low melting point).	High cost (5–20 times higher than conventional solvents). High viscosity. The preparation procedure involves the use of toxic compounds.	Medium	Low
Vegetable oils (e.g., olive oil)	Great solubility. Retards the oxidation time and degradation rate of the compounds. Eliminates the extra cost of evaporation.	High viscosity producing low diffusivity and consequently low extraction yield.	Low	Low
Deep eutectic solvents and natural deep eutectic solvents (e.g., menthol)	Their preparation is simple and cheap. Low toxicity and biodegradability. Recovery, recycling, and purification of the extracted compounds are easy. Wide range of combinations is estimated.	High viscosity that produces problems with mass transfers and limits the extraction process. The application at an industrial scale might be difficult.	Medium	Low
Terpenes (e.g., limonene)	Consider GRAS. High biocompatibility with apolar compounds. Recyclability.	Only can be used for lyophilic compounds.	Low	Medium

Table 2. Carotenoid recovery using green solvents in fruit and vegetable by-products.

Matrix	Solvent	Target Compounds	Yield	Extraction Method	Optimal Conditions	Isolation Process	Reference
Mango peel (*Mangifera indica* L.)	Supercritical CO_2	All-trans-β-carotene	6290 mg/100 g dry peel	SFE	25.0 MPa, 60 °C and 15% *w*/*w* ethanol	Not necessary, directly include in food product	[35]
Mango peel (*Mangifera indica* L.)	Supercritical CO_2	Total carotenoids	560.4 mg/100 g dry peel	SFE	30 MPa and 40 °C	NA	[36]
Passion fruit bagasse (*Passiflora edulis*)	Supercritical CO_2	Total carotenoids	5.3 mg/100 g dry sample	SFE	26 MPa and 60 °C	NA	[37]
		β -carotene	1.6 mg/100 g dry sample				
		β-cryptoxanthin	62.9 mg/100 g dry sample				
Orange peel (*C. sinensis* L. *Osbeck*)	Ionic liquid: 1-butyl-3-methylimidazolium chloride and ethanol (1:2)	Total carotenoids	3.208 mg/100 g dry sample	UAE	6 extraction repetitions of 5 min, 1:3 ratio	Resins (XAD-7HP)	[38]
Tomato wastes (*Solanum lycopersicum*)	Ionic liquid: hexafluorophosphate 1-butyl- 3-methylimidazolium	Lycopene	0.556 mg/100 g dry sample	Traditional extraction	NA	NA	[39]
Passion fruit peel (*Passiflora edulis*)	Olive oil	Total carotenoids	1.207 mg/100 g of dry sample	UAE	39 min, 47 °C, 30/100 ratio	NA	[40]
	Sunflower		1.185 mg/100 g of dry sample				

Table 2. *Cont.*

Matrix	Solvent	Target Compounds	Yield	Extraction Method	Optimal Conditions	Isolation Process	Reference
Pomegranate peels (*Punica granatum* L.)	Sunflower oil	Total carotenoids	0.613 mg/100 g of dry sample	UAE	30 min, 51.5 °C, 0.10 g/mL ratio	Not necessary, directly include in food product	[41]
Mandarin epicarp (*Citrus reticulata*)	Sunflower oil	β-carotene	0.140 mg/100 g of dry sample	UAE	60 min, 60 °C, 0.4 ratio	Not necessary, directly include in food product	[42]
Peach palm peels (*Bactris gasipaes*)	Sunflower oil	Total carotenoids	163.4 mg/100 g of dry sample	UAE	30 min, 35 °C, and 1528 W/m^2	NA	[18]
Mango pulp (Chausa variety)	Flaxseed oil	Total carotenoids	0.841 mg/100 g of dry sample	High shear dispersion	20,000 rpm, 4 min	Not necessary, directly include in food product	[43]
Carrot wastes (*Daucus carota*)	Flaxseed oil	Total carotenoids	77.48% of recovery	MAE	9.39 min, 8.06:1 ratio, 165 W power	Not necessary, directly include in food product	[44]
Pumpkin peel (*Cucurbita maxima*)	Corn oil	Total carotenoids	3.803 mg/100 g of dry peel	UAE	1:10 ratio, amplitude of 20%, 30 min, 22–25 °C	Not necessary, directly include in food product	[45]
			3.494 mg/100 g of dry peel	MAE			
Tomato wastes (*Solanum lycopersicum*)	Sunflower oil	Lycopene	91.4 mg/100 g dry sample	UAE	70 W/m^2 ultrasonic intensity and 10 min	Not necessary, directly include in food product	[46]
Carrot wastes (*Daucus carota* L.)	Flaxseed oil	Total carotenoids	3460 mg/100 g of dry sample	Magnetic stirrer	10 min	Not necessary, directly include in food product	[8]
Buriti peel (*Mauritia flexuosa* L.)	Choline chloride-based and ethanol	Total carotenoids	1043 mg/100 g of dry sample	NA	30 min, 50 °C, 0.1/2 ratio	NA	[47]
Apricot pulp ('Bebekos' cultivar)	Choline chloride and L (+)-tartaric acid (2:1) and Methanol 80:20 (*v*/*v*)	β-carotene	41.3 mg/100 g of dry sample	UAE	10 min and 35 °C	Not necessary, directly include in food product	[48]
			76.1 mg/100 g of dry sample	MAE	20 min and 52 °C		
Tomato by-products (*Solanum lycopersicum*)	DL-menthol and lactic acid (8:1)	Lycopene	1.446 mg/100 g of dry sample	UAE	70 °C, 120 mL/g ratio, 10 min	NA	[49]
Pumpkin by-products (*Cucurbita maxima*)	Caprylic acid: Capric acid (3:1)	β-carotene	15,141 mg/100 g	UAE	50 °C, 52.5 W/cm^3 ultrasonic power, 7 mL/g ratio and 10 min	Switching NADES polarity	[50]
Orange peel (*Citrus cinensis* L. *osbeck*)	D-limonene	β-carotene	1125 mg/100 g of dry peel	UAE	5 min, 20 °C, 1/10 ratio	Not necessary, directly include in food product	[13]

SFE: supercritical fluid extraction, UAE: ultrasound-assisted extraction, MAE: microwave-assisted extraction, NA: not available.

3.1. Supercritical Fluids

One of the most important advantages of SCF extraction is allowing the extraction of bioactive compounds from plant material at low temperatures, which reduces the degradation of thermolabile compounds. Additionally, this process avoids the use of hazardous solvents, and also SCF is easy to remove from the target compounds [12]. Usually, SCF extraction uses CO_2 because it possesses low critical constants (Tc = 31.1 °C, Pc = 7.38 MPa) and is denominated as a generally recognized as safe (GRAS) solvent, so it is safe for human consumption. Supercritical CO_2 is a very nonpolar solvent, compatible for the extraction of low-polar compounds with small molecular weight such as triglycerides, fatty acids, aromas, and carotenoids [35,51]. Other fluids can be used in a supercritical state for the extraction of bioactives in fruit and vegetable materials, for example, ethane, propane, and dimethyl ether. These solvents have critical points comparable to CO_2, and a higher polarity index, which makes them a better choice for the extraction of higher-polarity compounds [4,14,52].

SCFs have been recognized as successfully extracting carotenoids from different sources, such as the fruits, pulps, and wastes of passion fruit (*Passiflora edulis*), peach (*Bactris gasipaes*), apricot (*Prunus armeniaca*), banana (*Musa X paradisiaca*), etc. [36,51,53]. Sánchez-Camargo et al. [35] performed the extraction of carotenoids from mango (*Mangifera indica* L.) peel using CO_2 as a SCF and ethanol as a co-solvent; the optimum conditions were 25.0 MPa, 60 °C, and 15% *w*/*w* ethanol with a result of 6290 mg all-trans-β-carotene/100 g dry peel. They demonstrated that the extracted all-trans-β-carotene can be included directly in sunflower oil to protect it against lipid oxidation without the solvent elimination process. These results can be compared with the experiment of García-Mendoza et al. [36], who also analyzed carotenoids in mango peels and obtained 560.4 mg/100 g of total carotenoids in dry peels at the optimum conditions of 30 MPa at 40 °C. Another example using CO_2 as a SCF was performed by Filho et al. [54], who studied the carotenoids from the freeze-dried pulp waste of pitanga fruits (*Eugenia uniflora* L.). The optimization showed the best conditions to be 60 °C and 25 MPa, and they succeeded in extracting 55% of the total carotenoid content, including 74% of the rubixanthin and 78% of the lycopene from the pulp. In the case of Viganó et al. [37], they extracted carotenoid from passion fruit by-products with optimum conditions (40 °C, 35 MPa) and achieved a recovery of 94.5% compared to the traditional Soxhlet extraction. They showed the effectiveness of SCFs for carotenoid extractions and how varying the conditions of pressure and temperature can extract different carotenoids. The high number of carotenoids extracted using CO_2 could be related to molecular interactions between carotenoids and supercritical carbon dioxide, as well as its nonpolar character, which facilitates carotenoid diffusion from the vegetable membranes [37]. The use of supercritical CO_2 constitutes an important extraction strategy due to its efficiency in the recovery of bioactive compounds. Additionally, target compounds can be obtained with no traces of solvent and can be directly included in the final food product. Additionally, it is environmentally friendly and nontoxic to human health, but it could require a large investment for industry at the beginning [23,55].

3.2. Ionic Liquids

ILs are organic salts showing melting points lower than 100 °C and have been successfully utilized for green extraction of carotenoids. Studies report that ILs assist with the permeabilization of the cell wall, interrupting the hydrogen bond network of cellulose [1]. One of the advantages of using ILs is their unique physicochemical properties, which depend on the structure. In addition, their insignificant volatility, nonflammability, and stability make them more attractive for bioactive compound extractions [17,56].

ILs can be used as a substitute for conventional organic solvents in the extraction process of a variety of compounds. However, carotenoid extraction using ILs has not been extensively studied. Murador et al. [21] developed a carotenoid extraction process from orange peels using ILs combined with ultrasound technique. They tested four different ionic liquids: 1-butyl-3-methylimidazolium chloride ([BMIM][Cl]), 1-n-butyl-3-methylimidazolium hexafluorophosphate ([BMIM][PF6]), 1-n-butyl-3-methylimidazolium tetrafluoroborate ([BMIM][BF4]), and 1-hexyl-3-methylimidazolium chloride ([HMIM][Cl]). Furthermore, the extraction yield was compared to that from a traditional extraction with acetone. It was demonstrated that the most effective IL was [BMIM][Cl], with a total carotenoid extract of 32.0 μg/g, while the result using acetone was 7.8 μg/g of dry matter. It is important to mention that the optimum conditions were a solid–liquid ratio of 1:3, an IL/co-solvent proportion of 1:2, and six extraction repetitions of 5 min. The effectiveness of ILs can be attributed to the affinity of [BMIM][Cl] with carotenoid compounds, hydrophobic and hydrogen bond interactions, and the aforementioned cellulose wall disruption activity of ILs, which can favor the solvent interaction with the target compound. In another study, the extraction of lycopene from tomatoes was performed using [BMIM][PF6] and [BMIM][Cl]. The results were in good agreement with the previous study, and the greater yields were extracted with [BMIM][Cl]. The total recovered was 5.560 μg/g of lycopene from tomatoes, compared to 3.650 μg/g with acetone, showing that IL application in the extraction process is a viable alternative [39]. Another study used the conductor-like screening model for real solvents (COSMO-RS) to predict the extraction efficiency of ILs, and it reported similar results. It demonstrated that imidazolium-based ILs exhibited higher cell disruption capability compared to pyridinium and ammonium-based ILs [7]. Extraction with ILs is an efficient option to extract carotenoid compounds, and selecting the optimum conditions of solid–liquid ratio, extraction time, use of co-solvents, and the best IL for a specific matrix are fundamental to favor the best interaction between the ILs and the target compounds.

The separation of biocompounds from the ILs has been reviewed, and the extraction, purification, and recovery of several compounds (caffeine, hydrolyzable tannins, and gallic acid), as well as the recycling and reuse of ILs, has been considered [15]. However, in the literature there are few studies that describe integrated and full processes for the isolation and recovery of carotenoids from ILs. Back-extraction techniques and antisolvent precipitation of bioactives are the most commonly used methods. The use of ion-exchange resins, microporous resins, distillable ILs, and thermoresponsive polymeric ILs are some other alternatives. One example of a recovery process for orange peel carotenoids from ILs was developed by Murador et al. [21]. They used XAD-7HP resin with a recovery of the ILs from 59.5–63.8% and carotenoids from 52.2–58.7%, which raises the possibility of including the obtained extracts directly in food products as a natural pigment. Nevertheless, to achieve better results and determine IL effectiveness, more research is needed to optimize the recovery process.

3.3. *Vegetable Oils*

Carotenoids are oil-soluble pigments and the use of oils as solvents can be an excellent alternative to replace organic solvents. The use of vegetable oils is according to the principles of a green process, because they are environmentally friendly solvents and produce an extract without contaminants. One of the most important advantages of the use of oils for carotenoid extraction is that elimination of the solvent is not necessary, and the extracted compound can be included directly in final food products [40]. Additionally, oils play a role as an impediment against oxygen, retarding the oxidation time and degradation of the carotenoid extracts [41].

There are some studies showing the use of vegetable oils as an alternative for carotenoid extraction. The most popular oils used are flaxseed, olive, soy, sunflower, corn, and peanut. Elik et al. [44] studied the extraction of carotenoids from carrot juice wastes using flaxseed oils in microwave-assisted extraction, and they recovered 77.48% of carotenoids at the optimal conditions of 165 W, 9.39 min, and a 8.06:1 g/g oil-to-waste ratio. They also compared this result to the traditional extraction using n-hexane, ethanol, and acetone (2:1:1 $v/v/v$), where the recovery was about 87% of the carotenoids in 120 min. Traditional extraction had a higher extraction yield, but the elimination of the organic solvent takes a long time and is energy-consuming. It is important to mention that, in the previous case, the result was an enriched oil with carotenoids, phenolics, and antioxidants.

There is also the case of Goula et al. [41], who extracted carotenoids from pomegranate peels with sunflower oil and soy oil. Their results demonstrated that combining the use of green solvents with ultrasound successfully extracted 0.620 and 0.670 mg carotenoids/100 g of dry peels using sunflower oil and soy oil, respectively. Sharma and Bhat [45] compared the extraction of carotenoids in pumpkin peels using corn oil with that using hexane/isopropyl alcohol as a traditional solvent. The results showed that almost twice as much total carotenoid was extracted when employing a green solvent (33.70 μg/g) compared to that using conventional extraction (16.20 μg/g). One aspect in the selection of the right oil is the viscosity; a low viscosity is associated with a better migration through the matrix, and consequently, the extraction yield increases [57]. The extraction efficiency of carotenoids using oils can also depend on other factors including polarity, the amount of phospholipids, and chain length of fatty acids [58]. Oils with higher quantities of short-chain fatty acids and lower amounts of phospholipids can improve carotenoid extraction [59].

The selection of the optimal conditions for carotenoid extraction using vegetable oils is an important step, due to their easy degradation under exposure to light, heat, and oxygen for a long term. For this reason, some studies have optimized the extraction process and considered variables such as intensity or power employed, time, temperature, and solid–liquid ratio [60]. Ordoñez et al. [18] performed the extraction of carotenoids from peach palm fruit using ultrasound-assisted extraction and sunflower oil as a solvent. The optimal ultrasound-assisted extraction (UAE) conditions were obtained with an ultrasonic intensity of 1528 W/m^2, extraction temperature of 35 °C, and extraction time of 30 min. Under those conditions, carotenoid recovery was 163 mg/100 g of dry peel. Chutia and Mahanta [40] found similar trends using ultrasound and olive oil in passion fruit peels. The optimum conditions were 39 min treatment time, 47 °C temperature, and 30 g/100 mL solid–liquid ratio. It was shown that the extraction yields of carotenoids are time-dependent, and yields increased with the ultrasound time increasing from 10 to 40 min. Temperature is another important variable to consider in the optimization process. In these studies, it was shown that increasing the temperature to around 50 °C, but no higher than 60 °C, increases the extraction yield. Additionally, a higher temperature decreases oil viscosity; so, consequently, a greater fluidity facilitates the diffusion of extractable lipophilic compounds [18]. These results show that the use of vegetable oils could efficiently replace organic solvents because they can extract the same amount or even more than conventional solvents, and furthermore, they can be included in the final food products, avoiding the process of solvent elimination.

3.4. Deep Eutectic Solvents and Natural Deep Eutectic Solvents

Deep eutectic solvents (DESs) are fully compliant with all 12 principles of green chemistry; they are nontoxic, inexpensive, biodegradable, and environmentally friendly [23,34]. Additionally, they have gained interest in recent years due to their cost–benefit balance and easy preparation, since the cost of DESs is comparable to that of conventional solvents [12,61]. When natural components are used for the preparation of the eutectic mixture, they are considered natural deep eutectic solvents (NADESs). DESs and NADESs are mixtures of two or three components forming hydrogen bonds. The hydrogen bond acceptor is an organic salt (quaternary ammonium or phosphonium salt) and the hydrogen bond donors are sugar, alcohol, amino acid, organic acid, etc. [62]. These solvents are considered an interesting alternative because of their recycling potential and easy purification of the extracted bioactive compounds. The main drawback of these solvents is their high viscosity, which produces mass transfer problems and restricts the extraction process [63,64]. Studies recommend reducing the viscosity by adding water in the range of 5–30%, but this process can limit their capacity for extracting hydrophobic compounds [62,65]. DESs are also considered promising solvents for green extraction processes due to the formation of strong hydrogen bonds between their components and the extracted compounds, and high quantities of water can decrease this interaction [66,67].

The use of choline chloride-based deep eutectic solvents as a co-solvent for the extraction of carotenoids in Buriti fruit (*Mauritia flexuosa*) wastes was studied, and the recovery was 10.4 mg/g. It was demonstrated that it did not increase the ethanolic extraction yield because β-carotene, which is the carotenoid usually present in fruits and vegetables, does not have the functional group for interacting with the hydrophilic choline chloride-based DES [47]. Additionally, it is well-known that all DESs with concentrations ranging from 50–30% of water presents low carotenoid extraction yields. Another parameter that must be considered in DES extractions and optimization processes is viscosity. A higher solvent viscosity limits the solute migration from the solid matrix to the liquid medium and, in consequence, a lower yield is extracted [68]. In another study, with a different combination of components, choline chloride and tartaric acid in apricots, it was reported that these solvents extracted 41.30 mg/100 g of the dry sample while organic solvents extracted 11.50 mg/100 g [48]. In that study, ethanol was used as a co-solvent to decrease the viscosity and recover a higher yield of carotenoids.

The most characterized DESs are water-soluble; thus, their application to hydrophobic compounds is limited and there are few studies using DESs for carotenoid extraction [69]. Recently, the use of hydrophobic DESs has been studied to extract nonpolar phytochemicals [70]. The first publication about hydrophobic DESs involved using a fatty acid (decanoid acid) as a hydrogen-bond acceptor (HBD) and a quaternary ammonium salt as a hydrogen-bond donor (HBA) to extract volatile fatty acids (VFAs) [71]. However, their use to extract carotenoids from fruit and vegetable by-products is still narrow. Silva et al. [49] performed the extraction of lycopene from tomato waste using a hydrophobic eutectic mixture (HEM) compound of DL-menthol as HBA and lactic acid as HBD, and UAE was used. The optimum conditions were 70 °C, 8:1 mol HBA/mol HBD, 120 mL/g solvent: sample, and 10 min of extraction time. The results showed an excellent capacity for extracting lycopene, with a yield of 1446 mg/g. Stupar et al. [50] extracted β-carotene from pumpkin by-products using hydrophobic NADESs based on fatty acids. Caprylic acid: capric acid (3:1) was selected as the optimal NADES, obtaining 151.41 μg/mL yield of β-carotene at the optimum conditions of 50 °C, ultrasonic power of 60% (52.5 W/cm^3), and a solvent-to-solid ratio of 7 mL/g during 10 min of extraction.

The excellent extraction capacity of DESs and their low or nonexistent toxicity allow their employment in food without further isolation. However, an efficient method for separation and DES reusability must be developed before DESs can be employed in industrial applications as a substitute for conventional organic solvents [72]. Stupar et al. [50] proposed a sustainable method for the separation of carotenoids from hydrophobic NADESs. This consists of switching solvent polarity from hydrophobic to hydrophilic adding water and ammonium hydroxide. The pH of the solvents was easily changed, and as a result the polarity of the NADES extract also changed. They obtained precipitated carotenoids because of their low solubility in the hydrophilic media. After the separation of carotenoids, the solvent could be reused in hydrophilic form or switched back to hydrophobic using CO_2.

The use of DESs as an innovative green alternative to the traditional carotenoid extraction process for the revalorization of food wastes and by-products is increasing, but the number of studies is still limited. The variables, optimum conditions, and characteristics of the DESs could change for each matrix and more studies in this area are needed. Furthermore, their potential inclusion in food products directly is a promising alternative for industry, but the development of databases containing information on their physicochemical qualities and toxicity is needed.

3.5. Limonene

Limonene is a cyclic monoterpene and is one of the most important essential oils in citrus fruit skins. There are two isomers, D- and L-limonene, and D-limonene represents 90% of total citrus essential oils. D-limonene is recognized for its antimicrobial and antioxidant properties and can be exploited as an antioxidant agent in the food industry [73]. Limonene does not show any functional groups available for hydrolysis; therefore, it seems to be a highly lipophilic solvent [74]. Recently, due to its highly nonpolar properties, researchers have found this compound is highly suited to replace petroleum-based solvents.

The evidence for the use of D-limonene for carotenoid extractions is still limited, but the effectiveness of this compound as a solvent has been demonstrated. Boukroufa et al. [13] used D-limonene to extract carotenoids from orange peels and compared it with a traditional extraction using hexane. They obtained yields of 11.25 mg/ L at the optimal conditions of 208 W/cm^{-2}, 20 °C, and 5 min of extraction. They observed that the carotenoid content was practically the same for both solvents, which indicates that limonene can be used as a replacement for traditional organic solvents. Another important point in that study was regarding the recycling of the solvents used, and it was demonstrated that hexane allows the recovery of only 50% of the solvent against 90% for D-limonene. Moreover, the extracts obtained using D-limonene can be directly used in food products because it they are GRAS by the US Food and Drug Administration.

4. Combining Nonconventional Methods and Green Solvents

Energy consumption is a current concern for the industry, and it can be associated with environmental impact, increased production costs, and, in consequence, loss of profits. Traditional extraction processes for carotenoids such as distillation, agitation, shaking, centrifugation, and Soxhlet are well known for being energy-consuming [13,75]. One of the six principles of green extraction is to reduce energy consumption by energy recovery and using innovative technologies [76]. The use of nonconventional methods has the objective to diminish extraction time and energy consumption compared to the conventional processes. These methods are employed in a temperature-controlled environment or without heat, which makes these methods an excellent alternative for the extraction of thermolabile compounds such as carotenoids [77]. Another advantage of green techniques is the purity of the target compounds, and carotenoids extracted with the nonconventional methods have high purity and better yields [73]. The green extraction methods are free from hazardous organic solvents and, combined with the use of green solvents, can achieve more effective results. A brief overview of the nonconventional techniques (UAE and MAE) being combined with green solvents for carotenoid extraction in fruit and vegetable by-products will be described below.

4.1. Ultrasound-Assisted Extraction

UAE is a nonconventional technology that uses waves with a frequency above 10 MHz. Ultrasonic waves penetrate cell membranes and facilitate the interaction with the solvent; consequently, mass transfer and bioactive compound extraction improves [78]. UAE is also a non-thermal, fast, and efficient technology, but to obtain good extraction, small particle size (<50 μm) is essential. One advantage of this technique is that ultrasonic power, temperature, and solid–liquid ratio can be controlled for an efficient extraction [79]. The use of an optimized range of ultrasound power is a crucial parameter to obtain a better yield of carotenoids. Nevertheless, one disadvantage of this method is that high ultrasound intensity can contribute to the degradation of antioxidant compounds such as carotenoids, because of the formation and accumulation of OH and H radicals during the cavitation process [80,81]. An important advantage of this technology is that it is energy- and time-saving. The use of traditional stirring can consume around 4.536 × 106 J in 2 h, compared to UAE at 2.999 × 106 J in 10 min [82]. Mercado-Mercado et al. [83] used UAE to evaluate the bioaccessibility of carotenoids from mango by-products and the optimal conditions were 30 min, duty cycle of 0.8, sonication amplitude of 30%, and a solid-to-liquid ratio of 300 mg to 10 mL. It was demonstrated that β-cryptoxanthin content was higher (3.59 mg/g of dry matter) in UAE-treated peel that in the control peel (0.66 ± 0.03 mg/g of dry matter). In addition, the β-carotene content in the UAE peel and UAE paste was higher than in the control peel and control paste [83].

4.2. Microwave-Assisted Extraction

MAE is a simple, rapid, economical technique, and it also uses a low amount of solvents for carotenoid extraction [45]. The parameters that can influence the efficiency of MAE are power, frequency, extraction time, temperature, moisture of matrix, type of solvent, and the solid–liquid ratio [84]. MAE has been compared with Soxhlet extraction and, even though it has a lower extraction time, the yield is higher in the traditional extraction; this can be explained by the degradation of compounds in MAE. However, Soxhlet extraction is time-consuming and it uses large amounts of solvent, which enhances the extraction cost [85,86].

Baria et al. [43] studied an optimization for the extraction of carotenoids from mango pulp wastes comparing magnetic stirring, ultrasonication (US), microwave (MW), and high shear dispersion (HSD) techniques. It was shown that the maximum total carotenoid content was obtained using HSD with flaxseed oil as solvent at the optimal conditions of 20,000 rpm for 4 min. The MW technique resulted in poor extraction, which could be due to the presence of fewer polar groups in mango pulp, which reduces the penetration of microwaves in the matrix. The structure of the matrix is a very important parameter influencing the extraction of carotenoids [85]. In another study, it was demonstrated that UAE at the conditions of 9 min, 47 °C, and 30 g/100 mL of solid–liquid ratio can extract 91.4% of carotenoids, while MAE at the optimal conditions of 200 W power, 25 min, and 10 g/100 mL solid-to-liquid ratio can extract 86.9% of carotenoids in passion fruit peels using vegetable oils as solvents [40].

Other advantages of MAE are the low cost and its applicability in a variety of matrices. However, a negative aspect of this technique is that thermolabile compounds can be damaged and cis–trans isomerization of carotenoids can occur. For that reason, experiments with intermittent radiation of 3 min have been carried out and this method showed efficiency in the recovery of carotenoids [85].

5. Application of Green Extracts in Food Products and Toxicity Issues

Lipid pigments such as carotenoids possess a great variety of beneficial effects for human health, as described above; for this reason, they have been presented as food antioxidants and bioactive substances that might be used to enrich food products. Carotenoids can also be included in dietary supplements, cosmetics, and animal fodder due to their coloring and antioxidant properties [7]. Carotenoids are mostly extracted in the food sector to be used as colorants in fruit juices, pasta, drinks, sweets, margarine, cheeses, and sausages. Traditional extraction of carotenoids using organic solvents requires an extensive downstream operation for the purification of the target compounds, which means higher costs for industry. Obtaining extracts without the need of downstream purification, using green solvents that can be included directly in the final product, is one of the challenges for the food industry. In this regard, there are two options for the food industry: either including the obtained extracts directly in food products without the elimination of the solvents or developing a process to isolate the target compounds and recycle the solvents (Figure 3).

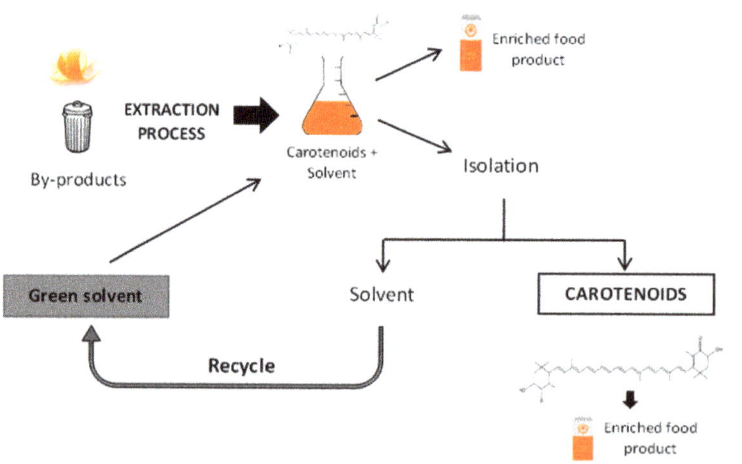

Figure 3. Process to obtain an enriched food product with carotenoids using green solvents.

The inclusion of carotenoids extracted with green solvents in food products should guarantee that solvents do not cause toxic effects for human consumers. For example, it has been demonstrated that very low concentrations of an IL do not exhibit toxicological effects, and traces of it can be found in market food products [87]. Furthermore, doses of 1-butyl-3-methyl-imidazolium at 10 mg/kg of body weight per day used in the extraction of carotenoids from tomatoes do not pose a toxicological risk [88]. Martins and De Rosso [88] studied the application of imidazolium-based ILs for extracting carotenoids used in the food industry and have filed a patent. NADESs have also been tested and shown to be safe for consumption in vitro by an antiproliferation assay in tumor cells. They demonstrated that extracts using NADESs could be considered ready-to-use in food products without requiring an expensive and long downstream process for solvent elimination. Additionally, the antioxidant properties of NADESs have been well studied and could be beneficial to extend the shelf life of food products [89–93]. However, the sensory properties in the final products should be analyzed in future research. The studies that have included carotenoids extracted with green solvents and used in food products are listed in Table 3 and will be discussed below.

Table 3. Applications of carotenoids extracted with green solvents in the food industry.

Solvent	Source	Products	Application	Reference
Sunflower oil	Mandarin epicarp (*Citrus reticulata*)	Cakes and bread	Enhance sensory attributes. Improve coloring properties	[42]
Ionic liquids 1-Decyl-trimethylammonium bromide $[N_{1,1,1,10}]$ Br	Palm heart fruit wastes (*Bactris gasipaes*)	Chitosan film	Ecological packaging Antioxidant	[94]
Flaxseed oil	Carrot waste (*Daucus carota* L.)	Encapsulated emulsion	Natural colorant Food fortification	[8]
Supercritical CO_2	Mango peel (*Mangifera indica* L.)	Sunflower oil	Reducing deterioration Antioxidant natural additive Retarding lipid oxidation	[35]

Vegetable oils used as solvents are completely safe and have been included in food products directly. For example, Ordoñez-Santos et al. [42] extracted carotenoids in mandarin epicarp with sunflower oil as solvent using UAE. Then, 20 g of the lipid extract was included in two kinds of samples, cake and bread, as a natural colorant resource and compared with a control with 0.0025 g of tartrazine. In comparison to the control, the samples made with the mandarin epicarp extract showed a substantial increase in all carotenoid pigments. Additionally, the product with the added lipid extract showed a better color than the control. This could be an excellent alternative to replace synthetic coloring additives such as tartrazine in bakery products. In another study, Tiwari et al. [8] developed an encapsulated emulsion based on green extracted carotenoids from carrot wastes using flaxseed oil as solvent. According to their chemical structure, carotenoids and oils are nonpolar; due to this characteristic, the authors formulated an emulsion adding proteins and lactose. They demonstrated that carotenoids extracted in flaxseed oil [95] were encapsulated in an emulsion-based delivery system using 10% whey protein concentrate, 5% lactose, and 1% gum Arabic, with the encapsulation efficiency being 92%. Furthermore, the authors applied the dried emulsion in the preparation of a flavored milk-based drink as a source of natural functional colorant [96]. It is important to mention that the final enriched emulsion had other bioactive compounds besides carotenoids, because flaxseed oil is one of the richest vegetable sources of omega-3 fatty acids.

CO_2 is also generally recognized as a safe (GRAS) solvent, and as a result the compounds extracted with supercritical CO_2 are safe for human consumption [12]. Sanchez-Camargo et al. [35] obtained carotenoids from mango peels using supercritical CO_2. Then, extracts were added to sunflower oil to prevent lipid oxidation in concentrations of 200, 500, and 1000 mg extract per kg of oil, equivalent to 6, 14, and 28 mg all-trans-β-carotene per kg of oil. The results showed that it was efficient in inhibiting sunflower oil lipid oxidation. The antioxidant activity of carotenoids in edible oils is highly dependent on oxygen pressure levels and carotenoid concentration. For this reason, it is important to consider that the amount of β-carotene should be from 50–300 ppm; otherwise, it can act as a pro-oxidant. In that study, β-carotene was employed at concentrations lower than 30 ppm and no effect was found [35].

The inclusion of green extracts in the food industry has also been explored in the area of food packaging. Bio-based materials with antioxidant properties are attracting considerable interest in industry because non-ecofriendly packages are considered one of the most important environmental problems. De Souza Mesquita et al. [94] developed a green extraction and purification of carotenoids in palm heart (*Bactris gasipaes*) fruit wastes using ionic liquids. After the extraction, the carotenoids were used as a component in a chitosan-based film usable for food packaging in three different doses of 0.025, 0.050, and 0.100% (w/w). The mechanical characteristics of the films were measured to assess the impact of carotenoid addition on the properties of the films. The study found positive effects for all the doses of added carotenoids on physical characteristics such as elasticity, wettability, and solubility. Additionally, the antioxidant capacity of the films was evaluated, and it persisted at the highest levels throughout the whole 20-day experiment period.

The inclusion of other bioactive compounds extracted with green solvents, such as polyphenols, has also been used for fortification in the food industry [6]. The determination of the RDI and daily upper limits for green solvents such as NADESs and ILs is one of the challenges remaining for their inclusion in food products. Recommendations in appropriate legislation would undoubtedly help to expand the use of green solvents in the food industry. Future studies should be directed to study cytotoxicity, recommended daily intake, and the sensory properties of bioactives extracted with green solvents.

6. Conclusions

The choice of solvent is a crucial step for an efficient green extraction, and it strongly depends on the composition of the target compound. The recovery of nutritive and valuable compounds derived from fruit and vegetable by-products is still an important challenge for food industries. Green extractions using GRAS solvents can be an effective alternative to improve extraction procedures and optimize carotenoid extraction. In order to follow the principles of green chemistry, such as Principle 2, the use of alternative solvents (principally water or agro-solvents), and Principle 4, the production of co-products instead of waste in the bio- and agro-refining industries, this study described the potential of various green solvents and emerging techniques for carotenoid extraction from fruit and vegetable by-products. The use of SCFs, limonene, ILs, NADESs, and vegetable oils has shown promising results in the different fruit and vegetable matrices mentioned in this review. Nevertheless, the best alternative for a green extraction depends on the characteristics of the bioactive compounds extracted and the optimization of the variables that can affect the interactions between solvents and target compounds. Green solvents are undoubtedly a good alternative to replace organic solvents due to their properties and their low toxicity. This work presents valuable information and future trends for the inclusion of carotenoids extracted with green solvents in food products. However, there is still a lot of work to do for the inclusion of green solvents in the food industry.

Author Contributions: A.V.-O., writing—original draft preparation; J.B., writing—reviewing and editing; D.L.-M., writing—reviewing and editing; M.J.E., conceptualization, supervision, and funding acquisition; A.F., writing—reviewing. All authors have read and agreed to the published version of the manuscript.

Funding: This work was financially supported by the Ministry of Science and Innovation (Spain) -State Research Agency (PID-2019-111331RB-I00/AEI/10.13039/501100011033) and by the "Generación Bicentenario" scholarship from the Ministry of Education of the Republic of Peru (PRONABEC).

Institutional Review Board Statement: Not applicable.

Informed Consent Statement: Not applicable.

Data Availability Statement: Not applicable for review.

Conflicts of Interest: The authors declare no conflict of interest.

References

1. Yu, J.; Liu, X.; Zhang, L.; Shao, P.; Wu, W.; Chen, Z.; Li, J.; Renard, C. An overview of carotenoid extractions using green solvents assisted by Z-isomerization. *Trends Food Sci. Technol.* **2022**, *123*, 145–160. [CrossRef]
2. Kultys, E.; Kurek, M.A. Green Extraction of Carotenoids from Fruit and Vegetable Byproducts: A Review. *Molecules* **2022**, *27*, 518. [CrossRef] [PubMed]
3. Luzardo-Ocampo, I.; Ramírez-Jiménez, A.K.; Yañez, J.; Mojica, L.; Luna-Vital, D.A. Technological Applications of Natural Colorants in Food Systems: A Review. *Foods* **2021**, *10*, 634. [CrossRef] [PubMed]
4. Martins, N.; Ferreira, I.C.F.R. Wastes and by-products: Upcoming sources of carotenoids for biotechnological purposes and health-related applications. *Trends Food Sci. Technol.* **2017**, *62*, 33–48. [CrossRef]
5. Razi Parjikolaei, B.; Bahij El-Houri, R.; Fretté, X.C.; Christensen, K.V. Influence of green solvent extraction on carotenoid yield from shrimp (*Pandalus borealis*) processing waste. *J. Food Eng.* **2015**, *155*, 22–28. [CrossRef]
6. Panić, M.; Drakula, S.; Cravotto, G.; Verpoorte, R.; Hruškar, M.; Radojčić Redovniković, I.; Radošević, K. Biological activity and sensory evaluation of cocoa by-products NADES extracts used in food fortification. *Innov. Food Sci. Emerg. Technol.* **2020**, *66*, 102514. [CrossRef]
7. Rajabi, M.S.; Moniruzzaman, M.; Mahmood, H.; Sivapragasam, M.; Bustam, M.A. Extraction of β-carotene from organic phase using ammonium based ionic liquids. *J. Mol. Liq.* **2017**, *227*, 15–20. [CrossRef]
8. Tiwari, S.; Upadhyay, N.; Malhotra, R. Three way ANOVA for emulsion of carotenoids extracted in flaxseed oil from carrot bio-waste. *J. Waste Manag.* **2021**, *121*, 67–76. [CrossRef]
9. Singh, A.; Ahmad, S.; Ahmad, A. Green extraction methods and environmental applications of carotenoids-a review. *RSC Adv.* **2015**, *5*, 62358–62393. [CrossRef]
10. Nagarajan, J.; Kay, H.P.; Arumugam, P.; Krishnamurthy, N.P.; Ramakrishnan, N.R.; Aldawoud, T.M.S.; Galanakis, C.M.; Wei, O.C. Valorisation of carrot peel waste by water-induced hydrocolloidal complexation for extraction of carotenoid and pectin. *Chemosphere* **2021**, *272*, 129919. [CrossRef]
11. Yara-Varón, E.; Fabiano-Tixier, A.S.; Balcells, M.; Canela-Garayoa, R.; Bily, A.; Chemat, F. Is it possible to substitute hexane with green solvents for extraction of carotenoids? A theoretical versus experimental solubility study. *RSC Adv.* **2016**, *6*, 27750–27759. [CrossRef]
12. Cvjetko Bubalo, M.; Vidović, S.; Radojčić Redovniković, I.; Jokić, S. New perspective in extraction of plant biologically active compounds by green solvents. *Food Bioprod. Process.* **2018**, *109*, 52–73. [CrossRef]
13. Boukroufa, M.; Boutekedjiret, C.; Chemat, F. Development of a green procedure of citrus fruits waste processing to recover carotenoids. *Resour. Effic. Technol.* **2017**, *3*, 252–262. [CrossRef]
14. Putnik, P.; Lorenzo, J.M.; Barba, F.J.; Roohinejad, S.; Režek Jambrak, A.; Granato, D.; Montesano, D.; Bursa, D. Novel Food Processing and Extraction Technologies of High-Added Value Compounds from Plant Materials. *Foods* **2018**, *7*, 106. [CrossRef]
15. Ventura, S.P.M.; Silva, F.A.; Quental, M.V.; Mondal, D.; Freire, M.G.; Coutinho, A.P. Ionic-Liquid-Mediated Extraction and Separation Processes for Bioactive Compounds: Past, Present, and Future Trends. *Chem. Rev.* **2017**, *117*, 6984–7052. [CrossRef]
16. Chanioti, S.; Tzia, C. Extraction of phenolic compounds from olive pomace by using natural deep eutectic solvents and innovative extraction techniques. *Innov. Food Sci. Emerg. Technol.* **2018**, *48*, 228–239. [CrossRef]
17. Cvjetko Bubalo, M.; Vidović, S.; Radojčić Redovniković, I.; Jokić, S. Green solvents for green technologies. *J. Chem. Technol. Biotechnol.* **2015**, *90*, 1631–1639. [CrossRef]
18. Ordóñez-Santos, L.E.; Pinzón-Zarate, L.X.; González-Salcedo, L.O. Optimization of ultrasonic-assisted extraction of total carotenoids from peach palm fruit (*Bactris gasipaes*) by-products with sunflower oil using response surface methodology. *Ultrason. Sonochem.* **2015**, *27*, 560–566. [CrossRef]
19. Satari, B.; Karimi, K. Citrus processing wastes: Environmental impacts, recent advances, and future perspectives in total valorization. *Resour. Conserv. Recycl.* **2018**, *129*, 153–167. [CrossRef]
20. Mahato, N.; Sharma, K.; Sinha, M.; Cho, M.H. Citrus waste derived nutra-/pharmaceuticals for health benefits: Current trends and future perspectives. *J. Funct. Foods* **2018**, *40*, 307–316. [CrossRef]

21. Murador, D.C.; Salafia, F.; Zoccali, M.; Martins, P.L.G.; Ferreira, A.G.; Dugo, P.; Mondello, L.; de Rosso, V.V.; Giuffrida, D. Green Extraction Approaches for Carotenoids and Esters: Characterization of Native Composition from Orange Peel. *Antioxidants* **2019**, *8*, 613. [CrossRef] [PubMed]
22. Sagar, N.A.; Pareek, S.; Sharma, S.; Yahia, E.M.; Lobo, M.G. Fruit and Vegetable Waste: Bioactive Compounds, Their Extraction, and Possible Utilization. *Compr. Rev. Food Sci. Food Saf.* **2018**, *17*, 512–531. [CrossRef] [PubMed]
23. Mäki-Arvela, P.; Hachemi, I.; Murzin, D.Y. Comparative study of the extraction methods for recovery of carotenoids from algae: Extraction kinetics and effect of different extraction parameters. *J. Chem. Technol. Biotechnol.* **2014**, *89*, 1607–1626. [CrossRef]
24. Ramos-Escudero, F.; González-Miret, M.L.; Viñas-Ospino, A.; Ramos Escudero, M. Quality, stability, carotenoids and chromatic parameters of commercial Sacha inchi oil originating from Peruvian cultivars. *Food Sci. Technol.* **2019**, *56*, 4901–4910. [CrossRef] [PubMed]
25. Mercadante, A.Z.; Rodrigues, D.B.; Petry, F.C.; Mariutti, L.R.B. Carotenoid esters in foods—A review and practical directions on analysis and occurrence. *Int. Food Res. J.* **2017**, *99*, 830–850. [CrossRef] [PubMed]
26. Watanabe, M.; Musumi, K.; Ayugase, J. Carotenoid pigment composition, polyphenol content, and antioxidant activities of extracts from orange-colored Chinese cabbage. *LWT—Food Sci. Technol.* **2011**, *44*, 1971–1975. [CrossRef]
27. Zhao, X.J.; Xing, T.T.; Li, Y.F.; Jiao, B.N. Analysis of phytochemical contributors to antioxidant capacity of the peel of Chinese mandarin and orange varieties. *Int. J. Food Sci. Nutr.* **2019**, *70*, 825–833. [CrossRef]
28. Arunkumar, R.; Gorusupudi, A.; Bernstein, P.S. The macular carotenoids: A biochemical overview. *Biochim. Biophys. Acta Mol. Cell Biol. Lipids* **2020**, *1865*, 158617. [CrossRef]
29. Etzbach, L.; Pfeiffer, A.; Weber, F.; Schieber, A. Characterization of carotenoid profiles in goldenberry (*Physalis peruviana* L.) fruits at various ripening stages and in different plant tissues by HPLC-DAD-APCI-MSn. *Food Chem.* **2018**, *245*, 508–517. [CrossRef]
30. Fuad, N.I.N.; Sekar, M.; Gan, S.H.; Lum, P.T.; Vaijanathappa, J.; Ravi, S. Lutein: A comprehensive review on its chemical, biological activities and therapeutic potentials. *Pharmacogn. J.* **2020**, *12*, 1769–1778. [CrossRef]
31. Mussagy, C.U.; Khan, S.; Kot, A.M. Current developments on the application of microbial carotenoids as an alternative to synthetic pigments. *Crit. Rev. Food Sci. Nutr.* **2021**, *62*, 6932–6946. [CrossRef] [PubMed]
32. Jiao, Y.; Reuss, L.; Wang, Y. β-Cryptoxanthin: Chemistry, Occurrence, and Potential Health Benefits. *Curr. Pharmacol. Rep.* **2019**, *5*, 20–34. [CrossRef]
33. Lefebvre, T.; Destandau, E.; Lesellier, E. Selective extraction of bioactive compounds from plants using recent extraction techniques: A review. *J. Chromatogr. A* **2021**, *1635*, 461770. [CrossRef] [PubMed]
34. Chemat, F.; Abert-Vian, M.; Fabiano-Tixier, A.S.; Strube, J.; Uhlenbrock, L.; Gunjevic, V.; Cravotto, G. Green extraction of natural products. Origins, current status, and future challenges. *TrAC—Trends Anal. Chem.* **2019**, *118*, 248–263. [CrossRef]
35. Sánchez-Camargo, A.d.P.; Gutiérrez, L.F.; Vargas, S.M.; Martinez-Correa, H.A.; Parada-Alfonso, F.; Narváez-Cuenca, C.E. Valorisation of mango peel: Proximate composition, supercritical fluid extraction of carotenoids, and application as an antioxidant additive for an edible oil. *J. Supercrit. Fluids.* **2019**, *152*, 104574. [CrossRef]
36. Garcia-Mendoza, M.P.; Paula, J.T.; Paviani, L.C.; Cabral, F.A.; Martinez-Correa, H.A. Extracts from mango peel by-product obtained by supercritical CO_2 and pressurized solvent processes. *LWT—Food Sci. Technol.* **2015**, *62*, 131–137. [CrossRef]
37. Viganó, J.; Coutinho, J.P.; Souza, D.S.; Baroni, N.A.F.; Godoy, H.T.; Macedo, J.A.; Martínez, J. Exploring the selectivity of supercritical CO_2 to obtain nonpolar fractions of passion fruit bagasse extracts. *J. Supercrit. Fluids* **2016**, *110*, 1–10. [CrossRef]
38. Murador, D.C.; Braga, A.R.C.; Martins, P.L.G.; Mercadante, A.Z.; de Rosso, V.V. Ionic liquid associated with ultrasonic-assisted extraction: A new approach to obtain carotenoids from orange peel. *Int. Food Res. J.* **2019**, *126*, 108653. [CrossRef]
39. Martins, P.L.G.; De Rosso, V.V. Carotenoids achieving from tomatoes discarded using ionic liquids as extracting for application in food industry. *XIV Saf. Health Environ. World Congr. Conf.* **2014**, 35–38. [CrossRef]
40. Chutia, H.; Mahanta, C.L. Green ultrasound and microwave extraction of carotenoids from passion fruit peel using vegetable oils as a solvent: Optimization, comparison, kinetics, and thermodynamic studies. *Innov. Food Sci. Emerg. Technol.* **2020**, *67*, 102547. [CrossRef]
41. Goula, A.M.; Ververi, M.; Adamopoulou, A.; Kaderides, K. Green ultrasound-assisted extraction of carotenoids from pomegranate wastes using vegetable oils. *Ultrason. Sonochem.* **2017**, *34*, 821–830. [CrossRef] [PubMed]
42. Ordóñez-Santos, L.E.; Esparza-Estrada, J.; Vanegas-Mahecha, P. Ultrasound-assisted extraction of total carotenoids from mandarin epicarp and application as natural colorant in bakery products. *LWT—Food Sci. Technol.* **2021**, *139*, 110598. [CrossRef]
43. Baria, B.; Upadhyay, N.; Singh, A.K.; Malhotra, R.K. Optimization of 'green' extraction of carotenoids from mango pulp using split plot design and its characterization. *LWT—Food Sci. Technol.* **2019**, *104*, 186–194. [CrossRef]
44. Elik, A.; Yanık, D.K.; Göğüş, F. Microwave-assisted extraction of carotenoids from carrot juice processing waste using flaxseed oil as a solvent. *LWT—Food Sci. Technol.* **2020**, *123*, 109100. [CrossRef]
45. Sharma, M.; Bhat, R. Extraction of Carotenoids from Pumpkin Peel and Pulp: Comparison between Innovative Green Extraction Technologies (Ultrasonic and Microwave-Assisted Extractions Using Corn Oil). *Foods* **2021**, *10*, 787. [CrossRef] [PubMed]
46. Rahimi, S.; Mikani, M. Lycopene green ultrasound-assisted extraction using edible oil accompany with response surface methodology (RSM) optimization performance: Application in tomato processing wastes. *Microchem. J.* **2019**, *146*, 1033–1042. [CrossRef]

47. Leite, P.I.P.; De Assis, C.F.; Silvino dos Santos, E.; Padilha, C.E.d.A.; Ferrari, M.; Sousa Junior, F.C. Choline chloride-based deep eutectic solvents do not improve the ethanolic extraction of carotenoids from buriti fruit (*Mauritia flexuosa* L.). *Sustain. Chem. Pharm.* **2021**, *20*, 100375. [CrossRef]
48. Koutsoukos, S.; Tsiaka, T.; Tzani, A.; Zoumpoulakis, P.; Detsi, A. Choline chloride and tartaric acid, a Natural Deep Eutectic Solvent for the efficient extraction of phenolic and carotenoid compounds. *J. Clean. Prod.* **2019**, *241*, 118384. [CrossRef]
49. Silva, Y.P.A.; Ferreira, T.A.P.C.; Jiao, G.; Brooks, M.S. Sustainable approach for lycopene extraction from tomato processing by-product using hydrophobic eutectic solvents. *J. Food Sci. Technol.* **2019**, *56*, 1649–1654. [CrossRef]
50. Stupar, A.; Šeregelj, V.; Ribeiro, B.D.; Pezo, L.; Cvetanović, A.; Mišan, A.; Marrucho, I. Recovery of β-carotene from pumpkin using switchable natural deep eutectic solvents. *Ultrason. Sonochem.* **2021**, *76*, 105638. [CrossRef]
51. De Andrade Lima, M.; Kestekoglou, I.; Charalampopoulos, D.; Chatzifragkou, A. Supercritical fluid extraction of carotenoids from vegetable waste matrices. *Molecules* **2019**, *24*, 466. [CrossRef] [PubMed]
52. Kehili, M.; Kammlott, M.; Choura, S.; Zammel, A.; Zetzl, C.; Smirnova, I.; Allouche, N.; Sayadi, S. Supercritical CO_2 extraction and antioxidant activity of lycopene and β-carotene-enriched oleoresin from tomato (*Lycopersicum esculentum* L.) peels by-product of a Tunisian industry. *Food Bioprod. Process.* **2017**, *102*, 340–349. [CrossRef]
53. Massa, T.B.; Stevanato, N.; Cardozo-Filho, L.; Da Silva, C. Pumpkin (*Cucurbita maxima*) by-products: Obtaining seed oil enriched with active compounds from the peel by ultrasonic-assisted extraction. *J. Food Process. Eng.* **2019**, *42*, e13125. [CrossRef]
54. Filho, G.; De Rosso, V.; Meireles, A.; Rosa, P.; Oliveira, A.; Mercadante, A.; Cabral, F. Supercritical CO_2 extraction of carotenoids from pitanga fruits (*Eugenia uniflora* L.). *J. Supercrit. Fluids* **2008**, *46*, 33–39. [CrossRef]
55. Chemat, F.; Vian, M.A.; Cravotto, G. Green extraction of natural products: Concept and principles. *Int. J. Mol. Sci.* **2012**, *13*, 8615–8627. [CrossRef] [PubMed]
56. Vieira, F.A.; Guilherme, R.J.R.; Neves, M.C.; Rego, A.; Abreu, M.H.; Coutinho, J.A.P.; Ventura, S.P.M. Recovery of carotenoids from brown seaweeds using aqueous solutions of surface-active ionic liquids and anionic surfactants. *Sep. Purif. Technol.* **2018**, *196*, 300–308. [CrossRef]
57. Portillo-López, R.; Morales-Contreras, B.E.; Lozano-Guzmán, E.; Basilio-Heredia, J.; Muy-Rangel, M.D.; Ochoa-Martínez, L.A.; Rosas Flores, W.; Morales-Castro, J. Vegetable oils as green solvents for carotenoid extraction from pumpkin (*Cucurbita argyrosperma* Huber) byproducts: Optimization of extraction parameters. *Food Sci.* **2021**, *86*, 3122–3136. [CrossRef]
58. Nour, V.; Corbu, A.R.; Rotaru, P.; Karageorgou, I.; Lalas, S. Effect of carotenoids, extracted from dry tomato waste, on the stability and characteristics of various vegetable oils. *Grasas Aceites* **2018**, *69*, 238. [CrossRef]
59. Konuskan, D.B.; Arslan, M.; Oksuz, A. Physicochemical properties of cold pressed sunflower, peanut, rapeseed, mustard and olive oils grown in the Eastern Mediterranean region. *Saudi J. Biol. Sci.* **2019**, *26*, 340–344. [CrossRef]
60. Chuyen, H.V.; Roach, P.D.; Golding, J.B.; Parks, S.E.; Nguyen, M.H. Optimisation of extraction conditions for recovering carotenoids and antioxidant capacity from Gac peel using response surface methodology. *Int. J. Food Sci. Technol.* **2017**, *52*, 972–980. [CrossRef]
61. Zhuang, B.; Dou, L.L.; Li, P.; Liu, E.H. Deep eutectic solvents as green media for extraction of flavonoid glycosides and aglycones from *Platycladi Cacumen*. *J. Pharm. Biomed. Anal.* **2017**, *134*, 214–219. [CrossRef]
62. Panić, M.; Radić Stojković, M.; Kraljić, K.; Škevin, D.; Radojčić Redovniković, I.; Gaurina Srček, V.; Radošević, K. Ready-to-use green polyphenolic extracts from food by-products. *Food Chem.* **2019**, *283*, 628–636. [CrossRef] [PubMed]
63. Gullón, P.; Gullón, B.; Romaní, A.; Rocchetti, G.; Lorenzo, J.M. Smart advanced solvents for bioactive compounds recovery from agri-food by-products: A review. *Trends Food Sci. Technol.* **2020**, *101*, 182–197. [CrossRef]
64. Manurung, R.; Siregar, A.G.A. Performance of menthol based deep eutectic solvents in the extraction of carotenoids from crude palm oil. *Int. J. GEOMATE* **2020**, *19*, 131–137. [CrossRef]
65. Oomen, W.W.; Begines, P.; Rianika Mustafa, N.; Wilson, E.G.; Verpoorte, R.; Choi, Y.H. Natural Deep Eutectic Solvent Extraction of Flavonoids of *Scutellaria baicalensis* as a Replacement for Conventional Organic Solvents. *Molecules* **2020**, *25*, 617. [CrossRef] [PubMed]
66. Bulut, O.; Akın, D.; Sönmez, Ç.; Öktem, A.; Yücel, M.; Öktem, H.A. Phenolic compounds, carotenoids, and antioxidant capacities of a thermo-tolerant *Scenedesmus* sp. (Chlorophyta) extracted with different solvents. *J. Appl. Phycol.* **2019**, *31*, 1675–1683. [CrossRef]
67. Liu, Y.; Friesen, J.B.; McAlpine, J.B.; Lankin, D.C.; Chen, S.N.; Pauli, G.F. Natural Deep Eutectic Solvents: Properties, Applications, and Perspectives. *J. Nat. Prod.* **2018**, *81*, 679–690. [CrossRef]
68. Abbott, A.P.; Boothby, D.; Capper, G.; Davies, D.L.; Rasheed, R.K. Deep Eutectic Solvents formed between choline chloride and carboxylic acids: Versatile alternatives to ionic liquids. *J. Am. Chem. Soc.* **2004**, *126*, 9142–9147. [CrossRef]
69. Chen, X.-Q.; Li, Z.-H.; Liu, L.-L.; Wang, H.; Yang, S.-H.; Zhang, J.-S.; Zhang, Y. Green extraction using deep eutectic solvents and antioxidant activities of flavonoids from two fruits of Rubia species. *LWT—Food Sci. Technol.* **2021**, *148*, 111708. [CrossRef]
70. Ricarte, G.N.; Coelho, M.A.Z.; Marrucho, I.M.; Ribeiro, B.D. Enzyme-assisted extraction of carotenoids and phenolic compounds from sunflower wastes using green solvents. *3 Biotech* **2020**, *10*, 405. [CrossRef]
71. Van Osch, D.J.G.P.; Zubeir, L.F.; Van Den Bruinhorst, A.; Rocha, M.A.A.; Kroon, M.C. Hydrophobic deep eutectic solvents as water-immiscible extractants. *Green Chem.* **2015**, *17*, 4518. [CrossRef]

72. Wang, L.T.; Yang, Q.; Cui, Q.; Fan, X.H.; Dong, M.Z.; Gao, M.Z.; Lv, M.J.; An, J.Y.; Meng, D.; Zhao, X.H.; et al. Recyclable menthol-based deep eutectic solvent micellar system for extracting phytochemicals from *Ginkgo biloba* leaves. *J. Clean. Prod.* **2020**, *244*, 118648. [CrossRef]
73. Zuin, V.G.; Ramin, L.Z.; Segatto, M.L.; Stahl, A.M.; Zanotti, K.; Forim, M.R.; Da Silva, M.F.d.G.F.; Fernandes, J.B. To separate or not to separate: What is necessary and enough for a green and sustainable extraction of bioactive compounds from Brazilian citrus waste. *Pure Appl. Chem.* **2020**, *93*, 13–27. [CrossRef]
74. Negro, V.; Mancini, G.; Ruggeri, B.; Fino, D. Citrus waste as feedstock for bio-based products recovery: Review on limonene case study and energy valorization. *Bioresour. Technol.* **2016**, *214*, 806–815. [CrossRef] [PubMed]
75. Virot, M.; Tomao, V.; Ginies, C.; Visinoni, F.; Chemat, F. Green procedure with a green solvent for fats and oils' determination Microwave-integrated Soxhlet using limonene followed by microwave Clevenger distillation. *J. Chromatogr. A* **2008**, *1196–1197*, 147–152. [CrossRef] [PubMed]
76. Rukavina, I.; Rodrigues, M.J.; Pereira, C.G.; Mansinhos, I.; Romano, A.; Ślusarczyk, S.; Matkowski, A.; Custódio, L. Greener Is Better: First Approach for the Use of Natural Deep Eutectic Solvents (NADES) to Extract Antioxidants from the Medicinal *Halophyte Polygonum maritimum* L. *Molecules* **2021**, *26*, 6136. [CrossRef]
77. Anticona, M.; Blesa, J.; Frigola, A.; Esteve, M.J. High biological value compounds extraction from citruswaste with non-conventional methods. *Foods* **2020**, *9*, 811. [CrossRef]
78. Montero-Calderon, A.; Cortes, C.; Zulueta, A.; Frigola, A.; Esteve, M.J. Green solvents and Ultrasound-Assisted extraction of bioactive orange (*Citrus sinensis*) peel compounds. *Sci. Rep.* **2019**, *9*, 16120. [CrossRef]
79. Tabaraki, R.; Heidarizadi, E.; Benvidi, A. Optimization of ultrasonic-assisted extraction of pomegranate (*Punica granatum* L.) peel antioxidants by response surface methodology. *Sep. Purif. Technol.* **2012**, *98*, 16–23. [CrossRef]
80. Kaderides, K.; Goula, A.M.; Adamopoulos, K.G. A process for turning pomegranate peels into a valuable food ingredient using ultrasound-assisted extraction and encapsulation. *Innov. Food Sci. Emerg. Technol.* **2015**, *31*, 204–215. [CrossRef]
81. Kumari, B.; Tiwari, B.K.; Hossain, M.B.; Brunton, N.P.; Rai, D.K. Recent Advances on Application of Ultrasound and Pulsed Electric Field Technologies in the Extraction of Bioactives from Agro-Industrial by-Products. *Food Bioproc. Technol.* **2018**, *11*, 223–241. [CrossRef]
82. Benmeziane, A.; Boulekbache-Makhlouf, L.; Mapelli-Brahm, P.; Khaled Khodja, N.; Remini, H.; Madani, K.; Meléndez-Martínez, A.J. Extraction of carotenoids from cantaloupe waste and determination of its mineral composition. *Int. Food Res. J.* **2018**, *111*, 391–398. [CrossRef] [PubMed]
83. Mercado-Mercado, G.; Montalvo-González, E.; González-Aguilar, G.A.; Alvarez-Parrilla, E.; Sáyago-Ayerdi, S.G. Ultrasound-assisted extraction of carotenoids from mango (*Mangifera indica* L. 'Ataulfo') by-products on in vitro bioaccessibility. *Food Biosci.* **2018**, *21*, 125–131. [CrossRef]
84. Soquetta, M.B.; Terra, L.d.M.; Bastos, C.P. Green technologies for the extraction of bioactive compounds in fruits and vegetables. *CyTA J. Food* **2018**, *16*, 400–412. [CrossRef]
85. Hiranvarachat, B.; Devahastin, S. Enhancement of microwave-assisted extraction via intermittent radiation: Extraction of carotenoids from carrot peels. *J. Food Eng.* **2014**, *126*, 17–26. [CrossRef]
86. Ho, K.K.H.Y.; Ferruzzi, M.G.; Liceaga, A.M.; San Martín-González, M.F. Microwave-assisted extraction of lycopene in tomato peels: Effect of extraction conditions on all-trans and cis-isomer yields. *LWT—Food Sci. Technol.* **2015**, *62*, 160–168. [CrossRef]
87. Martins, P.L.G.; Braga, A.R.; Vera De Rosso, V. Can ionic liquid solvents be applied in the food industry? *Trends Food Sci. Technol.* **2017**, *66*, 117–124. [CrossRef]
88. Martins, P.L.G.; De Rosso, V.V. Thermal and light stabilities and antioxidant activity of carotenoids from tomatoes extracted using an ultrasound-assisted completely solvent-free method. *Int. Food Res. J.* **2016**, *82*, 156–164. [CrossRef]
89. Panić, M.; Gunjević, V.; Radošević, K.; Bubalo, M.C.; Ganić, K.K.; Redovniković, I.R. Cosmotherm as an effective tool for selection of deep eutectic solvents based ready-to-use extracts from Graševina grape pomace. *Molecules* **2021**, *26*, 4722. [CrossRef]
90. Viñas-Ospino, A.; Panić, M.; Bagović, M.; Radošević, K.; Esteve, M.J.; Radojčić Redovniković, I. Green approach to extract bioactive compounds from orange peel employing hydrophilic and hydrophobic deep eutectic solvents. *Sustain. Chem. Pharm.* **2023**, *31*, 100942. [CrossRef]
91. Benlebna, M.; Ruesgas-Ramón, M.; Bonafos, B.; Fouret, G.; Casas, F.; Coudray, C.; Durand, E.; Cruz Figueroa-Espinoza, M.; Feillet-Coudray, C. Toxicity of Natural Deep Eutectic Solvent Betaine: Glycerol in Rats. *J. Agric. Food Chem.* **2018**, *66*, 6205–6212. [CrossRef] [PubMed]
92. Mitar, A.; Panić, M.; Kardum, J.P.; Halambek, J.; Sander, A.; Kučan, K.Z.; Redovniković, I.R.; Radošević, K. Physicochemical Properties, Cytotoxicity, and Antioxidative Activity of Natural Deep Eutectic Solvents Containing Organic Acid. *Chem. Biochem. Eng. Q.* **2019**, *33*, 1–18. [CrossRef]
93. Tang, X.; Liu, J.; Dong, W.; Li, P.; Li, L.; Lin, C.; Zheng, Y.; Hou, J.; Li, D. The Cardioprotective Effects of Citric Acid and L-Malic Acid on Myocardial Ischemia/Reperfusion Injury. *J. Evid. Based Complement. Altern. Med.* **2013**, *2013*, 820695. [CrossRef] [PubMed]
94. De Souza Mesquita, L.; Martins, M.; Maricato, E.; Nunes, C.; Quinteiro, P.; Dias, A.; Coutinho, J.; Pisani, L.; de Rosso, V.; Ventura, S. Ionic Liquid-Mediated Recovery of Carotenoids from the *Bactris gasipaes* Fruit Waste and Their Application in Food-Packaging Chitosan Films. *ACS Sustain. Chem. Eng.* **2020**, *8*, 4085–4095. [CrossRef]

95. Tiwari, S.; Upadhyay, N.; Singh, A.K.; Meena, G.S.; Arora, S. Organic solvent-free extraction of carotenoids from carrot bio-waste and its physico-chemical properties. *J. Food Sci. Technol.* **2019**, *56*, 4678–4687. [CrossRef]
96. Tiwari, S.; Upadhyay, N.; Singh, A.K. Stability assessment of emulsion of carotenoids extracted from carrot bio-waste in flaxseed oil and its application in food model system. *Food Biosci.* **2022**, *47*, 101631. [CrossRef]

foods

MDPI

Perspective

Advancing Health and Sustainability: A Holistic Approach to Food Production and Dietary Habits

Graça P. Carvalho [1,2,*], Eduardo Costa-Camilo [1,3] and Isabel Duarte [4,5]

1 Bioscience School of Elvas, Portalegre Polytechnic University, 7350-092 Elvas, Portugal; eduardocostacamilo@ipportalegre.pt
2 Research Centre for Endogenous Resource Valorization (VALORIZA), 7300-555 Portalegre, Portugal
3 Applied Biomolecular Sciences Unit (UCIBIO), NOVA School of Science and Technology, Universidade NOVA de Lisboa, 2829-516 Caparica, Portugal
4 Instituto Nacional de Investigação Agrária e Veterinária (INIAV), Apartado 6, 7350-951 Elvas, Portugal; isabel.duarte@iniav.pt
5 GeoBioTec, OVA School of Science and Technology (NOVA FCT), Universidade NOVA de Lisboa, 2829-516 Caparica, Portugal
* Correspondence: gpcarvalho@ipportalegre.pt

Abstract: Producing healthier food requires expertise in methods that yield significant benefits for human health, sustainability, economic growth, cultural heritage, and overall well-being. Investing in conscientious and sustainable food systems can improve individual and planetary quality of life by preventing diseases, delaying ageing, and enhancing well-being. While healthy eating habits begin at home, schools play a pivotal role in reinforcing them from an early age. Despite progress, challenges remain, underscoring the need for prioritizing food education and literacy across all ages. Knowledge of how sustainable food production impacts personal health and well-being is critical. A holistic approach is essential for addressing these complexities, considering physical, mental, social, and environmental factors to identify balanced and effective solutions. Such analyses examine how system components interact, guiding the development of sustainable practices. The DM4You project exemplifies this approach. It unites Portuguese partners to promote local food consumption, focusing on soups, an integral part of traditional Portuguese cuisine, made with diverse vegetables, legumes, olive oil, and sometimes by-products. DM4You monitors 80 healthy participants over three months to assess dietary habits, focusing on soup and fruit consumption. This crossover study evaluates the influence of diet on health, offering insights into sustainable and health-promoting dietary practices.

Keywords: healthy food production; sustainable agriculture; sustainable food systems; food literacy; local food

1. Transforming Food Systems for a Sustainable Future: An Integrated Approach

Food systems (FS) are defined as encompassing the entire spectrum of interconnected value-added activities involved in the production, aggregation, transformation, distribution, consumption, and disposal of food products derived from agriculture, forestry, or fisheries. These systems are situated within the broader economic, social, and natural environments. The food systems comprise several subsystems, including agricultural systems, waste management systems, and production factor supply systems, which interact with other fundamental systems, such as those concerned with energy, trade, and health.

Transformations in the structure of food systems may be precipitated by shifts in the configuration of other systems. To illustrate, a policy promoting organic food production would exert a significant impact on the entire food system. The European Commission has identified organic production as a pivotal component in achieving the objectives set forth in the European Green Deal, the "From Farm to Fork" strategy, and the biodiversity strategy. In consequence of the above, the European Commission has set an ambitious target of converting 25% of EU agricultural land to organic farming by 2030, together with a significant increase in organic aquaculture [1].

A sustainable food system strives to guarantee food security and nutrition for all, while preserving the economic, social, and environmental foundations essential for food security and nutrition, without jeopardizing the capacity of future generations to meet their own needs. The United Nations Sustainable Development Goals (SDGs), adopted in 2015, place sustainable food systems at the core of their agenda. The Sustainable Development Goals (SDGs) set forth a series of transformative changes in agriculture and food systems with the objective of eradicating hunger, achieving food security, and improving nutrition by 2030. To meet these goals, it is necessary to reshape the global food system in a way that makes it more productive, more inclusive of marginalized populations, more environmentally sustainable, more resilient, and capable of providing healthy and nutritious diets for all.

These challenges are complex and systemic in nature, necessitating a combination of interconnected actions at the local, national, regional, and global levels [2]. As emphasized by [3], the unavoidable demographic, economic and climate changes that are expected to occur in the coming decades will significantly exacerbate these issues unless the current systems are reformed. Moreover, food systems were accountable for 34% of the total global greenhouse gas emissions in 2015, thereby emphasizing the imperative for a sustainable transformation [4].

1.1. Building Sustainable Food Systems

In this context, innovation is a crucial factor in enabling concerted and coordinated efforts to transition towards healthier, more equitable, resilient, and sustainable food systems. To achieve such a transition, it is necessary to adopt a holistic approach to the analysis of food systems, encompassing the entire spectrum from farm to fork and extending to waste management. It is essential that policy options, incentives, and investments facilitate transformative changes across all domains of food systems, including food production, supply chains, food environments, and consumer behavior. The food systems approach emphasizes the importance of recognizing potential synergies, which necessitates an in-depth understanding of how policies implemented at one stage of the food supply chain can generate ripple effects throughout the entire system. It is of the utmost importance to identify strategic points for intervention along the supply chain to maximize the synergies and optimize the overall benefits [5].

Despite the global food system transitions that have made diets more accessible, there has not been a uniform improvement in nutritional outcomes, environmental sustainability, inclusivity, or equity [6,7]. It is not necessarily the case that increased accessibility equates to improved nutritional quality or the adoption of environmentally and socially sustainable practices.

There remains a clear correlation between access to recommended diets and the level of food system development within a given country, with more industrialized nations enjoying greater accessibility. However, emerging data indicate that in some countries, the consumption of processed foods is prevalent not only among urban households but also increasingly among rural populations [8]. This issue is highly relevant within the framework of the New Urban Agenda, as adopted by the United Nations General Assembly [9].

The impact of the COVID-19 pandemic on food systems, food production, and food security has been extensively documented by various scholars [3,10–15], alongside the repercussions of the war in Ukraine [16–18]. These events underscore the critical need to "build" more resilient food systems capable of adapting to shocks and disruptions while formulating policies that ensure food security during crises.

The development of targeted support policies can significantly enhance nutrition, reduce food loss and waste, optimize resource utilization, prevent deforestation, curb biodiversity loss, limit greenhouse gas emissions, reduce reliance on harmful chemicals, and provide support to farmers [19,20].

Key measures include incentivizing the adoption of sustainable agricultural practices local consumption, developing local markets with shorter supply chains, and improving rural–urban linkages [21]. Additionally, price regulation strategies that favor producers over intermediaries and efforts to reduce pre- and post-harvest losses and mitigate food waste are crucial components of a comprehensive approach to sustainable food system transformation [12,22].

1.2. Strategies for Advancing Sustainable Food Systems and Policy Alignment

The 2030 Agenda has underscored the pivotal role that transforming agri-food systems play in accelerating progress towards achieving numerous Sustainable Development Goals (SDGs). However, it has also highlighted the inherent complexities in driving transformative changes across the global food landscape [22–25]. According to FAO Director-General QU Dongyu, the uneven recovery from the global pandemic, compounded by the conflict in Ukraine, has disrupted access to nutritious foods and healthy diets [16–18]. This evolving "new normal" is characterized by the convergence of climate change, conflicts, and economic instability, pushing populations to the brink of food insecurity. In response, a mere maintenance of the status quo is untenable [26].

Reflecting this urgency, UN Deputy Secretary-General Amina Mohammed, at the UN Food Systems Summit + 2 Stocktaking Moment (UNFSS + 2) held in Rome from 24–26 July 2023, officially closed the summit by presenting the Secretary-General's Call to Action. This call emphasizes the need for all stakeholders to mobilize and commit to implementing six specific objectives aimed at accelerating the transformation of food systems. These objectives are critical to addressing the multifaceted challenges currently facing global food security:

1st—Integrate food system strategies into all national sustainable development policies;
2nd—Establish food system governance that brings together all stakeholders in society;
3rd—Invest in research, data, innovation, and technological capabilities;
4th—Promote the involvement and accountability of businesses in shaping the sustainability of food systems, recognizing their centrality in the food and agricultural ecosystem;
5th—Ensure full participation of women, farmers, youth, and indigenous peoples at the local level in all stages of implementation;
6th—Ensure long-term and favorable financing, investments, budget support, and debt restructuring [27].

In this context, the need for targeted support policies is imperative, both globally and within more developed nations, including Portugal. The alignment of public policies with European frameworks has resulted in the development of programs, financial support, and assistance for farmers, processors, retailers, and large-scale distributors. However, the responsibility also lies with individuals to make informed decisions that can contribute to challenging and changing the existing status quo [28].

Achieving healthy diets within the framework of sustainable food systems is a complex endeavor that necessitates a multi-dimensional approach. According to the Directorate-General for Food and Veterinary Affairs (DGAV), sustainable nutrition encompasses diets that are culturally accepted, accessible, and meet the physiological needs of the global population [28,29]. These diets must also ensure food safety, respect for biodiversity, and ecosystem integrity, while being economically viable and equitable. Furthermore, they should optimize the use of human and natural resources, actively combat food waste, and support the well-being of present and future generations [30]. To meet these goals, sustainable nutrition (SN) is founded on several core principles. It must

a. Be accessible to everyone;
b. Be culturally accepted;
c. Meet physiological needs;
d. Be safe;
e. Respect biodiversity and ecosystems;
f. Be fair and economically viable;

g. Utilize human resources effectively;
h. Utilize natural resources efficiently;
i. Combat waste.

Achieving these principles requires a concerted effort to strengthen urban–rural linkages, enhance product design, invest in innovative food systems, implement selective collection for alternative food waste uses, and promote behavior change within food environments. Recently, key performance indicators (KPI) were identified by analyzing legislation, guidelines and scientific literature to measure social sustainability in the Italian agri-food sector [31]. They are offering a set of 24 different KPIs, organized into macro-areas of interest such as employment and training, health and safety at work, human rights, territorial community, and health and safety in production, allowing farms to evaluate and improve their practices in line with social expectations and European policy directives [31].

The establishment and advancement of sustainable food systems must be founded upon a comprehensive and multifaceted approach to Food Education, which is vital for effecting change at multiple levels (Figure 1).

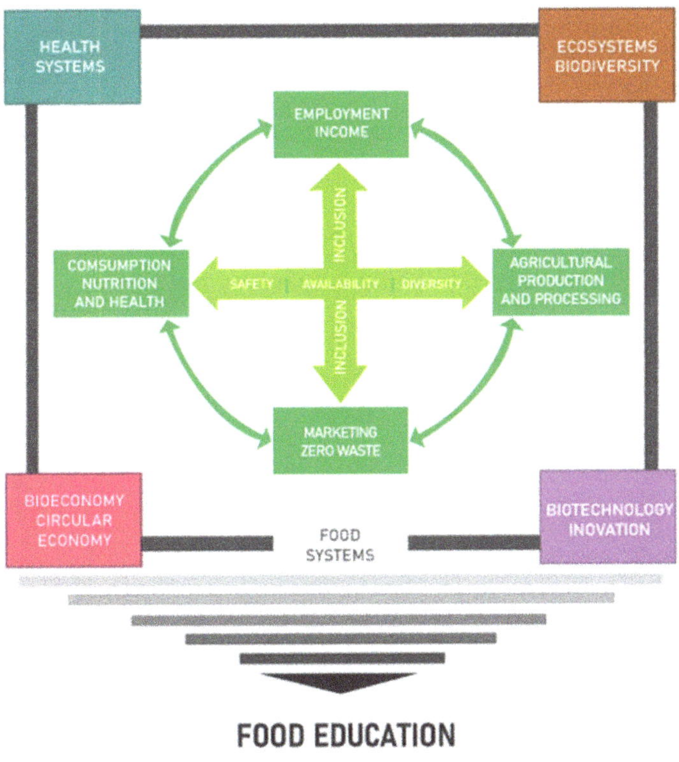

Figure 1. Food Education as Support and Foundation for Development (Adapted from [23]).

Food education serves as the cornerstone upon which these interactions are built, equipping individuals and communities with the requisite knowledge to make informed decisions about sustainable and nutritious food choices. It is thus imperative that an educational foundation be established to facilitate the creation of a well-informed society that can support and participate in sustainable food systems. This will ultimately result in the creation of healthier and more resilient communities and environments.

A recent study conducted in Istanbul, Turkey, involving 1600 volunteers with an average age of 28.2 $\pm$ 10.9 years, corroborates these findings [32]. The objective was to analyze the relationship between sociodemographic and anthropometric variables, nutritional knowledge and nutritional literacy. A total of 94.6% of respondents were deemed to possess adequate nutritional literacy [32]. The results indicated that age, gender, marital status, level of education, professional status, body mass index (BMI), and nutritional education were significantly associated with nutritional literacy [32]. A negative correlation was observed between body mass index (BMI) and age with nutritional literacy, whereas a positive correlation was evident between knowledge about nutrition and nutritional literacy [32].

Health literacy or the ability to access, understand, and use health information, has been recognized as an international public health goal. Nutrition literacy, a specific form of health literacy, has emerged as scholars explore the skills involved in being both food literate and health literate. This perspective suggests that a well-rounded concept of nutrition literacy should integrate core aspects of both health and food literacy [33].

Nutritional behavior is a highly complex phenomenon that is influenced by several factors, including nutritional knowledge. It is well documented that effective and continuous nutrition education plays an important role in the protection and development of health. It is also fundamental in changing the dietary habits of individuals of all ages. In the context of health-enhancing behaviors, health literacy is regarded as a pivotal predictive factor [34]. A lack of health literacy impairs the ability to identify and recognize health issues. Nutrition literacy, which constitutes an essential component of health literacy, denotes the capacity to comprehend, interpret, and access fundamental nutritional information to make informed decisions regarding nutritional matters. It is well established that an individual's knowledge, attitudes, skills, and behaviors concerning food and nutrition influence their dietary choices. A growing body of evidence indicates that most individuals encounter difficulties in interpreting the information presented on food labels [35]. Those with inadequate health literacy and/or mathematical abilities tend to exhibit poorer health outcomes [35].

2. Integrating Culture and Technology for Sustainable Food Systems: A Multidisciplinary Approach

In contemporary society, the food ecosystem is a multifaceted system that integrates environmental, social, economic, and ethical dimensions, reflecting the inter-connectedness of these factors in shaping sustainable and equitable food practices [36]. Achieving sustainable food systems requires a multidisciplinary approach that encompasses a broad spectrum of knowledge, encouraging change and fostering acceptance among all stakeholders while respecting cultural nuances. Cultural factors play a crucial role in determining the acceptance or rejection of various foods, such as insects, certain animals, and plants, which can vary significantly across sociocultural contexts [37]. It is also necessary to consider the lack of data on the chemical composition of (specifically) insect-based foods and toxicity studies, and potential allergenic implications must also be considered when determining whether to accept a foodstuff [38]. Also, we must consider how ethical decision-making within the food chain plays a crucial role in preventing food fraud and ensuring food safety, which in turn has wider benefits for global health and well-being. The reallocation of resources saved on combating fraud to address issues of hunger, oral health, and systemic health can facilitate the development of healthier communities and promote social justice [39]. Ethical conduct within the food chain reflects a responsibility to oneself, to others, and to the greater good, embodying values of compassion and respect for all life. Ultimately, this entails a commitment to human dignity and the interconnectedness of all beings [40].

In primary food production, it is essential to employ techniques and methods that not only respect the environment but also ensure the safety and quality of food products. Sustainable production practices such as organic farming, holistic management, syntropic agriculture, regenerative agriculture, and biodynamic agriculture represent a few of the many approaches that contribute to this goal. Agricultural product certification, such as Global G.A.P., provides a reliable and assured pathway for producers, processors, warehouse operators, and the entire food chain, ensuring that consumers receive safe and high-quality food [41].

In food processing, traditional thermal techniques remain widely used, relying on temperature changes (both high and low) to preserve and transform food while ensuring its safety [42]. However, ongoing research is focused on developing alternative non-thermal methods to achieve the same goals. These include High Pressure Processing (HPP), which allows for cold pasteurization using high pressure, as well as emerging technologies like microwave processing, ultrasonics, and food irradiation, among others [43].

It is imperative that the food industry prioritizes the promotion of sustainably produced foods and the development of strategies that can reduce food waste. This must be achieved by focusing on the protection of the environment and natural resources. To guarantee food safety, security and sustainability, it is essential that food production is increased, that the impact of this production on the environment is evaluated, and that innovative research, data, techniques and perspectives are developed [44].

Food technology encompasses any technological advancements that enhance food production, distribution, and supply, directly impacting how foods are produced, sold, and distributed. Companies are increasingly adopting the latest technologies to create jobs, reduce hunger, and promote responsible production and consumption throughout the entire food cycle [45].

In the domain of animal production, Artificial Intelligence (AI) has shown signific-cant efficacy, particularly in poultry farming [46]. AI systems can detect health issues in birds by analyzing the sounds they emit, and AI-powered robots are being deployed to perform tasks such as egg collection and assisting in the slaughtering process [46]. The objective is to redefine future agriculture. This is being achieved by utilizing Industry 4.0 applications, including artificial intelligence (AI), the Internet of Things (IoT), and cyber–physical systems. These are facilitating monitoring, precision forecasting, and providing comprehensive insights and system optimization, thereby transforming agricultural productivity [47].

The digitalization and integration of advanced technologies like automation, artificial intelligence, and machine learning have also revolutionized the healthcare sector. Patient data analysis enables the creation of personalized treatment plans tailored to individual needs, while large-scale data analysis enhances the quality of medical and epidemiological research [48].

In the field of medicine, immunotherapies are increasingly being used to treat various conditions, including certain cancers and degenerative diseases such as Alzheimer's. For instance, "Lecanemab", approved by the US Food and Drug Administration (FDA) in January 2023, is designed to remove amyloid protein buildup in the brains of individuals with early-stage Alzheimer's disease [49]. This approach signifies a shift from the traditional "one size fits all" model of medicine towards more individualized treatments. Telemedicine has experienced significant growth, providing remote medical care and patient monitoring, which is particularly valuable in remote areas or for patients with limited access to traditional healthcare [49]. However, the widespread adoption of technology in healthcare also presents challenges, including concerns over data privacy, cybersecurity, and the need for adequate training and empowerment of healthcare professionals. It is also critical to ensure that these technologies are applied ethically and in ways that prioritize patient welfare [50].

In biotechnology, there has been an exponential increase in research, yielding valuable insights that are being used to improve medical treatments, agricultural practices, and environmental solutions [51]. Environmental biotechnology, exemplified by bioremediation techniques using microorganisms to degrade plastic or the development of disease-resistant crops, highlights the potential of this field. Nonetheless, issues related to regulation and ethics, such as the use of genetically modified organisms (GMOs) and their release into the environment, necessitate rigorous safety assessments and responsible practices [52,53]. The complexities of modern supply chains and consumer preferences for safe food have driven the development of smart packaging, which encompasses active and intelligent technologies [52–54]. This innovation is becoming increasingly significant as it enhances the safety and quality of food products [54].

More recently, in a study on the association between ultra-processed foods (UPFs) and the increased prevalence of obesity and its complications, the authors state that promoting the consumption of whole and minimally processed foods and implementing nutrition education programs in schools are crucial steps to counteract the prevalence of obesity, which has become a global health problem [55]. They also point out that interdisciplinary and intersectoral collaboration will be essential to develop comprehensive solutions and improve public health outcomes worldwide.

The intersection of sustainable development and environmental biotechnology holds significant potential for addressing environmental challenges and fostering a healthier planet for all. This involves meeting present needs without compromising the ability of future generations to meet theirs.

In Portugal, the National Strategy for Food and Nutritional Security (ENSANP) was established, drawing on contributions from various governmental sectors and the results of extensive consultations with relevant entities [56,57]. This initiative also expanded the membership of the National Food and Nutritional Security Council (CON-SANP), initially created by Resolution of the Council of Ministers No. 103/2018, on July 26. This integrated approach underscores the importance of collaborative efforts in ensuring food security and nutritional well-being on a national scale.

Food Waste

It has been estimated that food losses due to the processing and production operations of the agri-food sector represent between 30 and 80% of overall yield [58]. On 30 October 2023, a pivotal report was presented, highlighting that food waste "occurs primarily at the household level". This finding underscores the need to reassess the National Strategy for Food and Nutritional Security, which has been traditionally focused on operators but now requires a stronger emphasis on consumer behavior. The shift away from traditional Portuguese eating habits, particularly the Mediterranean Diet (MD), is increasingly evident among the younger population, driven by a lack of awareness regarding the critical role of healthy eating in promoting long-term health and well-being [59].

Currently, over half of the adult population in Portugal is overweight, contributing significantly to the high prevalence of non-communicable diseases such as coronary heart disease, obesity, and diabetes. In the European Union, approximately 87.6 million tonnes of food are lost or wasted annually, with domestic food waste now estimated at twice the previous figures, reaching 74 kg per capita in 2021 [60].

In Portugal alone, around 1 million tons of food are wasted every year, with households responsible for 31% of this total [61]. To address this, it is crucial to implement practices and tools that enable the conversion of wasted food into consumable products. The European Union has committed to halving per capita food waste by 2030, supported by policies and instruments introduced in 2020 under the Circular Economy Action Plan, the Farm to Fork Strategy, and the Biodiversity Strategy [62].

On September 29th, on the International Day of Awareness of Food Loss and Waste, at a conference organized by the FAO, Andersen stated the following: "Progress towards each of the Sustainable Development Goals (SDGs) is slowing or regressing. But if there's one thing we know, it's that crises can bring out the best in people. We must all show these qualities now—particularly those who have the power to effect change in areas of great impact, such as agri-food systems" [63].

Regarding the importance of agri-food systems in the setbacks we are witnessing, the same author states the following: "We need a sustainable transition of the agri-food system that provides healthy and nutritious food for all [61,62]. That slows climate change and strengthens resilience. That improves livelihoods. And that protects the health of people, animals, and ecosystems. Halving food waste and losses, as advocated in SDG target 12.3, would bring enormous gains: Food security, Climate change mitigation, Protection of nature and biodiversity, a lighter burden on water—which is crucial in a world facing growing climate disruption" [64].

The global food system is currently confronting a series of formidable challenges, including biodiversity loss, depletion of natural resources, food insecurity, climate change, malnutrition, and other health-related issues [65].

A literature review encompassing 160 studies on market dynamics, information dissemination, regulatory frameworks, and incentives aimed at improving food systems, revealed key insights: (i) Less intrusive policy instruments, such as information-based strategies and behavioral nudges, are more popular but less effective compared to more stringent market-based and regulatory measures; (ii) consumers tend to rely on information-based tools to make sustainable food choices and are willing to pay a premium for sustainable products; (iii) sociodemographic factors, particularly gender (with females showing greater awareness) and education level (with higher education correlating with more sustainable choices), play a critical role in shaping these decisions [66].

There is a growing social interest in the interplay between food and health, which increasingly influences food choices. However, this interest is not uniform across society-ty, varying significantly with the economic capacity of individuals or households. This variation is more pronounced among higher economic strata, where awareness and education are generally more advanced, as corroborated by the study [66].

The successful achievement of these goals hinges on the dissemination of knowledge and the promotion of food literacy, an essential concept for informed decision-making. Food choices, which shape consumption patterns, are highly individualized and, in the case of children, are often influenced by family feeding models [67].

For adolescents, the systematic implementation of food and nutrition education in schools is crucial. Such education supports adolescents in making decisions that not only safeguard their current health but also ensure long-term well-being [68]. It has been observed that improving the quality of students' diets can enhance both their lifestyle behaviors and academic performance, with reciprocal benefits [68].

A study further explores the association between diet quality and lifestyle behaviors among higher education students, emphasizing the need to establish healthy campus committees to plan targeted interventions and educational programs focused on diet and lifestyle [69].

Moreover, there is a positive relationship between sustainable consumption and purchase intentions among higher education students. This indicates that environmental concern can indeed translate into effective purchasing decisions when aligned with product characteristics [70].

3. Promoting Healthy Eating in Portugal: National Strategies and the Role of the Mediterranean Diet

In Portugal, the National Plan for Healthy Eating (PNPAS) was established as a central platform for disseminating knowledge on nutrition and healthy eating. Functioning under the Directorate-General of Health (DGS), PNPAS serves as both an information hub and an activation platform, engaging a broad range of stakeholders including the health sector, local authorities, international institutions, associations, social projects, and consumers [71].

Complementing this initiative, the Integrated Strategy for the Promotion of Healthy Eating (EIPAS) is designed to ensure the implementation of its defined measures across various state services and bodies, both direct and indirect [72]. The monitoring and evaluation of these measures are overseen by an interministerial Working Group led by the DGS, ensuring a coordinated approach to advancing public health nutrition.

Recently, PNPAS, in collaboration with the European region and the World Health Organization (WHO), assessed Portuguese children's exposure to digital marketing of food and beverages [73]. The study revealed that girls aged 13–16 from lower socioeconomic backgrounds are the most exposed demographic, with 81% of foods advertised to children through digital platforms being unhealthy [73].

Portugal benefits from numerous national and European support structures aimed at enhancing food literacy. Among these, the Mediterranean Diet stands out as a historically rooted dietary pattern, recognized as the best diet globally by U.S. News and World Report for six consecutive years [74]. It has also been classified as an Intangible Cultural Heritage of Humanity by UNESCO since 2013 [74]. In support of this, the Mediterranean Diet Competence Center was established in Tavira in 2018, followed by the formation of the Network of Higher Education Institutions for the Safeguarding of the Mediterranean Diet (RIESDM) in 2019, to which we belong, alongside the Regional Council for the Mediterranean Diet [75].

The Mediterranean Diet, structured around 10 core principles, serves as a guide for consumers to adopt healthier and more sustainable eating habits, considering the three pillars of sustainability—economic, environmental, and social [76,77]. A recent study by the University of South Australia highlights the diet's economic benefits, demonstrating that a family of four can save $28 per week, or $1456 annually, by adhering to the Mediterranean diet compared to a typical Western diet [78]. Moreover, studies conducted in the UK have shown that following the Mediterranean diet is associated with a reduced risk of dementia, even among individuals with a genetic predisposition [78].

These findings underscore the significance of integrating traditional dietary practices like the Mediterranean Diet into national strategies, not only for improving public health but also for promoting sustainability and economic resilience [78].

DM4You

DM4You—Potential of the Mediterranean Diet in Increasing Quality of Life: +sustainability + health is led by the Polytechnic Institute of Portalegre and involves 12 other national partners as well as some international partners.

With a duration of 3 years, the project aims to be a driving force to reinforce, strengthen, disseminate, and educate about the importance of the principles of the Mediterranean Diet, specifically the consumption of soup and fruit and their effects on the health and well-being of populations. In addition to training and food education, the project embraces a more ambitious goal: it proposes to study the effects of soup consumption, through a clinical trial, on inflammatory processes associated with aging. In addition to blood, saliva collection may also allow for the identification of protein biomarkers in a less invasive manner, facilitating the continuation of the studies.

4. Conclusions

The promotion of sustainable food systems and healthier dietary habits requires a multifaceted approach encompassing education, technological innovation, cultural integration, and policy alignment. By focusing on holistic strategies, such as those demonstrated by the DM4You project, it is possible to enhance public awareness and encourage the adoption of practices that are both health-promoting and environmentally sustainable.

The DM4You initiative illustrates the potential of traditional dietary elements, such as soups and fruits, in improving well-being and reducing the impacts of ageing, while respecting cultural heritage. These efforts align with broader objectives like reducing food waste, combating climate change, and fostering resilient food systems capable of withstanding crises.

Moreover, integrating food education into public health initiatives and fostering literacy in nutrition and sustainability across all societal levels are critical for empowering informed food choices. The adoption of evidence-based practices and policies, supported by collaborative efforts, is essential to overcome current challenges.

Ultimately, transforming food systems requires sustained commitment at local, national, and global levels, ensuring that they are equitable, inclusive, and resilient for present and future generations. By prioritizing sustainability and well-being, it is possible to create food systems that serve as a foundation for healthier lives and a more sustainable planet.

Author Contributions: Conceptualization, G.P.C.; writing—original draft preparation, G.P.C. and E.C.-C.; writing—review and editing, G.P.C., E.C.-C. and I.D.; funding acquisition, G.P.C. All authors have read and agreed to the published version of the manuscript.

Funding: This work is supported by DM4You. "Potencial da Dieta Mediterrânica no aumento da qualidade de vida:+ saúde + sustentabilidade" by Agenda de Inovação para a Agricultura 20 | 30 «Terra Futura», PRR-C05-i03-I-000152 LA1.3./ LA1.4. and The APC was funded by national funds through the Fundação para a Ciência e a Tecnologia, I.P. (Portuguese Foundation for Science and Technology) by the project UIDB/05064/2020 (VALORIZA—Research Centre for Endogenous Resource Valorization).

Conflicts of Interest: The authors declare that the research was conducted in the absence of any commercial or financial relationships that could be construed as a potential conflict of interest.

References

1. Comissão Europeia. 2023. Available online: https://agriculture.ec.europa.eu/farming/organic-farming/organic-action-plan_pt (accessed on 12 July 2023).
2. FAO. 2018. Available online: https://www.fao.org/3/ca2079en/CA2079EN.pdf (accessed on 12 July 2023).
3. Barrett, C.B.; Benton, T.; Fanzo, J.; Herrero, M.; Nelson, R.J.; Bageant, E.; Buckler, E.; Cooper, K.; Culotta, I.; Fan, S.; et al. Socio-Technical Innovation Bundles for Agri-Food Systems Transformation. 2020. Available online: https://www.nature.com/documents/Bundles_agrifood_transformation.pdf (accessed on 12 July 2023).
4. Crippa, M.; Solazzo, E.; Guizzardi, D.; Monforti-Ferrario, F.F.N.T.; Leip, A. Food systems are responsible for a third of global anthropogenic GHG emissions. *Nat. Food* **2021**, *2*, 198–209. [CrossRef] [PubMed]
5. FAO; UNEP; UNDP. *A Multi-Billion-Dollar Opportunity–Repurposing Agricultural Support to Transform Food Systems*; FAO; UNDP; UNEP: Rome, Italy, 2021. [CrossRef]
6. Barrett, C.B.; Fanzo, J.; Herrero, M.; Mason-D'Croz, D.; Mathys, A.; Thornton, P.; Wood, S.; Benton, T.G.; Fan, S.; Lawson-Lartego, L.; et al. COVID-19 pandemic lessons for agri-food systems innovation. *Environ. Res. Lett.* **2021**, *16*, 101001. [CrossRef]
7. Ambikapathi, R.; Schneider, K.; Davis, B.; Herrero, M.; Winters, P.; Fanzo, J.C. Global food systems transitions have enabled affordable diets but had less favorable outcomes for nutrition, environmental health, inclusion and equity. *Nat. Food* **2022**, *3*, 764–779. [CrossRef]
8. FAO; IFAD; UNICEF; WFP; WHO. Urbanization, agrifood systems transformation and healthy diets across the rural–urban continuum. In *Brief to The State of Food Security and Nutrition in the World 2023*; FAO: Rome, Italy, 2023. [CrossRef]
9. Reardon, T.; Tschirley, D.; Liverpool-Tasie, L.S.O.; Awokuse, T.; Fanzo, J.; Minten, B.; Vos, R.; Dolislager, M.; Sauer, C.; Dhar, R.; et al. The processed food revolution in African food systems and the double burden of malnutrition. *Glob. Food Secur.* **2021**, *28*, 100466. [CrossRef]
10. Laborde, D.; Martin, W.; Vos, R. *Estimating the Poverty Impact of COVID-19, the MI-RAGRODEP and POVANA Frameworks*; International Food Policy Research Institute: Washington, DC, USA, 2020. [CrossRef]
11. Fan, S.; Chen, K.; Wei, S.; Johan, S. *Rethinking Agrifood Systems for the Post-COVID World*; Academy of Global Food Economics and Policy (AGFEP): Beijing, China, 2021.
12. Campos, A. Soberania Alimentar é Necessária para Garantir a Segurança Alimentar e a Sustentabilidade no Planeta. Available online: https://www.cna.pt/news/show/361 (accessed on 21 June 2021).
13. Okolie, C.; Ogundeji, A. Effect of COVID-19 on agricultural production and food security: A scient metric analysis. *Humanit. Soc. Sci. Commun.* **2022**, *9*, 1–13. [CrossRef]
14. Ali, J. Mapping scientific knowledge discovery on COVID-19 pandemic and agriculture: A bibliometric analysis and future research directions. *Environ. Sci. Pollut. Res.* **2023**, *30*, 95155–95171. [CrossRef] [PubMed]
15. Machado, B.; Pinto, E.; Silva, M.; Veiga, E.; Sá, C.; Silva, P.O.; Pimenta, A.; Gomes, A.; Almeida, A.; Sá, L. Impact of the COVID-19 pandemic on the mental and physical health and overall wellbeing of university students in Portugal. *PLoS ONE* **2023**, *15*, e0285317. [CrossRef]
16. Comissão Europeia. Ações da UE para Reforçar a Segurança Alimentar Mundial (2019–2024). Available online: https://commission.europa.eu/strategy-and-policy/priorities-2019-2024/stronger-europe-world/eu-actions-enhance-global-food-security_pt (accessed on 12 July 2023).
17. Ozili, P.K. Global economic consequences of Russian invasion of Ukraine. In *Dealing with Regional Conflicts of Global Importance*; IGI Global: Hershey, PA, USA, 2024; pp. 195–223. [CrossRef]
18. Agboklou, K.; Özkan, B.; Gujrati, R. Challenges to achieving zero hunger by 2030: The impact of armed conflict on global food security with a focus on the Russo-Ukrainian War. *J. Lifestyle SDG'S Rev.* **2024**, *4*, e01595. [CrossRef]
19. Ordovás, J. Balancing Public Health/Population Nutrition and Precision Nutrition in the Development of Dietary Guidelines. In *Precision Nutrition*; Academic Press: Cambridge, MA, USA, 2024; pp. 425–438. [CrossRef]
20. Silvestri, C.; Silvestri, L.; Piccarozzi, M.; Ruggieri, A. Toward a framework for selecting indicators of measuring sustainability and circular economy in the agri-food sector: A systematic literature review. *Int. J. Life Cycle Assess.* **2024**, *29*, 1446–1484. [CrossRef]

21. Piñeiro, V.; Arias, J.; Dürr, J.; Elverdin, P.; Ibáñez, K.A.; Opazo, C.M.; Owoo, N.; Page, J.R.; Prager, S.D.; Torero, M. A scoping review on incentives for adoption of sustainable agricultural practices and their outcomes. *Nat. Sustain.* **2020**, *3*, 809–820. [CrossRef]
22. Delgado, L.; Schuster, M.; Torero, M. Quantity and quality food losses across the value Chain: A Comparative analysis. *Food Policy* **2021**, *98*, 1–16. [CrossRef]
23. Laborde, D.; Torero, M. Modeling Actions for Transforming Agrifood Systems. In *Science and Innovations for Food Systems Transformation*; Springer: Cham, Switzerland, 2023; pp. 105–132. [CrossRef]
24. Varzakas, T.; Smaoui, S. Global food security and sustainability issues: The road to 2030 from nutrition and sustainable healthy diets to food systems change. *Foods* **2024**, *13*, 306. [CrossRef]
25. Martin, W.; Vos, R. *The SDGs and Food System Challenges: Global Trends and Scenarios Toward 2030*; International Food Policy Research Institute: Washington, DC, USA, 2024. Available online: https://hdl.handle.net/10568/138940 (accessed on 12 July 2023).
26. WHO. 122 Million More People Pushed into Hunger Since 2019 Due to Multiple Crises, Reveals UN Report. Available online: https://www.who.int/news/item/12-07-2023-122-million-more-people-pushed-into-hunger-since-2019-due-to-multiple-crises--reveals-un-report (accessed on 12 July 2023).
27. UNEP U.N. Deputy Secretary-General's Closing Remarks and Introduction of the SG's Call to Action: Food Systems Summit. Available online: https://www.un.org/sg/en/content/deputy-secretary-general/statement/2023-07-26/deputy-secretary-generals-closing-remarks-and-introduction-of-the-sg%E2%80%99s-call-action-food-systems-summit (accessed on 26 July 2023).
28. Clark, M.; Macdiarmid, J.; Jones, A.D.; Ranganathan, J.; Herrero, M.; Fanzo, J. The role of healthy diets in environmentally sustainable food systems. *Food Nutr. Bull.* **2020**, *41* (Suppl. S2), 31S–58S. [CrossRef]
29. Carlsson, L.; Callaghan, E. The social license to practice sustainability: Concepts, barriers and actions to support nutrition and dietetics practitioners in contributing to sustainable food systems. *J. Hunger. Environ. Nutr.* **2024**, *19*, 198–216. [CrossRef]
30. DGAV. Available online: https://www.dgav.pt/wp-content/uploads/2021/10/Brochura_DiaAlimentacao20.pdf (accessed on 16 October 2023).
31. Briamonte, L.; Pergamo, R.; Salerno, C.; Uliano, A. Measuring Social Sustainability in the Italian Agri-Food Sector: Proposed Key Performance Indicators. *Foods* **2024**, *13*, 2849. [CrossRef]
32. Sanlier, N.; Kocaay, F.; Kocabas, S.; Ayyildiz, P. The Effect of Sociodemographic and Anthropometric Variables on Nutritional Knowledge and Nutrition Literacy. *Foods* **2024**, *13*, 346. [CrossRef] [PubMed]
33. Velardo, V. The nuances of health literacy, nutrition literacy, and food literacy. *J. Nutr. Educ. Behav.* **2015**, *47*, 385–389. [CrossRef] [PubMed]
34. Carrara, A.; Schulz, P. The role of health literacy in predicting adherence to nutritional recommendations: A systematic review. *Patient Educ. Couns.* **2018**, *101*, 16–24. [CrossRef]
35. Draper, A.K.; Adamson, A.J.; Clegg, S.; Malam, S.; Rigg, M.; Duncan, S. Front-of-pack nutrition labelling: Are multiple formats a problem for consumers? *Eur. J. Public Health* **2013**, *23*, 517–521. [CrossRef]
36. El Bilali, H.; Strassner, C.; Ben Hassen, T. Sustainable Agri-Food Systems: Environment, Economy, Society, and Policy. *Sustainability* **2021**, *13*, 6260. [CrossRef]
37. Luderer, C.; Calado, V.H. Sustentabilidade Alimentar: Do Futuro Incerto à Resposta Certa. In *Os Desafios (Geográficos) da Governação Territorial*; U.Minho: Braga, Portugal, 2023; pp. 135–160.
38. Tavares, P.P.L.G.; dos Santos Lima, M.; Pessôa, L.C.; de Andrade Bulos, R.B.; de Oliveira, T.T.B.; Assis, D.d.J.; Morte, E.S.d.B.; Ribeiro, C.V.D.M.; de Souza, C.O. Innovation in alternative food sources: A review of a tech-nological state-of-the-art of insects in food products. *Foods* **2022**, *11*, 3792. [CrossRef] [PubMed]
39. Safavi, A.; Thrastardottir, R.; Thorarinsdottir, R.I.; Unnthorsson, R. Insect Production: A Circular Economy Strategy in Iceland. *Sustainability* **2024**, *16*, 9063. [CrossRef]
40. Varzakas, T.; Antoniadou, M. A Holistic Approach for Ethics and Sustainability in the Food Chain: The Gateway to Oral and Systemic Health. *Foods* **2024**, *13*, 1224. [CrossRef]
41. Global G.A.P. Available online: https://www.globalgap.org/pt/ (accessed on 6 October 2023).
42. Issa, A.; Mekanna, A.N.; Doumit, J.; Bou-Mitri, C. Redefining food safety: The confluence of Web 3.0 and AI technologies in the meat supply chain—A systematic review. *Int. J. Food Sci. Technol.* **2024**, *59*, e1–e14. [CrossRef]
43. Palamae, S.; Temdee, W.; Saetang, J.; Patil, U.; Suyapoh, W.; Yingkajorn, M.; Fan, X.; Zhang, B.; Benjakul, S. Impact of high-pressure processing on hemolymph, color, lipid globular structure and oxidation of the edible portion of blood clams. *Food Chem.* **2024**, *447*, 138948. [CrossRef]
44. Çakmakçı, S.; Polatoğlu, B.; Çakmakçı, R. Foods of the Future: Challenges, Opportunities, Trends, and Expectations. *Foods* **2024**, *13*, 2663. [CrossRef]
45. Prasanna, S.; Verma, P.; Bodh, S. The role of food industries in sustainability transition: A review. *Environ. Dev. Sustain.* **2024**, 1–21. [CrossRef]
46. Liu, H.; Wang, Y.; Yan, Z. Artificial Intelligence and Food Processing Firms Productivity: Evidence from China. *Sustainability* **2024**, *16*, 5928. [CrossRef]
47. Galanakis, C. The Future of Food. *Foods* **2024**, *13*, 506. [CrossRef]
48. Ahmed, Z.; Mohamed, K.; Zeeshan, S.; Dong, X. Artificial intelligence with multi-functional machine learning platform development for better healthcare and precision medicine. *Database* **2020**, *2020*, baaa010. [CrossRef]

49. Alzheimer's Society. Available online: https://www.alzheimers.org.uk/blog/lecanemab-new-drug-early-stage-alzheimers-disease (accessed on 5 October 2023).
50. Vilas-Boas, J.L.; Rodrigues, J.J.; Alberti, A.M. Convergence of Distributed Ledger Technologies with Digital Twins, IoT, and AI for fresh food logistics: Challenges and opportunities. *J. Ind. Inf. Integr.* **2023**, *31*, 100393. [CrossRef]
51. Ghag, S.B. Genetically modified organisms and their regulatory frameworks. In *Global Regulatory Outlook for CRISPRized Plants*; Academic Press: Cambridge, MA, USA, 2024; pp. 147–166. [CrossRef]
52. Stazi, A. Biotechnology, Genetically Modified Organisms and Millennium Development Goals. In *GMOs, Food Traceability and RegTech: Genetically Modified Food, Traceability Systems and Block-Chain as a Regulatory Technology*; Springer: Cham, Switzerland, 2024; pp. 1–14.
53. Young, E.; Mirosa, M. A systematic review of consumer perceptions of smart packaging technologies for food. *Front. Sustain. Food Syst.* **2020**, *4*, 63. [CrossRef]
54. Monda, A.; de Stefano, M.; Villano, I.; Allocca, S.; Casillo, M.; Messina, A.; Monda, V.; Moscatelli, F.; Dipace, A.; Limone, P.; et al. Ultra-Processed Food Intake and Increased Risk of Obesity: A Narrative Review. *Foods* **2024**, *13*, 2627. [CrossRef] [PubMed]
55. Diário da República. Presidência do Conselho de Ministros-Resolução do Conselho de Ministros n.º 145/2021. Available online: https://files.dre.pt/1s/2021/11/21300/0003000034.pdf (accessed on 3 November 2021).
56. Graça, P.; Gregório, M.J.; Freitas, M.D.G. A decade of food and nutrition policy in Portugal (2010–2020). *Port. J. Public Health* **2021**, *38*, 94–118. [CrossRef]
57. Read, Q.; Muth, M. Cost-effectiveness of four food waste interventions: Is food waste reduction a "win–win"? *Resour. Conserv. Recycl.* **2021**, *168*, 105448. [CrossRef]
58. Mendonça, N.; Gregório, M.J.; Salvador, C.; Henriques, A.R.; Canhão, H.; Rodrigues, A.M. Low adherence to the Mediterranean diet is associated with poor socioeconomic status and younger age: A cross-sectional analysis of the EpiDoC cohort. *Nutrients* **2022**, *14*, 1239. [CrossRef]
59. UNEP (United Nations for Environment Programme). *Food Waste Index Report 2021*; United Nations: Nairobi, Kenya, 2021.
60. Kiran, B.R.; Prasad, M.N.V.; Mohan, S.V. Farm to fork: Sustainable agrifood systems. In *Sustainable and Circular Management of Resources and Waste Towards a Green Deal*; Elsevier: Amsterdam, The Netherlands, 2023; pp. 25–38. [CrossRef]
61. Conti, C.; Zanello, G.; Hall, A. Why are agri-food systems resistant to new directions of change? A systematic review. *Glob. Food Secur.* **2021**, *31*, 100576. [CrossRef]
62. Andersen, I. Food for Nought: Time to End Loss and Waste. Available online: https://www.unep.org/news-and-stories/speech/food-nought-time-end-loss-and-waste (accessed on 29 September 2023).
63. Breitenmoser, L.; Behner, D.; Baertsch, A.; Mondardini, M.R.; Hugi, C. Citizen Science-Based Waste Diaries: An Exploratory Case Study of Household Waste in Switzerland. *Sustainability* **2024**, *16*, 3800. [CrossRef]
64. Sousa, J. *Literacia Alimentar: Decisão Informada*; Fundação Francisco Manuel dos Santos, Faculdade de Medicina da Universidade de Lisboa: Lisboa, Portugal, 2023.
65. Ammann, J.; Arbenz, A.; Mack, G.; Nemecek, T.; El Benni, N. A review on policy instruments for sustainable food consumption. *Sustain. Prod. Consum.* **2023**, *36*, 338–353. [CrossRef]
66. Ruggiero, C.F.; Trexberg, K.M.; Moore, A.M.; Savage, J.S. Applying the family stress model to responsive feeding and early obesity prevention. *Appetite* **2024**, *200*, 107515. [CrossRef]
67. Botelho, G.; Lameiras, J. Adolescentes e obesidade: Considerações sobre a importância da educação alimentar. *Acta Port. Nutr.* **2018**, *15*, 30–35. [CrossRef]
68. Doak, S.; Kearney, J.M.; McCormack, J.M.; Keaver, L. The relationship between diet and lifestyle behaviors in a sample of higher education students; a cross-sectional study. *Clin. Nutr. ESPEN* **2023**, *54*, 293–299. [CrossRef] [PubMed]
69. Montreuil Carmona, L.; Barretto, A. Does intention translate into actual purchase? Analysis of inducers of sustainable consumption in higher education students. *Young Consum* **2023**. *ahead of print*. [CrossRef]
70. Pinto, P.C. At the crossroads: Human rights and the politics of disability and gender in Portugal. *Alter* **2011**, *5*, 116–128. [CrossRef]
71. Graça, P.; Gregório, M.J.; de Sousa, S.M.; Brás, S.; Penedo, T.; Carvalho, T.; Bandarra, N.M.; Lima, R.M.; Simão, A.P.; Goiana-Da-Silva, F.; et al. A new interministerial strategy for the promotion of healthy eating in Portugal: Implementation and initial results. *Health Res. Policy Syst.* **2018**, *16*, 1–16. [CrossRef] [PubMed]
72. Figueira, M.; Bica, M.; Vicente, B.; Gaspar, M.; Gregório, M. *Exposição das Crianças Portu-Guesas ao Marketing Digital de Alimentos e Bebidas: Estudo Piloto da Ferramenta CLICK da OMS Europa*; Direção-Geral da Saúde: Lisboa, Portugal, 2023; ISBN 978-972-675-340-7.
73. Andrade, V.; Tagarro, M.; Jorge, R.; Lúcia da Mata Silvério Ruivo, P.; García-Conesa, M.T.; Philippou, E.; Pinto, P. Associações Entre Dieta Mediterrânica, Fatores de Estilo de Vida, e Bem-Estar Subjetivo. In 1ª Congresso da Rede das Instituições do Ensino Superior para a Salvaguarda da Dieta Mediterrânica. Rede das Instituições do Ensino Superior para a Salvaguarda da Dieta Mediterrânica (RIESDM). Available online: http://hdl.handle.net/10400.15/3525 (accessed on 12 July 2021).
74. Romano, A.; Mateus, M.P. Mediterranean Diet: A multidisciplinary approach to a new territorial strategy. In Proceedings of the Worldwide Perspectives on Geographical Indications, Montpellier, Frence, 5–8 July 2022; p. 133.
75. Qazi, A.; Al-Mhdawi, M.K.S. Exploring the relative importance of sustainable competitiveness pillars. *J. Clean. Prod.* **2024**, *443*, 140986. [CrossRef]
76. Qazi, A. Exploring the impact of global competitiveness pillars on sustainable development. *Environ. Impact Assess. Rev.* **2024**, *105*, 107404. [CrossRef]

77. Murphy, K. The Mediterranean Diet Good for Your Health and Your Hip Pocket. Available online: https://www.unisa.edu.au/media-centre/Releases/2023/the-mediterranean-diet-good-for-your-health-and-your-hip-pocket/ (accessed on 24 May 2023).
78. Shannon, O.R. Mediterranean diet adherence is associated with lower dementia risk, independent of genetic predisposition: Findings from the UK Biobank prospective cohort study. *BMC Med.* **2023**, *21*, 81. [CrossRef]

foods

MDPI

Systematic Review

Discovering and Mapping Colloquial Terminologies Describing Underutilized and Neglected Food Crops—A Comprehensive Review

Szymon Wojciech Lara, Amalia Tsiami * and Peter Cross

London Geller College of Hospitality and Tourism, University of West London, St Mary's Road, Ealing, London W5 5RF, UK; szymon.lara@uwl.ac.uk (S.W.L.); peter.cross@uwl.ac.uk (P.C.)
* Correspondence: amalia.tsiami@uwl.ac.uk

Abstract: Global levels of biodiversity and dietary diversity are decreasing, leading to food and nutrition insecurity. This is partially due to the homogenization of the global food supply with commodity crops. The reintroduction or introduction of neglected and underutilized species, minor, forgotten, and indigenous crops and landrace varieties to the wider food systems and further diversification have been outlined as the future strategies for tackling the above by the United Nations and the Food and Agriculture Organization in their policy frameworks. Most of the above species/crops are marginalized and only used across local food systems and in research. With over 15,000 different seed banks and repositories worldwide, information transparency and communication are crucial for database searching and their effective utilization. Much confusion persists around the true nature of those plants, and this prohibits the efficient utilization of their economic potential. A linguistic corpus search and a systematic literature review were conducted using the six most popular collocates to the above terms, which were as follows: ancient, heirloom, heritage, traditional, orphan, and the more distinct term 'landrace'. The results were interpreted using the Critical Discourse Analysis method. The definitions' findings show that heirloom, heritage, and ancient are mainly used in the United Kingdom and USA, where they are used to describe 'naturalized' and 'indigenized' or 'indigenous' food crops with a strong affiliation to 'family' and the 'act of passing seeds down from generation to generation'. Orphan crops, on the other hand, are often described as being 'overlooked' by growers and 'underfunded' by researchers. Landrace is most strongly affiliated with 'locality', 'biocultural diversity', and 'indigenous', and with genomics literature, where the characteristics are often discussed in the context of genetics and population biology. Contextualizing, most of the terms were found to be 'arbitrary' and 'undefinable' due to their continuing evolution in the socially accepted form of language, perhaps apart from landrace. The review has retrieved 58 definitions for the mentioned 6 terms, together with the primary key terms creating a tool to facilitate a better inter-sector communication and aid in policy.

Keywords: underutilized food crops; heirloom; landrace; heritage; orphan crops; ancient crops

1. Introduction

According to the World State of Plant Fungi Report, the global levels of biodiversity are decreasing at a worrying rate, where two in five plants are threatened with extinction [1]. The agricultural sector, together with aquaculture, are the major contributors to the loss of biodiversity, with 32.8% combined impact to extinction threats [1–3]. This is partially due to high-input intensive agricultural practices, based on exhaustive methods, such as monocropping, which are grounded in the "cheaper food" design paradigm [4]. This paradigm has been the primary motivator for the unsustainable practices across the entirety of food systems, making populations vulnerable to future food and nutrition insecurity, especially in the eve of the looming Climate Crisis, leading to poor resilience of production systems and livelihoods [4,5].

Despite the negative outcomes, these agricultural methods are still crucial for maintaining a sufficient and stable production of food, in both developed and underdeveloped countries [6]. There are two major factors contributing to this: predictability and price. It is argued that the drastic elimination of these practices from global food systems is unfeasible and would lead to further problems, such as hunger and poverty, and contribute to a decrease in dietary diversity [1–4]. Simultaneously, researchers are warning of the prolonged consequences of a lack of action, where future populations could also be put at risk of the above issues [7–9]. It is, therefore, necessary to consider different strategies for a sustainable change in food systems, where agrobiodiversity-oriented farming methods can be introduced feasibly, without damage to existing food provenance guarantees, with considerations for efficiency, sustainability, nutritional quality, dietary diversity, and economic prosperity [7–9].

The Food and Agriculture Organization (FAO) and the United Nations (UN) have stated that the process of diversification of agricultural practices across food systems, amongst others, is a good strategy for tackling the induced food and nutrition insecurity. The process of crop diversification has been linked to a decrease in the aspects of "low food diversity", "monoculture", and "genetic erosion" and correlated with an improvement in the overall "sustainability" of food production, consumption, nutritional quality, and dietary diversity [7,10–16].

Crop commodity diversification also corresponds with an increased resistance to other threats, such as plant disease, climate change, and soil depletion [7,14–16]. Furthermore, the importance of minor crops within cultural and food heritage paradigms, and the major role of this heritage in their long-term conservation, are also significant. As explained by Ryan et al. [17], some minor crops, which tend to be of high local importance, have their own stories and "timing of change, being additionally connected to transitions in the ways they are processed and transformed into foods and to shifting food preferences".

A key characteristic of many minor crops lies in their ongoing popularity within local food systems, as most can be classed as underutilized, neglected, forgotten, or minor in the grand scheme of regionalized and globalized agriculture. In other contexts where agri-development has already displaced the local varieties, some rural communities deliberately pursue low-input farming with neglected and underutilized species (NUS) and landrace varieties of crops. However, the overall tendency to lean towards the homogenization of the food systems, as well as at the local and indigenous levels, is progressive, leading to varietal and genetic loss [18].

Seed banks act as the major institutions for preserving crop seeds and associated information collection; however, this is often limited to the English/Latin colloquial and taxonomic species names, which can include agri-developed cultivars and landrace (LR) species, which means that it can be difficult to distinguish the two. Moreover, the cultural context of LRs [19,20] is 'culturally/people'-based, as opposed to cultivars strictly referring to 'true to type' propagation, which is often commercially driven. These parameters, together with accession year, and sometimes with unspecified collection location, make their popularization—that could otherwise draw on perceptions of heritage and local environment suitability—challenging. This scenario has developed, especially as seedbanks' primary purpose is seed/genetic preservation and crop development, not revival [19,20].

In both the scientific and non-scientific literature, these crops are often described with the following overarching English terms: neglected, underutilized, minor, forgotten, and indigenous, which are frequently seen as interchangeable and are very often used to describe "uncommercial" and "wild edibles" or "semi-domesticated" varieties of crops, and sometimes localized and cultural cultivars of commodity crops or landraces [7,13,21,22].

Furthermore, the literature often use NUS to describe neglected and underutilized crops, as well as wild species that have potential for domestication or commercial cultivation [23,24], as opposed to minor crops and forgotten crops, which refer specifically to domesticated crops and sometimes specific cultivars and are used interchangeably with other terms, namely the following: (1) orphan, (2) heirloom, (3) heritage, (4) ancient,

(5) traditional, and sometimes (6) landrace crops. A common trend is observed with the above terms being used solely in a descriptive format and coexisting in the scientific literature as the direct collocates of NUS, minor, and forgotten, and vice versa [1,7,13,21–27]. Landrace is a unique term in this debate, as it is used colloquially alongside the other terminology relating to neglected crops; however, landrace definitions are based more closely on a botanical framework, as discussed further below [27]. The additional terms (noted above 1–6) can also be found across a range of online sources, such as blogs, community websites and gazettes, seed bank databases, and other grey literature, and are adapted by gardeners, allotment holders, and crop 'seed guardians' across the English-speaking world [25–27].

The successful utilization of these crops for the fight against food and nutrition insecurity depends greatly on their visibility to potential growers, researchers, policy makers and investors across the value chains and markets. A foundational challenge lies in the genre of definitions, or their absence.

The understanding of the meanings embodied within these terms varies greatly across the literature, sometimes leading to problems in intersectoral communications, for example, between the researchers, the seed banks, and the growers [28]. Across the literature from multiple disciplines, these terms are used in various contexts and their usage is predetermined to subjective variables, such as the scientific discipline, the language, and the form, such as in food science or product development, where many so-called 'novel foods/ingredients' are in fact common across other sectors of the food systems, sometimes being commodity-like in one region of the world but completely neglected in other.

The terms are rarely defined, and most often one-word lexemes are used, in the form of connotations assigned as synonyms to provide a description and not a definition [25].

Villa et al. [27] shows information on the etymology of the term *landrace*, together with a review of pre-existing definitions and descriptions, including popular synonyms, such as 'farmers varieties', highlighting the sophisticated nature of the word and that it may never be suitable for a strict definition as the characteristics of the hypothetical *landrace crop* have also not been well-defined. Landrace definitions also refer to genetic aspects and population biology, and there are differences in how the terminology may fit different groups of plants [27]. A deeper lack of strict definitions and boundaries is present for the remaining five terminologies, as those are used interchangeably and are more categorial.

Based on the existing meanings and descriptions of the above terminologies, we aim to assess the similarities and differences in which these words are received across the scientific and grey literature, and to present the meaning in an accessible tabular format in order to form a descriptive tool for future researchers encountering terminology issues, limiting the uniformity and, therefore, the transparency of data across the fields.

2. Materials and Methods

The search strategy consisted of a primary literature review using the following initially predetermined terms: *neglected*, *underutilized*, *minor*, *forgotten*, and *indigenous* to showcase the language inventory used for such edible plants. The identified key words were (1) *orphan*, (2) *landrace*, (3) *heirloom*, (4) *heritage*, (5) *ancient*, and (6) *traditional*. These 6 terms represent entries into different terminology groupings and were, therefore, considered to be key for the scoping review, as they had the potential to be developed into separate categories, based on embodied meanings from the published literature.

The data search was conducted in 2 stages, firstly through a systematic literature review and secondly through a linguistic corpus search. The conceptual framework developed for this review study was based on the work of Jabareen [29], titled "Building a Conceptual Framework: Philosophy, Definitions, and Procedure".

2.1. Stage 1—Systematic Literature Review

The review was conducted in accordance with the Joanna Briggs Institute's guidance for Systematic Reviews for Effective Data Synthesis (Joanna Briggs Institute, 2022), with the steps defined below. The literature search was conducted electronically on the following academic databases: (1) Science Direct, (2) Emerald Insight, (3) ProQuest, (4) PubMed, and (5) Google Scholar. An additional search was conducted using an electronic institutional database/library called Summon Search Engine, with access provided on behalf of the University of West London [30]. A total of 2 separate searches were conducted across each database using 2 Boolean Codes with the 6 key words embodied within. These were as follows: ((*landrace crop definition*) OR (*orphan crop definition*) OR (*heritage crop definition*) OR (*heirloom crop definition*) OR (*ancient crop definition*)) OR (*traditional crop definition*)) and: *landrace definition* OR *orphan definition* OR *heritage definition* OR *heirloom definition* OR *ancient definition* OR *traditional definition*.

Inclusion and Exclusion Criteria

The searches were limited to peer-reviewed and published journal articles only, although some non-peer-reviewed material was also considered for inclusion. Grey literature was not considered for inclusion at this stage. The searches were not limited by any date restraints, the geographical location of the published articles, or the area of study, as those criteria did not seem relevant and could have led to discrimination against valuable articles. The articles were required to be written in English, with the limitation of this highlighted throughout this review article. In order to be included in the review, the article had to show significant correlation to the topic under investigation, with the key words mentioned somewhere throughout the text.

2.2. Stage 2—LinguisticCorpora Search

In addition to the scoping review, we also ran online searches using the British National Corpus (BNC). The BNC is a renowned tool, frequently used by researchers from the field of English language and linguistics to determine the frequency of occurrence of specific terms and their contexts—"BNC is a 100-million-word collection of samples of written and spoken language from a wide range of sources, designed to represent a wide cross-section of British English from the later part of the 20th century, both spoken and written" [31–33]. The selected 6 terms were individually inserted into the search engine on the BNC and run to automatically generate findings. The database also created a detailed analysis of the findings, including the major themes, which were grouped according to the nature of resources available. Further to that, the BNC also showed the major (usually top 10) synonyms, constructs, and other language features relating to the inserted key word. The findings generated through this search method were used as support for the main findings retrieved through the scoping review. The inclusion criteria for these findings were based on the relevance to the topic, meaning that constructs with no straight correlation to the area under investigation (food systems, food, and nutrition security) were excluded from the final findings, for example: *art theme*. Some relevance to the mentioned topics had to be present, for example: *botany theme*, *agriculture theme*, and *gardening theme*. The key constructs/definitions from both searches (Table 1) were combined and are presented in Table 2. Furthermore, for the term *landrace*, an additional resource was used, the American English National Corpus. This was due to lack of entries available on the BNC, and it has been acknowledged to be equivalent to the BNC.

Table 1. This table presents the articles retrieved through the scoping review that contain intentional definitions (19 out of 33 shortlisted articles). The remaining articles with ostensive definitions can be found in Appendix A. The article numbers correspond with the chronological occurrence in the text.

No.	Articles with Intentional Definitions Identified Through the Scoping Review (Citations)	Terminology
[27]	(Villa et al., 2005)	Landrace
[34]	(Armstead et al., 2009)	Orphan
[35]	(Benlioğlu and Adak, 2019)	Landrace
[36]	(Brouwer et al., 2016)	Heirloom
[37]	(Casañas et al., 2017)	Landrace
[38]	(Chiurugwi et al., 2018)	Orphan
[39]	(Dwivedi, Goldman, and Ortiz, 2019)	Heirloom Landrace Traditional
[40]	(Epping and Laibach, 2020)	Orphan
[41]	(Giambanelli et al., 2013)	Ancient
[42]	Glover and Stone, 2017	Ancient Heirloom
[43]	(Jordan, J., 2007)	Heirloom
[44]	(Mabhaudhi, et al., 2019)	Orphan
[45]	(Oldfield and Alcorn., 1987)	Traditional
[46]	(Sogbohossou et al., 2018)	Orphan
[47]	(Varshney et al., 2012)	Orphan
[48]	(Wendin et al., 2020)	Heritage
[49]	(Zeven, 1998)	Landrace
[50]	(Tadele, Z., 2009)	Orphan
[51]	(Naylor et al., 2004)	Orphan

Table 2. In this table, the 58 extracted definitions (most relevant international and ostensive) have been presented together with the 6 key terms. The definitions have been listed according to frequency of use, from the most common to least common, across for the 6 key terms. The green boxes indicate definition occurrence of the selected key term with the presented definition somewhere in the retrieved literature (see Appendix A for a full list of shortlisted articles) or in the BNC/AENC. This table could be used to increase the accuracy of future literature searches and to help differentiate between the key terms.

Key Definitions Identified	Orphan	Traditional	Ancient	Heirloom	Heritage	Landrace
Neglected	Yes	Yes	Yes	Yes	Yes	Yes
Underutilized	Yes	Yes	Yes	Yes	Yes	Yes
Regional	Yes	Yes	Yes	Yes	Yes	Yes
Indigenous	Yes	Yes	No	Yes	Yes	Yes
Tradition	No	Yes	Yes	No	Yes	Yes
Generations	No	Yes	Yes	Yes	Yes	No
Older Cultivars	Yes	No	Yes	Yes	No	Yes
Under-represented	Yes	No	No	Yes	Yes	Yes
Low Yielding	Yes	Yes	Yes	No	No	No
From a long time ago	No	Yes	Yes	No	Yes	No
Culture	No	Yes	Yes	No	Yes	No

Table 2. *Cont.*

Key Definitions Identified	Orphan	Traditional	Ancient	Heirloom	Heritage	Landrace
Ancestors	No	No	Yes	Yes	Yes	No
Passed down for generations	No	No	Yes	Yes	Yes	No
Traditional Cultivars	No	Yes	No	Yes	No	Yes
Wild	Yes	No	Yes	Yes	No	No
Heritage Tomatoes	No	Yes	No	Yes	Yes	No
Representing the Communities	Yes	Yes	No	No	No	Yes
Natural and Cultural Environments	Yes	No	No	No	Yes	Yes
Remote	Yes	No	No	Yes	No	Yes
Local	Yes	No	No	Yes	No	Yes
Family	No	No	No	Yes	Yes	Yes
Primitive	Yes	Yes	Yes	No	No	No
Small Landholders	Yes	No	No	Yes	No	Yes
Natural	Yes	Yes	No	No	No	Yes
Conserved	Yes	No	No	Yes	No	Yes
Grown for 50 years +	No	Yes	No	Yes	No	No
Historic	No	Yes	No	No	Yes	No
Wild Ancestors	Yes	No	Yes	No	No	No
Genetically Distinct	No	No	No	Yes	No	Yes
Traditional Agriculture	Yes	No	No	No	Yes	No
Adopted	Yes	No	No	No	No	Yes
Variability	No	No	Yes	No	No	Yes
Underexploited	Yes	No	No	No	No	No
Overlooked	Yes	No	No	No	No	No
Unfunded	Yes	No	No	No	No	No
Unresearched	Yes	No	No	No	No	No
Underdeveloped	Yes	No	No	No	No	No
Valuable Traits	Yes	No	No	No	No	No
Promising	Yes	No	No	No	No	No
With Some Importance	Yes	No	No	No	No	No
Authenticity	No	Yes	No	No	No	No
Historic Significance	No	No	Yes	No	No	No
Polen-Fertilized	No	No	No	Yes	No	No
Open-Pollinated	No	No	No	Yes	No	No
Anti-Commodity	No	No	No	Yes	No	No
Organic	No	No	No	Yes	No	No
Non-commercial	No	No	No	Yes	No	No
Non-hybrid	No	No	No	Yes	No	No
Historical Importance	No	No	No	No	Yes	No
Traditional Society	No	No	No	No	Yes	No
Older Varieties	No	No	No	No	Yes	No
Distinct Identity	No	No	No	No	No	Yes
Lacking Importance	No	No	No	No	No	Yes
Tolerates Biotic/Abiotic Stress	No	No	No	No	No	Yes
High Yield Stability	No	No	No	No	No	Yes
Complex Nature	No	No	No	No	No	Yes
Medicinal	No	No	No	No	No	Yes
Sustainable	No	No	No	No	No	Yes

2.3. Shortlisting, Screening, and Selection of Evidence

Through the scoping review, we identified a total of 517 records using the Boolean codes for all databases combined. The records were imported into a software program called Zotero 6.0.16, where the duplicates were removed, resulting in 496 records (see the PRISMA flow chart in Figure 1).

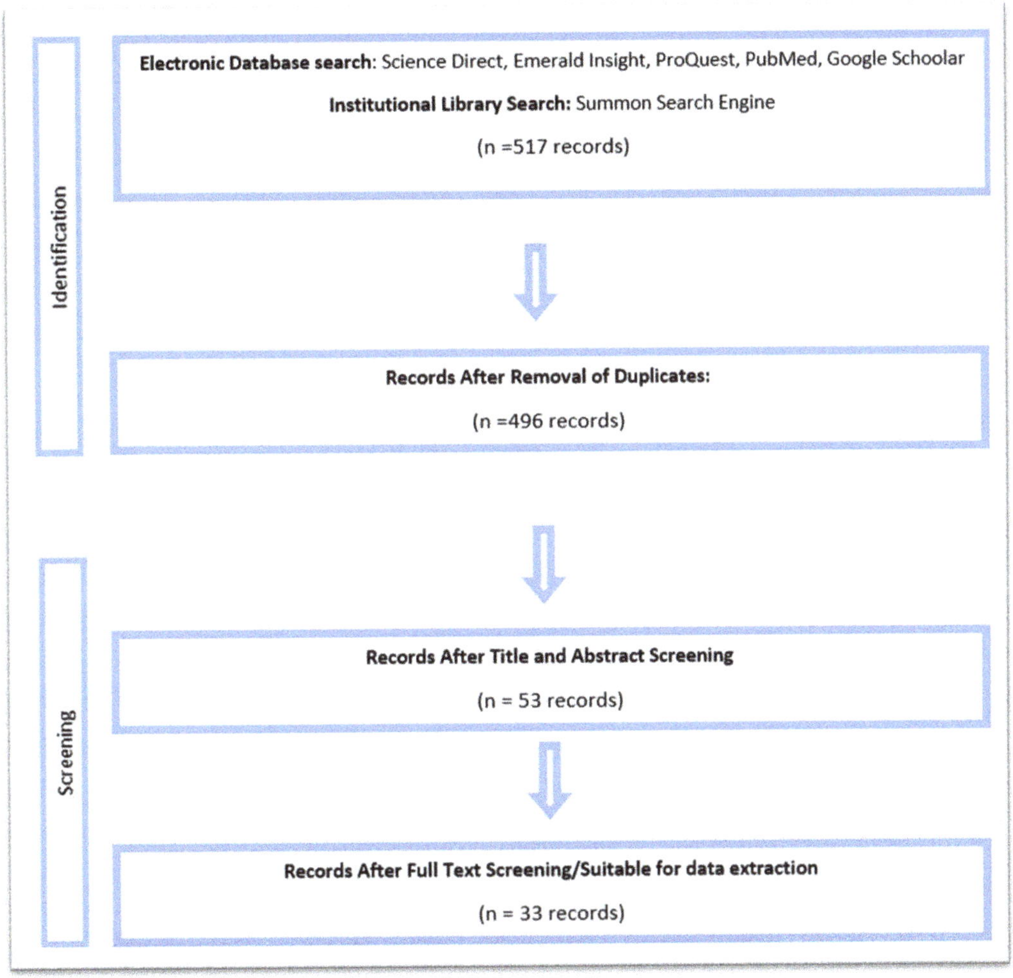

Figure 1. PRISMA flow chart displaying the search and shortlisting processes, as defined in Section 2.3, and the screening process, as defined in Section 2.4.

Title and abstract screening of the records was then undertaken. We searched for the previously mentioned 6 key words, and articles showing some relevance to the review were considered for further inclusion. Other articles that did not contain relevant key words but showed some indirect relevance to the area of study were considered for further screening. The removal of irrelevant articles resulted in 53 records. A full-text screening of these records was then undertaken, where, in addition to the previously mentioned key words, we also looked at the connectors used in the definitions across the literature, which were as follows: *"is"*, *"may"*, *"can"—be defined "as" "defined"*, and *"described"*.

2.4. Data Classification and Analysis

For the findings from the Stage 1 scoping reviews, the shortlisted articles were analyzed using the NVivo 12 Pro software program. Highlighted fragments of text were divided into the following 2 sections: (1) *intentional definitions* and (2) *ostensive definitions*, meaning complete definitions and descriptors, respectively. In the instance of a definition of one of the 6 terminologies, the relevant fragment of the article was selected and imported into the *intentional definitions* section of results. In order for the text to be classed as an intentional definition, it had to (1) express the overall meaning of the term and (2) be of a determinant nature to the term. All other relevant sections of the texts were classed as ostensive definitions, as the meaning of the term was usually (1) touched upon briefly in a secondary importance and supported with existing examples, more in a (2) description-like format, often for the purpose of providing some background information in the answer to an overarching question.

The data from both stages were analyzed using the Critical Discourse Analysis (CDA) method by Fairclough [33]. CDA was chosen as the appropriate theoretical framework for this research as different texts use linguistic strategies to reflect their ideological positions and CDA decodes these ideologies to show the power structures constructed in the discourse [52]. We analyzed the findings at the textual dimension to critically understand the choices and patterns in vocabulary, such as wording and chosen metaphors, to investigate the ideologies, the grammar, cohesion, and text structure for each of the key words.

2.5. Data Presentation

The data from both stages are presented and discussed according to Fairclough's [33] three dimensions of CDA. Fairclough sees discourse as being three dimensional, including the textual dimension, the discourse practice dimension, and the social practical dimension.

The selected fragments of text from Stage 1 are quoted and discussed simultaneously. Whereas the words identified through the BNC for Stage 2 were thematically analyzed and discussed. It is important to note that identifying encoded ideologies is an artificial process and the found ideologies can be separated imprecisely in different ways, as they are all intertwined. The analysis was conducted on the 6 terms and was organized alphabetically, in separate stages. The total findings from both stages were presented tabularly.

3. Results

This section shows the overall results from both the qualitative (scoping review) and the quantitative (BNC) studies. The systematic scoping review approached a yield of 517 records. Out of those, only 33 were shortlisted for final data extraction, as shown in the PRISMA flow chart in Figure 1. The citations of the identified articles (19 out of 33) are presented in Table 1, as well as the terminology of the crop discussed in each publication. The dominant (key) terms were selected, and those were orphan, traditional, ancient, heirloom, heritage, and landrace. The definitions have been listed according to their frequency of use, from the most common to the least common. The green boxes indicate the definition occurrence of the selected key term, with the presented definition somewhere in the retrieved literature or in the BNC, as presented in Table 2. A full list of the shortlisted articles is available in Appendix A.

The sections below show the qualitative and quantitative data retrieved through the systematic review of the literature and the BNC search, respectively. Each key word is presented and discussed separately. The results below show the intentional definitions and the BNC lexeme for the selected key words, presented in qualitative and quantitative formats.

3.1. Term—Ancient

The word 'ancient' has been loosely defined as *"of or from a long time ago, having lasted for a very long time"*, and this construct is the dominant definition of the word [53]. The findings indicated that *ancient* seems to be often used to promote alternative grains for flour production, baking, and can be found on some restaurant menus. Furthermore, we have identified two articles (see Table 1 for no [41,42]) that provided intentional definitions and zero ostensive definitions for the term. The focus was drawn onto the aspect of historical significance through words such as *ancestors* and *generations*. There are some similarities to *heirloom*, which also has been defined through a similar 'historical/family' lens, but more as *interpersonal* and *intergenerational* relations, as follows:

> *"Ancient grains are represented by populations of primitive grains, which were not subject to any modern breeding or selection, and which retained characters of wild ancestors, such as large individual variability, ear height, brittle rachis, and low harvest index"*—[41]

> *"Ancient cultivars are those that have been passed down for the generations without alteration"*—[42]

In addition to the definition search, we conducted a corpus search on the BNC database. There were 4846 entries across the entire database, and the analysis revealed that the dominant set of topics was from the field of *ancient history* and *archaeology*, with the 3 most popular collocates being: (1) *history*, (2) *culture*, and (3) *tradition*. Overall, the analysis showed that the topics, collocates, synonyms, and clusters did not resemble any significant correlation between the word *ancient* and any relevant disciplines related to food systems and food and nutrition security. On the other hand, the analysis of the ostensive definitions found that the most popular synonyms were (1) *traditional*, (2) *historical*, and (3) *heritage*, which were found in clusters describing the importance of *ancient farming* and *cultivars*. Based on these outcomes, the existing definitions of ancient are sufficient in meaning; however, the word can perhaps be drawn towards the aspect of edible grains, especially when derived from the grey literature. This may lead to some possible confusion amongst the general public, as many so-called 'ancient grains', such as spelt, are often found to be domesticated and popularized significantly later than wheat or barley, which are all considered as more *'modern'* [53]. Therefore, the term 'ancient' reflects more on older landrace/orphan crops versus hybrids and other improved cultivars.

3.2. Term—Landrace

According to the Cambridge Dictionary [53], the dominant definition for 'landrace' is as follows: *"a variety (= type) of a crop or a breed of an animal that has developed over time to suit the conditions of a particular local area"*. The word is also widely used in the botanical context and does not seem to have any other contradictory definitions or meanings.

We have identified 44 references in 12 different articles, and 5 of those were classed as intentional definitions (article numbers: [27,35,37,39,49]), and the remaining were classed as ostensive definitions, with the key journal article titled *'Landraces: A review of definitions and classifications'* authored by Villa et al. [27]. The document includes various existing definitions of the term from the pre-World War One era, when the term was vaguely used. It was only after the Second World War that it started to be recognizable across food system practitioners [21]. As previously mentioned, Villa et al. [27] discusses the etymology of the term *landrace*, together with pre-existing definitions and descriptions, highlighting the sophisticated nature of the word and that it may never be suitable for a strict definition, as the characteristics of the hypothetical *landrace crop* are difficult to define well because of the complexity of the genetic aspects and the population biology.

The proposed definition presented by Villa et al. [27] is a working definition, as outlined in the article; nevertheless, it is still the most widely accepted version, especially across plant sciences.

> *"Dynamic population(s) of a cultivated plant that have historical origin, distinct identity and lacks formal crop improvement, as well as often being genetically diverse, locally adapted and associated with traditional farming systems"*—[27]

The other identified definitions include the following:

"Each crop landrace has a specific local name assigned to it, highlighting its features and importance to the particular habitat and representing the class of humans inhabiting that area"—[35,37]

"An autochthonous landrace is a variety with a high capacity to tolerate biotic and abiotic stress resulting in a high yield stability and an intermediate yield level under a low input agricultural system"—[49]

Another definition, which was proposed more recently, shows some correlation between the terms *landrace* and *traditional*, and is synonymized with *heirloom.*

"Landraces–defined as traditional cultivars developed over time after adapting to both natural and cultural environments– or heirloom cultivars"—[39]

We have found that many journal articles do not provide their own definitions of the term. This could have been caused by the complex nature and continuing evolution of the word. This has been outlined in the Villa et al. [27] publication, in the following way,: *"As landraces have a rather complex nature it is not possible to give an all-embracing definition as it would result in a description"*.

We conducted a search for the term 'landrace' using the BNC, however, no entries were found. We then conducted the same search using the AENC, and identified 24 entries; however, analysis was unavailable. The AENC was treated as a secondary data source and only in the instance of lack of information on the BNC. The lack of data on the BNC could indicate the scarcity of this particular term amongst general websites and mean that this word (in its colloquial meaning) is used more widely across the literature in the USA rather than in the UK, hence, it was absent from the BNC. It is important to note that the AENC search could be limited by geographical bias, as certain entries might originate from sources outside of the USA but be written in American English. The term 'landrace' seems to be more recent in use across Britain and was likely derived from the USA agricultural sector at some point in the twentieth century [27].

The ostensive definitions showed that the most popular synonyms for the word were as follows: (1) *traditional*, followed by (2) *primitive*, and then (3), *remote, local, medicinal, natural, old, sustainable, conserved, unstandardized, adapted, under-represented, undocumented, misidentified, wild, native,* and *heritage.* Based on these findings, as well as the initial literature review, we have concluded that there is no unitary definition for the term 'landrace'. Different communities of practice possess different understandings of the meaning of the term. Sometimes, *landraces* are referred to as *heritage varieties,* or vice versa. This often occurs amongst professional chefs who prefer the use of the term *heritage* and in plant sciences [27,54,55]. The issue of misidentification is also common, as sometimes there are no 'reliable' tools to define between *landraces* and other cultivars amongst communities of growers, unless the growers save farm seed as opposed to buying commercial cultivars [22,27].

The term 'landrace' refers exclusively to cultivated plants, with a focus on the elements of locality, local adaptation, genetic diversity within populations, and the possession of some historical/heritage elements that determine the crop's uniquity, often not just in botanical contexts but also social contexts. *Landre-crop* is yet another lexeme that is frequently used across the literature and is used equivalently to the other six terms as well as NUS, forgotten, and *minor*, resulting in a slightly different meaning when compared to plain *landrace,* which is more of a definition itself, in an attempt to make the plant categories more botanical (scientific) rather than colloquial.

3.3. Term—Heirloom

For the term *'heirloom'*, we have identified 35 references in 9 various journal articles. Four of these references [36,39,42,43] were classed as intentional definitions and the remaining were classed as ostensive definitions, where the word has been synonymized with *landrace* on three occasions. The dominant dictionary definition for the term is as follows: "a valuable object that has been given by older members of a family to younger members of the same family over many years" and the food-related definition is: "a fruit, plant, or seed of a type that has existed for many years" [53]. The definitions indicate that the determining factor for naming a crop *heirloom* is its historical significance, to pair with the social construct of the family's heritage. Restaurateurs tend to use this term in menu creation and advertisement, but, interestingly, this is often practiced only in relation to tomato-based dishes [43].

> *"Heirloom varieties, commonly defined as having been grown for over 50 years, have been grown and harvested for multiple generations"*—[36]

> *"Heirloom cultivars are generally characterized as traditional or older cultivars that are open pollinated, passed down from gardener to gardener or handed down in families, and often not used in large-scale agricultural enterprises. The definition of heirloom varies, and the term does not carry a precise scientific designation"*—[39]

> *"The term 'heirloom' itself generally applies to varieties that are capable of being pollen-fertilised and that existed before the 1940s, when industrial farming spread dramatically in the USA and the variety of species grown commercially was significantly reduced"*—[43]

> *"HRP defines heirloom rice as cultivars that have been 'handed down for several generations through family members and grown by small landholders in their ancestral farms"*—[42]

> *"Heirloom rice as a social construction that facilitates the cooperation between the CHRP and HRP partners and enables the commodification of what was formerly an anti-commodity"*—[42]

The British National Corpus search resulted in 45 entries across 21 topics related to the field of food systems, with the 5 most popular being *seed, tomato, variety, garden,* and *plant.* The collocates were similar, as they included the above words and several additional words such as *family, grow, inherit, organic,* and *open-pollinated.* The clusters were also identical, as they included the above words.

The analysis of the data from the scoping review showed that the most common ostensive definitions involved the following words: *anti-commodity, non-commercial, traditional, culture, family, local, landrace, organic, open-pollinated, wild, regional, non-hybrid,* and *genetically distinct.* It seems that the *open-pollinated* characteristic is a determining factor for the heirloom classification of plants. The identified definitions varied greatly, with some touching on the aspect of the decreasing popularity in naming these crops as *commodities.* The commodity status seems to be replaced with societal relations in the form of family and the "handing down" of crops through generations. In conclusion, the term 'heirloom' resembles some relevance to *culture, community,* and *the act of food production,* all of which refer to the overall indigenous food system of rural settlements, which can prove to be a valuable source of new ingredients, flavors, and culinary concepts [19,56].

3.4. Term—Heritage

Not a single intentional definition for the term 'heritage' has been identified through the review. This is likely to be caused by the insignificance and broadness of the term, indicating its 'common' usage. Therefore, the meaning of this word is obvious and there is no major need to redefine it. Although, one journal article [48] on *cereals* stated the following: *"in this paper the term (heritage cereals) is meant to include ancient cereals, landraces and older varieties"* [48], but no strict definition has been found. The Cambridge Dictionary possesses a definition of the word, but in a broader context, as follows: *"heritage features belonging to the culture of a particular society, such as traditions, languages, or buildings, that were created in the past and still have historical importance"*; however, no correlation to botany or gastronomy has been identified. Furthermore, it is important to mention that heritage crops, most often tomatoes, are correlated with traditional agriculture and society, similarly to the previously mentioned concept of heirloom crops and are often cited in policy literature [57,58].

On the other hand, the BNC has identified 1937 inserts of the word across the entire database. The 20 most popular topics were of a similar nature to the previously described *ancient*, as these are mainly orientated around the aspects *of historical significance, museums, monuments*, and *landmarks*. The collocates and synonyms also relate to the above topics, and the word was synonymized with the term *traditional.*

3.5. Term—Traditional

We have identified five references across four different journal articles (No: [14,22,25,49]) that provided some ostensive definitions for the term 'traditional'; however, no intentional definitions were stated (apart from 41 and 47), as with the term 'heritage'. 'Traditional' is also used in an ostensive manner to describe foods that are of certain *"historical value and significance"* [56]. The term 'traditional' has been used in a broad form to describe heirloom crops as follows: *"heirloom crop cultivars are traditional cultivars grown for a long time (>50 years)"* [37]. It may seem logical to establish this term as significantly broader in meaning when compared to heritage, and the remaining terminologies are more specific in the meaning, making *traditional* an overarching term that encompasses the remaining five terms.

The BNC search resulted in 9599 entries, with the dominant sources being of an academic nature. The 20 most common topics were of various natures, with the most relevant to the field of study being: *tradition*, *culture*, and *authenticity*, but with no strong correlation to food systems or FNS.

The analysis of the scoping review did not reveal any significant ostensive definitions. Traditional crops were described as *"often low yielding"* by Oldfield and Alcorn [45]; however, some orphan crops tend to have higher yield ratios when compared to their improved relatives [50]. It is important to note that the term *traditional* is frequently used by various organizations, such as the FAO and the UN, but it is less popular amongst some academic disciplines. For example, in anthropology and ethnobotany searches, it is argued that *traditional* resembles 'static-ness' and does not reflect the true form of the indigenous food systems, which are believed to be often evolving; therefore, terms such as 'customary practices' are used instead [59]. From the gastronomical perspective, the word *traditional* is often used as a descriptor of dishes that have some historical background, often related to ethnic communities or regions.

3.6. Term—Orphan

The search resulted in 23 references from 8 different articles [34,38,40,44,46,47,51,60]. We shortlisted four definitions for the term, where some were found multiple times across six various documents. The main concept that can be extracted from the below definitions is the *economic importance*, together with two key words, *underutilized* and *neglected*. It seems that *orphan crops* are perceived as *less known* and *under-researched* with *underexploited value* due to *insignificant investment and academic attention.*

> *"So-called orphan crops are underutilized species that have local significance, especially for small-scale farmers, but are neglected on a global scale"*—[40,44]

> *"Orphan crops are those which are grown as food, animal feed or other crops of some importance in agriculture, but which have not yet received the investment of research effort or funding required to develop significant public bioinformatics resources. Where an orphan crop is related to a well-characterised model plant species, comparative genomics and bioinformatics can often, though not always, be exploited to assist research and crop improvement"*—[34]

> *"What is an orphan crop? In this context, we define it as a plant species which is grown as a food, animal feed or other crop of some importance in agriculture, but which has not received the investment of research effort—or of funding, which often amounts to the same thing—required to develop significant public bioinformatics resources."*—[34]

> *"Orphan crops are comparatively underexploited or underutilised food plants characterized as having relatively low or no perceived economic importance or agricultural significance in advanced economies, meaning they receive relatively little research and development attention"*—[38,46,47]

The BNC search resulted in 153 entries, mainly from *fiction* and *media* sources. The top 20 topics did not resemble any significant correlation to the field of study. The most popular collocates were *adopted, widowed, rescued,* and *abandoned.* These words could be interpreted from the botanical perspective, especially when looked at through the lens of the above definitions, where 'orphan' is linked to *locality* and *low economic significance.*

The scoping review search identified the following key words: *underutilized, unfinanced, neglected,* and *native.* Additionally, the following ostensive definitions were also identified:

> *"They (orphan crops) are often overlooked by researchers, despite valuable traits that are promising for emerging markets"*—[50]

> *"A huge difference exists on the way orphan or underutilized crops received financial investments for research and development"*—[51]

Based on the above results, it can be seen that *orphan* is often used as a reference to *underutilized and neglected crops,* minor, forgotten, and indigenous, including *cultivars,* as well as wild edible plants. This seems to be of significant value when compared to the remaining terminologies, as those have a stronger tendency to lean toward cultivars and are used more in agri-literature.

4. Discussion

As outlined by Villa et al. [27], the term 'landrace' has been built on from previous definitions—originally developed at the beginning of the 20th century by Von Rünker (1908) and Mansholt (1909).

The defining entries for the term 'orphan' have shown a tendency to lean towards the practicality and usability of the hypothetical *orphan* plant. However, it seems that the plants possess promising attributes, but are yet to be utilized due to a lack of funding and research, as explained by Varshney et al. [47]. Sogbohossou et al. [46], and Chiurugwi et al. [38]. Further to that, *orphan* appears to be used interchangeably with NUS or just *underutilized, minor, forgotten, lost,* and *priority crops,* mainly across the botanical sciences and agri-sciences [61]. Whereas, *heirloom* plants seem to be understood as social constructs, where family and community and the non-commercial and traditional food aspects are the determining factors, which are found more often in grey literature. Within *heirloom,* the human elements refer strongly to the act of *passing down,* which could mean that these are human cultivars that are being preserved, but not the wild varieties. This goes along with the existing definitions found in the plant sciences literature [39,42]. Moreover, *heirloom* plants have been referred to as *locality* more often than *orphan.* The term *landrace* seems to be multidisciplinary in the way that it is described. The descriptions span from *locality, communities,* and *remoteness* to *medicinal properties* and *sustainability.* There is some correlation between *landrace* and *orphan,* as shown through the *under-represented* factors, as both types of plant seem to possess that characteristic.

All of these foods are also sometimes referred to as *indigenous,* which botanically refers to the crops' original regions of domestication and diversity. When referring to the identified papers, discussions are oriented on the act of initial crop domestication, planting, harvest, storage, and reformulation, together with cultural aspects such as gastronomy and beliefs. Often, the term 'indigenous' refers more specifically to foods within indigenous peoples' food systems, which can then include both indigenous crops that originated in those parts of the world and other traditionally consumed foods that were introduced in the past [56]. Crops that are grown as landraces still within rural and indigenous food systems can include crops grown for millennia, or else more recent historic introductions, but essentially, they have been grown for long enough to become locally adapted and to have been adopted into the food systems for enough generations to be considered as *traditional* [13,62]. However, it is important to note that *landrace* is often outlined as a separate category in common UK legislature [20,56]. Seedbanks play a crucial role here, functioning as the repositories of seeds and genetic material that can used in research but are not designed to facilitate the actual crop revival into the food systems [20,63–66].

The meaning of *traditional* is much broader when compared to the remaining five terms. The information embodied in the description for the term was also multidisciplinary and often lacked clarity. This could also be due to the much wider usage of that word in other contexts, as shown through the corpus search, where the number of identified entries on the BNC for *traditional* reached 9599, showing the abundance of that word. There is some overlap with other terms, such as *heritage, ancient,* and *heirloom.* The historical attributes seem to be dominant. *Ancient* is also correlated with *history, tradition,* and *culture,* but also encompasses *primitiveness* and *the act of passing something down from generation to generation,* which is also emphasized through *ancestry* or *heirloom. Heritage* is narrower, however, in the sense that it most often refers to specific cultivars, especially those of tomatoes and oats, which has been found multiple times throughout the shortlisting process for this review.

Further research should be carried out using other non-English terms that show linkage with some of the identified terms through this systematic review. Moreover, it is worth noting that some terms could have been missed due to the literature publication formats, a lack of digitalization, or document accessibility through modern search engines. A good example of this is the Series of Volumes of Lost Crops of Africa, where the term 'lost-crops' is frequently used [64–66]. Further English search terms could include indigenous, lost, minor, and forgotten and an analysis of the meanings embodied there could be used in the promotion of forgotten crops along the food supply chains, through their popularization along the value chains, directing investments, markets and focusing attention of policy makers [67–69].

5. Conclusions

A scoping literature review was conducted on the existing definitions of 6 key terms (*orphan, heirloom, heritage, ancient, traditional,* and *landrace*), according to the Joanna Brigs Institute's guidance and the PRISMA model, using 33 shortlisted articles. Out of those, 58 definitions have been identified as linking to the above key terms. The key term *landrace* has a botanical and taxonomic framing, and is used to describe crops that are uncommercial, localized, and indigenous. In addition, sometimes *traditional* and *landrace crops* are also determined in genetic and population biology characteristics. *Orphan, heirloom,* and *heritage* are used mainly in the UK and the USA and can refer to the minor cultivars and sometimes-forgotten *landraces*. *Orphan* is also commonly used in the agri-development literature globally. *Ancient* generally refers to old varieties/cultivars, species, or wild edibles and cultivated wild plants, and is often used for grains. Some of the above terms can be categorized under *NUS* (neglected and underutilized species), which includes all crops and cultivable wild species that are not major/significant in relation to food systems or NUC (c = crops), which includes all minor crops but not the wild species. *Landrace crops* are a separate category, consisting of both minor and major species.

Highlighting the ever-changing nature of some of the above terms, alongside a more precise application in legislature, has the potential to improve communication in different disciplines and stakeholders from across food systems. Better framing in both the scientific and non-scientific literature could enhance the transparency and accessibility of knowledge to non-experts in the light of growing popularity of interdisciplinary research. All species, crops, and landraces that are categorized under any of the mentioned terms have the potential to help with the increase in agrobiodiversity, facilitating a better dietary diversity and contributing to the elimination of food and nutrition insecurity caused by many factors, including looming climate change. Future research could focus on the identification and definition of terms from foreign languages. Policy development in the above fields should be open to using a variety of terminologies, including new emerging trends, such as the use of *forgotten crops,* as opposed to prioritizing some terms for all beneficiaries, and consider their complex meanings and linkage to various natural and social aspects, such as geography, language, culture, and development level, in order to increase the level of mutual understanding between researchers, policy makers, growers, and consumers, beyond the English-speaking world.

Author Contributions: Conceptualization, S.W.L.; methodology, S.W.L. and A.T.; software, S.W.L. and A.T.; validation, S.W.L., A.T. and P.C.; formal analysis, S.W.L., A.T. and P.C.; investigation, S.W.L., A.T. and P.C.; resources, S.W.L. and A.T.; data curation; S.W.L., A.T. and P.C.; writing-review and editing; S.W.L., A.T. and P.C.; visualization, S.W.L. and A.T.; supervision, S.W.L., A.T. and P.C.; project administration, S.W.L.; funding acquisition, A.T. All authors have read and agreed to the published version of the manuscript.

Funding: This research received no external funding and the APC was funded by the publisher.

Data Availability Statement: Data is contained within the article. The data presented in this study are available in the provided tables, references and appendices. Please use the stated article names and DOIs for data searching.

Conflicts of Interest: The authors declare no conflict of interest.

Appendix A

Table displaying all of the 33 shortlisted articles (numbered) and the occurrence of ostensive/intensive definitions.

No	Ostensive Definition	Intentional Definition
[27]	Yes	Landrace
[31]	Yes	No
[34]	Yes	Orphan
[35]	Yes	Landrace
[36]	Yes	Heirloom
[37]	Yes	Landrace
[38]	Yes	Orphan
[39]	Yes	Heirloom Landrace Traditional
[40]	Yes	Orphan
[41]	Yes	Ancient
[42]	Yes	Ancient Heirloom
[43]	Yes	Heirloom
[44]	Yes	Orphan
[45]	Yes	Traditional
[46]	Yes	Orphan
[47]	Yes	Orphan
[48]	Yes	Heritage
[49]	Yes	Landrace
[57]	Yes	No
[58]	Yes	No
[50]	Yes	Orphan
[51]	Yes	Orphan

No	Ostensive Definition	Intentional Definition
[70]	Yes	No
[71]	Yes	No
[72]	Yes	No
[73]	Yes	No
[74]	Yes	No
[75]	Yes	No
[76]	Yes	No
[77]	Yes	No
[75]	Yes	No
[73]	Yes	No
[74]	Yes	No

References

1. Antonelli, A.; Smith, R.J.; Fry, C.; Simmonds, M.S.; Kersey, P.J.; Pritchard, H.W.; Abbo, M.S.; Acedo, C.; Adams, J.; Ainsworth, A.M.; et al. State of the World's Plants and Fungi. Ph.D. Thesis, Royal Botanic Gardens (Kew), Sfumato Foundation, Richmond, UK, 2020.
2. Williams, B.A.; Grantham, H.S.; Watson, J.E.M.; Alvarez, S.J.; Simmonds, J.S.; Rogéliz, C.A.; Da Silva, M.; Forero-Medina, G.; Etter, A.; Nogales, J.; et al. Minimising the loss of biodiversity and ecosystem services in an intact landscape under risk of rapid agricultural development. *Environ. Res. Lett.* **2020**, *15*, 014001. [CrossRef]
3. Thrupp, L.A. Linking agricultural biodiversity and food security: The valuable role of agrobiodiversity for sustainable agriculture. *Int. Aff.* **2000**, *76*, 265–281. [CrossRef] [PubMed]
4. Tsiafouli, M.A.; Thébault, E.; Sgardelis, S.P.; de Ruiter, P.C.; van der Putten, W.H.; Birkhofer, K.; Hemerik, L.; de Vries, F.T.; Bardgett, R.D.; Brady, M.V.; et al. Intensive agriculture reduces soil biodiversity across Europe. *Glob. Chang. Biol.* **2015**, *21*, 973–985. [CrossRef] [PubMed]
5. Rauf, S.; da Silva, J.T.; Khan, A.A.; Naveed, A. Consequences of plant breeding on genetic diversity. *Int. J. Plant Breed.* **2010**, *4*, 1–21.
6. Giampietro, M. Socioeconomic constraints to farming with biodiversity. *Agric. Ecosyst. Environ.* **1997**, *62*, 145–167. [CrossRef]
7. Henle, K.; Alard, D.; Clitherow, J.; Cobb, P.; Firbank, L.; Kull, T.; McCracken, D.; Moritz, R.F.A.; Niemelä, J.; Rebane, M.; et al. Identifying and managing the conflicts between agriculture and biodiversity conservation in Europe—A review. *Agric. Ecosyst. Environ.* **2008**, *124*, 60–71. [CrossRef]
8. Sarkar, A.; Sensarma, S.R. (Eds.) *Sustainable Solutions for Food Security: Combating Climate Change by Adaptation*; Springer: Berlin, Germany, 2019; pp. 3–23.
9. Frison, E.A.; Cherfas, J.; Hodgkin, T. Agricultural biodiversity is essential for a sustainable improvement in food and nutrition security. *Sustainability* **2011**, *3*, 238–253. [CrossRef]
10. Fess, T.L.; Kotcon, J.B.; Benedito, V.A. Crop breeding for low input agriculture: A sustainable response to feed a growing world population. *Sustainability* **2011**, *3*, 1742–1772. [CrossRef]
11. Galluzzi, G.; Eyzaguirre, P.; Negri, V. Home gardens: Neglected hotspots of agro-biodiversity and cultural diversity. *Biodivers. Conserv.* **2010**, *19*, 3635–3654. [CrossRef]
12. Mustafa, M.A.; Mayes, S.; Massawe, F. Crop diversification through a wider use of underutilised crops: A strategy to ensure food and nutrition security in the face of climate change. In *Sustainable Solutions for Food Security: Combating Climate Change by Adaptation*; Springer: Cham, Switzerland, 2019; pp. 125–149. [CrossRef]
13. Padulosi, S.; Thompson, J.; Rudebjer, P.G. *Fighting Poverty, Hunger and Malnutrition with Neglected and Underutilized Species: Needs, Challenges and the Way Forward*; Bioversity International: Rome, Italy, 2013.
14. Kahane, R.; Hodgkin, T.; Jaenicke, H.; Hoogendoorn, C.; Hermann, M.; Keatinge, J.D.H.; d'Arros Hughes, J.; Padulosi, S.; Looney, N. Agrobiodiversity for food security, health and income. *Agron. Sustain. Dev.* **2013**, *33*, 671–693. [CrossRef]
15. Medina-Lozano, I.; Bertolín, J.R.; Diaz, A. Nutritional value of commercial and traditional lettuce (*Lactuca sativa* L.) and wild relatives: Vitamin C and anthocyanin content. *Food Chem.* **2021**, *359*, 129864. [CrossRef] [PubMed]
16. van der Merwe, J.D.; Cloete, P.C.; van der Hoeven, M. Promoting food security through indigenous and traditional food crops. *Agroecol. Sustain. Food Syst.* **2016**, *40*, 830–847. [CrossRef]
17. Ryan, P.; Kordofani, M.; Saad, M.; Hassan, M.; Dalton, M.; Cartwright, C.; Spencer, N. Nubian agricultural practices, crops and foods: Changes in living memory on Ernetta Island, northern Sudan. *Econ. Bot.* **2022**, *76*, 250–272. [CrossRef]
18. Jackson, L.E.; Pascual, U.; Hodgkin, T. Utilizing and conserving agrobiodiversity in agricultural landscapes. *Agric. Ecosyst. Environ.* **2007**, *121*, 196–210. [CrossRef]

19. Gupta, A.; Sood, S.; Agrawal, P.K.; Bhatt, J.C. Under-utilized food crops of Himalayan region: Utilization and prospective. In *Newer Approaches to Biotechnology*; Narendra Publishing House: Delhi, India, 2013; pp. 101–120.
20. Maxted, N.; Scholten, M.; Ford-Lloyd, B.; Allender, C.; Astley, D.; Vincent, H.; Kell, S.P. *Landrace Conservation Strategy for the United Kingdom*; The University of Birmingham: Birmingham, UK, 2014.
21. Durst, P.B.; Bayasgalanbat, N. *Promotion of Underutilized Indigenous Food Resources for Food Security and Nutrition in Asia and the Pacific*; FAO: Rome, Italy, 2014.
22. Massawe, F.J.; Mayes, S.; Cheng, A.; Chai, H.H.; Cleasby, P.; Symonds, R.; Ho, W.K.; Siise, A.; Wong, Q.N.; Kendabie, P.; et al. The potential for underutilised crops to improve food security in the face of climate change. *Procedia Environ. Sci.* **2015**, *29*, 140–141. [CrossRef]
23. Hunter, D.; Borelli, T.; Beltrame, D.M.O.; Oliveira, C.N.S.; Coradin, L.; Wasike, V.W.; Wasilwa, L.; Mwai, J.; Manjella, A.; Samarasinghe, G.W.L.; et al. The potential of neglected and underutilized species for improving diets and nutrition. *Planta* **2019**, *250*, 709–729. [CrossRef]
24. Ulian, T.; Diazgranados, M.; Pironon, S.; Padulosi, S.; Liu, U.; Davies, L.; Howes, M.J.R.; Borrell, J.S.; Ondo, I.; Pérez-Escobar, O.A.; et al. Unlocking plant resources to support food security and promote sustainable agriculture. *Plants People Planet* **2020**, *2*, 421–445. [CrossRef]
25. Joshi, B.K.; Shrestha, R.; Gauchan, D.; Shrestha, A. Neglected, underutilized, and future smart crop species in Nepal. *J. Crop Improv.* **2020**, *34*, 291–313. [CrossRef]
26. Thompson, J.; Hodgkin, T.; Atta-Krah, K.; Jarvis, D.; Hoogendoorn, C.; Padulosi, S. Biodiversity in agroecosystems. In *Farming with Nature: The Science and Practice of Ecoagriculture*; Scherr, S.J., McNeely, J.A., Eds.; Island Press: Washington, DC, USA, 2007; pp. 46–60.
27. Villa, T.C.C.; Maxted, N.; Scholten, M.; Ford-Lloyd, B. Defining and identifying crop landraces. *Plant Genet. Resour.* **2005**, *3*, 373–384. [CrossRef]
28. Jaenicke, H.; Lengkeek, A. Marketing the products of underutilized crops-challenges and opportunities for pro-poor economic development. In *XXVII International Horticultural Congress-IHC2006: International Symposium on Cultivation and Utilization of Asian*; ISHS: Belgium, Brussels, 2006; Volume 770, pp. 87–94.
29. Jabareen, Y. Building a conceptual framework: Philosophy, definitions, and procedure. *Int. J. Qual. Methods* **2009**, *8*, 49–62. [CrossRef]
30. UWL Summon. 2022. Available online: https://www.uwl.ac.uk/current-students/library/finding-and-using-information/using-libsearch (accessed on 2 January 2022).
31. Sha, G. Using Google as a super corpus to drive written language learning: A comparison with the British National Corpus. *Comput. Assist. Lang. Learn.* **2010**, *23*, 377–393. [CrossRef]
32. Burnard, L. About the British National Corpus. 2022 (bnc). *Natcorp.ox.ac.uk.* Available online: http://www.natcorp.ox.ac.uk/corpus/#:~:text=The%20British%20National%20Corpus%20BNC,century%2C%20both%20spoken%20and%20written (accessed on 24 February 2022).
33. Fairclough, D.L. *Design and Analysis of Quality of Life Studies in Clinical Trials*; CRC Press: Boca Raton, FL, USA, 2010.
34. Armstead, I.; Huang, L.; Ravagnani, A.; Robson, P.; Ougham, H. Bioinformatics in the orphan crops. *Brief. Bioinform.* **2009**, *10*, 645–653. [CrossRef] [PubMed]
35. Benlioğlu, B.; Adak, M.S. Importance of crop wild relatives and landraces genetic resources in plant breeding programmes. *J. Exp. Agric. Int.* **2019**, *37*, 1–8. [CrossRef]
36. Brouwer, B.; Winkler, L.; Atterberry, K.; Jones, S.; Miles, C. Exploring the role of local heirloom germplasm in expanding western Washington dry bean production. *Agroecol. Sustain. Food Syst.* **2016**, *40*, 319–332. [CrossRef]
37. Casañas, F.; Simó, J.; Casals, J.; Prohens, J. Toward an evolved concept of landrace. *Front. Plant Sci.* **2017**, *8*, 145. [CrossRef] [PubMed]
38. Chiurugwi, T.; Kemp, S.; Powell, W.; Hickey, L.T. Speed breeding orphan crops. *Theor. Appl. Genet.* **2019**, *132*, 607–616. [CrossRef] [PubMed]
39. Dwivedi, S.; Goldman, I.; Ortiz, R. Pursuing the potential of heirloom cultivars to improve adaptation, nutritional, and culinary features of food crops. *Agronomy* **2019**, *9*, 441. [CrossRef]
40. Epping, J.; Laibach, N. An underutilized orphan tuber crop—Chinese yam: A review. *Planta* **2020**, *252*, 1–9. [CrossRef]
41. Giambanelli, E.; Ferioli, F.; Koçaoglu, B.; Jorjadze, M.; Alexieva, I.; Darbinyan, N.; D'Antuono, L.F. A comparative study of bioactive compounds in primitive wheat populations from Italy, Turkey, Georgia, Bulgaria and Armenia. *J. Sci. Food Agric.* **2013**, *93*, 3490–3501. [CrossRef]
42. Glover, D.; Stone, G.D. Heirloom rice in Ifugao: An 'anti-commodity' in the process of commodification. *J. Peasant Stud.* **2018**, *45*, 776–804. [CrossRef]
43. Jordan, J.A. The heirloom tomato as cultural object: Investigating taste and space. *Sociol. Rural.* **2007**, *47*, 20–41. [CrossRef]
44. Mabhaudhi, T.; Chimonyo, V.G.P.; Hlahla, S.; Massawe, F.; Mayes, S.; Nhamo, L.; Modi, A.T. Prospects of orphan crops in climate change. *Planta* **2019**, *250*, 695–708. [CrossRef] [PubMed]
45. Oldfield, M.L.; Alcorn, J.B. Conservation of traditional agroecosystems. *BioScience* **1987**, *37*, 199–208. [CrossRef]

46. Sogbohossou, E.O.D.; Achigan-Dako, E.G.; Maundu, P.; Solberg, S.; Deguenon, E.M.S.; Mumm, R.H.; Hale, I.; Van Deynze, A.; Schranz, M.E. A roadmap for breeding orphan leafy vegetable species: A case study of *Gynandropsis gynandra* (Cleomaceae). *Hortic. Res.* **2018**, *5*, 2. [CrossRef] [PubMed]
47. Varshney, R.K.; Ribaut, J.-M.; Buckler, E.; Tuberosa, R.; Rafalski, J.A.; Langridge, P. Can genomics boost productivity of orphan crops? *Nat. Biotechnol.* **2012**, *30*, 1172–1176. [CrossRef]
48. Wendin, K.; Mustafa, A.; Ortman, T.; Gerhardt, K. Consumer awareness, attitudes and preferences towards heritage cereals. *Foods* **2020**, *9*, 742. [CrossRef]
49. Zeven, A.C. Landraces: A review of definitions and classifications. *Euphytica* **1998**, *104*, 127–139. [CrossRef]
50. Tadele, Z. Role of orphan crops in enhancing and diversifying food production in Africa. *Afr. Technol. Dev. Forum J.* **2009**, *6*, 9–15.
51. Naylor, R.L.; Falcon, W.P.; Goodman, R.M.; Jahn, M.M.; Sengooba, T.; Tefera, H.; Nelson, R.J. Biotechnology in the developing world: A case for increased investments in orphan crops. *Food Policy* **2004**, *29*, 15–44. [CrossRef]
52. Machin, D.; Caldas-Coulthard, C.R.; Milani, T.M. Doing critical multimodality in research on gender, language and discourse. *Gend. Lang.* **2016**, *10*, 301–308. [CrossRef]
53. Cambridge Dictionary. Available online: https://dictionary.cambridge.org/ (accessed on 20 December 2022).
54. Standage, T. *An Edible History of Humanity*; Bloomsbury Publishing: New York, USA, 2009.
55. Pereira, L.M.; Calderón-Contreras, R.; Norström, A.V.; Espinosa, D.; Willis, J.; Lara, L.G.; Khan, Z.; Rusch, L.; Palacios, E.C.; Amaya, O.P. Chefs as change-makers from the kitchen: Indigenous knowledge and traditional food as sustainability innovations. *Glob. Sustain.* **2019**, *2*, e16. [CrossRef]
56. Maffi, L. Biocultural diversity and sustainability. In *The SAGE Handbook of Environment and Society*; Sage: Newcastle upon Tyne, UK, 2007; pp. 267–277.
57. Abdoulaye, T.; Sanders, J.H. New technologies, marketing strategies and public policy for traditional food crops: Millet in Niger. *Agric. Syst.* **2006**, *90*, 272–292. [CrossRef]
58. Dawson, I.K.; Powell, W.; Hendre, P.; Bančič, J.; Hickey, J.M.; Kindt, R.; Hoad, S.; Hale, I.; Jamnadass, R. The role of genetics in mainstreaming the production of new and orphan crops to diversify food systems and support human nutrition. *New Phytol.* **2019**, *224*, 37–54. [CrossRef]
59. Lugo-Morin, D.R. Indigenous communities and their food systems: A contribution to the current debate. *J. Ethn. Foods* **2020**, *7*, 6. [CrossRef]
60. Tadele, Z. Orphan crops: Their importance and the urgency of improvement. *Planta* **2019**, *250*, 677–694. [CrossRef]
61. Leakey, R.R. From ethnobotany to mainstream agriculture: Socially modified Cinderella species capturing 'trade-ons' for 'land maxing'. *Planta* **2019**, *250*, 949–970. [CrossRef]
62. Gaisberger, H.; Deletre, M.; Gaiji, S.; Bordoni, P.; Padulosi, S.; Hermann, M.; Arnaud, E. *Diversity of Neglected and Underutilized Plant Species (NUS) in Perspective*; Bioversity International: Rome, Italy, 2014.
63. Dierig, D.; Blackburn, H.; Ellis, D.; Nesbitt, M.A.R.K. Curating seeds and other genetic resources for ethnobiology. In *Curating Biocultural Collections: A Handbook*; Royal Botanic Gardens, Kew: Richmond, UK, 2014; pp. 107–125.
64. National Research Council. *Lost Crops of Africa: Volume I: Grains*; National Academies Press: Washington, DC, USA, 1996.
65. National Research Council. *Lost Crops of Africa: Volume II: Vegetables*; National Academies Press: Washington, DC, USA, 2006.
66. National Research Council. *Lost Crops of Africa: Volume III: Fruits*; National Academies Press: Washington, DC, USA, 2008.
67. Scarano, A.; Semeraro, T.; Chieppa, M.; Santino, A. Neglected and underutilized plant species (Nus) from the apulia region worthy of being rescued and re-included in daily diet. *Horticulturae* **2021**, *7*, 177. [CrossRef]
68. Coral, L.L.; Cusimamani, E.F. An Andean ancient crop, Chenopodium quinoa Willd: A review. *Agric. Trop. Et Subtrop.* **2014**, *47*, 142–146. [CrossRef]
69. Negri, V.; Torricelli, R. *The Italian Landrace Conservation Strategy*; The University of Birmingham: Perugia, Italy, 2015; pp. 15–17.
70. Barker, A.V.; Eaton, T.E.; Meagy, J.; Jahanzad, E.; Bryson, G.M. Variation of mineral nutrient contents of modern and heirloom cultivars of cabbage in different regimes of soil fertility. *J. Plant Nutr.* **2017**, *40*, 2432–2439. [CrossRef]
71. Bartel, C.; Tadele, Z.; Paterson, A.; Bhattacharjee, R.; Gedil, M.; Esfeld, K.; Dubois, T.; Plaza, S.; Lambein, F.; Dominique, D. African orphan crops: Their significance and prospects for improvement. *ADTF J.* **2009**, *6*, 1–82.
72. Boukid, F.; Folloni, S.; Sforza, S.; Vittadini, E.; Prandi, B. Current trends in ancient grains-based foodstuffs: Insights into nutritional aspects and technological applications. *Compr. Rev. Food Sci. Food Saf.* **2018**, *17*, 123–136. [CrossRef] [PubMed]
73. Chapagain, T.; Riseman, A. Evaluation of heirloom and commercial cultivars of small grains under low input organic systems. *Am. J. Plant Sci.* **2012**, *3*, 10–4236. [CrossRef]
74. Elia, A.; Santamaria, P. Biodiversity in vegetable crops: A heritage to save. The case of the Puglia region. *Ital. J. Agron.* **2013**, *8*, 21–34. [CrossRef]
75. Grassbaugh, E.M.; Regnier, E.E.; Bennett, M.A. Comparison of organic and inorganic mulches for heirloom tomato production. In Proceedings of the XXVI International Horticultural Congress: Sustainability of Horticultural Systems in the 21st Century, Toronto, ON, Canada, 11–17 August 2002; Volume 638, pp. 171–176.

76. Jain, S.M.; Gupta, S.D. *Biotechnology of Neglected and Underutilized Crops*; Springer: Berlin, Germany, 2013.
77. Mir, R.A.; Sharma, A.; Mahajan, R. Crop landraces: Present threats andopportunities for conservation. In *Rediscovery of Genetic and Genomic Resources for Future Food Security*; Springer: Singapore, 2020; pp. 335–349.

MDPI AG
Grosspeteranlage 5
4052 Basel
Switzerland
Tel.: +41 61 683 77 34

Foods Editorial Office
E-mail: foods@mdpi.com
www.mdpi.com/journal/foods

www.ingramcontent.com/pod-product-compliance
Lightning Source LLC
Chambersburg PA
CBHW061605220926
51850CB00005B/432